D1653078

Catalytic Chemical Vapor Deposition

Catalytic Chemical Vapor Deposition

Technology and Applications of Cat-CVD

Hideki Matsumura
Hironobu Umemoto
Karen K. Gleason
Ruud E.I. Schropp

Authors

Prof. Hideki Matsumura
JAIST (Japan Advanced Institute of Science and Technology)
1-1 Asahidai, Nomi
923-1292 Ishikawa
Japan

Prof. Hironobu Umemoto
Shizuoka University
3-5-1 Jyohoku, Nakaku
Hamamatsu
432-8561 Shizuoka
Japan

Prof. Dr. Karen K. Gleason
Massachusetts Inst. of Technology
77 Massachusetts Ave
MA
United States

Prof. Ruud E.I. Schropp
University of the Western Cape
Department of Physics and Astronomy
Robert Sobukwe Road, Belville 7535
South Africa

Cover: © Stanislaw Pytel/Getty Images, © Jena Ardell/Getty Images, © Jamie Brand/EyeEm/Getty Images

Library of Congress Card No.:
applied for

British Library Cataloguing-in-Publication Data
A catalogue record for this book is available from the British Library.

Bibliographic information published by the Deutsche Nationalbibliothek
The Deutsche Nationalbibliothek lists this publication in the Deutsche Nationalbibliografie; detailed bibliographic data are available on the Internet at <http://dnb.d-nb.de>.

Print ISBN: 978-3-527-34523-6
ePDF ISBN: 978-3-527-81864-8
ePub ISBN: 978-3-527-81866-2
oBook ISBN: 978-3-527-81865-5

Cover Design: Wiley
Typesetting SPi Global, Chennai, India
Printing and Binding C.O.S. Printers Pte Ltd, Singapore

Printed on acid-free paper

10 9 8 7 6 5 4 3 2 1

Contents

Preface

Chemical vapor deposition (CVD) is one of the key technologies in the present semiconductor industry. Among various CVD systems, CVD using heated metal filaments is a new and unique method to obtain device quality thin films at substrate temperatures lower than 300 °C. The method bears several different names. For instance, H. Matsumura and his coworkers have been using the term "catalytic chemical vapor deposition (Cat-CVD)" for more than 30 years since 1985, and probably, this may be the first name for this CVD method. However, other groups like to use the term "hot-wire CVD" and sometimes "hot filament CVD." In addition, recently, a new concept has been proposed in polymer synthesis by CVD using heated filaments with initiators. The groups dealing with this polymer deposition like to call it "initiated CVD (*i*CVD)."

In this book, as the principal author, H. Matsumura, has been using the term "Cat-CVD," and as there is general agreement that catalytic cracking is the major decomposition mechanism of molecules in CVD using heated filaments, the term "Cat-CVD" is used in this book instead of using hot-wire CVD or hot filament CVD, although *i*CVD has some other concepts involved in its deposition mechanism.

In the Cat-CVD method, molecules of source gases are decomposed by the catalytic cracking reaction with heated catalyzers, and such decomposed species are transported to a substrate in a vacuum chamber to form thin films on it. The method has been already industrially implemented and some consumer products using Cat-CVD in their fabrication process have been sold in the market, although many companies are not publicly revealing their processing method.

The adoption of Cat-CVD appears successful in the fabrication of various devices. Therefore, Cat-CVD technology is collecting more attention for its future feasibility. Because of these developments, I have been asked many times whether there is a proper book to study the technology systematically, probably since I have worked in this field for a long time as mentioned above. Thus, I thought of the publication of comprehensive book about Cat-CVD and related technologies. As far as I know, this may be the first book summarizing Cat-CVD (hot-wire CVD) technology and other related technologies with a wide systematic view.

For better understanding of Cat-CVD technology, this book starts with an explanation of the physics in a vacuum chamber. Then, the features of Cat-CVD are explained in comparison with other thin film technologies such as plasma-enhanced chemical vapor deposition (PECVD). The book deals with a wide scope of Cat-CVD and related technologies, from fundamentals to applications including the design of a Cat-CVD apparatus. Some of the techniques used in the observation of Cat-CVD-related phenomena are introduced to understand the concepts of such techniques. For instance, the laser-induced fluorescence, deep ultraviolet absorption, and other methods are briefly introduced as tools to determine the decomposed species in Cat-CVD process. Some of the fundamental physics of film materials are explained to understand the quality of films prepared by Cat-CVD. For instance, the physics of amorphous materials such as amorphous silicon (a-Si) is briefly mentioned. However, the explanation of detailed techniques and the deep physics concerned with them is beyond the scope of this book. If readers like to know more detailed information on them, they are referred to other books or references. I have arranged for understanding the total story and concepts of Cat-CVD and related technologies by this single book, excluding detailed stories in regarding individual parts.

This book is also written with the help of Professor Hironobu Umemoto at Shizuoka University, Japan, who is an expert of chemical analysis and who has revealed a lot of chemical phenomena concerned with Cat-CVD technology. Through his works, catalytic cracking has been confirmed in many cases for the decomposition mechanism of CVD using heated filaments. Chapters 3 and 4, in which the radical detection techniques and the mechanism of molecular decomposition are described, are mainly written by him. He also helped to check the validity of explanation in this book as a whole. The story of polymer synthesis by *i*CVD is written by Professor Karen K. Gleason at Massachusetts Institute of Technology, USA, who is a pioneer to use hot filament CVD for the synthesis of polymers and an inventor of *i*CVD. Chapter 6 is mainly written by her. Device application of Cat-CVD and related technologies are written with the help of Professor Ruud E.I. Schropp, a former Professor at Utrecht University, the Netherlands, presently Extraordinary Professor at University of the Western Cape, South Africa, who is an expert of device physics and has contributed to showing various feasibilities of Cat-CVD, although he has chosen to use the term "hot-wire CVD" in his papers instead of Cat-CVD. Major parts of Chapter 8 are written by him. He also helps to check the validity of explanations and expressions in this book as a whole. This book is completed by the help of all these experts in individual fields.

The purpose of this book is to be helpful for researchers, engineers, and students who begin to study on or wish to summarize the study on Cat-CVD, hot-wire CVD, *i*CVD, and related technologies. Comments and criticisms from users of this book will be received with many thanks.

December, 2018

On behalf of all authors
Hideki Matsumura
Emeritus Professor of School of Materials Science, at JAIST
(Japan Advanced Institute of Science and Technology)

Abbreviations

A list of Symbols or Abbreviations used in Several Chapters of this Book

CVD	chemical vapor deposition
APCVD	atmospheric pressure chemical vapor deposition
LPCVD	low-pressure chemical vapor deposition
Cat-CVD	catalytic chemical vapor deposition
HWCVD	hot-wire chemical vapor deposition
*i*CVD	initiated chemical vapor deposition
ALD	atomic layer deposition
PECVD	plasma-enhanced chemical vapor deposition
RF	radio frequency
VHF	very high frequency
ICP	inductive coupled plasma
D_{CS}	distance between catalyzer and substrate
T_{cat}	temperature of catalyzer
T_g	temperature of gas.
T_s	temperature of substrate
T_{holder}	temperature of substrate holder
RT	room temperature
P_g	pressure of gas during deposition
n_g	density of gas molecules
FR(X)	flow rate of gas X
S_{cat}	total surface area of catalyzer
V_{ch}	inner volume of deposition chamber
S_{ch}	inner surface area of chamber wall.
k	Boltzmann constant (1.38×10^{-23} J/K = 8.62×10^{-5} eV/K)
v_{th}	thermal velocity of gas molecule
λ	mean free path
λ_A	mean free path of molecule A
t_{col}	interval time between collisions
N_{col}	number of collisions of molecule with a unit area of solid per unit time.
$N_{col\text{-}space}$	number of collisions among gas molecules in space of chamber
t_{res}	residence time of molecules in chamber
V_{sheath}	sheath voltage of plasma

AES	Auger electron spectroscopy
AFM	atomic force microscopy
FTIR	Fourier transform infrared spectroscopy
LEED	Low-energy electron diffraction
LIF	laser-induced fluorescence
RBS	Rutherford backscattering
SIMS	secondary ion mass spectroscopy
SEM	scanning electron microscope
STEM	scanning transmission electron microscope
TEM	transmission electron microscope
XRD	X-ray diffraction
WVTR	water vapor transmission rate
SRV	surface recombination velocity
IR	infrared absorption
n	refractive index
σ_{p}	photoconductivity
σ_{d}	dark conductivity
$\sigma_{\mathrm{p}}/\sigma_{\mathrm{d}}$	photosensitivity
TFT	thin film transistor
FET	field effect transistor
HEMT	high electron mobility transistor
OLED	organic light-emitting diode
ULSI	ultralarge-scale integration (integrated circuits)

The symbols and abbreviations used only in a single chapter are not listed above.

The symbols such as α, σ, γ, and A are sometimes used with different meanings in different chapters. In such cases, the meaning of symbols is explained when they are used.

A list of chemical symbols used in some chapters in this book.

SiH_4	silane
GeH_4	germane
CH_4	methane
PH_3	phosphine
B_2H_6	diborane
NH_3	ammonia
HMDS	hexamethyldisilazane
HMDSO	hexamethyldisiloxane
HFPO	hexafluoropropylene oxide
PET	poly-ethylene terephthalate
PTFE	polytetrafluoroethylene
PGMA	poly glycidyl methacrylate
TMA	trimethylaluminum
TBPO	tert-butyl peroxide

1

Introduction

This chapter presents the outline of thin film technologies currently used and describes the relationship among catalytic chemical vapor deposition (Cat-CVD) and other conventional thin film technologies. In this chapter, the history of Cat-CVD and its related technologies is also briefly reviewed. Finally, the structure of this book is explained for easier reading.

1.1 Thin Film Technologies

Most industrial consumer products are coated with various thin films. Some products may be covered by painting or plating films. Thin film coating is also seen in modern electronic products. In such cases, quite often, the quality of the coating films determines the performance of the electronic products themselves. For instance, in liquid crystal displays (LCDs) or organic electroluminescent displays, transistors made of semiconductor thin films are working as a key device to control the brightness and colors of pictures. In ultralarge-scale integrated circuits (ULSI), used in computers as a key device, many thin films are contained, and the quality of such thin films strongly determines the performance of ULSI. In solar cells, the quality of thin films also determines their energy conversion efficiency.

So far, many thin film technologies have been invented. Coating technology on various tools was invented at least more than 10 000 years ago. It is well known that the painted drawing on the walls of stone caves can have a history of more than 40 000 years.

Here, let us depict a family tree of thin film technologies. It is very useful to make a family tree of related research for taking a view point of our own research in the tree to judge the value of the research itself. It can sometimes lead to new ideas and helps to define what we should do in future for our own technology from a deeper understanding of the position of research. The family tree is shown in Figure 1.1. It demonstrates that the thin film technologies are divided to three major technologies.

The first one is the pasting of thin films on solid substrates, where such thin films are prepared elsewhere. As a historical technology, pasting of gold leaf on solids is known. Because gold bullion can be stretched and transformed to a sheet

Catalytic Chemical Vapor Deposition: Technology and Applications of Cat-CVD,
First Edition. Hideki Matsumura, Hironobu Umemoto, Karen K. Gleason, and Ruud E.I. Schropp.

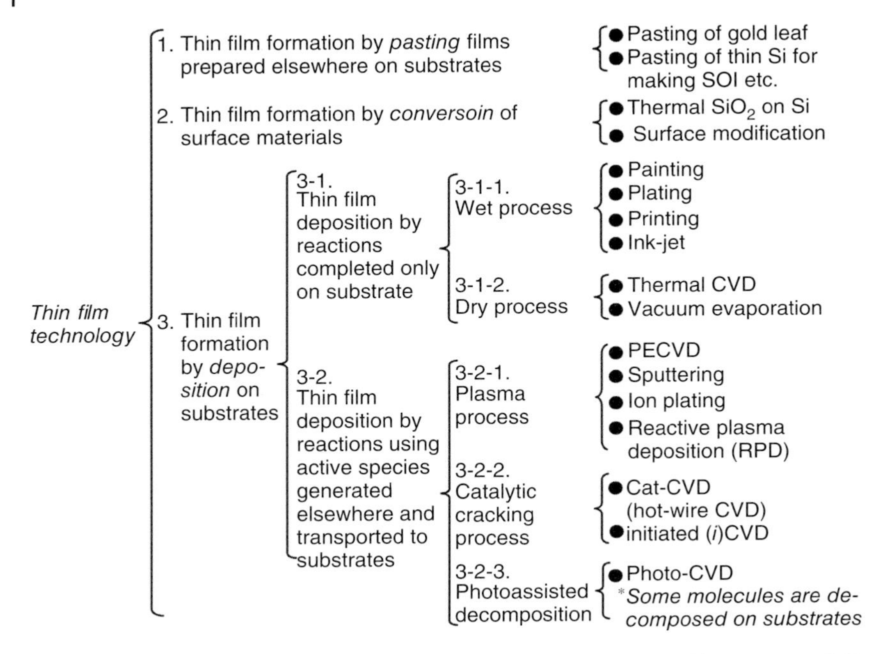

Figure 1.1 A family tree of thin film technologies. SOI and PECVD are abbreviations of silicon on insulator and plasma-enhanced chemical vapor deposition, respectively.

by beating, a thin gold leaf with a thickness less than 100 nm can be made and attached strongly on the surface of solids. The second is the formation of thin films by converting the thin surface layer of a solid to a layer of different materials by chemical reactions of the solid surface with active gases. For instance, the thermal oxidation of crystalline silicon (c-Si) wafers to form a silicon dioxide (SiO_2) layer at the surface is a widely spread technique within this category. The third is the formation of thin films by deposition on solid substrates. Films are formed at the surface of substrates by deposition of species supplied from outside of the substrate. This technology is divided into two groups: one is the method in which the film forming process is completed on the surface of substrates and the other is the method in which molecules are decomposed in advance remotely from the substrates, and such activated species are used in the reactions of film formation at the surface of substrates. Usually, by using species activated in advance, the high-quality films can be obtained at relatively low substrate temperatures, typically lower than 300 °C.

The methods utilizing activated species created outside of the substrates are further divided into three groups. One of the methods uses plasma for the generation of active species, the second method uses catalytic cracking reactions, and the third one uses the energy of radiation, such as photochemical vapor deposition (Photo-CVD). In Photo-CVD, there is discussion on the possibility of direct excitation of adsorbed molecules on the surface of substrates rather than in the gas phase outside of substrates. In this case, the method fits better in the group of 3-1-2.

Plasma-enhanced chemical vapor deposition (PECVD) is the first method, in which molecules of source gases are decomposed by the collisions with energetic electrons generated in plasma. The method using catalytic cracking reactions to decompose molecules is Cat-CVD. In Cat-CVD, as heated metal wires are usually used, the method is often called hot-wire CVD or hot filament CVD. In Cat-CVD, no plasma is needed.

A similar technology using heated metal wires is known as initiated chemical vapor deposition (*i*CVD). In this newly developed method, activation of initiators is due to heated metal wires, and the activated species are inducing polymerization of adsorbed monomers at the surface of substrates, resulting in high-quality organic thin films. This method is suitable for the preparation of device quality organic polymer films.

Looking at the family tree, one may notice that there is plenty of spaces for the first divided groups in which the films are prepared in advance, outside of the substrates. For instance, in the family tree of creatures, at first, the creatures are divided into plants and animals, and there is a roughly equal number of branches in the two groups. However, in the case of the family tree of thin film technology, the first and second divided groups do not have as many branches or methods compared with the third method of thin film deposition. This means that there is a big potential to invent new thin film technologies in the first and second groups. For instance, if patterned thin films, made apart from the substrates, are pasted on substrates, the thin film formation process becomes much more cost effective. The prototype of this idea is already seen commercially for pasting painted sheets on walls of air planes, trains, and buses for advertisement. If thin films are more sophisticated, this idea will be used in industrial electronic devices, in particular, large area devices. By looking at the family tree, you can enjoy more to create new ideas.

1.2 Birth of Cat-CVD

Catalytic cracking is a well-known phenomenon. It was already known in the 1910s that hydrogen (H) atoms are simply generated by catalytic cracking reactions of a hydrogen molecule (H_2) with heated tungsten (W) wires [1]. Since the 1970s, H. Matsumura et al. concentrated on making high-quality and thermally stable amorphous silicon (a-Si) films by using fluorine (F) atoms as dangling bond terminators. Particularly, they used silicon difluoride (SiF_2) molecules as a source for the formation of fluorinated a-Si (a-Si:F) films. By using SiF_2, such a-Si:F films were formed by simple plasma-less thermal CVD, although the quality of the films was slightly less than that of hydrogenated a-Si (a-Si:H) films prepared by PECVD. Thus, to improve film quality, Matsumura attempted to introduce H atoms into such a-Si:F films. As the films were obtained without the use of plasma, he pursued the use of H atoms created by catalytic cracking reaction of H_2 molecules at heated W wires to complete the development of a plasma damage-free deposition system. The work had been carried out from 1983 to 1985. The result was very encouraging. The property of such films appeared better than those of PECVD a-Si:H films. Matsumura named the method “Cat-CVD” in 1985.

The film quality was excellent; however, the results could not make the industry adopt this method, as they would not add halogen gases to their system because their systems preparing a-Si:H films had just been constructed in many companies. Then, he attempted to deposit a-Si:H by cracking of silane (SiH_4) gas with heated W wires and succeeded in obtaining device quality a-Si:H films in the period from 1985 to 1986. This makes the birth of Cat-CVD technology. Matsumura succeeded in obtaining device quality a-Si films without the assistance of plasmas for the first time.

1.3 Research History of Cat-CVD and Related Technologies

The combination of thermal CVD with catalysis of metals was first reported in 1970 by S. Yamazaki et al. at Doshisha University, Japan. He installed catalysts such as platinum and nickel oxide just near the substrates in a quartz tube of a conventional atmospheric pressure thermal CVD apparatus for preparing silicon nitride (SiN_x) films [2]. This catalyzer was just put near the substrates and it was unheated. He discovered that the temperature for forming SiN_x films on the substrates could be reduced from 700 to 600 °C by the effect of the presence of catalysts nearby. The role of catalysts and the mechanism for lowering the deposition temperature were not clearly explained in the report. This also did not show the invention of a low pressure and low temperature deposition method, although metal wires were used as an example of catalysis.

Later, in 1979, H. Wiesmann et al., at the Brookhaven National Laboratory, USA, reported their discovery that a-Si films could be formed when heated W wires and carbon foils were exposed to silane (SiH_4) gas in a low-pressure chamber [3]. This is the first report on the discovery of a-Si formation using heated W wires. However, as the quality of the a-Si films was less than that of PECVD a-Si:H and therefore the method appeared non-attractive, this work did not collect much attention and had been forgotten for several years until the work of Cat-CVD by Matsumura was reported. In addition, as they believed that the process for the decomposition of SiH_4 molecules was simply pyrolytic, the work did not lead to further understanding of the deposition mechanism based on catalytic reactions on the surface of metals.

In 1982, S. Matsumoto et al., at the National Institute for Research in Inorganic Materials, Japan, reported that diamond-like carbons (DLCs) could be obtained from methane (CH_4) gas when substrates were put in a conventional high-temperature thermal CVD chamber at temperatures between 800 and 1000 °C [4]. The substrates were heated further by heated W wires placed near them. At that time, the role of W wires was not clear, except for the role of additional heating of the substrates. Although they attempted to lower the temperatures later, this did not lead to the invention of a low-temperature deposition system at that time. Later, there were a number of reports discussing the role of W wires. In 1985, A. Sawabe and T. Inuzaka at Aoyama Gakuin University, Japan, reported the effect of electrons emitted from heated W wires on the growth of carbon films to enhance the crystallization of DLC films [5].

In 1985 and 1986, Matsumura et al. at the University of Hiroshima, Japan (later he moved to JAIST), reported the success of obtaining device quality hydrofluorinated a-Si (a-Si:F:H) [6, 7] and a-Si:H [8] at low temperatures, and they presented the concept of catalytic cracking playing a major role in the decomposition of material gases. In 1987, Matsumura also succeeded in making device quality amorphous silicon–germanium (a-SiGe) films [9] and in 1989, he also succeeded in obtaining device quality SiN_x films [10]. The success of formation of SiN_x films by Cat-CVD was also reported by K. Yasui et al., at the Nagaoka University of Technology, Japan, in 1990 [11]. In their case, they used SiH_4 and mono-methylamine as source gases and also used a W filament heated at 2400 °C.

Matsumura continued the research of Cat-CVD, and in 1991, he succeeded in obtaining even polycrystalline Si (poly-Si) or microcrystalline silicon (μc-Si) by adjusting the deposition parameters from those used in a-Si formation to mixture of SiH_4 with a large amount of H_2 [12]. In addition, these successes encouragingly demonstrated the big feasibility of this method as a new thin film technology.

After the success of preparation of a-SiGe, in 1988, Doyle et al. at Colorado University, USA, demonstrated the results of further advanced studies on Cat-CVD a-Si films by showing the feasibility to obtain a-Si films with excellent properties, although they named the method "evaporative surface decomposition (ESD)" [13].

In 1991, A.H. Mahan et al., at Solar Energy Research Institute (SERI, presently, National Renewable Energy Laboratories, NREL), USA, reported, by detailed comparison of two deposition methods, the superiority of Cat-CVD a-Si to that deposited by PECVD, although they named the method at that time "hot-wire CVD" [14]. Their scientifically well-elaborated reports contributed to the wide expansion of Cat-CVD-related research in the world. Since then, the number of researchers working on Cat-CVD or hot-wire CVD has increased.

Actually, in 1992, J. L. Dupuie and E. Gulari, University of Michigan, USA, reported the success of obtaining high-quality aluminum nitride (AlN) films by Cat-CVD using ammonia (NH_3) and trimethylaluminum (TMA) with a W catalyzer heated at about 1750 °C [15]. In addition, in 1992, J. L. Dupuie et al. also succeeded in confirming the formation of device quality SiN_x films by using Cat-CVD technology [16].

There have been many reports on various examples of application of Cat-CVD a-Si films and SiN_x films in the 1990s to 2000s. For instance, in 1995, R. Hattori et al. succeeded in obtaining high-quality SiN_x coating films for gallium–arsenide (GaAs) high-frequency transistors by using Cat-CVD and reported the superiority of device performance of GaAs transistors using Cat-CVD films to that using PECVD films [17]. In 1997, R.E.I. Schropp et al. succeeded in preparing state-of-the-art solar cells [18] and thin film transistors (TFTs) [19]. They also succeeded in preparing SiN_x films used as passivation films [20], gate dielectrics [21], and chemical barrier films [22] for various devices.

Concerned with the application to solar cells, there are also many reports from the NREL group. The progress of their group was summarized in Ref. [23]. For instance, in 1993, at first, E. Iwaniczko and coworkers succeeded in fabricating p-i-n a-Si solar cells by using i-a-Si layer prepared by Cat-CVD with a deposition rate of 0.9 nm/s. The deposition rate was very fast, compared with PECVD a-Si at that time. After various efforts, in 1998, A.H. Mahan et al. succeeded in obtaining full Cat-CVD p-i-n a-Si solar cells with an efficiency of 9.8%, at a deposition rate of 1.6 nm/s for the i-layer. The deposition rate was improved to 1.8 nm/s keeping the efficiency at 9.8% in 1999 by Q. Wang and coworkers.

Apart from these movements, many research studies on Cat-CVD a-Si and microcrystalline silicon (μc-Si), which includes crystallites in a-Si, had been carried out by the group of B. Schroeder et al. at the University of Kaiserslautern, Germany [24], by the group of J.E. Bouree and coworkers at Ecole Polytechnique, France [25], and some other groups.

Apart from these applications to electronic devices, from the middle of 1990s, chemistry groups mainly led by K.K. Gleason and coworkers at the Massachusetts Institute of Technology (MIT), USA, started to use this CVD technology using heated filaments for the formation of completely different materials such as polymer films and also started to think of applications, although they did not use the term "Cat-CVD" because there was no verification of catalytic reactions in their systems at that time. Probably, the research might be encouraged by the work of DLC films using a hot metal filament, as they like to use the term of "hot filament CVD" in the initial stage.

In 1996, S.J. Limb et al. of K. Gleason's group reported the success of obtaining polytetrafluoroethylene (PTFE, widely known by its commercial name "Teflon") films by using hexafluoropropylene oxide (HFPO) gas and heated nickel–chrome (NiCr) wires [26]. The temperature of the NiCr wires was about 325 to 535 °C, much lower than that of the W catalyzer during deposition of a-Si films. They revealed the superior property of PTFE films prepared by their hot filament CVD.

In 2001, H.G. Pryce Lewis et al., belonging to the same group, discovered and reported that the deposition rates of such PTFE films could be easily increased from 40 to 1000 nm/min, for instance, by adding perfluorooctane sulfonyl fluoride (PFOSF, $CF_3(CF_2)_7SO_2F$) gas as an initiator of reactions to HFPO as the source gas [27]. Since then, they had started to study on the mechanism of organic film formation and the role of an initiator. Finally, they reached a conclusion on the film formation process, revealing the importance of the selection of a proper initiator, and named the method "initiated chemical vapor deposition (*i*CVD)" in 2005 [28].

According to their explanation, in *i*CVD, at first, vapors of a monomer and an initiator are fed into the vacuum CVD chamber, and secondly, the initiator is activated in the gas phase in the hot zone near the heated filament. They claim that the activation is carried out by the heat transferred from the filament rather than by catalytic cracking. Subsequently, the monomer and the activated initiator are adsorbed on cooled substrates near to room temperature, and finally, polymerization starts on the surface of the substrates and growth of polymer films is started.

As the temperature of the filament in *i*CVD is usually kept below 500 °C, which is below the temperature for CH_3 decomposition, the filament metal is not carburized (carbided), and thus, the filament can be used for a very long time. Also, in *i*CVD, as gas phase reactions are important, the gas pressure during deposition is usually much higher than that in Cat-CVD.

Although the apparatus of *i*CVD is quite similar to that of Cat-CVD, the deposition mechanism may be markedly different, and so, they used a different name for their deposition method. As a new synthesizing method of high-quality organic polymer, further progress of the *i*CVD research is expected.

In this book, the deposition mechanism of Cat-CVD (hot-wire CVD) is explained in detail, and in addition, *i*CVD and its application are briefly introduced.

1.4 Structure of This Book

This book aims to present comprehensive overviews of all CVD methods using heated filaments, such as Cat-CVD technologies. It includes fundamental physics of vacuum and molecular dynamics in the vacuum chambers to provide easier understanding of the relevant background to the reader. It also includes a brief explanation of the conventional PECVD method, to compare it with Cat-CVD and understand the difference from Cat-CVD. An analysis of chemical reactions is given in detail, as this is a key to understand the deposition mechanism of Cat-CVD. This, particularly in relation to CVD using heated filaments, has not been systematically described in any other book as far as the authors know. The explanation is put forward as simple as possible, for easier understanding.

Fundamental physics and fundamental processing concepts of Cat-CVD are explained in Chapter 2, along with the explanation of the differences from PECVD, after first describing the features of PECVD. The detailed explanation of various radical detection techniques, which reveal the deposition mechanism of Cat-CVD, is described in Chapter 3. In Chapter 4, the deposition mechanism and chemical reactions in Cat-CVD are described based on the experimental results.

In Chapter 5, the properties of inorganic films prepared by Cat-CVD are mentioned. In this chapter, the meaning of physical parameters that are obtained in Cat-CVD films is explained, particularly for a-Si as a typical example of amorphous materials. The film formation process is explained by using various models. To explain the phenomena in a straightforward manner, the model is considerably simplified. Obviously, this does require sacrificing a detailed and accurate explanation, for which we refer to dedicated publications.

In Chapter 6, the preparation of organic films, based on processes similar to Cat-CVD, is presented. Particularly, *i*CVD and its application are explained.

In Chapter 7, the physics used for designing a Cat-CVD apparatus is elaborated. Gas flow, the factors deciding film uniformity, the effect of thermal radiation from the heated catalyzer, as well as methods to suppress it, and the key points for designing mass production machines are explained and discussed. The issue of contamination released from the heated wires and methods for chamber cleaning are also discussed. The lifetime of catalyzing wires is one of the challenges in Cat-CVD. This topic is dealt with in this chapter, along with various possible methods for extension of the lifetimes of catalyzing wires.

In Chapter 8, various applications of Cat-CVD technologies are introduced. A broad variety of applications have been developed.

In Chapters 9 and 10, the radical species generated in the Cat-CVD apparatus, the properties, and the applications of such radicals are summarized. Particularly, in Chapter 10, a new impurity doping technology, named "Cat-doping," is introduced. In this new technology, boron and phosphorus atoms are introduced into crystalline silicon at temperatures as low as 80 °C and demonstrated to be active dopants.

Throughout this book, the reader will appreciate that CVD using heated filaments, such as Cat-CVD and related technologies, has a promising future.

References

1 Langmuir, I. (1912). The dissociation of hydrogen into atoms. *J. Am. Chem. Soc.* 34: 860–877.
2 Yamazaki, S., Wada, K., and Taniguchi, I. (1970). Silicon nitride prepared by the SiH_4–NH_3 reaction with catalysts. *Jpn. J. Appl. Phys.* 9: 1467–1477.
3 Wiesmann, H., Ghosh, A.K., McMahon, T., and Strongin, M. (1979). a-Si:H produced by high-temperature thermal decomposition of silane. *J. Appl. Phys.* 50: 3752–3754.
4 Matsumoto, S., Sato, Y., Kamo, M., and Setaka, N. (1982). Vapor deposition of diamond particles from methane. *Jpn. J. Appl. Phys.* 21: L183–L185.
5 Sawabe, A. and Inuzaka, T. (1985). Growth of diamond thin films by electron assisted chemical vapor deposition. *Appl. Phys. Lett.* 46: 146–147.
6 Matsumura, H. and Tachibana, H. (1985). Amorphous silicon produced by a new thermal chemical vapor deposition method using intermediate species SiF_2. *Appl. Phys. Lett.* 47: 833–835.
7 Matsumura, H., Ihara, H., and Tachibana, H. (1985). Hydro-fluorinated amorphous-silicon made by thermal CVD (chemical vapor deposition) method. In: *Proceedings of the 18th IEEE Photovoltaic Specialist Conference, Las Vegas, USA, October 21–25, 1985*, 1277–1282.
8 Matsumura, H. (1986). Catalytic chemical vapor deposition (CTC-CVD) method producing high quality hydrogenated amorphous silicon. *Jpn. J. Appl. Phys.* 25: L949–L951.
9 Matsumura, H. (1987). High-quality amorphous silicon germanium produced by catalytic chemical vapor deposition. *Appl. Phys. Lett.* 51: 804–805.
10 Matsumura, H. (1989). Silicon nitride produced by catalytic chemical vapor deposition method. *J. Appl. Phys.* 66: 3612–3617.
11 Yasui, K., Katoh, H., Komaki, K., and Kaneda, S. (1990). Amorphous SiN films grown by hot-filament chemical vapor deposition using monomethylamine. *Appl. Phys. Lett.* 56: 898–900.
12 Matsumura, H. (1991). Formation of polysilicon films by catalytic chemical vapor deposition (Cat-CVD) method. *Jpn. J. Appl. Phys.* 30: L1522–L1524.
13 Doyle, J., Robertson, R., Lin, G.H. et al. (1988). Production of high-quality amorphous silicon films by evaporative silane surface decomposition. *J. Appl. Phys.* 64: 3215–3223.
14 Mahan, A.H., Carapella, J., Nelson, B.P. et al. (1991). Deposition of device quality, low H content amorphous silicon. *J. Appl. Phys.* 69: 6728–6730.

15 Dupuie, J.L. and Gulari, E. (1992). The low temperature catalyzed chemical vapor deposition and characterization of aluminum nitride thin films. *J. Vac. Sci. Technol., A* 10: 18–28.
16 Dupuie, J.L., Gulari, E., and Terry, F. (1992). The low temperature catalyzed chemical vapor deposition and characterization of silicon nitride thin films. *J. Electrochem. Soc.* 139: 1151–1159.
17 Hattori, R., Nakamura, G., Nomura, S. et al. (1997). Noise reduction of pHEMTs with plasma-less SiN passivation by catalytic CVD. In: *Technical Digest of 19th Annual IEEE GaAs IC Symposium, held at Anaheim, California, USA* (12–15 October 1997), 78–80.
18 Schropp, R.E.I., Feenstra, K.F., Molenbroek, E.C. et al. (1997). Device-quality polycrystalline and amorphous silicon films by hot-wire chemical vapour deposition. *Philos. Mag. B* 76: 309–321.
19 Meiling, H. and Schropp, R.E.I. (1997). Stable amorphous-silicon thin-film transistors. *Appl. Phys. Lett.* 70: 2681–2683.
20 van der Werf, C.H.M., Goldbach, H.D., Löffler, J. et al. (2006). Silicon-nitride at high deposition rate by hot wire chemical vapor deposition as passivating and antireflection layer on multicrystalline silicon solar cells. *Thin Solid Films* 501: 51–54.
21 Stannowski, B., Rath, J.K., and Schropp, R.E.I. (2001). Hot-wire silicon nitride for thin-film transistors. *Thin Solid Films* 395: 339–342.
22 Spee, D., van der Werf, K., Rath, J., and Schropp, R.E.I. (2012). Excellent organic/inorganic transparent thin film moisture barrier entirely made by hot wire CVD at 100 °C. *Phys. Status Solidi RRL* 6: 151–153.
23 Nelson, B.P., Iwaniczko, E., Mahan, A.H. et al. (2001). High-deposition rate a-Si:H n-i-p solar cells grown by HWCVD. *Thin Solid Films* 395: 292–297.
24 Schroeder, B., Weber, U., Sceitz, H. et al. (2001). Current status of the thermo-catalytic (hot-wire) CVD of thin silicon films for photovoltaic application. *Thin Solid Films* 395: 298–304.
25 Niikura, C., Kim, S.Y., Drévillon, B. et al. (2001). Growth mechanisms and structural properties of microcrystalline silicon films deposited by catalytic CVD. *Thin Solid Films* 395: 178–183.
26 Limb, S.J., Labelle, C.B., Gleason, K.K. et al. (1996). Growth of fluorocarbon polymer thin films with high CF_2 fractions and low dangling bond concentrations by thermal chemical vapor deposition. *Appl. Phys. Lett.* 68: 2810–2812.
27 Pryce Lewis, H.G., Caulfield, J.A., and Gleason, K.K. (2001). Perfluorooctane sulfonyl fluoride as an initiator in hot filament chemical vapor deposition of fluorocarbon thin films. *Langmuir* 17: 7652–7655.
28 Chan, K. and Gleason, K.K. (2005). Initiated chemical vapor deposition of linear and cross-linked poly(2-hydroxyethyl methacrylate) for use as thin film hydrogels. *Langmuir* 21: 8930–8939.

2

Fundamentals for Studying the Physics of Cat-CVD and Difference from PECVD

For better understanding of the deposition mechanism in catalytic chemical vapor deposition (Cat-CVD), in this chapter, the dynamics of molecules in a vacuum chamber are briefly summarized. As this is not a book on vacuum physics, explanations are simplified just to show the background physics needed for a basic understanding of Cat-CVD. In this chapter, the properties or features of the conventional plasma enhanced chemical vapor deposition (PECVD) method are also summarized to more clearly understand the features of Cat-CVD. Finally, the unique features of Cat-CVD are demonstrated.

2.1 Fundamental Physics of the Deposition Chamber

2.1.1 Density of Molecules and Their Thermal Velocity

When molecules are introduced into a vacuum chamber, they are moving around in the chamber with velocities determined by the temperature of the gas molecules. If the chamber walls and other components inside the chamber have the same temperature, the molecules inside the chamber are in thermal equilibrium after many collisions with the walls or other components in the chamber. In this case, the gas temperature is simply the same as the chamber walls. However, if there is a heated surface inside the chamber, the temperature of the gas molecules attains a distribution. Here, we start to consider the fundamental factors to understand what is going on inside the chamber.

The relationship among the density of gas molecules n_g at gas pressure P_g and gas temperature T_g is expressed by Eq. (2.1) by following a simple formula of an ideal gas, where k refers to Boltzmann constant, $k = 1.38 \times 10^{-23}$ J/K $= 8.62 \times 10^{-5}$ eV/K.

$$n_g = \frac{P_g}{kT_g} \tag{2.1}$$

The deviation from the ideal gas law is minor at low pressures and high temperatures, such as below one atmosphere and over room temperature. Just for giving a quick idea of the numerical values involved, the typical values of n_g are summarized in Table 2.1 for various P_g's and T_g's. The P_g's and T_g's taken here

Catalytic Chemical Vapor Deposition: Technology and Applications of Cat-CVD,
First Edition. Hideki Matsumura, Hironobu Umemoto, Karen K. Gleason, and Ruud E.I. Schropp.

Table 2.1 Typical values of density of gas molecules n_g for various gas temperatures T_g's and gas pressures P_g's.

T_g	$P_g = 1$ Pa	$P_g = 10$ Pa	$P_g = 100$ Pa	$P_g = 101\,325$ Pa (760 Torr)
0 °C (273 K)	$2.65 \times 10^{14}/cm^3$	$2.65 \times 10^{15}/cm^3$	$2.65 \times 10^{16}/cm^3$	$2.69 \times 10^{19}/cm^3$
27 °C (300 K)	$2.42 \times 10^{14}/cm^3$	$2.42 \times 10^{15}/cm^3$	$2.42 \times 10^{16}/cm^3$	$2.45 \times 10^{19}/cm^3$
250 °C (523 K)	$1.39 \times 10^{14}/cm^3$	$1.39 \times 10^{15}/cm^3$	$1.39 \times 10^{16}/cm^3$	$1.40 \times 10^{19}/cm^3$
1000 °C (1273 K)	$5.69 \times 10^{13}/cm^3$	$5.69 \times 10^{14}/cm^3$	$5.69 \times 10^{15}/cm^3$	$5.77 \times 10^{18}/cm^3$
1800 °C (2073 K)	$3.50 \times 10^{13}/cm^3$	$3.50 \times 10^{14}/cm^3$	$3.50 \times 10^{15}/cm^3$	$3.54 \times 10^{18}/cm^3$
2000 °C (2273 K)	$3.19 \times 10^{13}/cm^3$	$3.19 \times 10^{14}/cm^3$	$3.19 \times 10^{15} cm^3$	$3.23 \times 10^{18}/cm^3$

are used in the discussion on the mechanism of film deposition in Cat-CVD later in this book.

In most of the Cat-CVD processes, P_g is about 1–100 Pa, and T_g is higher than room temperature (RT) but lower than 2000 °C because the temperature of the catalyzing wires T_{cat}, which is usually the highest temperature inside the chamber, is 1800–2000 °C for silicon (Si) film deposition, for instance. When silane (SiH_4) gas molecules are decomposed on a tungsten (W) catalyzer of 2000 °C in Si film deposition, decomposed species are emitted from the catalyzer with a temperature of about 1000 °C, as explained in Section 4.4. In addition, the substrate temperature T_s is often kept at 250 °C during deposition. Therefore, the values for $T_g = 250$ and 1000 °C are also included in Table 2.1.

The thermal velocity of gas molecules v_{th}, moving around inside the chamber, is evaluated by following the Boltzmann relation of kinetic gas theory, as shown in Eq. (2.2), where m refers to the mass of a molecule.

$$v_{th} = \sqrt{\frac{8kT_g}{\pi m}} = 5.93 \times 10^{-10} \sqrt{\frac{T_g(\mathrm{K})}{m(\mathrm{kg})}} \ (\mathrm{cm/s}) \tag{2.2}$$

The velocity of gas molecules has a distribution, called Maxwell distribution. We use the *average velocity* defined by Eq. (2.2) as a representative. Typical values of thermal velocity are also summarized in Table 2.2 at T_g between 0 and 2000 °C for various species such as hydrogen (H), nitrogen (N), and Si atoms and hydrogen molecules (H_2), ammonia molecules (NH_3), and SiH_4 molecules. These species are important for the deposition of Si films and silicon nitride (SiN_x) films, which are typical applications of Cat-CVD technology. Here, the masses of H, N, Si, H_2, NH_3, and SiH_4 used in the calculation are 1.67×10^{-27}, 2.33×10^{-26}, 4.66×10^{-26}, 3.45×10^{-27}, 2.83×10^{-26}, and 5.33×10^{-26} kg, respectively.

These two tables tell us that when P_g is about 1 Pa, the molecules with a density of 10^{13} to $10^{14}/cm^3$ are moving around inside the chamber with a velocity of about 1 km/s. In a solid Si crystal, the density of Si atoms is about $5 \times 10^{22}/cm^3$, and in atmospheric pressure of gas with temperature of RT, it is about $2 \times 10^{19}/cm^3$. The density of molecules or atoms in a low-pressure chamber of about 1 Pa is $1/10^6$

Table 2.2 Values of thermal velocity v_{th} for various T_g's for H, N, and Si atoms and H_2, NH_3, and SiH_4 molecules.

	Thermal velocity, v_{th} (cm/s)					
	H	N	Si	H_2	NH_3	SiH_4
0 °C (273 K)	2.40×10^5	6.42×10^4	4.54×10^4	1.67×10^5	5.82×10^4	4.24×10^4
27 °C (300 K)	2.51×10^5	6.73×10^4	4.76×10^4	1.75×10^5	6.11×10^4	4.45×10^4
250 °C (523 K)	3.32×10^5	8.88×10^4	6.28×10^4	2.31×10^5	8.06×10^4	5.87×10^4
1000 °C (1273 K)	5.18×10^5	1.39×10^5	9.80×10^4	3.60×10^5	1.26×10^5	9.16×10^4
1800 °C (2073 K)	6.61×10^5	1.77×10^5	1.25×10^5	4.60×10^5	1.60×10^5	1.17×10^5
2000 °C (2273 K)	6.92×10^5	1.85×10^5	1.31×10^5	4.81×10^5	1.68×10^5	1.22×10^5

to $1/10^5$ of that of a gas at atmospheric pressure and $1/10^9$ of that of atoms in a solid. These ratios give an idea of the relative densities of atoms or molecules.

2.1.2 Mean Free Path

2.1.2.1 Equation Expressing the Mean Free Path

Molecules in the reactor chamber are colliding with the chamber walls and also with the molecules themselves in space. The mean free path, λ, defined as the mean traveling distance of a molecule between the first collision and the second collision with another molecule of the same type is expressed by Eq. (2.3), where σ refers to the diameter of the molecule.

$$\lambda = \frac{1}{\sqrt{2}n_g\pi\sigma^2} = \frac{kT_g}{\sqrt{2}P_g\pi\sigma^2} \tag{2.3}$$

As molecules are rotating freely, they are assumed to be spherical in shape. This equation is valid for the case when only one type of molecules exists inside the chamber. However, in many cases, many kinds of molecules exist in the chamber at the same time. If there are two kinds of molecules inside the chamber, molecule A and molecule B, the mean free path of molecule A, λ_A, is expressed by Eq. (2.4),

$$\lambda_A = \frac{1}{\sqrt{2}n_A\pi{\sigma_A}^2 + \pi{\sigma_{AB}}^2 n_B\sqrt{1 + (m_A/m_B)}} \tag{2.4}$$

where n_A, n_B, σ_A, σ_{AB}, m_A, and m_B refer to the density of molecule A, the density of molecule B, the diameter of molecule A, the sum of the radius of molecule A and the radius of molecule B, the mass of molecule A, and the mass of molecule B, respectively.

In most of the cases, we have to use Eq. (2.4) to determine the mean free path of molecules exactly. However, for a quick understanding of the basic collisions among molecules, we often use Eq. (2.3) to estimate the mean free path roughly. In addition, although we refer to the mean free path of molecules, the equation also applies to atoms or other species such as electrons if such species receive no other energies apart from the thermal energy inside the chamber.

2.1.2.2 Estimation of Diameter of Molecules or Species

In the estimation of the mean free path, we have to know the diameter σ of the colliding species. There are several ways to estimate the diameter of molecules or species. σ can be evaluated experimentally or theoretically from the data of viscosity, density, and atomic distances in a molecule that is calculated by van der Waals bond length or covalent bond length. The estimated value of σ varies widely. For instance, σ for H_2 molecules varies from 128 pm estimated simply from the theoretical covalent bond radius to 275 pm estimated experimentally from viscosity and also to 419 pm estimated from density. This variation induces a large ambiguity in the estimation of mean free path. If we rely on only the experimental data, it is not easy to determine the various σ's and the mean free paths correctly.

Recently, the covalent bond radius for various elements in the periodic table has been calculated for a single bond, a double bond, and a triple bond [1–4]. The results for H, which has an atomic number of 1 in the periodic table, to argon (Ar), having an atomic number of 18, are summarized in Figure 2.1. In the figure, the atomic number and the radius of the covalent single bond, double bond, and triple bond are all described. From the combination of these covalent bond radii, we can roughly estimate the diameter of molecules composed by covalent bonds of the elements under the assumption of a sphere for a molecule.

The diameter is sometimes estimated from van der Waals radius or ionic bond radius apart from the covalent bond radius shown in Figure 2.1. The diameter of species depends on the bonding type. However, to estimate the physical values of species in the chamber, we have to use one of the known values just for obtaining rough figures. Thus, we estimate the diameters with some ambiguities. For instance, the diameter of H atoms is evaluated 64 pm from the covalent bond

Covalent bond radius

Element	H	He
Atomic number	1	2
Single bond radius (pm)	32	46
Double bond radius (pm)	—	—
Triple bond radius (pm)	—	—

Element	Li	Be	B	C	N	O	F	Ne
Atomic number	3	4	5	6	7	8	9	10
Single bond radius (pm)	133	102	85	75	71	63	64	67
Double bond radius (pm)	124	90	78	67	60	57	59	96
Triple bond radius (pm)	—	85	73	60	54	53	53	—

Element	Na	Mg	Al	Si	P	S	Cl	Ar
Atomic number	11	12	13	14	15	16	17	18
Single bond radius (pm)	155	139	126	116	111	103	99	96
Double bond radius (pm)	160	136	113	107	102	94	95	107
Triple bond radius (pm)	—	127	111	102	94	95	93	96

Figure 2.1 Covalent bond radius for various elements in the periodic table. Source: The table is newly constructed by using the data in Refs. [1–4].

Table 2.3 Estimated diameters of H, N, and Si atoms and H_2, NH_3, and SiH_4 molecules.

Species	Estimated values (pm)	Method of derivation	Value used here (pm)
H	64–106	Covalent bond radius in Figure 2.1 and Bohr radius	106
N	142	Covalent bond radius	142
Si	150–232	Bond length in the Si crystal observed by STEM and covalent bond radius	232
H_2	128–419	Covalent bond radius and radius estimated from viscosity or density	275
NH_3	206–374	Covalent bond radius and van der Waals radius	290
SiH_4	326–416	Si—H bond length and covalent bond radius	371

STEM: scanning transmission electron microscope.

radius in Figure 2.1; however, the diameter evaluated from Bohr's radius, 106 pm, which is the value of classical atomic model, is also still widely used.

The estimated diameters of H, N, and Si atoms, as well as H_2, NH_3, and SiH_4 molecules, are summarized in Table 2.3. In the table, the diameters used in the following calculations in this book are described in the right column. Average values within the range of estimated values are adopted for H_2, NH_3, and SiH_4. The Bohr radius for H and the covalent bond radius for N and Si are also adopted here. These values appear reasonable, considering the quantitative ranking of species in terms of the diameter.

2.1.2.3 Examples of Mean Free Path

Here, we estimate the mean free path λ for species used in a-Si and SiN_x deposition. Table 2.4 shows the values of λ of H, N, and Si atoms and H_2, NH_3, and SiH_4

Table 2.4 The estimated values of mean free path for H, N, and Si atoms and H_2, NH_3, and SiH_4 molecules for $P_g = 1$ Pa, using $\sigma_H = 1.06 \times 10^{-8}$ cm, $\sigma_{Si} = 2.32 \times 10^{-8}$ cm, $\sigma_{H_2} = 2.75 \times 10^{-8}$ cm, $\sigma_{SiH_4} = 3.71 \times 10^{-8}$ cm, $\sigma_N = 1.42 \times 10^{-8}$ cm, and $\sigma_{NH_3} = 2.90 \times 10^{-8}$ cm.

	$P_g = 1$ Pa					
T_g	H (cm)	N (cm)	Si (cm)	H_2 (cm)	NH_3 (cm)	SiH_4 (cm)
0 °C (273 K)	7.55	4.21	1.58	1.12	1.01	0.616
27 °C (300 K)	8.29	4.62	1.73	1.23	1.11	0.677
250 °C (523 K)	14.5	8.06	3.02	2.15	1.93	1.18
1000 °C (1273 K)	35.2	19.6	7.35	5.23	4.70	2.87
1800 °C (2073 K)	57.3	31.9	12.0	8.51	7.66	4.68
2000 °C (2273 K)	62.8	35.0	13.1	9.34	8.39	5.13

molecules for T_g between 0 and 2000 °C and for $P_g = 1$ Pa. Table 2.4 shows only the case of $P_g = 1$ Pa; however, it will be very easy to obtain λ for any value of P_g. As indicated in Eq. (2.3), λ is simply proportional to T_g and inversely proportional to P_g. For instance, λ of a SiH_4 molecule of temperature of 0 °C is 0.616 cm, but if T_g is elevated to 2000 °C, it becomes 5.13 cm. By elevating T_g, from 0 to 2000 °C, the mean free path increases by almost 1 order of magnitude.

As already briefly mentioned and also explained in detail later in Chapter 4, when a SiH_4 molecule impinges on the W catalyzer heated at 2000 °C, it is decomposed to Si + 4H with a high probability and desorbed from the surface of the catalyzer with a temperature of about 1000 °C. After touching the heated W surface, if SiH_4 molecules are not decomposed and recoiled back to the space, the mean free path of such SiH_4 molecules may be several centimeters for $P_g = 1$ Pa, although the exact estimation of the mean free path is not easy when various molecules and atoms exist in the same space. However, if they are decomposed, the decomposed H and Si atoms can travel much longer, before next collisions. The area within several centimeters from the catalyzer is a zone with thermal gradient in this case. This gives a rough idea what is going on near the heated catalyzer in the Cat-CVD chamber of $P_g = 1$ Pa.

2.1.2.4 Interval Time Between the First Collision and the Second Collision

The time interval between the first collision and the second collision, t_{col}, is simply evaluated from Eqs. (2.3) and (2.4), as expressed by Eq. (2.5). In addition, this t_{col} is also summarized in Table 2.5 for the case of only a single type of species existing in the Cat-CVD chamber.

$$t_{col} = \frac{\lambda}{v_{th}} \tag{2.5}$$

As shown in Table 2.5, approximately every 10^{-5} seconds, each species in the chamber collides with other species under 1 Pa.

Table 2.5 Examples of interval time between the first collision and the next collision, t_{col}, for H, N, and Si atoms and H_2, NH_3, and SiH_4 molecules for $P_g = 1$ Pa.

	t_{col} for $P_g = 1$ Pa (s)					
T_g	H	N	Si	H_2	NH_3	SiH_4
0 °C (273 K)	3.15×10^{-5}	6.56×10^{-5}	3.48×10^{-5}	6.71×10^{-6}	1.74×10^{-5}	1.45×10^{-5}
27 °C (300 K)	3.30×10^{-5}	6.86×10^{-5}	3.63×10^{-5}	7.03×10^{-6}	1.82×10^{-5}	1.52×10^{-5}
250 °C (523 K)	4.37×10^{-5}	9.08×10^{-5}	4.81×10^{-5}	9.31×10^{-6}	2.39×10^{-5}	2.01×10^{-5}
1000 °C (1273 K)	6.80×10^{-5}	1.41×10^{-4}	7.50×10^{-5}	1.45×10^{-5}	3.73×10^{-5}	3.13×10^{-5}
1800 °C (2073 K)	8.67×10^{-5}	1.80×10^{-4}	9.60×10^{-5}	1.85×10^{-5}	4.79×10^{-5}	4.00×10^{-5}
2000 °C (2273 K)	9.08×10^{-5}	1.89×10^{-4}	1.00×10^{-4}	1.94×10^{-5}	4.99×10^{-5}	4.20×10^{-5}

2.1.3 Collisions with a Solid Surface

2.1.3.1 Collisions with a Solid Surface

The molecule also collides with the solid surface such as the chamber wall. Here, let us evaluate the number of collisions of molecules with a unit area per unit time N_{col}. It is known that this N_{col} can be simply expressed by Eq. (2.6).

$$N_{col} = \frac{1}{4} n_g v_{th} \tag{2.6}$$

This equation, which is important when considering the collision of molecules with a heated catalyzer, is derived as follows.

At first, we consider a hemisphere with a radius r, which is shorter than the mean free path of the molecules, and also consider a small area dS at the center of it. Polar coordinates are set around the center as shown in Figure 2.2. Here, we consider a very thin shell with a thickness dr on the surface of the hemisphere, and a small volume at the position (r, θ, ϕ), surrounded between θ and $\theta + \mathrm{d}\theta$, ϕ and $\phi + \mathrm{d}\phi$, and r and $r + \mathrm{d}r$ in such a shell, as also indicated in Figure 2.2. Using the density of molecules n_g, the number of molecules existing in such a surrounded small volume is expressed by $(n_g \times r\sin\theta\ \mathrm{d}\phi\, r\ \mathrm{d}\theta \times \mathrm{d}r)$. Among such molecules, only the molecules moving toward the origin of coordinates can collide with the small area dS. Actually, such a small area dS is seen as $\mathrm{d}S' = \mathrm{d}S \cos\theta$ from the position (r, θ, ϕ), as the molecules are coming with an angle θ. The number of molecules moving toward dS' is $(n_g \times r\sin\theta\ \mathrm{d}\phi\ r\,\mathrm{d}\theta \times \mathrm{d}r) \times (\mathrm{d}S'/4\pi r^2) = (n_g \times \sin\theta \cos\theta\ \mathrm{d}\theta\ \mathrm{d}\phi\ \mathrm{d}r) \times (\mathrm{d}S/4\pi)$, as all molecules are freely moving over 360° space angles (4π steradian) and the probability having the direction toward dS' is $(\mathrm{d}S'/4\pi r^2)$.

All molecules colliding with the small area dS with thermal velocity v_{th} and within a time dt are the molecules existing inside the hemisphere of $r < v_{th}\mathrm{d}t$. Thus, the number of such molecules colliding from the angles of θ and $\theta + \mathrm{d}\theta$, ϕ

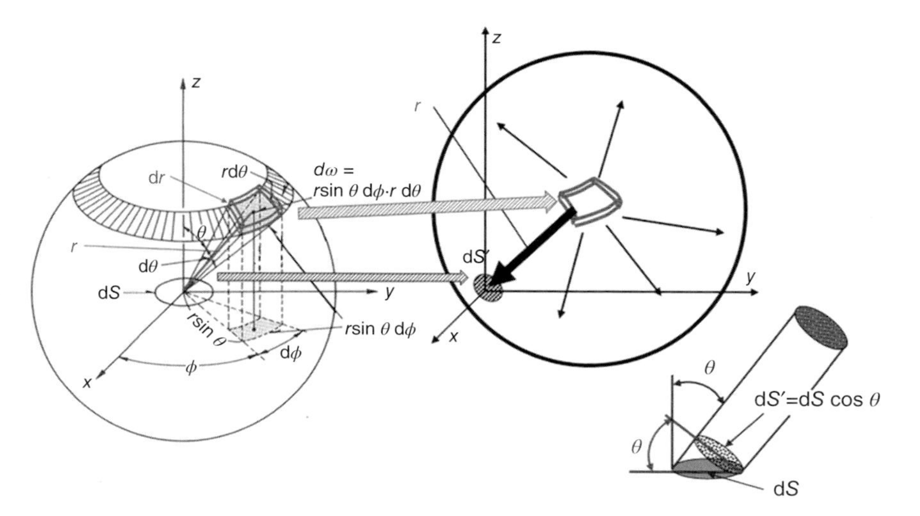

Figure 2.2 Schematic view of molecules colliding with the area dS.

and $\phi + \mathrm{d}\phi$ is evaluated by

$$\int_0^{v_{\mathrm{th}}dt} \frac{n_{\mathrm{g}}\mathrm{d}S}{4\pi} \sin\theta \cos\theta \mathrm{d}\theta \mathrm{d}\phi \mathrm{d}r = \frac{n_{\mathrm{g}}\mathrm{d}S}{4\pi}(v_{\mathrm{th}}\mathrm{d}t) \sin\theta \cos\theta \mathrm{d}\theta \mathrm{d}\phi$$

Thus, the total number of molecules colliding with the area dS within the time dt from any direction of a hemisphere, $N_{\mathrm{col}}\mathrm{d}S\mathrm{d}t$, is evaluated by taking the integral of θ and ϕ as follows:

$$N_{\mathrm{col}}\mathrm{d}S\mathrm{d}t = \frac{n_{\mathrm{g}}v_{\mathrm{th}}}{4\pi}\mathrm{d}S\mathrm{d}t \int_0^{\pi/2} \sin\theta \cos\theta \mathrm{d}\theta \int_0^{2\pi} \mathrm{d}\phi = \frac{1}{4}n_{\mathrm{g}}v_{\mathrm{th}}\mathrm{d}S\mathrm{d}t$$

From this relation, Eq. (2.6) can be derived.

Two physical parameters, the mean free path of molecules or species and the number of molecules or species colliding with the solid surface, are important values to consider the process of Cat-CVD. The number of colliding molecules per unit area and per second is simply estimated from Eq. (2.6). Typical areal collision frequencies are summarized in Table 2.6, for H_2, NH_3, and SiH_4 molecules.

2.1.3.2 Comparison of Collisions of Molecules in Space with Collisions at Chamber Wall

Here, let us estimate the number of collisions among molecules moving around inside the chamber with an inner volume of V_{ch}. Each molecule collides with another molecule once per t_{col}. Within this t_{col}, on average, $V_{\mathrm{ch}}n_{\mathrm{g}}/2$ times collisions should occur in total inside the chamber. As collisions occur between two molecules, the total number of molecules existing inside the chamber $V_{\mathrm{ch}}n_{\mathrm{g}}$ should be divided by 2. Thus, per unit time, the number of collisions inside the chamber $N_{\mathrm{col\text{-}space}}$ becomes Eq. (2.7),

$$N_{\mathrm{col\text{-}space}} = \frac{V_{\mathrm{ch}}n_{\mathrm{g}}}{2t_{\mathrm{col}}} = \frac{V_{\mathrm{ch}}n_{\mathrm{g}}}{2}\frac{v_{\mathrm{th}}}{\lambda} = \frac{\pi}{\sqrt{2}}V_{\mathrm{ch}}n_{\mathrm{g}}^2\sigma^2 v_{\mathrm{th}} \tag{2.7}$$

For instance, if we assume a cylindrical chamber with an inner diameter of 50 cm and an inner height of 50 cm, V_{ch} becomes approximately 1×10^5 cm^3.

Table 2.6 Number of collisions of H_2, NH_3, and SiH_4 molecules with a solid surface in a unit time and a unit area, for the case of $P_{\mathrm{g}} = 1$ Pa.

	Collisions on solid surface for $P_{\mathrm{g}} = 1$ Pa (times/cm^2 s)		
T_{g}	H_2	NH_3	SiH_4
0 °C (273 K)	1.11×10^{19}	3.86×10^{18}	2.81×10^{18}
27 °C (300 K)	1.06×10^{19}	3.70×10^{18}	2.69×10^{18}
250 °C (523 K)	8.03×10^{18}	2.80×10^{18}	2.04×10^{18}
1000 °C (1273 K)	5.12×10^{18}	1.79×10^{18}	1.30×10^{18}
1800 °C (2073 K)	4.03×10^{18}	1.40×10^{18}	1.02×10^{18}
2000 °C (2273 K)	3.84×10^{18}	1.34×10^{18}	9.73×10^{17}

When P_g in the chamber is 1 Pa, the density of SiH_4 molecules is 2.65×10^{14} cm^{-3} at T_g of 0 °C as shown in Table 2.1 and also t_{col} is 1.45×10^{-5} second as shown in Table 2.5. Thus, the total number of collisions among SiH_4 molecules inside the chamber is evaluated to be 9.14×10^{23} times per second.

On the other hand, the inner area of the chamber wall, S_{ch}, is calculated to be about 1.178×10^4 cm^2 in this case, and thus, the number of collisions of SiH_4 molecules with the chamber wall at $T_g = 0$ °C is estimated to be 3.31×10^{22} times/s from Table 2.6. The ratio of collisions with the chamber wall to collisions in the space inside the chamber is about 3.6%.

The above calculations are for the case of $V_{ch} = 1 \times 10^5$ cm^3. If a small experimental chamber with an inner diameter of 20 cm and an inner height of 20 cm is considered, V_{ch} becomes 6.280×10^3 cm^3 and S_{ch} 1.885×10^3 cm^2, and the total collisions of SiH_4 molecules inside the chamber is 5.72×10^{22} times/s and the number of collisions with the chamber wall is 5.29×10^{21} times/s. The ratio of collisions with the chamber wall to collisions in the chamber volume is about 9.2%. Thus, this ratio increases when a smaller chamber is used for deposition. That is, the film deposition is to be a matter of collisions on the walls in a smaller chamber. As the species can often stay long on the chamber wall, the increase of the effect of chamber wall in reactions involved in forming films becomes sometimes non-negligible. By changing the chamber size, we have to carefully adjust the residence time of species, as described in the following section. In addition to that, we have also taken into consideration the reactions at the chamber wall in some cases.

2.1.4 Residence Time of Species in Chamber

When we think about the dynamics of molecules or atoms in a low-pressure deposition chamber, we have to pay attention to one more important physical value, that is, the residence time t_{res}, which expresses the time a molecule stays inside the chamber before evacuation. This value is sometimes strongly related to the quality of films deposited in the chamber. When t_{res} is long, a molecule has many chances to be decomposed or involved in film formation process. Species existing inside the chamber are rather "old" species, and they have experienced many reaction events. However, when t_{res} is short, the chamber is filled with many "fresh" molecules, and most of the species still have a simple configuration because of the small number of reactions during their short t_{res}. When we design a deposition apparatus or when we enlarge the chamber based on the experimental data obtained from a small size experimental chamber, it is important to adjust this t_{res} to obtain similar quality films after changing the chamber.

First, let us consider a chamber with a volume of V_{ch}, and a flow of source gas with a rate of Q_0 introduced into the chamber, as illustrated in Figure 2.3. The t_{res} is expressed by Eq. (2.8),

$$t_{res}(\text{s}) = 5.92 \times 10^{-4} \frac{V_{ch}(\text{cm}^3) P_g(\text{Pa})}{Q_0(\text{sccm})} \tag{2.8}$$

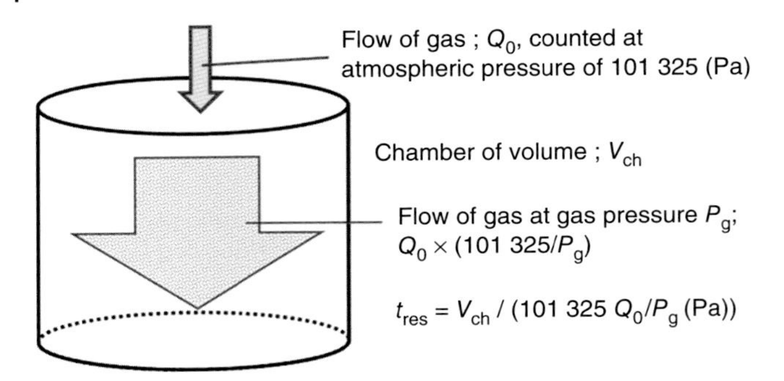

Figure 2.3 Schematic illustration of gas flow in a low-pressure chamber.

In Eq. (2.8), t_{res}, V_{ch}, and Q_0 use the unit of (s), (cm^3), and (sccm), respectively. The unit (sccm) expresses the volume of gas flow at standard atmospheric pressure in cubic centimeter per minute.

The derivation of this equation is quite simple. When a gas of volume Q_0 in atmospheric pressure, 101 325 Pa, is introduced into the chamber of the pressure P_g (Pa) per minute, it expands in the chamber to the volume of $Q_0 \times (101\,325/P_g)$. This volume of gas is supplied per minute; that is, the volume $Q_0 \times (101\,325/60P_g)$ of gas is supplied per second. As the volume of chamber is V_{ch}, the gas molecules having the flow of $Q_0 \times (101\,325/60P_g)$ can reside in the chamber for $V_{ch}/[Q_0 \times (101\,325/60P_g)]$ seconds. Here, $60/101\,325 = 5.92 \times 10^{-4}$ explains the prefactor in Eq. (2.8). This derivation assumes that the temperature in the chamber is not greatly different from room temperature. This assumption is not precarious even when the wires are heated. The average gas temperature in the chamber is usually below 400 K.

When $V_{ch} = 1 \times 10^5\ cm^3$ as already mentioned, for instance, then t_{res} is estimated to be 1.2 seconds when the flow rate is 50 sccm and the pressure is 1 Pa. That is, t_{res} is usually in the order of a second.

2.2 Difference Between Cat-CVD and PECVD Apparatuses

Figure 2.4 demonstrates the schematic views of a diode-type PECVD apparatus and a Cat-CVD apparatus. The structural difference between two deposition methods is apparent.

There are various types of apparatuses in PECVD, such as inductively coupled plasma (ICP) apparatus, diode-type apparatus with parallel two electrodes as shown here, and the diode-type apparatus with an array of many hollow cathodes for large-area deposition as shown later in Figure 2.11. Among them, the most widely used system in the industry is the diode-type one.

In PECVD, as the electric potential has to be sustained at the electrodes, the electric insulation has to be carefully taken. In addition, usually, as the radio-frequency (RF) power of 13.56 MHz is applied for generating the plasma,

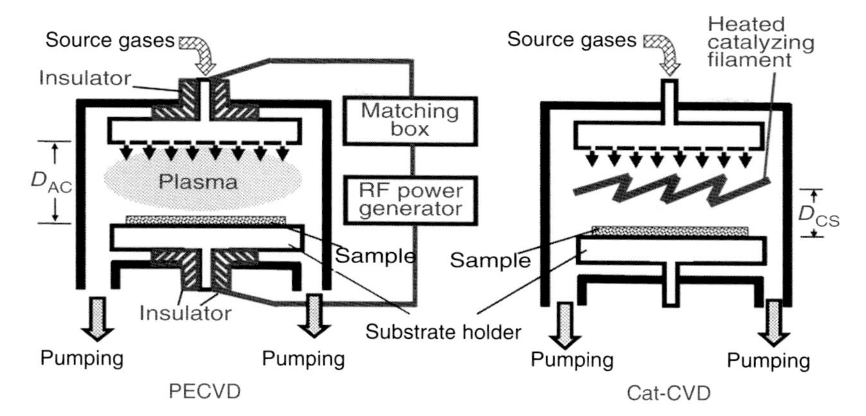

Figure 2.4 Schematic diagram of apparatuses for plasma-enhanced chemical vapor deposition (PECVD) and catalytic chemical vapor deposition (Cat-CVD).

the power matching has to be carefully controlled by a matching box for the efficient transfer of the RF power from a power generator to the chamber.

On the other hand, in Cat-CVD, a heated metal wire is placed inside the chamber, and this wire performs almost the same function as that the plasma does in PECVD. The structure is very simple, and the quality of films prepared by Cat-CVD is often superior to those by PECVD, as is described in Chapter 5 and other chapters in this book.

2.3 Fundamental Features of PECVD

2.3.1 Birth of PECVD

The fundamental features of PECVD are briefly explained, here, although many useful books have been already published for study on PECVD [5]. Readers who like to know more about PECVD are referred to these books.

PECVD is a common technique for preparing device quality thin films at low substrate temperatures below 400 °C. It is widely used in various fields such as liquid crystal display (LCD), ultralarge scale integrated circuits (ULSI), solar cell industries, and film coating industries. However, the history of PECVD is not a long one. The research on PECVD started in 1960s, and the technology was gradually implemented in industry in 1970s.

At first, H.F. Sterling and R.C.G. Swann established their patents on the invention of the plasma process, or PECVD, in 1964–1965 [6, 7], and later published their findings in a scientific paper [8]. The invention was an outcome of the research on Si epitaxial growth in a vertical quartz tube. Inside the tube, a carbon or molybdenum susceptor was placed, and it was heated remotely by the RF energy transmitted by inductive coupled coils mounted outside of the tube. At that time, this heating method using RF power supplied from the outside of the chamber was a new technology meant to heat up the sample. During the study on this thermal system, they discovered that the inorganic films could be obtained at low substrate temperatures when the glow discharge was generated in their quartz tube by the RF coil. After discovering the film deposition, various films have been prepared by this glow discharge method. The preparation of silicon nitride (SiN_x) films was reported in 1967 by R.G. Swan et al. [9] just after the report of sputtering SiN_x films by S.M. Hu [10]. The preparation of amorphous silicon (a-Si) films was reported in 1969 by R.C. Chittick et al. [11]. Since then, the technology began to be used for many purposes.

2.3.2 Generation of Plasma

In the diode-type PECVD apparatus, the chamber is filled with source gas molecules to a gas pressure P_g of 1–100 Pa. Plasma is generated in the gas-filled space between two parallel electrodes, anode and cathode, by applying a voltage on the two electrodes. Here, the distance between the two electrodes is described as D_{AC}.

When direct current (DC) voltages or low-frequency alternative current (AC) voltages are applied on the electrodes, and also when the applied voltage itself is larger than a certain threshold voltage, gas molecules are decomposed by the collisions with electrons that are accelerated by the electric field in the space between the two electrodes. Some of the molecules are ionized. The number of ions produced by a collision with a single electron, moving in a unit distance toward the anode, is defined as Townsend's first ionization coefficient, α_{ion}. These ions produced in the space between the electrodes are accelerated by the same electric field in the direction opposite to that of the electrons, and they finally collide with the cathode electrode, causing emission of many electrons from the electrode. The number of electrons emitted from the electrode by the collision with a single ion is defined as Townsend's second ionization coefficient, γ. These emitted electrons are accelerated toward the anode and in turn also take part in decomposing molecules. Thus, when α_{ion} and γ are larger than 1, an avalanche of ions and electrons are produced in the space between the electrodes and the plasma is generated. This situation is schematically explained in Figure 2.5.

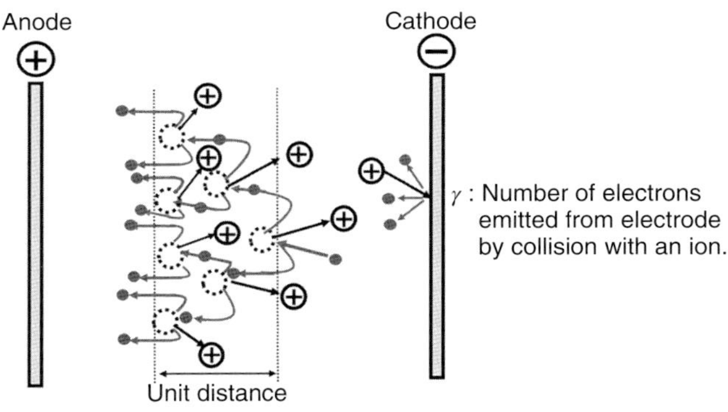

Figure 2.5 Schematic illustration of plasma formation with Townsend's ionization coefficients.

2.3.3 DC Plasma to RF Plasma

For easier formation of an electric discharge or easier generation of a plasma and also for eliminating the influence of electrodes, RF AC voltages of 13.56 MHz are very often applied to the electrodes. Sometimes, a very high frequency (VHF) AC voltage, higher than 27 MHz, is applied for obtaining a higher density of electrons in the plasma, as mentioned later. When AC voltage is applied on the two electrodes of the diode system, the polarity of electrodes changes quickly. Charged particles travel toward an electrode, but in the next half cycle of AC voltage, they have to change the direction. If we neglect the inertia for the movement of charged particles owing to the electric forces from the other charged particles, the following simple consideration can be made as a first-order approximation.

We consider that the electric field $E_0 \sin(\omega t + \theta)$ is applied in the space between two electrodes by applying AC voltage on them with an angular frequency ω or a frequency f ($\omega = 2\pi f$), where E_0, t, and θ refer to the amplitude of applied alternative electric field, the time, and the phase of the alternative field, respectively. It is known that the product of the mobility of a charged particle, μ, and the electric field is equal to the velocity of the particle, v. The velocity of the charged particle located at position x should be equal to $\mu E_0 \sin(\omega t + \theta)$. Thus, $dx/dt = v = \mu E_0 \sin(\omega t + \theta)$, then $x = (\mu E_0/\omega)\cos(\omega t + \theta) + C$, where C refers to an integral constant. Here, the x-axis is taken along the direction from an electrode to another electrode. That is, the amplitude A of the oscillation of the charged particles is simply expressed by

$$A = \frac{\mu E_0}{\omega} = \frac{\mu E_0}{2\pi f} \quad (2.9)$$

This equation tells us that an increase of ω or f causes a decrease of A and that when ω or f is large enough, $2A$ cannot reach D_{AC}, and thus, the charged particles are confined in the space between two electrodes. As the mobility of an electron, μ_e, is much larger than the mobility of an ion, μ_i, because of the large difference of mass, A for electrons is much larger than A for ions.

In most cases, when the frequency exceeds over several hundred kilohertz, this confinement happens to ions at first. Thus, in the most of RF-PECVD systems, Townsend's second ionization coefficient, γ, does not have any meaning, as ions cannot reach the electrodes any more. The influence of electrodes is eliminated in the generation of plasma, and only the value of α_{ion} becomes important. In the frequency range from several hundred kilohertz to about several megahertz, the threshold voltage for generation of plasma is higher than that for DC plasma generation, as electrons are not emitted from the electrode by ion bombardment.

When RF of 13.56 MHz is used, A becomes very small even for electrons, and they cannot reach the electrodes. The electron density increases enormously to make many ions and electrons in the confined space. When the frequency is in the range from about several megahertz or 10 MHz to several hundred megahertz, the threshold voltage required for generating a plasma decreases with respect to that for DC plasma generation, and therefore, the generation of a plasma becomes easier in RF than DC.

2.3.4 Sheath Voltage

In RF plasma, ions and electrons are oscillating within a small limited space. Then, the densities of electrons and ions in such a region become so high that both electrons and ions spread out toward the outside of plasma. That is, the electrons and ions collide with the electrodes at first. As indicated by Eq. (2.6), the number of such collisions is proportional to the density and the velocity of electrons or ions. However, the velocity of electrons, shown in Table 2.7, is about 3 orders of magnitude larger than that of ions. The thermal velocities of ions are the same order as those of neutral species listed in Table 2.2. Thus, the flow of electrons from plasma to the electrodes becomes much larger than that of ions, and actually, electron fluxes toward electrodes are induced. Here, if the electrodes are floating or electrically almost isolated, the electrodes become negatively charged, and the electric potential of the plasma becomes positive with respect to the electrodes. This induced potential is called the sheath voltage, V_{sheath}. The potential assumes a value so as to balance the electron flux and the ion flux at the electrodes. Based on this mechanism, the V_{sheath} is approximately derived as

$$V_{\text{sheath}} = \frac{kT_{\text{e}}}{2e} \ln\left(\frac{8M}{\pi m_{\text{e}}}\right) \tag{2.10}$$

where T_{e}, e, M, and m_{e} refer to the electron temperature, the charge of an electron, the mass of an ion, and the mass of an electron, respectively.

This situation is schematically illustrated in Figure 2.6a. This equation holds only for electrically isolated electrodes. However, even in the actual PECVD apparatus, although the electrodes are not completely isolated or floated, Eq. (2.10) can be still used approximately, as the currents supplied to the electrodes from the outside during the deposition is not so large compared with the inner currents induced by the electron collisions at the electrodes.

In addition, in most of the cases, two electrodes are not identical for RF signals. If there is a slight difference of areas in two electrodes, one electrode is biased to positive against the other electrode unintentionally. Actually, one electrode is often grounded, thus, such an electrode has a larger area than the other one by addition of the area of a chamber wall. Therefore, in RF-PECVD, there are anode and cathode electrodes like DC-PECVD. Of course, sometimes, this bias voltage is intentionally controlled by applying an external voltage. Then, the actual potential in most of the RF-PECVD systems is as shown in Figure 2.6b.

The figure shows that the potential difference between plasma and the electrode is larger at the cathode than at the anode. For keeping plasma, the loss of electrons from the electrically neutral plasma is compensated by new electrons provided

Table 2.7 Relation among kinetic energy, temperature T_{e}, and thermal velocity $v_{\text{th(e)}}$ of electrons.

Energy (eV)	1	3	5	10	20	30
T_{e} (K)	7.74×10^3	2.32×10^4	3.87×10^4	7.74×10^4	1.55×10^5	2.32×10^5
$v_{\text{th(e)}}$ (cm/s)	5.93×10^7	1.03×10^8	1.33×10^8	1.88×10^8	2.65×10^8	3.25×10^8

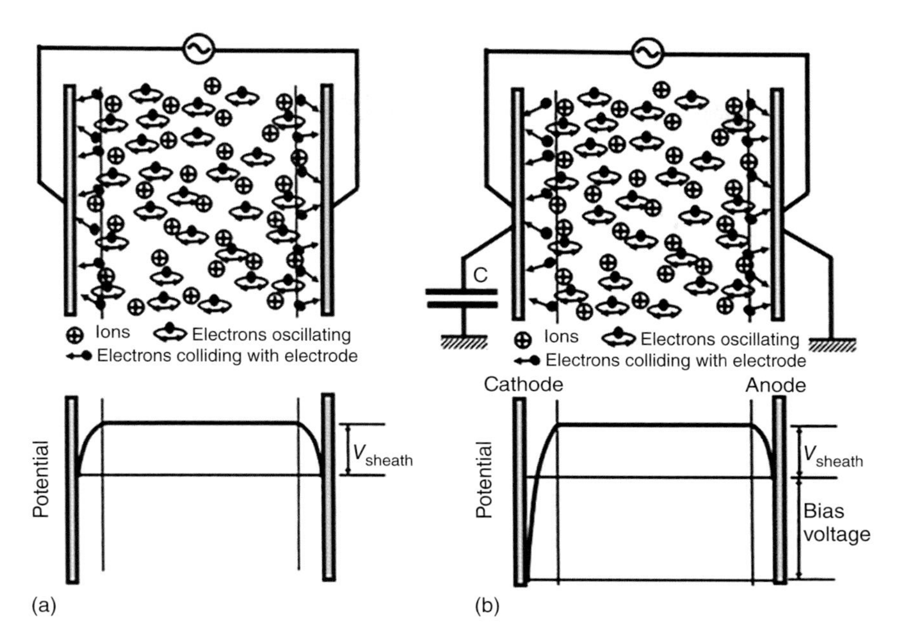

Figure 2.6 Movement of ions and electrons in AC electric field and potential in space between two electrodes. (a) Almost floating electrodes. (b) Biased electrodes.

from the electrode. In this way, currents start to flow through the plasma. The currents and the voltage between two electrodes should be maintained by the current supply from an RF power generator through an RF matching.

It is worth noticing an important point in RF-PECVD from the above explanation of the figure. By applying a bias voltage over the two electrodes, we can impose a voltage difference between two electrodes; however, the potential difference between plasma and the electrode cannot be eliminated. The generation of the sheath voltage is substantially originated by the difference between the electron velocity and the ion velocity. The existence of the sheath voltage is inevitable in PECVD, and we cannot escape from the influence of this sheath voltage.

If we consider the case of a-Si deposition by using SiH_4 gas, V_{sheath} is found to be in the range from 10 V to several tens volts, depending on P_g and RF power. The P_g and RF power determine T_e through the alternating electric field between two electrodes and the acceleration period of electrons. Here, we must notice that ions are accelerated by this potential toward the electrode, i.e. toward samples mounted on the electrode. Ions are finally implanted into samples with the energy of eV_{sheath} and contribute to the growth of the film.

2.3.5 Density of Decomposed Species in PECVD

2.3.5.1 Number of Collisions Between Electrons and Gas Molecules

When the collisions between electrons and source gas molecules are considered, we can use Eq. (2.4) for derivation of the mean free path of electrons. We assume that species A is an electron and species B is a gas molecule in Eq. (2.4). The size of an electron is negligibly small, and σ_A is approximated as 0. As the mass of an electron is much smaller than that of a source gas molecule, m_A/m_B is also approximated as 0. The term σ_{AB} should be actually the radius of a gas molecule B, and $\pi\sigma_{AB}^2$ means the scattering cross section for the collision between an electron and a molecule. Here, such a scattering cross section is described $\Sigma_{e\text{-}B}$, and hence, the mean free path of an electron, λ_e, is expressed by

$$\lambda_e = \frac{1}{\Sigma_{e-B} n_B}. \tag{2.11}$$

The interval time, counted from the first collision between an electron and molecule B to the second collision between the electron and another molecule B, $t_{\text{col-e}}$, is also derived as Eq. (2.12), similarly to Eq. (2.5),

$$t_{\text{col-e}} = \frac{\lambda_e}{v_{\text{th(e)}}}, \tag{2.12}$$

where $v_{\text{th(e)}}$ refers to the thermal velocity of the electron. Then, we can estimate the number of collisions within a unit time, the collision frequency, by $1/t_{\text{col-e}}$.

The kinetic energy of electrons, ε_k, can be expressed in two ways: one is the electron temperature, T_e, and the other is the thermal velocity. The relation is simply expressed by Eq. (2.13).

$$\varepsilon_k = \frac{1}{2} m_e {v_{\text{th(e)}}}^2 = \frac{3}{2} k T_e, \tag{2.13}$$

where m_e is the mass of an electron, which is equivalent to m_A in Eq. (2.4). Examples of these values are summarized in Table 2.7, to provide an idea of the movement of electrons.

If the volume of plasma-generated space, V_{plasma}, is filled with electrons with a density of n_e, a total number of electrons existing in such a plasma space should be $V_{\text{plasma}} n_e$. Thus, similar to Section 2.1.3.2 and Eq. (2.7), the number of collisions that occurs in the plasma space per unit time, $N_{\text{col-plasma}}$, becomes Eq. (2.14), considering that the factor (1/2) in Eq. (2.7) is not necessary in this case.

$$N_{\text{col-plasma}} = \frac{\mathrm{V}_{\text{plasma}} n_e}{t_{\text{col-e}}} = V_{\text{plasma}} n_e n_B \Sigma_{e-B} v_{\text{th(e)}} \tag{2.14}$$

2.3.5.2 Number of Decomposed Species in PECVD

Here, we estimate the number of decomposed species in PECVD. As an example of calculation, we here choose SiH_4 decomposition, as it is widely studied and useful for Si film deposition.

There are a number of reports on the density of electrons in PECVD. The electron density appears larger in a plasma generated in Ar gas [12] than in a SiH_4 plasma [13, 14]. It is about $10^{10}/cm^3$ in an Ar plasma and about $10^9/cm^3$ in a SiH_4 plasma. Of course, it is also depending on the RF power. Thus, here, it is approximated as in the order of $10^{10}/cm^3$ for RF-PECVD, and the value is likely to increase several times for increasing plasma frequency from 13.56 to 87 MHz [14].

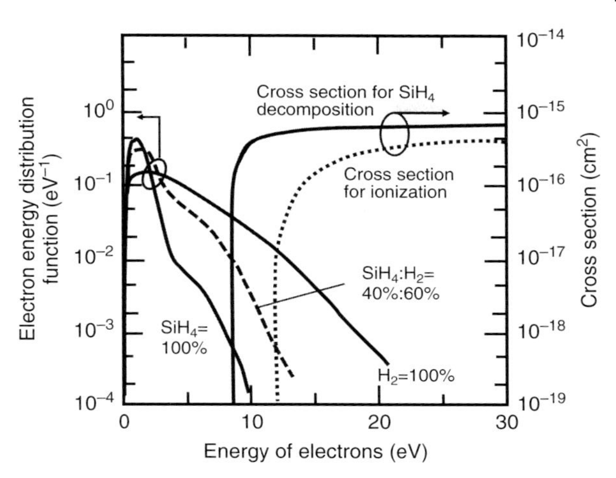

Figure 2.7 The electron energy distribution function (EEDF) and the cross sections for silane (SiH_4) decomposition and for ionization, as a function of electron energy, for radio-frequency plasma-enhanced chemical vapor deposition (RF-PECVD). Source: Data from Godyak et al. 1992 [12] and Gabriel et al. 2014 [13].

The distribution of electron energy, the electron energy distribution function (EEDF), and the cross section for collisions between electrons and SiH_4 molecules are also studied in RF-PECVD. Figure 2.7 shows EEDF for a system with only SiH_4 or H_2 gas and a system with a SiH_4 and H_2 mixture in RF-PECVD conditions. The EEDF depends on the plasma conditions, and it changes depending on the mixing ratio of SiH_4 and H_2. In PECVD, even if a pure SiH_4 gas is introduced into the chamber initially, H atoms are generated in the chamber by the decomposition of SiH_4, and thus, the chamber is soon filled with a plasma containing a SiH_4 and H_2 mixture.

The EEDF shown in the figure with a unit of (eV^{-1}) was derived from the data of the electron energy probabilistic function (EEPF), calculated by T. Shirafuji [15] with a unit of ($eV^{-3/2}$), by using the relation that EEDF is $(\varepsilon_k)^{1/2} \times$ EEPF. The expression using EEDF with a unit (eV^{-1}) is useful. For instance, the integral of EEDF from 0 to 8.75 eV of electron energy simply corresponds to the ratio of the number of electrons with the energies lower than 8.75 eV to the total number of electrons in plasma.

In the figure, the cross sections for SiH_4 decomposition and for ionization by electron impact are demonstrated. As shown in the figure, an energy of electrons over 8.75 eV is required for SiH_4 decomposition. However, as shown in the same figure of EEDF or its integral, the number of electrons with an energy in excess of this threshold value is not so large, only $1/10^3$ of the total number of electrons for the case of pure SiH_4 decomposition. Actually, SiH_4 and H mixture exists in the chamber. However, as we compare the decomposition of pure SiH_4 in PECVD with that in Cat-CVD later, here, we simply consider the phenomena by using pure SiH_4.

The figure also shows that the cross section of the collision of electrons with the energies over the threshold energy for SiH_4 decomposition is about 8×10^{-16} cm^2. In this case, the average energy of electrons in RF-PECVD apparently appears less than 5 eV as deduced from the figure. In the evaluation of the electron velocity, however, the electron energy with more than the threshold energy, 8.75 eV, must be used. Thus, employing the value for 10 eV, the velocity should be 1.88×10^8 cm/s, as shown in Table.2.7.

That is, the value of $n_e n_B \Sigma_{e\text{-}B} v_{th(e)}$ in eq. (2.14) becomes $10^{10} \times (1/10^3) \times 8 \times 10^{-16} \times 1.88 \times 10^8 \times n_B = 1.50 \times n_B$/s. If $P_g = 1$ Pa and the gas temperature is about 27 °C (300 K), as $n_B = 2.42 \times 10^{14}$/cm^3 as shown in Table 2.1, the value becomes $1.50 \times 2.42 \times 10^{14} = 3.63 \times 10^{14}$ cm^{-3}/s. This shows the density of collisions between electrons and SiH_4 molecules for SiH_4 decomposition in a unit volume in the RF-PECVD chamber. Later in Chapter 4, this is compared with Cat-CVD, and the collision will be shown to be much smaller than the case of Cat-CVD.

2.4 Drawbacks of PECVD and Technologies Overcoming Them

2.4.1 Plasma Damage

The plasma damage in PECVD has been studied by various approaches [16]. Here, we explain the phenomena as a low-energy ion implantation.

The range of implanted ions in a solid, which means the total length of the traveling trajectory of implanted ions, is not so clear for such low-energy implantation, with an acceleration voltage of the order of several tens volts. However, here, as a first-order approximation, we evaluate the trajectory range $\langle R \rangle$ of implanted Si and H atoms into crystalline silicon (c-Si), using a simple ion implantation theory, the so-called LSS (Lindhard–Scharff–Schiott) theory [17], to obtain a rough idea of the depth of the damage introduced in a solid by the impact of ions in PECVD. In the low-energy region, sputtering becomes dominant, and thus, LSS theory is usually not used. However, it is still useful for obtaining an image of the behavior of ions introduced into c-Si with a certain energy.

Ions implanted into c-Si lose their energy by collisions of many Si atoms. When the energy of implanted species decreases to a value smaller than the threshold energy for creating Si defects, the species travel deeper without creating defects anymore and finally stop. The threshold energy for creating Si defects in c-Si is not clear; however, it is often assumed to be about 15 eV or more [18]. During the evaluation of $\langle R \rangle$, we can also estimate the trajectory range of species for which the energy is larger than 15 eV and define it as the range of defect depth $\langle R_{defect} \rangle$.

Figure 2.8 shows the calculated range $\langle R \rangle$ of Si atoms and H atoms and also $\langle R_{defect} \rangle$ for the defects created by both Si and by H atoms (See Appendix 2.A). This is not the projected range $\langle R_p \rangle$, which is usually used for estimation of implanted ion depth. However, as the trajectory is so short, $\langle R \rangle$ has at least the same order as $\langle R_p \rangle$ and can be used for rough estimation. From the figure, if V_{sheath} is about 50 V, for instance, $\langle R \rangle$ and $\langle R_{defect} \rangle$ for Si atoms are about 1.5 and 0.8 nm, respectively. For H atoms, even if V_{sheath} is only 20 V, $\langle R \rangle$ and $\langle R_{defect} \rangle$ are 30 and 3 nm, respectively. The defects created by H atoms may be eliminated by making bonds with such H atoms, and this may make the phenomena a little bit complicated. However, it is clear that the acceleration of ions by V_{sheath} is enough to produce damage below the surface of c-Si, when V_{sheath} is larger than the threshold energy for creating Si defects. This is the meaning of the so-called "plasma damage."

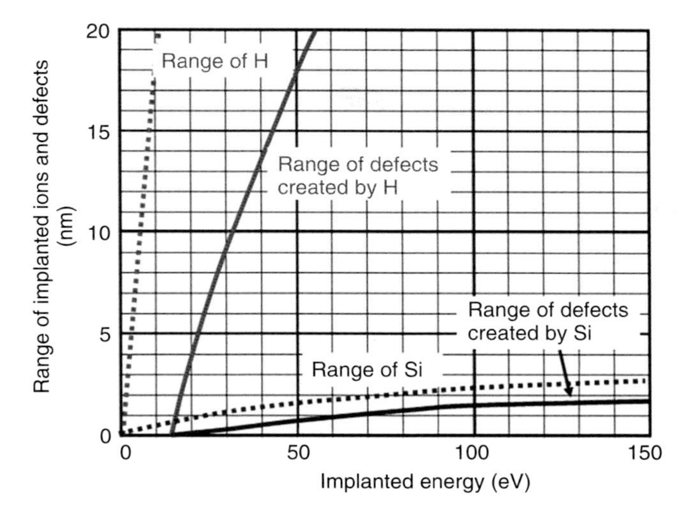

Figure 2.8 Range of trajectory for H and Si atoms and range for defect depth, as a function of accelerated voltage or sheath voltage of plasma.

Here, the real results of plasma damage can be seen in the case of a-Si deposition on a c-Si substrate by PECVD. Figure 2.9 demonstrates the atomic scale image at the interface between a-Si and c-Si, formed by (a) PECVD and also by (b) Cat-CVD. These are observed by a scanning transmission electron microscope (STEM) with an ultrahigh spatial resolution of 0.08 nm, from the (110) direction [19]. The Si–Si atomic pair with a distance of 0.15 nm can be seen in the image. The PECVD a-Si layer was made with an RF power density of 0.13 W/cm^2 using pure SiH_4, and the Cat-CVD a-Si layer was made by the conditions described later.

If you look at the interface, the image of c-Si atomic arrays fades gradually to dark a-Si region with a width of the transition layer of about 1.8 nm in PECVD a-Si/c-Si interface, but the width of transition layer is about only 0.6 nm in Cat-CVD a-Si/c-Si interface. In the figure, the signal intensity from Si atoms in STEM, measured for the area surrounded by dotted lines in STEM images, is also demonstrated to show the width of transition layers clearly in the middle of figure. When Si atoms are arranged in arrays, the intensity is emphasized largely. In STEM observation, the images are formed by the electrons transmitted through the samples with a thickness of 10 to 20 nm. That is, the width of transition layer reflects the fluctuation of interface. The image of roughness is also shown at the bottom of the figure. In the case of PECVD, the surface of c-Si is roughened with a scale of 1.8 nm. This value is not so different from the calculated depth of damages created by Si ion impact, shown in Figure 2.8. In the case of PECVD, this roughness may be inevitable. However, at the same time, in PECVD, H atoms are also implanted. The effect of defect elimination by H atoms has to be also taken into account, as it is known that ion-implanted damages are easily recovered with H atoms [20]. The implanted H atoms may play a role to ease the plasma damage, when H-related source gases are used.

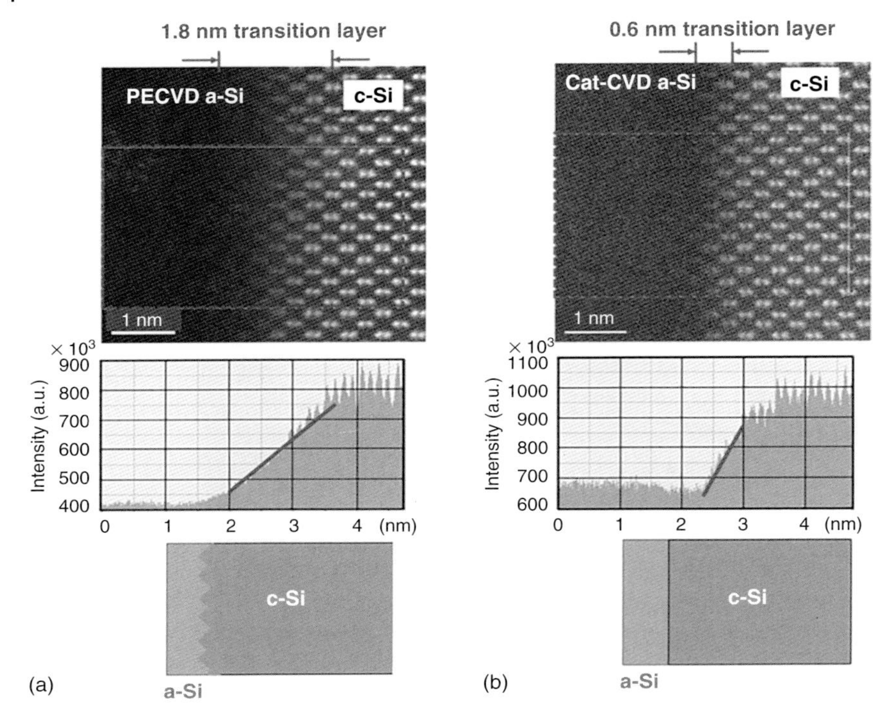

Figure 2.9 Top figures demonstrate atomic scale observation of the interface between amorphous silicon (a-Si) and crystalline silicon (c-Si) by a scanning transmission electron microscope (STEM), middle figures show signal intensity at the area surrounded by dotted lines in the STEM images as a function of position, and the bottom figures illustrate the interface roughness. Figure (a) shows the case of plasma-enhanced chemical vapor deposition (PECVD) a-Si and (b) shows catalytic chemical vapor deposition (Cat-CVD) a-Si. Source: Matsumura et al. 2015 [19]. Reproduced with permission from American Vacuum Society.

2.4.2 Increase of Frequency in PECVD

One of the simple methods to reduce plasma damage is the lowering of T_e, that is, to lower the velocity of electrons, v_e, in order to lower V_{sheath}, as shown in Eq. (2.10). As v_e is proportional to the strength of the applied electric field, E_0, as explained in Section 2.3.3, the RF power used for generating plasma must be lowered. However, this causes the reduction of number of decomposed molecules and a decrease of the deposition rates. To avoid this, another approach is considered, that is, the increase of both gas density and electron density. To increase the electron volume density, A in Eq. (2.9) is lowered by increasing f from 13.56 MHz to VHF domain, as at higher frequencies, the electrons are confined to even smaller regions and the probability of collision with gas molecules increases.

Increasing the frequency f sometimes helps to increase the deposition rate. The increased decomposition rate of gas molecules probably leads to an increase of the H atomic density in the gas phase when H-related source gases are used for film deposition. These H atoms may be effective in improving the film quality. Later in Chapter 5, similar effects are shown in Cat-CVD.

2.4.3 Power Transferring System

RF-PECVD and VHF-PECVD suffer from some difficulties. One is the method of transferring RF or VHF power into the chamber from a power generator. If the power is applied on the electrodes with DC or low frequency, the transfer of electric power is simple. However, when the frequency is in the RF range, we have to take care of the power transfer in the system of distributed constant circuits. The length of lines and impedance of the system becomes important. We have to take care of the reflection of input power from the chamber. To transfer the power properly into the electrodes inside the chamber, we need well-designed matching elements to adjust the impedance for avoiding the reflection of power.

Figure 2.10 shows a schematic view of power transfer system in RF-PECVD. Usually, the power matching is taken by adjusting inductance L and capacitance C, the so-called LC circuits. However, when the films are deposited on all surfaces inside the chamber, including the electrodes, the impedance of the chamber changes and matching has to be adjusted again. Even if this is done automatically, finally, the impedance of the chamber has to be recovered. That is, the chamber must be cleaned after a certain time. For instance, in the case of fabrication of thin film transistors (TFTs) used in LCD industry, the chamber used in mass production has to be cleaned every three to five deposition runs; although, in most cases, a-Si and SiN_x films are successively deposited in the same chamber.

For VHF-PECVD, the cleaning is even more important. Usually, gases used for cleaning are rather expensive and sometimes cause concerns in view of the greenhouse effect. The direction of development to increase the frequency in PECVD does not appear easy and does not provide a rosy future in mass production systems, without highly sophisticated and hard effort.

2.4.4 Large Area Uniformity for Film Deposition

As mentioned above, one of the reasons making the power transfer into the deposition chamber difficult is that the RF power has a property of a wave in transmitting lines. Similarly, the RF power shows wave characteristics inside the chamber, when the size of the chamber is large enough for large area deposition.

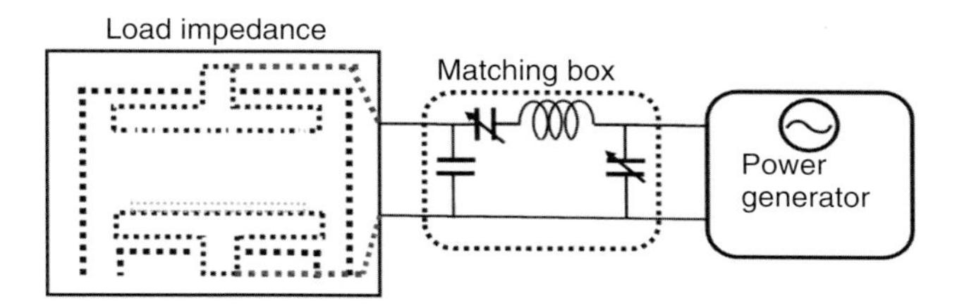

Figure 2.10 Power transfer system for radio-frequency plasma-enhanced chemical vapor deposition (RF-PECVD).

For instance, the wavelength for the 13.56 MHz frequency is about 22 m in vacuum. The length between the position of peak power and that of 0 should be one-fourth of the wavelength to form a standing wave, and this is about 5 m in vacuum. This length is likely to be slightly shortened when the chamber is filled by source gases, as the velocity of electromagnetic wave decreases slightly in a medium or plasma. When a size of the sample tray is about 2 m in a mass production machine, the value appears no longer negligibly small compared to one-fourth of the wavelength and uniform deposition is no longer achieved.

If the plasma frequency increases to 60 MHz to increase the deposition rate, it is even harder to realize uniform deposition over large area. As known in LCD industry, for the productivity improvement by reducing the film preparation cost, the enlargement of deposition area is an inevitable approach. The situation can also be seen in the solar cell industry. The achievement of this large area uniformity encounters a very high barrier for RF-PECVD and, particularly, for VHF-PECVD.

Here, we introduce some of the useful and effective attempts to overcome this drawback of high-frequency PECVD. One idea is the introduction of a "hollow cathode" system, as used in PECVD machines that are designed for fabrication of TFT of LCD. In a large electrode plate, many small holes with a diameter less than 10 mm and a depth also about 10 mm are built in. In each small hole, a small gas inlet and insulated small electrode are constructed. A glow discharge is generated inside the small holes. This type of discharge is called "hollow cathode discharge." If such small hollow cathodes are arranged in arrays in the electrode plate with a pitch of 100 mm or less, large area deposition becomes possible without the challenge of standing wave problems, although sophisticated technology is required to keep uniform glow discharges in these hollow cathode arrays. Using this technology, systems for 3 m size deposition, the so-called "10th generation PECVD machine," or even larger size systems have already been installed in LCD production lines. The hollow cathode principle is schematically illustrated in Figure 2.11 [21], although the exact design of the apparatus is not publically available yet.

Even with this technology, maintaining a stable plasma is still a big issue. When films are deposited on the electrodes near the hollow cathodes, plasma conditions are changed. Thus, frequent cleaning of the chamber and also of the areas near

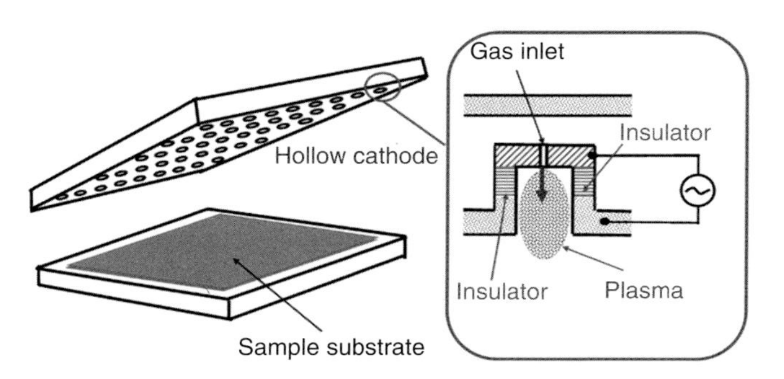

Figure 2.11 Schematic view of large area plasma-enhanced chemical vapor deposition (PECVD) apparatus with an array of hollow cathodes.

the electrodes is necessary. In a mass production system, etching gases such as nitrogen trifluoride (NF_3) are excited by microwave plasma outside of the chamber and the active species are transferred into the hollow cathodes through the gas inlet holes, in order to clean the hollow cathodes and the components surrounding them. This type of production system is quite successful in the LCD industry; however, it should be noticed that the cost of NF_3 is not negligible. In addition, when we use halogenated gases such as NF_3, the vacuum system should be specifically prepared for it. All these aspects are the reason of the high cost of this system. In the case of the LCD industry, the added cost can be absorbed by combining them with other costs. However, in solar cell industry, the increased cost is very serious, as much lower costs per square meter of product is required, requiring consistent minimization of production cost.

2.5 Features of Cat-CVD as Technology Overcoming Drawbacks of PECVD

The features of Cat-CVD are mentioned in this section. In Cat-CVD, the following advantages can be easily identified:

(1) **There is no worry about plasma damage:** This is the first advantage of Cat-CVD. As already demonstrated in Figure 2.9, the interface between the Cat-CVD film and the substrate is apparently different from that of a PECVD film and the substrate. In the case of film deposition on compound semiconductors such as gallium arsenide (GaAs), whose surface is more sensitive to damage than c-Si, Cat-CVD clearly demonstrates superiority to PECVD [22, 23]. In the case of deposition on c-Si, probably as H atoms implanted in c-Si during deposition play a key role in eliminating some of the Si defects created by ion bombardment, the superiority of Cat-CVD may be masked when source gases including H atoms are used. This is contrast to GaAs, as H atoms in GaAs do not simply function as terminators of defects and sometimes change the property of GaAs itself [24] by increasing the resistivity. Therefore, H incorporation may not be effective in recovering the quality by passivation of ion bombarded states. However, as the difference exists as shown in Figure 2.9, the superiority of Cat-CVD will be recognized even in deposition on c-Si when device becomes more delicate.

(2) **Source gas molecules can be decomposed at any gas pressure:** In glow discharge PECVD, gas pressure is limited to a range where a plasma can be maintained, whereas in Cat-CVD, any gas pressure can be used for the decomposition of gas molecules. Although the properties of the films depend on gas pressure, this gives wider freedom for film deposition and adjustment of film properties.

(3) **As there are no special insulators at the electrodes, the mechanical system for mass production can be easily installed at the substrate holders or other parts. The cost of apparatus can be lowered because of its simple structure:** As shown in Figure 2.4, the deposition system of Cat-CVD is much simpler than that of PECVD. As we need not take care of electric potential,

the installation of moving components such as moving rollers becomes simple, and thus, the design of mass production machine becomes easy as well. These factors make the cost of apparatus low, in addition to simple structure. This is a clear advantage of Cat-CVD.

(4) **As we need not worry about anomalous glow discharges, the film can be deposited uniformly on samples with various shapes including sharp edge:** Contrary to PECVD, thin films can be deposited on substrates of any shape, as we need not worry about abnormal discharges. This advantage is used for the deposition of SiN_x insulators on mechanical components with various shapes and for deposition of polytetrafluoroethylene (PTFE, or Teflon, by its commercial name) films on razor blades with sharp edges to make their surface smooth.

(5) **Large area deposition is easy by simply expanding the area spanned by catalyzing wires. In addition, when the catalyzing wires are hung vertically, the films can be deposited on both sides of the catalyzer and the productivity doubles:** In Cat-CVD, by enlarging the area spanned by catalyzing wires, the deposition area can be expanded without any worry about nonuniformity because of standing wave of RF power, which is seen in simple PECVD. For this expansion of the spanned area, vertical setting of catalyzing wires appears more realistic, as thermal expansion of catalyzing wires does not affect the distance between the catalyzer and substrates.

Figure 2.12 shows a schematic diagram of a vertical type Cat-CVD apparatus with catalyzers hung vertically in the deposition chamber, although this does not show the exact design of commercial machines. In the figure, two sample trays are placed parallel to each other at both sides of hanging catalyzing wires. This structure adds another advantage to Cat-CVD. As the most of areas inside the chamber are covered with substrate trays, the chamber wall or other surfaces of the chamber are free from undesirable film coatings. In addition, as no glow discharge is used, and thus, the change of impedance of the deposition system due to film deposition does not cause trouble, the films deposited on undesirable areas do not affect seriously the next deposition. That is, the cleaning cycle of the Cat-CVD chamber can be prolonged. For instance, in the case of a-Si deposition for solar cell production, after continuous deposition of a-Si films for a month, the hanging parts of catalyzing wires and the mechanical parts for moving trays are all slid out from the deposition chamber, and these parts are cleaned chemically, elsewhere outside of the chamber. Equivalent parts that had been cleaned and dried already are reused and newly slid back to the top of the chamber. We need not use expensive halogen cleaning gases and also the chamber downtime can be much shortened in this way. This is a big advantage of Cat-CVD.

(6) **The efficiency of source gas use is by 5–10 times higher than that of the conventional PECVD:** In PECVD, the molecules of source gases are decomposed by physical collisions with energetic electrons as mentioned earlier. This means that the collisions of a point with another point in three-dimensional space must occur in PECVD, as illustrated in Figure 2.13, left side. The probability of such collisions is not high, and many of the molecules are evacuated without having reacted. On the other hand, in

Cat-CVD, the molecules collide with a two-dimensional solid surface; that is, the collisions of a point with a two-dimensional area is utilized, as shown in Figure 2.13, right. The probability of collisions becomes high and, in general, under typical deposition conditions, a molecule collides with the surface of a catalyzer more than six times before evacuation, as is discussed in Section 4.1.1 in more detail. Thus, the efficiency of gas use in Cat-CVD is much higher than that of PECVD.

The data of gas efficiency in the factory has not been made public by manufacturing companies. However, it can be estimated that the gas efficiency in PECVD is only several percent to obtain high-quality films with a cost-effective deposition rate, while the gas efficiency in Cat-CVD reaches 30% or more in a mass production system. In addition, in Cat-CVD, films with the same quality as PECVD films are obtained at almost double deposition rates.

When SiH_4 gas is used as a source gas, the advantage of high gas utilization rates becomes even greater, particularly in large area deposition. The total amount of exhaust gas after deposition can be reduced and also the amounts of dangerous SiH_4 gas stored at the factory can also be much reduced. These advantages make the film deposition cost low. The superiority of Cat-CVD to PECVD is clear, particularly when large area deposition is carried out.

In Figure 2.13, we have intuitively explained the differences in decomposition of source gas molecules in Cat-CVD from that in PECVD. The explanation is quite sketchy, in order to provide a quick grasp of the principles. Detailed discussion is presented in Chapter 4, by using the concepts described for PECVD in Section 2.3.5.

The typical features and the advantages of Cat-CVD all demonstrate the superiority of Cat-CVD over PECVD. However, although any gas molecule can be decomposed in PECVD, as gas molecules are decomposed by physical collisions with energetic electrons, in Cat-CVD, we sometimes have to find alternative gases and proper catalyzers suitable to crack the feedstock molecules. Nevertheless, if we find a proper combination of gas, catalyzer material, and appropriate catalyzing temperatures, we can make high-quality film of any material using a simple apparatus.

2.A Rough Calculation of Ranges $\langle R \rangle$ of Si and H Atoms and Defect Range $\langle R_{defect} \rangle$ Created by Si and H Atoms Implanted with Very Low Energy

A particle implanted into a solid loses its energy through both the elastic collisions with nuclei of atoms in the solid and the inelastic collision with electron clouds surrounding the nuclei. In these energy loss mechanisms, the elastic collisions create defects in the solid. In the very low energy region, such as the energy obtained by acceleration in the sheath voltage, the influence of inelastic energy loss is negligible. The energy loss due to the elastic collisions with the nuclei of

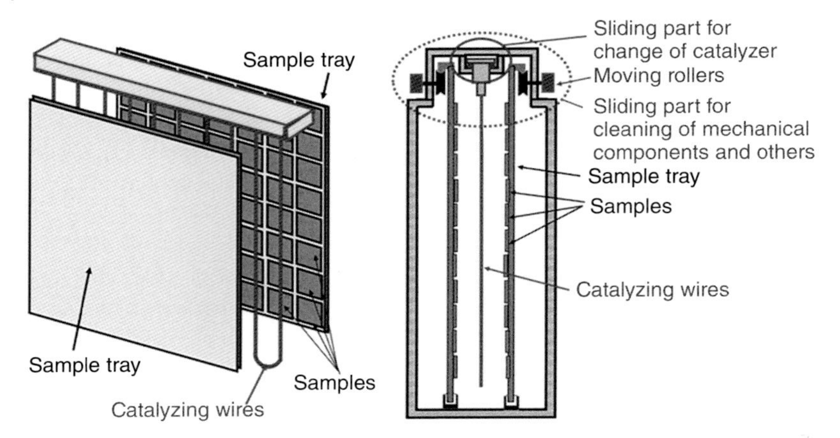

Figure 2.12 Schematic view of vertical type catalytic chemical vapor deposition (Cat-CVD) apparatus. This system is used in large area mass production machines.

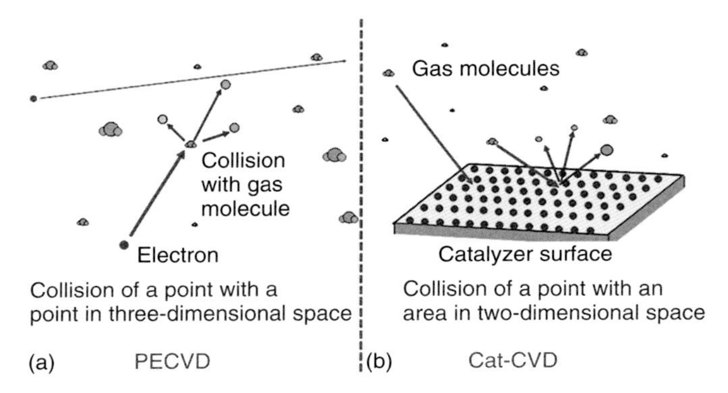

Figure 2.13 Collisions in plasma-enhanced chemical vapor deposition (PECVD) (a) and in catalytic chemical vapor deposition (Cat-CVD) (b).

atoms is called "nuclear stopping power, S_n." This S_n is defined as dE/dx, where E and x refer to the energy of the particle and the position of it, respectively.

E_0, E_{defect}, and R refer to the initial energy of an implanted atom, a threshold energy for creating defects in the solid by the implanted atom, and the total length of the trajectory of the atom, respectively. The total traveling length of the atom before losing the energy from E_0 to E_{defect}, R_{defect}, is expressed by

$$R_{defect} = -\int_{E_0}^{E_{defect}} (1/S_n)dE \tag{2.A.1}$$

In general, S_n and the energy factors are determined by the combination of implanted atoms and atoms composing the solid. Thus, in general, S_n and the energies are all described by normalized formula. It is known that the normalized S_n is expressed by Eq. (2.A.2) when the normalized energy $\varepsilon < 0.01$.

$$S_n = 1.593 \times \varepsilon^{1/2} \tag{2.A.2}$$

Here, the normalized energy, ε, is expressed by

$$\varepsilon = \frac{32.53 \times M_2 \times E\,(\text{keV})}{Z_1 Z_2 (M_1 + M_2)({Z_1}^{2/3} + {Z_2}^{2/3})^{1/2}} \tag{2.A.3}$$

where M_1, M_2, Z_1, Z_2, and E refer to the mass number of the implanted atom, the mass number of atoms composing the solid, the atomic number of the implanted atom, the atomic number of the atoms of the solid, and the energy of the implanted atom, respectively.

When the normalized S_n is converted to the conventional unit of energy loss (eV/[10^{15} atoms/cm^2]), the normalized S_n must be multiplied with a constant C_{cov} described in Eq. (2.A.4).

$$C_{cov} = \frac{8.462 \times M_1 Z_1 Z_2}{(M_1 + M_2)(Z_1^{2/3} + Z_2^{2/3})^{1/2}} \tag{2.A.4}$$

When Si atoms are implanted into Si solids, $M_1 = M_2 = 28$, $Z_1 = Z_2 = 14$ and then,

$$\varepsilon = 2.435 \times 10^{-5} \times E\,(\text{eV}) \tag{2.A.5}$$

and when H atoms are implanted into Si solids, by inserting $M_1 = 1$, $M_2 = 28$, $Z_1 = 1$, $Z_2 = 14$,

$$\varepsilon = 8.598 \times 10^{-4} \times E\,(\text{eV}) \tag{2.A.6}$$

Thus, $\varepsilon = 0.01$ is equivalent to $E = 411$ eV for Si implantation into Si solid, and $E = 11.6$ eV for H implantation into Si solid.

Here, considering the Si atomic density in crystalline Si solid of 4.995 (=5.00) $\times 10^{22}$ atoms/cm^3, the unit (eV/[10^{15} atoms/cm^2]) is converted to

$$S_n\,(\text{eV/nm}) = 9.562 \times (E\,[\text{eV}])^{1/2}\ (\text{valid for } E < 411\text{ eV}) \tag{2.A.7}$$

for Si implantation into Si solid and

$$S_n\,(\text{eV/nm}) = 0.3656 \times (E\,[\text{eV}])^{1/2}\ (\text{valid for } E < 11.6\text{ eV}) \tag{2.A.8}$$

for H implantation into Si solid.

If the threshold for defect creation E_{defect} is assumed to be 15 eV, for Si atoms implanted with the energy of 30 eV, $R_{defect} = 0.335$ nm, and for H atoms with 30 eV, $R_{defect} = 8.78$ nm. In the calculation for H implantation, as the equation holds only for energies less than 11.6 eV, the calculation using Eq. (2.A.6) is not correct any more. The value is the estimation of an order of magnitude.

If we insert 0 eV for E_{defect}, R_{defect} becomes the total traveling distance along the trajectory of implanted atoms, R. The R for Si implantation with 30 eV is 1.15 nm and that for H implantation with 30 eV is 30.0 nm. The results are summarized in Figure 2.8.

This is the estimation of R of implanted atoms, but not the estimation of projected range, which is measured along the depth. The calculation is only an order of magnitude estimation of the possible defect depth induced in PECVD. However, the results imply that even if the sheath voltage is small, if it is larger than the voltage that accelerates ions to an energy that causes the creation of defects, the defects will be created to a depth of several nanometers by H implantation.

In many cases, source gases used in PECVD includes H atoms, and thus, defects up to several nanometers of depth are always created in Si solids. Although H atoms may also function to terminate defects, a solid outer surface always suffers from the defects created by accelerated atoms in PECVD.

References

1 Pyykkö, P., Riedel, S., and Patzschke, M. (2005). Triple-bond covalent radii. *Chem. Eur. J.* 11: 3511–3520.
2 Pyykkö, P. and Atsumi, M. (2009). Molecular single-bond covalent radii for elements 1-118. *Chem. Eur. J.* 15: 186–197.
3 Pyykkö, P. and Atsumi, M. (2009). Molecular double-bond covalent radii for elements Li-E112. *Chem. Eur. J.* 15: 12770–12779.
4 Pyykkö, P. (2012). Refitted tetrahedral covalent radii for solids. *Phys. Rev. B* 85: 024115-1–024115-7.
5 Chapman, B. (1980). *Glow Discharge Processes*. New York: Wiley. ISBN: 0471-07828-X.
6 Sterling, H.F. and Swann, R.C.G. (1966). Perfectionnements aux methods de formation de couches. French Patent 1,442,502, filed at 05 August 1964, published at 17 June 1966.
7 Sterling, H.F. and Swann, R.C.G. (1968). Improvements in or relating to a method of forming a layer of an inorganic compound. British Patent 1,104,935, filed at 07 May 1965, published at 06 March 1968.
8 Sterling, H.F. and Swann, R.C.G. (1965). Chemical vapour deposition promoted by R.F. discharge. *Solid State Electron.* 8: 653–654.
9 Swann, R.G., Mehta, R.R., and Cauge, T.P. (1967). The preparation and properties of thin film silicon-nitrogen compounds produced by a radio frequency glow discharge reaction. *J. Electrochem. Soc.* 114: 713–717.
10 Hu, S.M. (1966). Properties of amorphous silicon nitride films. *J. Electrochem. Soc.* 113: 693–698.
11 Chittick, R.C., Alexander, J.H., and Sterling, H.F. (1969). The preparation and properties of amorphous silicon. *J. Electrochem. Soc.* 116: 77–81.
12 Godyak, V.A., Piejak, R.B., and Alexandrovich, B.M. (1992). Measurement of electron energy distribution in low-pressure RF discharges. *Plasma Sources Sci. Technol.* 1: 36–58.
13 Gabriel, O., Kirner, S., Klick, M. et al. (2014). Plasma monitoring and PECVD process control in thin film silicon-based solar cell manufacturing. *EPJ Photovoltaics* 5: 55202-1–55202-9.
14 Takatsuka, H., Noda, M., Yonekura, Y. et al. (2004). Development of high efficiency large area silicon thin film modules using VHF-PECVD. *Sol. Energy* 77: 951–960.
15 Shirafuji, T. (2011). Chemical reaction engineering of plasma CVD. *J. High Temp. Soc.* 37: 281–288. (in Japanese).
16 Gallagher, A. (1987). Apparatus design for glow-discharge a-Si:H film-deposition. *Int. J. Solar Energy* 5: 311–322.

17 Lindhard, J., Scharff, M., and Schiøtt, H.E. (1963). Range concepts and heavy ion ranges. *Mat. Fys. Medd. Dan. Vid. Selsk* 33: 1–42.
18 Vepřek, S., Sarrott, F.-A., Rambert, S., and Taglauer, E. (1989). Surface hydrogen content and passivation of silicon deposited by plasma induced chemical vapor deposition from silane and the implications for the reaction mechanism. *J. Vac. Sci. Technol., A* 7: 2614–2624.
19 Matsumura, H., Higashimine, K., Koyama, K., and Ohdaira, K. (2015). Comparison of crystalline-silicon/amorphous-silicon interface prepared by plasma enhanced chemical vapor deposition and catalytic chemical vapor deposition. *J. Vac. Sci. Technol., B* 33: 031201-1–031201-4.
20 Thi, T.C., Koyama, K., Ohdaira, K., and Matsumura, H. (2016). Defect termination on crystalline silicon surfaces by hydrogen for improvement in the passivation quality of catalytic chemical vapor-deposited SiN_x and SiN_x/P catalytic-doped layers. *Jpn. J. Appl. Phys.* 55: 02BF09-1–02BF09-6.
21 Sun, S., Takehara, T., and Kang, I.D. (2005). Scaling up PECVD system for large-sized substrate processing. *J. Soc. Inf. Disp.* 13: 99–103.
22 Hattori, R., Nakamura, G., Nomura, S. et al. (1997). Noise reduction of pHEMTs with plasmaless SiN passivation by catalytic CVD. In: *Technical Digest of 19th Annual IEEE GaAs IC Symposium, Held at Anaheim, California, USA* (12–15 October 1997), 78–80.
23 Higashiwaki, M., Matsui, T., and Mimura, T. (2006). AlGaN/GaN MIS-HFETs with f_T of 163 GHz using Cat-CVD SiN gate-insulating and passivation layers. *IEEE Electron Device Lett.* 27: 16–18.
24 Murphy, R.A., Lindley, W.T., Peterson, D.F. et al. (1972). Proton-guarded GaAs IMPATT diode. In: *Proceedings of Symposium on GaAs*, 224–230.

3

Fundamentals for Analytical Methods for Revealing Chemical Reactions in Cat-CVD

In this chapter, methods to reveal the chemical reactions important in catalytic chemical vapor deposition (Cat-CVD) processes are described. Many radical species play key roles in the chemical processes in chemical vapor deposition (CVD). Information on the behaviors of these radical species is essential to understand the process mechanisms. Because radical species are highly reactive and are removed rapidly, their densities in the gas phase cannot be high. Then, highly sensitive techniques are required to detect radicals. The individual detection techniques are explained concretely.

3.1 Importance of Radical Species in CVD Processes

In Cat-CVD, source gas molecules are decomposed on hot catalyzer surfaces to produce free radicals. Free radicals are atoms and molecules with unpaired electrons and shall be called just *radicals* in this book. These radicals may directly deposit on the substrate surfaces to grow films, whereas nonradical stable species may hardly deposit. The situation is similar in plasma enhanced chemical vapor deposition (PECVD), although the radical production processes are different. Radical species may also initiate chemical reactions in the gas phase to produce other radicals. For example, H atoms produced on heated catalyzer surfaces may react with SiH_4 to produce SiH_3 [1, 2]. On substrate surfaces, radical species may not only deposit but also abstract atoms to generate surface radicals [3, 4]. The surface radicals thus produced may be terminated by other radicals. As such termination processes are always exothermic, the substrate surfaces are heated locally. Local heating may induce not only surface migration of adsorbed species but also cleavage of weak chemical bonds.

In short, radical species in the gas phase play key roles in the CVD processes, and it is important to know what kinds of radicals are produced to understand the underlying mechanisms and to control the processes. In PECVD, as already explained in Chapter 2, radical species are mainly produced by electron bombardment. Electronically excited species are often produced in such processes, and optical emission spectroscopy is widely used to identify the radical species in plasma. On the other hand, electronically excited species are not produced in catalytic decomposition, and this technique cannot be employed. In any case, emission spectroscopy cannot provide direct information on the densities of radical

Catalytic Chemical Vapor Deposition: Technology and Applications of Cat-CVD,
First Edition. Hideki Matsumura, Hironobu Umemoto, Karen K. Gleason, and Ruud E.I. Schropp.

species in the ground state, which are more abundant than excited species and much more important in deposition processes.

Detection of radical species adsorbed on solid surfaces is another important target to make clear the deposition processes. However, *in situ* identification is more difficult for adsorbed species. It should be remembered that CVD processes are usually carried out in low or medium vacuum, not in ultrahigh vacuum. Many surface analysis techniques, which utilize ions and electrons, are difficult to be applied in CVD processes, as it is impossible to accelerate charged particles without collisions. Then, only gas-phase diagnosis techniques shall be discussed in this chapter.

Two review articles for the detection of radical species have been published recently by one of the present authors [5, 6]. These should be useful to review the former data. Besides these, several excellent reviews are available for gas-phase diagnoses used not only in Cat-CVD but also in plasmas [7–13].

In this chapter, for the convenience of explanation, the definitions of some concepts are treated flexibly. SiH_2 and BH are not radicals in a strict sense because their spin states are singlet and they do not have unpaired electrons. However, considering their high chemical reactivity, we would like to include these species in radicals.

3.2 Radical Detection Techniques

In general, radical densities are much lower than those of stable nonradical species, such as material gas molecules and end products because of their high reactivity. The typical densities of radicals, which contribute to deposition, are between 10^{10} and 10^{13} cm^{-3}. Then, highly sensitive techniques are required to detect them. Laser spectroscopic and mass spectrometric techniques are common in the detection of such low-density species in the gas phase.

In the mass spectrometric detection, radical species are ionized, and the ions produced are detected after mass selection. In mass selection, ions must fly several centimeters or more without colliding. Under typical CVD conditions, the pressure is more than 1 Pa and the mean free path is in the order of centimeters as shown in Table 2.4. In other words, mass spectrometric detection is not easy without sampling, and thus, *in situ* detection is difficult. Usually, a sampling hole is placed between the CVD and ionization chambers, and the ionization and mass selection parts are differentially pumped.

In contrast to mass spectrometric techniques, many photon-in photon-out type laser techniques, such as laser-induced fluorescence (LIF), can be applied in low-vacuum systems, typical in CVD. By using these techniques, *in situ*, highly sensitive, quantitative, state-specific, real-time, and nonintrusive detection is possible. In some techniques, such as two-photon LIF, high spatial resolution can also be expected. Conventional light sources can also be used to detect radicals [14, 15], but lasers are more suitable because lasers are intense, tunable in wavelength, coherent, monochromatic, directional, and less divergent. Outputs in ultraviolet (UV) and vacuum ultraviolet (VUV), with wavelengths shorter than 190 nm, regions can be easily obtained with pulsed lasers by using nonlinear optical media. One point to be noted is that, in laser spectroscopic detection, the detectable species are limited compared to those by mass spectrometric techniques because the first step, photoabsorption, is species and state specific. Synchrotron radiation dispersed by a monochromator is another choice [16] but is less handy.

There are many laser techniques to detect radical species in the gas phase, but none of them are almighty. In order to elucidate the detailed chemical kinetics in CVD processes, a number of techniques must be combined. In the following, the principles of one- and two-photon LIF, VUV laser absorption, cavity ringdown, resonance-enhanced multiphoton ionization (REMPI), and tunable diode laser absorption will be introduced.

3.3 One-Photon Laser-Induced Fluorescence

3.3.1 General Formulation

LIF is one of the most widely used techniques to detect small radical species, such as atoms and diatoms. In this technique, lower state species, usually ground-state species, are excited to one of the excited states, and spontaneous emission is observed. From this nature, an excited state, which is accessible by phototransition, must be present. In addition, this technique cannot be applied to predissociative species, such as CH_3 and SiH_3, which dissociate and do not fluoresce after photoabsorption. As this technique is indirect compared to simple emission or absorption spectroscopy, there are possibilities to lead artifacts. This point will be discussed in the following subsections. The LIF scheme is illustrated in Figure 3.1.

The firsts step in LIF is the absorption of a photon to excite the lower state, state 1, to an upper state, state 2. The transition probability is given by

$$W_{12} = B_{12}\rho(\nu) \tag{3.1}$$

where B_{12} is the Einstein B coefficient, the proportional constant for the photoabsorption, and $\rho(\nu)$ is the energy density of photons at the frequency of ν.

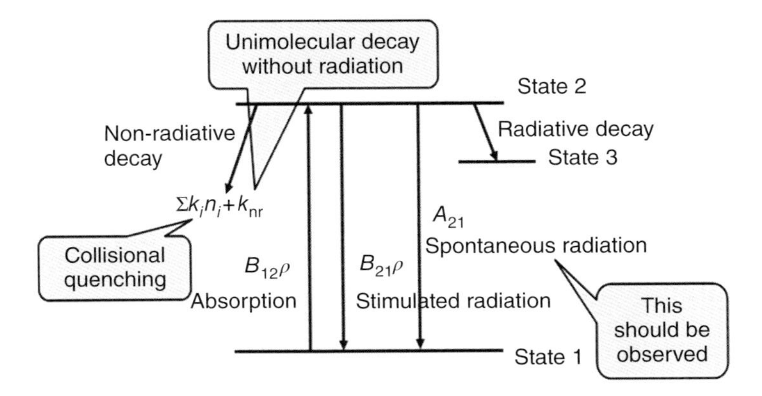

Figure 3.1 Schematic diagram representing the principle of LIF detection.

This absorption process is selective and, in general, just one specific state can be excited. The transition probability from state 2 to state 1 is

$$W_{21} = B_{21}\rho(\nu) + A_{21} \tag{3.2}$$

The first and second terms correspond to stimulated and spontaneous radiation, respectively. A_{21} is the Einstein A coefficient, which is equivalent to the inverse of the radiative lifetime of state 2 if radiative decay to other states, such as state 3, can be neglected. B_{21} is the proportional constant for the stimulated emission relevant to the two states. The following relationships are satisfied among the coefficients [17]:

$$g_1 B_{12} = g_2 B_{21} \tag{3.3}$$

$$A_{21} = \frac{8\pi h\nu^3}{c^3} B_{21} \tag{3.4}$$

where g_1 and g_2 are the statistical weights of states 1 and 2, respectively, whereas h and c are the Planck constant and the speed of light, respectively.

Here, we assume a two-state system ignoring the presence of state 3. In general, there may be multiple decay processes if just one excited state is populated by laser radiation. However, a two-state system is usually enough to model LIF as will be discussed in Section 3.3.2. The time differential of the upper-state population, $N_2(t)$, should be expressed by

$$\begin{aligned}\frac{\mathrm{d}N_2(t)}{\mathrm{d}t} &= \rho(\nu)B_{12}N_1(t) - \left\{\rho(\nu)B_{21} + A_{21} + k_{\mathrm{nr}} + \sum k_i n_i\right\} N_2(t)\\ &= \rho(\nu)B_{12}N - \left\{\rho(\nu)(B_{12} + B_{21}) + A_{21} + k_{\mathrm{nr}} + \sum k_i n_i\right\} N_2(t)\end{aligned} \tag{3.5}$$

where $N_1(t)$ is the lower-state population and $N(=N_1(t) + N_2(t))$ is the population of the lower state before excitation, which is what we would like to know. The rate constant for the collisional quenching (de-excitation) of the upper state and the number density of the quencher molecules are represented by k_i and n_i, whereas k_{nr} represents the rate constant for the unimolecular nonradiative decay, such as predissociation. The quencher density may be assumed to be time independent, as the density of excited species is much smaller. In other words, pseudo-first-order processes can be assumed.

Just for simplicity, we assume rectangular pulse excitation with a duration time of Δ. Then, by solving the differential equation, Eq. (3.5), the upper-state population just after the laser pulse terminated, at $t = \Delta$, is calculated to be

$$N_2(\Delta) = \frac{\rho(\nu)B_{12}N\left(1 - \exp\left[-\left\{\rho(\nu)(B_{12} + B_{21}) + A_{21} + k_{\mathrm{nr}} + \sum k_i n_i\right\}\Delta\right]\right)}{\rho(\nu)(B_{12} + B_{21}) + A_{21} + k_{\mathrm{nr}} + \sum k_i n_i} \tag{3.6}$$

The time-integrated LIF intensity should be proportional to

$$I_{\mathrm{LIF}} \propto N_2(\Delta)\phi = \frac{\rho(\nu)A_{21}B_{12}N\left(1 - \exp\left[-\left\{\rho(\nu)(B_{12} + B_{21}) + A_{21} + k_{\mathrm{nr}} + \sum k_i n_i\right\}\Delta\right]\right)}{\left\{\rho(\nu)(B_{12} + B_{21}) + A_{21} + k_{\mathrm{nr}} + \sum k_i n_i\right\}\left(A_{21} + k_{\mathrm{nr}} + \sum k_i n_i\right)} \tag{3.7}$$

Here, ϕ is the quantum yield of the fluorescence. If the laser is intense enough to saturate the transitions, the LIF intensity saturates against $\rho(\nu)$ and should be

$$I_{\mathrm{LIF}} \propto \frac{g_2 A_{21} N}{(g_1 + g_2)\left(A_{21} + k_{\mathrm{nr}} + \sum k_i n_i\right)} \tag{3.8}$$

and independent of $\rho(\nu)$. If the laser intensity is weak, that should be

$$I_{\mathrm{LIF}} \propto \frac{\rho(\nu) A_{21} B_{12} N \left[1 - \exp\left\{-\left(A_{21} + k_{\mathrm{nr}} + \sum k_i n_i\right)\Delta\right\}\right]}{\left(A_{21} + k_{\mathrm{nr}} + \sum k_i n_i\right)^2} \tag{3.9}$$

and proportional to $\rho(\nu)$. In both cases, the LIF intensity is proportional to N, and it is easy to evaluate the *relative populations* under various conditions. The situation is similar when the temporal profile of the laser is not rectangular or under continuous wave (CW) excitation instead of pulse excitation.

3.3.2 Validity of the Assumption of a Two-State System

When we detect molecular species, there may be many rotational levels, both in the ground state and in the excited state. When just one rotational level is excited, there may be multiple radiative decay processes. As an example, we would like to show the detection of BH(X $^1\Sigma^+$) by LIF. The symbols such as X $^1\Sigma^+$ are explained in Appendix 3.A. When we excite BH(X $^1\Sigma^+$, $v'' = 0, J''=2$) to (A $^1\Pi$, $v' = 0, J' = 3$), fluorescence not only to the original rotational level, $J'' = 2$, but also to the $J'' = 3$ and 4 levels is expected, according to the selection rule in the optical transition, J'-$J'' = 0, \pm 1$ in this system. Here, v and J are the quantum numbers for vibration and total angular momentum, respectively. In molecular species, the total angular momentum is determined mainly by the rotational motion. Collisional rotational level mixing to populate rotational levels other than $J' = 3$ may take place before radiative decay. From BH(A $^1\Pi$, $v' = 0$), fluorescence not only to the vibrational ground state, BH(X $^1\Sigma^+$, $v'' = 0$), but also to the vibrationally excited states, BH(X $^1\Sigma^+$, $v'' \geq 1$), is possible. Then, many rotational and vibrational levels should be involved. Even though, this system can be regarded as a two-state one during laser excitation, under some conditions. The radiative lifetime of BH(A $^1\Pi$, $v' = 0$) is 159 ns [18], which is much longer than the duration of nanosecond lasers, commonly used in gas-phase diagnoses. Then, the spontaneous radiative decay during the excitation can be ignored. The situation is similar for other di- and triatomic species, which are common in CVD processes. The transition wavelengths and the radiative lifetimes of typical di- and triatomic hydride molecules are summarized in Table 3.1.

The collisional level mixing is minor during the laser pulse when the pressure is less than 10^2 Pa. As has been discussed in Chapter 2, the interval time between collisions is 10^{-7} s at 10^2 Pa. This interval is much longer than the duration time of nanosecond lasers. In addition, not all the collisions are inelastic. Collisional level mixing may take place during the lifetime of the upper state, but that is not a problem in the density measurements. This is because the wavelength shift due to the rotational level mixing is minor and we can detect all the fluorescence by using a suitable optical filter and a photomultiplier tube (PMT). A monochromator is

Table 3.1 Wavelengths to detect di- and triatomic hydride radicals by one-photon LIF and the radiative lifetimes of the upper states.

Radical	Transition[a)]	Wavelength (nm)	Radiative lifetime	References
BH	$A\ ^1\Pi - X\ ^1\Sigma^+$	433[b)]	159 ns[c)]	[18]
CH[d)]	$A\ ^2\Delta - X\ ^2\Pi$	431[b)]	534 ns[c)]	[18]
NH	$A\ ^3\Pi - X\ ^3\Sigma^-$	336[b)]	404 ns[c)]	[18]
OH	$A\ ^2\Sigma^+ - X\ ^2\Pi$	309[b)]	693 ns[c)]	[18]
SiH	$A\ ^2\Delta - X\ ^2\Pi$	413[b)]	0.7 μs[c)]	[18]
PH	$A\ ^3\Pi - X\ ^3\Sigma^-$	342[b)]	0.45 μs[c)]	[18]
NH_2	$\tilde{A}\ ^2A_1 - \tilde{X}\ ^2B_1$	598[e)]	10.0 μs	[19]
SiH_2[d)]	$\tilde{A}\ ^1B_1 - \tilde{X}\ ^1A_1$	580, 610[e)]	1.0 μs	[20, 21]
PH_2	$\tilde{A}\ ^2A_1 - \tilde{X}\ ^2B_1$	454, 474[e)]	0.65, 1.44 μs	[22]

a) The superscripts represent the spin states, whereas the capital Greek letters represent the magnitude of the electronic orbital angular momentum along the internuclear axis. A_1 and B_1 specify the symmetry of wave functions of polyatomic molecules. Refer to Appendix 3.A for details.
b) Band origin of the (0,0) band.
c) Lifetime of the $v' = 0$ state.
d) CH and SiH_2 have not been detected by LIF in catalytic decomposition, although SiH_2 has been detected in plasma processes [21, 23].
e) One or two of the strongest bands.

not usually used. The vibrational relaxation is much slower than the rotational mixing, and the vibrational relaxation can usually be ignored even during the lifetime of the upper state.

The radiative lifetimes of excited atomic species are, in general, shorter than those of molecular species, as is shown in Table 3.2, and the spontaneous decay during the laser pulse cannot be ignored. This will be a problem when we measure the *absolute densities* by comparing the intensity of induced fluorescence with that of Rayleigh scattering (see Section 3.3.8 for details). Fortunately, as for Si atoms, it is possible to regard the system to be a true two-state one when we excite $Si(3s^23p^2\ ^3P_1)$ to $Si(3s^23p4s\ ^3P_0)$, as the optical transitions from $Si(3s^23p4s\ ^3P_0)$ to $Si(3s^23p^2\ ^3P_0)$ and $Si(3s^23p^2\ ^3P_2)$ are strictly forbidden. (Refer to Appendix 3.A for the explanation of the symbols, such as $3s^23p^2\ ^3P_1$.) The situation is a little different in the B atom system, which cannot be regarded as a two-state one. This is because the upper state, $B(2s^23s\ ^2S_{1/2})$, fluoresces not only to $B(2s^22p\ ^2P_{3/2})$ but also to $B(2s^22p\ ^2P_{1/2})$, and some uncertainty cannot be avoided in the absolute density measurement. This difficulty, however, cannot be a problem as long as we measure the *relative densities* by changing the experimental conditions, such as catalyzer temperature or partial pressure. As for H, N, and O atoms, the absolute densities can be determined by absorption spectroscopy as will be described in Section 3.5, whereas the absolute P atom densities have been estimated from the imprisonment lifetime, which is longer than the natural radiative lifetime and depends on the density of the ground-state atoms [22]. As for C atoms, LIF measurements have not been reported in the catalytic decomposition.

Table 3.2 Wavelengths to detect atomic radicals by one-photon LIF and the radiative lifetimes of the upper states.

Radical	Transition[a)]	Wavelength (nm)	Radiative lifetime (ns)	References
H	$2p\ ^2P_{3/2,1/2}$–$1s\ ^2S_{1/2}$	121.6	1.7	[24]
B	$2s^23s\ ^2S_{1/2}$–$2s^22p\ ^2P_{3/2}$	249.8	6.0	[24]
C	$2s^22p3s\ ^3P_2$–$2s^22p^2\ ^3P_2$	165.7	3.8	[24]
N	$2s2p^4\ ^4P_{5/2}$–$2s^22p^3\ ^4S_{3/2}$	113.5	6.9	[24]
N	$2s^22p^23s\ ^4P_{5/2}$–$2s^22p^3\ ^4S_{3/2}$	120.0	2.5	[24]
O	$2s^22p^33s\ ^3S_1$–$2s^22p^4\ ^3P_2$	130.2	2.9	[24]
Si	$3s^23p4s\ ^3P_0$–$3s^23p^2\ ^3P_1$	252.4	4.5	[24]
P	$3s^23p^24s\ ^4P_{5/2}$–$3s^23p^3\ ^4S_{3/2}$	177.5	4.6	[24]

a) $^2S_{1/2}$ means that the spin multiplicity is doublet and the orbital angular momentum quantum number is zero, whereas the total angular momentum quantum number is 1/2. Refer to Appendix 3.A for details.

3.3.3 Anisotropy of the Fluorescence

The induced fluorescence from atomic species is not necessarily isotropic, and this anisotropy can be a potential problem. The rotational period of molecules is in the order of picoseconds and much shorter than the radiative lifetimes. Then, the induced fluorescence from molecular species can be assumed to be isotropic. On the other hand, the radiative lifetimes of atomic species are, in some cases, shorter than the time needed for collisional depolarization. Fortunately, however, in many systems including those of Si and B atoms, the depolarization coefficients are zero when the excitation laser is linearly polarized, and it is not necessary to take into account the effect of anisotropy [25]. This problem can also be avoided by introducing collision partners, such as rare gas atoms, to the system.

3.3.4 Correction for Nonradiative Decay Processes

The effect of collisional quenching of the upper states must be taken into account when the radiative lifetime is long or the quencher pressure is high. The LIF intensity, $I_{\mathrm{LIF}}(t)$, decays single exponential after pulse excitation:

$$I_{\mathrm{LIF}}(t) \propto \exp\left\{-\left(A_{21} + k_{\mathrm{nr}} + \sum k_i n_i\right) t\right\} \tag{3.10}$$

and the time-integrated intensity depends on the values of k_i and n_i. The rate constants for the quenching, k_i, of many excited atomic and molecular species have already been reported. Table 3.3 summarizes the rate constants at room temperature for six hydride molecules [26]. These data would be helpful, but it is desired to measure the rate constants in each individual case, as they depend on the surrounding temperature. It is easy to determine the rate constant by measuring the decay profiles as a function of the partial quencher pressure.

Table 3.3 Rate constants for the quenching of excited CH(A $^2\Delta$), NH(A $^3\Pi$), OH(A $^2\Sigma^+$), SiH(A $^2\Delta$), PH(A $^3\Pi$), and NH_2(Ã $^2A_1$) in cm^3/s at room temperature [26].

	CH(A $^2\Delta$)	NH(A $^3\Pi$)	OH(A $^2\Sigma^+$)	SiH(A $^2\Delta$)[a)]	PH(A $^3\Pi$)[b)]	NH_2(Ã $^2A_1$)
H_2	1.0×10^{-11}	8.4×10^{-11}	1.3×10^{-10}		8.8×10^{-11}	4.7×10^{-10}
N_2	2.4×10^{-13}	3×10^{-14}	2.4×10^{-11}		3.9×10^{-12}	8×10^{-11}
O_2	1.7×10^{-11}	5.0×10^{-11}	1.5×10^{-10}	1.3×10^{-10}	8.3×10^{-11}	
CO	5.2×10^{-11}	1×10^{-10}	3.7×10^{-10}		2.1×10^{-10}	5×10^{-10}
NO	1.3×10^{-10}		4×10^{-10}	4.6×10^{-10}		
H_2O	9×10^{-11}	3.7×10^{-10}	5.9×10^{-10}		4.1×10^{-10}	
CO_2	4.3×10^{-13}	6.9×10^{-12}	3.4×10^{-10}		3.4×10^{-12}	
N_2O	3.4×10^{-12}	1×10^{-11}	4×10^{-10}		8.0×10^{-12}	
NH_3	3.4×10^{-10}	5.9×10^{-10}	7×10^{-10}		7.6×10^{-10}	4.5×10^{-10}
CH_4	2.3×10^{-11}	8.0×10^{-11}	2.6×10^{-10}			3.7×10^{-10}
C_2H_2	1.7×10^{-10}					
C_2H_4	1.9×10^{-10}	2×10^{-10}	7×10^{-10}			3×10^{-10}
C_2H_6	1.2×10^{-10}	2.3×10^{-10}	5.8×10^{-10}		1.5×10^{-10}	
C_3H_8	2×10^{-10}	3.7×10^{-10}	1.1×10^{-9}			

a) Reference [27].
b) Reference [28].
Source: From Umemoto 2004 [26].

When atomic species, such as H, B, C, N, O, Si, and P, are detected by one-photon LIF, the collisional quenching of the upper states can usually be ignored as the radiative lifetimes are short, as is shown in Table 3.2, as long as radiation trapping can be ignored. When the ground-state density is high, the resonance radiation can be easily trapped and the apparent radiative lifetime becomes long [29]. Attention is needed in such cases [22, 30].

Predissociation can be a problem in the determination of absolute densities of molecular species, but, fortunately, this is rather minor for many diatomic hydride molecules common in CVD processes, including BH [31], CH [32], NH [33], OH [34], SiH [35], and PH [36], as long as they are excited to the low rotational levels of the vibrational ground states. As long as we treat small molecules, other nonradiative unimolecular decay processes, internal conversion and intersystem crossing, are minor.

3.3.5 Spectral Broadening

The LIF intensity depends on the absorption coefficient, which varies with the absorption spectral profile. There are three types of spectral broadening: natural, Doppler, and pressure broadenings. Among them, the Doppler broadening, caused by the translational motion of the absorbing species, is the most important under typical CVD conditions. When the Doppler width of the absorption spectrum changes, such as by the changes in wire temperature, the wavelength-integrated LIF intensity must be measured by scanning the exciting wavelength. However, when the distance between the wire and the detection zone is more than 10 cm and the pressure is higher than 10 Pa, such an effect can be ignored because of rapid collisional thermalization, as will be discussed in Section 3.3.7. Under such conditions, the translational as well as the rotational temperature depends little on the wire temperature. When the distance between the wire and the detection zone is short, on the other hand, the Doppler profile depends not only on the wire temperature but also on the distance from the wire. In other words, it is possible to determine the translational temperature distribution from the Doppler profile measurements. This subject shall be commented in Sections 3.6.1 and 4.4 in Chapter 4. When the LIF intensity is measured by varying the total pressure, the pressure broadening may be another potential problem, but such an effect is minor below 10^2 Pa.

3.3.6 Typical Apparatus for One-Photon LIF and the Experimental Results

Figure 3.2 shows a schematic diagram of a typical experimental setup for one-photon LIF. Baffled arms, Brewster windows, and a collimating lens system are usually used to reduce the stray light. Nanosecond tunable pulsed lasers, such as Q-switched Nd^{3+}:YAG laser-pumped dye lasers and optical parametric oscillators, are widely used. The output of the laser can be doubled in frequency with a nonlinear optical crystal, such as β-BaB_2O_4 (BBO), when necessary, such as in the detection of NH, OH, and PH. The induced fluorescence is usually monitored at an angle perpendicular to the laser beam with a PMT. A monochromator is not necessarily required, and interference filters or cutoff (or band pass) filters are enough to isolate the fluorescence. The photomultiplier signals can be processed with an oscilloscope or a boxcar averager. It is possible to reduce the effect of background signals, such as blackbody radiation from hot wires, by monitoring the signals just after the laser pulse. CW lasers are not usually used because of the difficulty in the generation of UV and VUV outputs

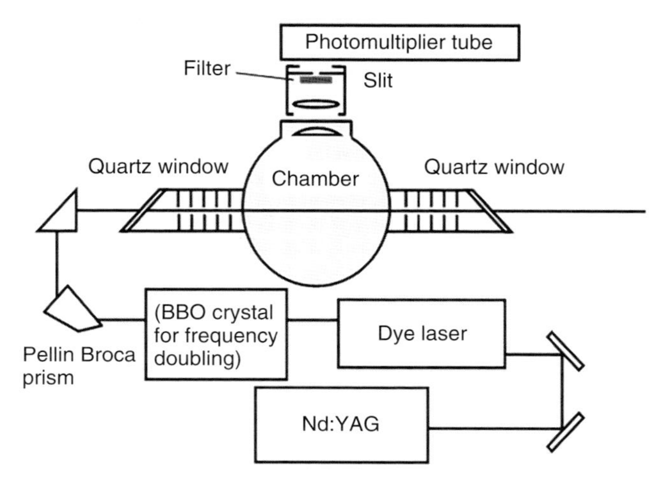

Figure 3.2 Schematic diagram of the apparatus for the LIF detection.

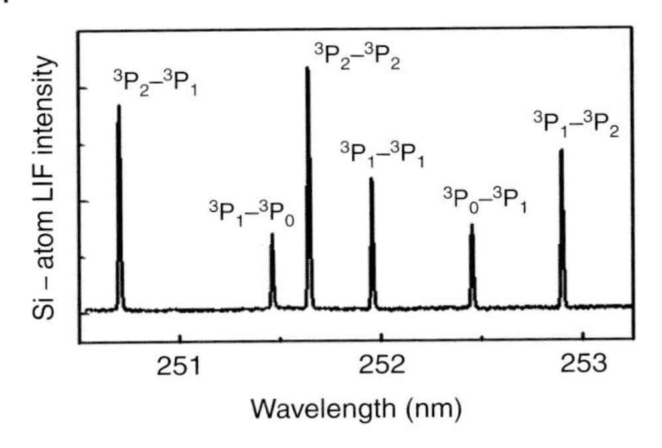

Figure 3.3 LIF spectrum of Si($3s^23p^2\,^3P_{0,1,2}$) formed in the catalytic decomposition of SiH_4. The SiH_4 flow rate was 0.5 sccm, whereas the pressure was 4 mPa. The W wire temperature was 2.30×10^3 K. Source: Nozaki et al. 2000 [38]. Reprinted with permission of American Institute of Physics.

as well as that in the elimination of background signals. However, as CW lasers have better wavelength resolution, they are advantageous in the Doppler profile measurements of heavy species, such as SiH_2 [37].

Figure 3.3 shows an LIF spectrum of Si atoms formed by the catalytic decomposition of SiH_4 on a heated W (tungsten) wire at low pressure conditions where collisional processes in the gas phase can be ignored [38]. The laser intensity was low enough not to saturate the transitions. Si atoms in the $3s^23p^2\,^3P_{0,1,2}$ states were excited to the $3s^23p4s\ ^3P_{0,1,2}$ states, and the fluorescence from these excited states were monitored. Both the lower and the upper states have three spin–orbit states, and there are nine combinations. Because of a selection rule, however, only six transitions can be observed. From the relative populations of $3s^23p^2\,^3P_0$, $^3P_1$, and $^3P_2$ spin–orbit states, the electronic temperature of Si atoms just after the formation was determined to be 1.3×10^3 K, 1.0×10^3 K lower than the wire temperature. The absolute densities can also be estimated from the spectral peak heights, as will be discussed in Section 3.3.8.

Figure 3.4 shows an LIF spectrum of BH formed in the catalytic decomposition of $B_2H_6/He/H_2$ [39]. Many spectral lines appear because molecular species have freedom of rotation and vibration. This spectrum corresponds to the (0,0) and (1,1) bands of the A $^1\Pi$ – X $^1\Sigma^+$ transition. (0,0) means the transition between $v' = 0$ and $v'' = 0$ vibrational levels, whereas (1,1) means the transition between $v' = 1$ and $v'' = 1$. The selection rule for the rotational transition is fairly strict and the total angular momentum quantum number, J, may not change or may change

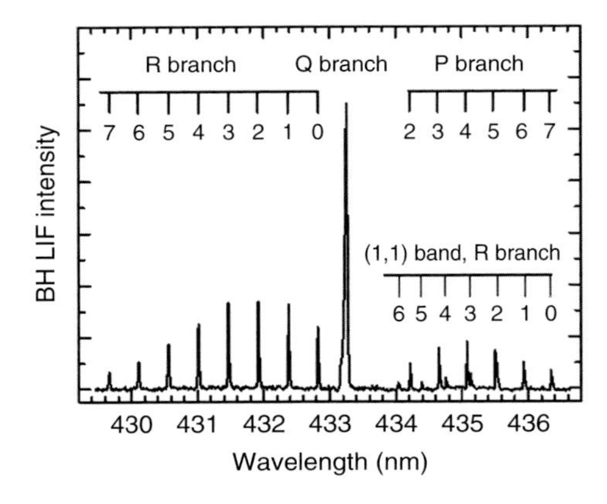

Figure 3.4 LIF spectrum of BH(X $^1\Sigma^+$, $v'' = 0,1$) formed in the catalytic decomposition of $B_2H_6/He/H_2$. The flow rates of B_2H_6/He (2.0% dilution) and H_2 were 10 and 20 sccm, respectively, whereas the total pressure was 3.9 Pa. The W wire temperature was 2.05×10^3 K. The numbers are the total angular momentum quantum numbers of the lower X $^1\Sigma^+$ state. Source: Umemoto et al. 2014 [39]. Reprinted with permission of American Chemical Society.

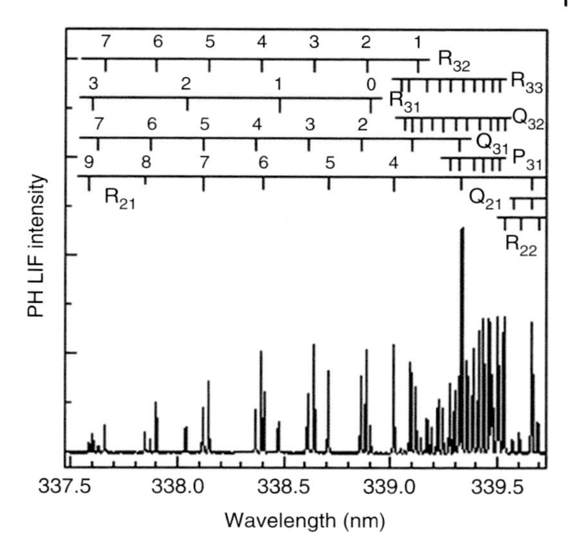

Figure 3.5 LIF spectrum of PH (X $^3\Sigma^-$, $v'' = 0$) formed in the catalytic decomposition of PH_3/He/H_2. The flow rates of PH_3/He (2.0% dilution) and H_2 were 10 and 20 sccm, respectively, whereas the total pressure was 4.0 Pa. The W wire temperature was 2.15×10^3 K. The numbers are the total angular momentum quantum numbers exclusive of electron spin of the lower X $^3\Sigma^-$ state. Source: Umemoto et al. 2012 [22]. Copyright 2012: The Japan Society of Applied Physics.

just one; $\Delta J = 0, \pm 1$. The transition without changes is called the Q branch. In the R branch, the total angular momentum quantum number of the upper state, J', is larger than that of the lower state, J'', by unity; $J' = J'' + 1$. In the P branch, J' must be $J'' - 1$. In Figure 3.4, the rotational lines of the Q branch are not resolved, but those of the R branch are well resolved, and the rotational state distribution can be determined.

Figure 3.5 illustrates an LIF spectrum of PH formed in the catalytic decomposition of PH_3/He/H_2 [22]. The spectrum, corresponding to the A $^3\Pi_i$ – X $^3\Sigma^-$ transition, is much more complex compared to that of BH because the spin multiplicities of both upper and lower states are triplet. Fortunately, the spectral assignment has been carried out by successive spectroscopists, as will be mentioned in Section 3.3.7. Vibrationally excited PH was not identified in this case.

In the detection of H [5, 40, 41], N [42, 43], O [30, 44], and P [22] atoms by one-photon LIF, VUV radiation is required, as shown in Table 3.2. Lyman-α radiation at 121.6 nm, needed to detect H atoms, can be obtained by a frequency tripling technique by using Kr (krypton) as a nonlinear optical medium. In the detection of O and P atoms, a four-wave mixing technique must be used to produce appropriate VUV outputs: 130.2 and 177.5 nm, respectively. Figure 3.6 shows a schematic diagram of the experimental apparatus to detect O atoms, where two tunable lasers are used. A collimating lens made of MgF_2 (magnesium fluoride) or LiF (lithium fluoride) and a solar-blind PMT are used to detect the induced fluorescence. In the detection of N atoms, VUV radiation at 113.5 nm can be obtained by frequency tripling by Kr, while four-wave mixing must be employed to generate the output at 120.0 nm. Hg (mercury) vapor is used as a mixing medium. A NO (nitric oxide) cell can be used to confirm the generation of VUV radiation. NO can be easily ionized by VUV radiation, while that is not ionized by unfocused visible or near UV radiation. Figure 3.7 illustrates the LIF spectra of O atoms formed in the catalytic decomposition of a H_2/O_2 mixture

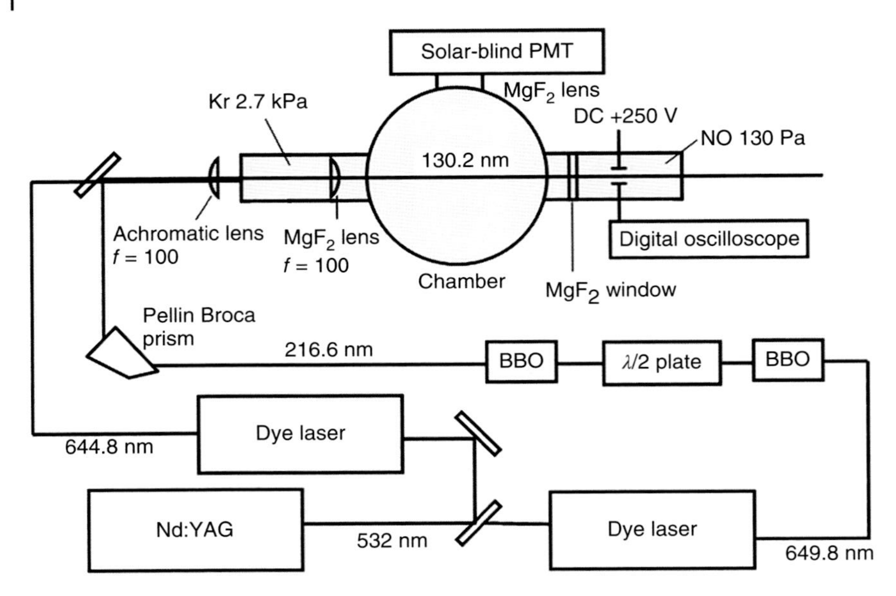

Figure 3.6 Schematic diagram of the apparatus for four-wave mixing to detect O atoms.

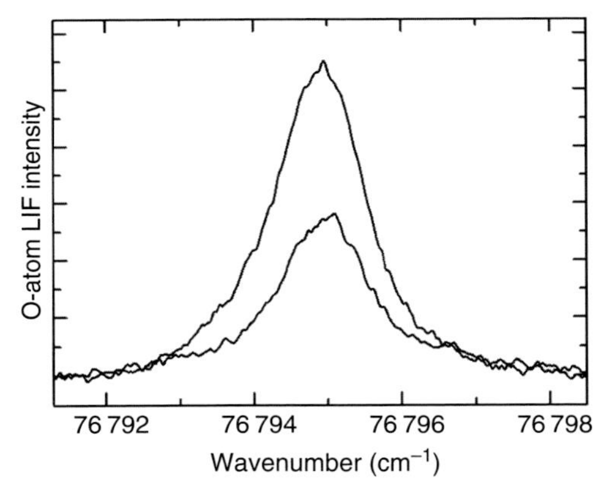

Figure 3.7 VUV LIF spectra of O atoms formed in the catalytic decomposition of a H_2/O_2 mixture on a heated W wire. The H_2 flow rate was 100 sccm, whereas the O_2 flow rate was 1.20 (upper) or 0.60 (lower) sccm. The total pressure was 17 Pa, and the catalyzer temperature was 2.00×10^3 K. Source: Data from Umemoto and Moridera 2008 [30].

on a heated W wire. The spectrum corresponds to the $2s^22p^33s\ ^3S_1–2s^22p^4\ ^3P_2$ transition.

Finally, it should be noted that one-photon LIF as well as VUV laser absorption, which will be discussed in Section 3.5, cannot be used when the atomic density is too high. For example, when the H atom density is more than 10^{11} cm^{-3}, the system becomes optically too thick to allow the laser beam at 121.6 nm to reach the central part of the chamber. In such cases, a two-photon LIF technique is useful, which will be mentioned in Section 3.4.

3.3.7 Determination of Rotational and Vibrational State Distributions of Molecular Radicals

The rotational temperature can be evaluated from the rotational state distribution as long as the distribution is Boltzmann-like, which is usual in

Cat-CVD chambers. As both rotational and translational relaxations are fast, the translational temperature may be assumed to be the same as the rotational one. Such information is essential in the evaluation of absolute molecular densities from pressures. Information on the temperature is also important when we figure out the reaction rates in the gas phase, especially for reactions with large activation energies.

It is possible to determine the rotational state distributions from the LIF spectra, as shown in Figures 3.4 and 3.5. The spectral assignments have been reported for many diatomic molecules, including BH [45], CH [46], NH [47], OH [48], SiH [49], and PH [50]. The Hönl-London factor, which is proportional to the product of the Einstein B coefficient, B_{12}, and the rotational degeneracy factor, $2J + 1$, is given as a function of J for various types of transitions [51]. J is the total angular momentum quantum number, which is almost determined by the rotational motion. The correction for the nonradiative decay processes is often necessary, but the J dependences of the radiative lifetime, which is equivalent to $1/A_{21}$, and the quenching rate constant, k_i, are usually small. The rate constant for the nonradiative decay, k_{nr}, is usually much smaller than the radiative decay rate for species detected by LIF.

The rotational temperature of BH determined from the spectrum shown in Figure 3.4 is 340 K. This temperature is similar to those obtained for other hydride radicals, such as NH, OH, SiH, and PH, produced in similar procedures and typical when the distance from the wire is more than several centimeters and the pressure is more than several pascals [22, 30, 38, 41].

The vibrational state distribution of diatomic species can also be determined if Franck–Condon factors, the squares of the vibrational overlap integrals and proportional to the transition probabilities, are known, but vibrationally excited radicals have not been identified in catalytic decomposition except for HCN and B_2H_6 systems [39, 52]. In the catalytic decomposition of B_2H_6, taking into account the difference in the Franck–Condon factors [53], the density of BH($v = 1$) was estimated to be ~15% of BH($v = 0$). In the HCN system, the CN($v = 1$)/CN($v = 0$) population ratio was determined to be 23%. In these estimations, the contribution of off-diagonal transitions is ignored. The probability of off-diagonal transitions from the $v' = 0$ state to $v'' \geq 1$ states is much smaller than the diagonal one for many diatomic molecules, including the A–X transitions of BH [53], CH [54], NH [55], OH [56], SiH [57], and PH [58].

3.3.8 Estimation of Absolute Densities in One-Photon LIF

It is possible to estimate the absolute densities of radical species by comparing the LIF intensity with that of Rayleigh scattering, which is the elastic scattering of light caused by particles much smaller than the wavelength, such as atoms and molecules. The laser intensity may be weak enough not to saturate the transition or strong enough to saturate the transition. In the former case, information on the absolute value of the Einstein B coefficient, B_{12}, is necessary, whereas in the latter case, information on the absolute pulse energy of the laser is required. CW lasers can also be used, but we restrict the discussion only to pulsed laser systems. The absolute pulse energy can be easily measured with a joule meter.

In the following, formulation under saturated conditions shall be presented [21], but similar formulation is possible for proportional conditions [12, 59].

In LIF measurements, excited species are produced along the optical path, but fluorescence only from a small volume, V, can be detected. The fluorescence is usually isotropic and that in a solid angle, Ω, is detected. The absolute sensitivity of the detector, η, is also necessary to evaluate the absolute number of photons emitted. Fortunately, the product of these three parameters, $V\Omega\eta$, can be estimated by measuring the Rayleigh scattering intensity caused by rare gas atoms, such as Ar (argon). Under saturated conditions, the time-integrated LIF intensity after pulse excitation is given by

$$I_{\rm LIF} = \frac{g_2 A_{21} N h\nu}{(g_1 + g_2)\left(A_{21} + \sum k_i n_i\right)} V\eta \frac{\Omega}{4\pi} \tag{3.11}$$

in energy unit, when nonradiative decay, corresponding to $k_{\rm nr}$, can be ignored. The Rayleigh scattering intensity is expressed by

$$I_{\rm Rayleigh} = \frac{N_0 \sigma_{\rm R} E_{\rm L}}{S_{\rm L}} V\Omega\eta \tag{3.12}$$

N_0 is the density of rare gas atoms, $E_{\rm L}$ is the laser pulse energy, $S_{\rm L}$ is the cross-sectional area of the laser beam, and $\sigma_{\rm R}$ is the differential cross section for Rayleigh scattering, which can be calculated based on the refractive index [59].

$$\sigma_{\rm R} = \frac{9\pi^2}{\lambda^4 N_0^{\,2}} \left(\frac{n^2 - 1}{n^2 + 2}\right)^2 \approx \frac{4\pi^2}{\lambda^4 N_0^{\,2}} (n-1)^2 \tag{3.13}$$

Here, λ is the wavelength and n is the refractive index when the gas density is N_0. It is assumed that the polarization direction of the laser is perpendicular to the direction of observation. Then, the absolute density of radical species can be evaluated by the following equation:

$$N = \frac{g_1 + g_2}{g_2} \frac{A_{21} + \sum k_i n_i}{A_{21}} \frac{4\pi}{h\nu} \frac{N_0 \sigma_{\rm R} E_{\rm L}}{S_{\rm L}} \frac{I_{\rm LIF}}{I_{\rm Rayleigh}} \tag{3.14}$$

The first part of Eq. (3.14), the ratio of the statistical weights, is given by a function of the total angular momentum quantum numbers of the upper and the lower states, J' and J''. The degeneracy factors are, in general, given by $2J+1$, but, depending on the conditions, a minor modification is necessary. For example, when we excite BH(X $^1\Sigma^+$, $v'' = 0$, $J'' = 2$) to (A $^1\Pi$, $v' = 0$, $J' = 3$) with an intense linearly polarized radiation, the value of $(g_1 + g_2)/g_2$ should be 2, instead of $(2J'' + 1 + 2J' + 1)/(2J' + 1) = 12/7$ [25]. Rayleigh scattering is not isotropic when the laser is linearly polarized due to the conservation of the angular momentum. A half-wave plate (or a double Fresnel rhomb) should be used depending on the optical geometry. Strictly speaking, the wavelength of the induced fluorescence and that of Rayleigh scattering may be different. However, as has been mentioned in Section 3.3.2, the wavelength shift in LIF is usually minor and the change in the detection sensitivity may be ignored. In the detection of molecular species, we usually select one isolated spectral line, which does not overlap with others. To evaluate the total population, information on the rotational state distribution, discussed in Section 3.3.7, is necessary. As for H, O, and N atoms, absorption spectroscopy can also be used to determine the absolute densities, which will be discussed in Section 3.5.

3.4 Two-Photon Laser-Induced Fluorescence

Ground-state H atoms, H(1s $^2S_{1/2}$), can be two-photon excited to H(3s $^2S_{1/2}$) and H(3d $^2D_{5/2,3/2}$) via a virtual state by focused laser radiation at 205.1 nm. Radiation at 205.1 nm can be obtained by tripling the output of a tunable laser by two BBO crystals and a half-wave plate. The experimental setup for two-photon LIF is schematically drawn in Figure 3.8. Baffled arms are not used to prioritize tight focusing. H(3p $^2P_{3/2,1/2}$) cannot be produced directly because of the selection rule on the azimuthal quantum number, l, associated with two-photon absorption, $\Delta l = 0, \pm 2$, but can easily be produced by collisional mixing. H(3s $^2S_{1/2}$), H(3p $^2P_{3/2,1/2}$), and H(3d $^2D_{5/2,3/2}$) fluoresce Balmer-α at 656.3 nm, which can be monitored to evaluate the ground-state H atom densities. This technique is less sensitive compared to one-photon LIF but can be used in the presence of a large amount of H atoms, 10^{14} cm^{-3} or more [5]. As an example of observation, a typical two-photon LIF spectrum of H atoms is shown in Figure 3.9 [40].

The absolute density calibration is possible by measuring the two-photon LIF signal intensity of a known amount of Kr following excitation at 204.2 nm [60, 61]. Besides H atoms, similar two-photon LIF techniques can be applied to detect ground-state O [62], N [63], and C atoms [64]. As for O and N atoms, and possibly for C atoms, however, the densities obtained by catalytic decomposition are too low to be measured by this technique.

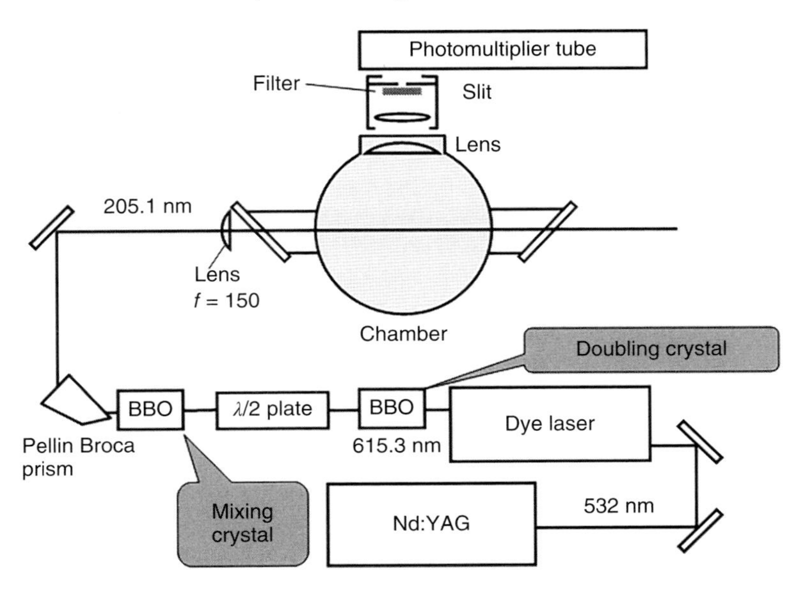

Figure 3.8 Schematic diagram of the apparatus for two-photon LIF to detect H atoms at 205.1 nm.

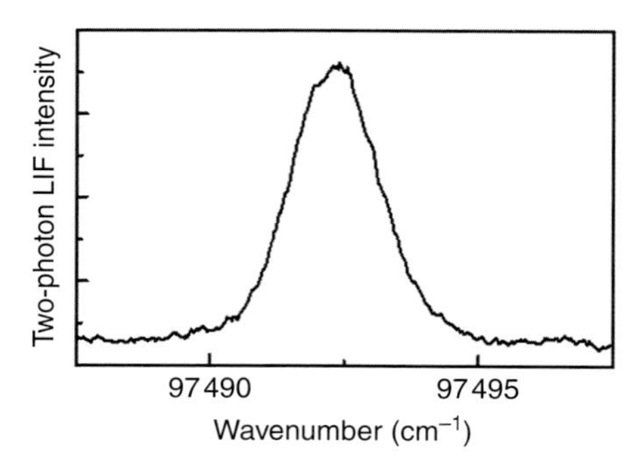

Figure 3.9 Two-photon LIF spectrum of H atoms formed in the catalytic decomposition of H_2. The H_2 flow rate was 150 sccm, whereas pressure was 5.6 Pa. The W wire temperature was 2.20×10^3 K. Source: Umemoto et al. 2002 [40]. Reprinted with permission of American Institute of Physics.

Two-photon LIF at 205.1 nm to detect H atoms is subject to collisional quenching because the radiative lifetimes of H(3s $^2S_{1/2}$), H(3p $^2P_{3/2,1/2}$), and H(3d $^2D_{5/2,3/2}$) are much longer (158, 45, and 15 ns, respectively) than that of H(2p $^2P_{3/2,1/2}$), 1.7 ns [24], whereas the rate constants for the quenching of these states are extremely large; the rate constant for H_2 is as large as 2×10^{-9} cm^3/s [60, 65]. This point must be kept in mind when the material gas pressure is high. It should also be noted that this technique cannot be applied in the presence of material gases, which can be two-photon decomposed to produce H atoms, such as NH_3 and Si_2H_6.

Ground-state H atoms can also be two-photon excited to H(2s $^2S_{1/2}$) [5, 30]. The exciting wavelength is 243.1 nm. H(2s $^2S_{1/2}$) is metastable and does not fluoresce, but that can be easily collisionally relaxed to the near-by 2p $^2P_{3/2,1/2}$ states, which fluoresce Lyman-α. This two-photon excitation technique, together with the 2 + 1 REMPI technique, which will be focused in Section 3.6.1, is the best for measuring the Doppler profiles because this transition has no fine structures. The situation is more complex in one-photon LIF at 121.6 nm or the two-photon LIF at 205.1 nm. For example, the spin–orbit splitting for H(2p $^2P_{3/2,1/2}$) is 0.365 cm^{-1} and the Lyman-α lines corresponding 2p $^2P_{3/2}$–1s $^2S_{1/2}$ and 2p $^2P_{1/2}$–1s $^2S_{1/2}$ transitions cannot be resolved at moderate temperatures.

3.5 Single-Path Vacuum Ultraviolet (VUV) Laser Absorption

VUV laser absorption is one of the most common techniques to evaluate the absolute densities of some atomic species, such as H and O atoms [5, 40, 41, 44]. The wavelengths are the same as those used in the one-photon LIF. One problem is that the well-known Beer–Lambert law, $\mathrm{Tr} = \exp(-kl)$, cannot be applied. Here, Tr is the transmittance, k is the absorption coefficient, and l is the optical path length. This relationship can only be used when the absorption spectrum is much broader than the bandwidth of the incident radiation. In VUV absorption by atoms, the absorption spectrum is too sharp.

As has been mentioned in Section 3.3.5, under typical CVD conditions (below 10^2 Pa and over room temperature), the natural and pressure broadenings are much narrower than the Doppler one caused by the translational motion of the absorbing species and the absorption spectral profile can be approximated by the following Gaussian function [17]:

$$k(\nu - \nu_0) = \left(\frac{m}{2\pi k_B T}\right)^{1/2} \frac{g_2 c^3 N}{8\pi g_1 {\nu_0}^3 \tau} \exp\left\{-\frac{mc^2(\nu - \nu_0)^2}{2k_B T {\nu_0}^2}\right\} \tag{3.15}$$

whereas the Doppler width (full width at half maximum) is represented by

$$\Delta\nu_D = 2\nu_0\left(\frac{2\ln 2\ k_B T}{mc^2}\right)^{1/2} \tag{3.16}$$

Here, ν is the frequency of the incident radiation, ν_0 is the frequency at the absorption peak, k_B is the Boltzmann constant, c is the speed of light, m is the mass of the absorbing species, τ is the radiative lifetime of the upper state, T is the absolute translational temperature, and N is the density of absorbing species, whereas g_1 and g_2 are the statistical weights of the lower and upper states. The general representation of the transmittance is given by [17]:

$$\mathrm{Tr}(\nu_L) = \frac{\int I(\nu - \nu_L)\exp\{-k(\nu - \nu_0)l\}d\nu}{\int I(\nu - \nu_L)d\nu} \tag{3.17}$$

Here, ν_L and $I(\nu - \nu_L)$ represent the central frequency and the spectral profile of the incident laser radiation. In most cases, the exact profile of $I(\nu - \nu_L)$ is not known, and Gaussian functions, $\exp\{-(\nu - \nu_L)^2/\alpha^2\}$, are usually assumed, where α is a parameter to represent the spectral width of the laser.

In the detection of H atoms by the absorption of Lyman-α at 121.6 nm, a little modification is necessary because Lyman-α is an unresolved doublet because of spin–orbit interaction [5]. In this case, the optical density, $k(\nu - \nu_0)l$, must be expressed by a sum of two terms and the transmittance should be

$$\mathrm{Tr}(\nu_L) = \frac{\int_0^\infty \exp\left\{-\left(\frac{\nu-\nu_L}{\alpha}\right)^2\right\} \exp\left\{-k_0 l \exp\left(-\left(\frac{\nu-\nu_0}{\beta_0}\right)^2\right) - k_1 l \exp\left(-\left(\frac{\nu-\nu_1}{\beta_1}\right)^2\right)\right\} d\nu}{\int_0^\infty \exp\left\{-\left(\frac{\nu-\nu_L}{\alpha}\right)^2\right\} d\nu} \tag{3.18}$$

The subscripts 0 and 1 stand for the transitions to 2p $^2P_{1/2}$ and 2p $^2P_{3/2}$, respectively. The frequencies at the absorption peaks are represented by ν_0 and ν_1. The values of β_0, β_1, k_0, and k_1 are given by

$$\beta_0 = \left(\frac{2k_B T}{m}\right)^{1/2} \frac{\nu_0}{c}, \beta_1 = \left(\frac{2k_B T}{m}\right)^{1/2} \frac{\nu_1}{c}, k_0 = \left(\frac{m}{2\pi k_B T}\right)^{1/2} \frac{c^3 N}{8\pi {\nu_0}^3 \tau}, \text{ and}$$

$$k_1 = \left(\frac{m}{2\pi k_B T}\right)^{1/2} \frac{c^3 N}{4\pi {\nu_1}^3 \tau}.$$

The value of α can be determined from the absorption profile under completely thermalized conditions. Figure 3.10 illustrates typical absorption spectra of H atoms produced by the catalytic decomposition of H_2 [40]. A combination of a VUV monochromator and a solar-blind PMT can be used to evaluate the transmitted laser intensity, whereas the measurement of the NO^+ ion current is another convenient technique to evaluated the transmitted laser intensity.

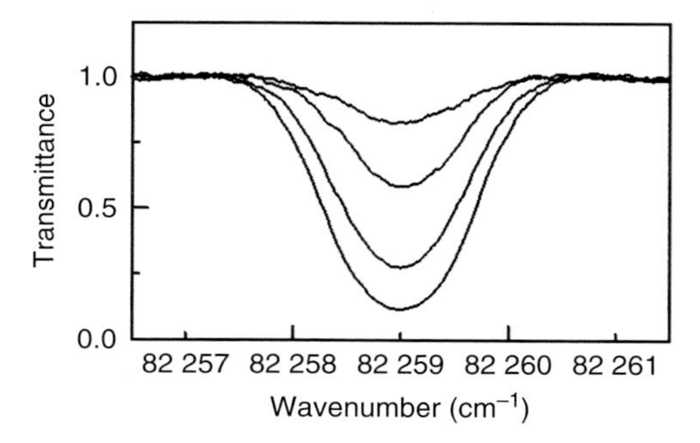

Figure 3.10 Vacuum ultraviolet absorption spectra of H atoms formed by catalytic decomposition of H_2. The H_2 flow rate was 150 sccm, whereas the pressure was 5.6 Pa. The path length was 45 cm. The W wire temperatures were 1.27×10^3, 1.32×10^3, 1.37×10^3, and 1.41×10^3 K from top to bottom. The corresponding H atom densities are 1.5×10^{10}, 4.3×10^{10}, 1.0×10^{11}, and 1.8×10^{11} cm^{-3}, respectively. Source: Umemoto et al. 2002 [40]. Reprinted with permission of American Institute of Physics.

NO is not ionized by unfocused radiation at 364.7 nm, the wavelength before frequency tripling to produce Lyman-α.

As the absolute values of the absorption coefficients, k_0 and k_1 in Eq. (3.18), and the path length, l, are known, it is easy to evaluate the absolute densities by this technique. By scanning the wavelength, it is possible to eliminate the contribution of broad background absorption by source and product molecular species, such as SiH_4 and CH_4. This technique is insensitive to collisional quenching of the upper states. The problem is that the dynamic range is rather narrow. For precise measurements, the absorbance should be between 10% and 90%, considering the fluctuation in laser pulse energy. When the path length is 10 cm, the density to be evaluated is in the order of 10^{11} cm^{-3} for H atoms and 10^{12} cm^{-3} for O atoms. As for N atoms, the density obtained by the catalytic decomposition of N_2 or NH_3 was too low to be detected by absorption, but it was possible to observe the absorption in microwave discharge [41].

Figure 3.11 compares the H atom detection techniques, one-photon LIF, two-photon LIF, VUV laser absorption, and 2 + 1 REMPI. As for REMPI, see Section 3.6.1 for details. As is true with any techniques, these have merits and limitations. For instance, one-photon LIF has a high sensitivity, whereas two-photon LIF can be used in the presence of an excess amount of H atoms. Absolute density can be determined by VUV absorption, while 2 + 1 REMPI can be used in the presence of strong background emission and is the best to determine the translational temperatures.

3.6 Other Laser Spectroscopic Techniques

Although LIF is one of the most sensitive techniques to detect atomic and molecular species, that cannot be applied to predissociative species, such as CH_3 and SiH_3, from which no fluorescence is expected. To detect such species, other techniques must be employed. REMPI, cavity ringdown spectroscopy (CRDS), and tunable diode laser absorption spectroscopy (TDLAS) are useful

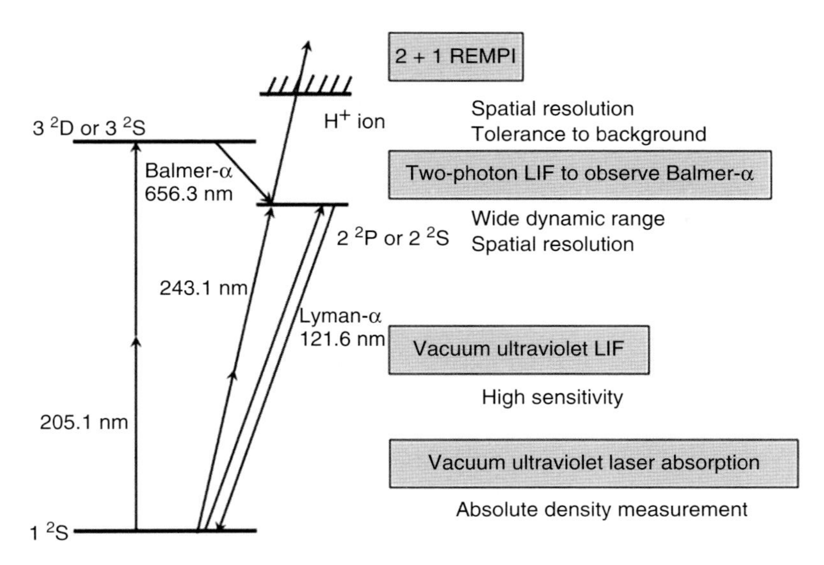

Figure 3.11 Comparison of the four techniques to detect H atoms.

to detect such species. With the latter two techniques, absolute densities can also be determined. These techniques can be used in the presence of strong background emission, such as blackbody radiation from hot filaments. There are some other minor techniques, which are not affected by background emission, such as amplified spontaneous emission (ASE), two-photon polarization, and third-harmonic generation (THG) techniques.

3.6.1 Resonance-Enhanced Multiphoton Ionization

REMPI involves a resonant single- or multiple-photon absorption to an electronically excited state followed by the absorption of another photon to ionize the intermediate state. By detecting ions or electrons produced, it is possible to evaluate the density of ground-state species. When m photons are needed to excite resonantly and n photons to ionize, that is called $m + n$ REMPI. $2 + 1$ REMPI is the most popular, but sometimes, $3 + 1$ or $2 + 2$ REMPI techniques are utilized. In $2 + 1$ REMPI, the signal intensity is usually proportional to the square of the laser intensity, as the final ionization step can be easily saturated. As a focused laser beam is used and the ionization volume is small, spatial resolution is expected. REMPI is extremely sensitive when combined with a time-of-flight mass spectrometric technique. On the other hand, when used without mass selection, high vacuum is not required, and this technique can be used under typical CVD conditions. In such cases, care must be paid to the choice of applied voltage. It is necessary to collect all electrons or ions produced, but an electron avalanche must be avoided. There are many examples of the detection by REMPI applied to Cat-CVD processes, including H atoms [5], Si atoms [66], B atoms [67], and CH_3 radicals [68–71]. $SiH_3(\tilde{X}\ ^2A_1)$ [72] and $CH_2(\tilde{X}\ ^3B_1)$ [73, 74] have also been detected by REMPI, although application to catalytic decomposition has not been made. Table 3.4 summarizes the wavelengths and transitions to

Table 3.4 Wavelengths to detect H, B, and Si atoms and hydride radicals by REMPI.

Radical	Resonant transition	Wavelength (nm)	$n+m$	References
H	2s ^{2}S–1s ^{2}S	243.1	2+1	[5]
H	2p ^{2}P–1s ^{2}S	364.7	3+1	[5]
B	2s^{2}4p ^{2}P–2s^{2}2p ^{2}P	346.1	2+1	[67]
CH	E′ $^2\Sigma^+$–X $^2\Pi$	291	2+1	[75]
CH	D $^2\Pi$–X $^2\Pi$	311	2+1	[75]
CH_2	4d $^3A_2$–$\tilde{X}$ $^3B_1$	392	3+1	[73]
CH_2	H (3p)–$\tilde{X}$ $^3B_1$	312	2+1	[74]
CH_3	$3p_z$ $^2A''_2$–$\tilde{X}$ $^2A''_2$	333	2+1	[71]
Si	3s^{2}3p4s $^3P_2$–3s^{2}3p^2 $^3P_2$	251.6	1+1	[66]
SiH	F $^2\Pi$–X $^2\Pi$	428	2+1	[76]
SiH_3	E $^2A''_2$ (4p)–$\tilde{X}$ $^2A_1$	350–415	2+1	[72]

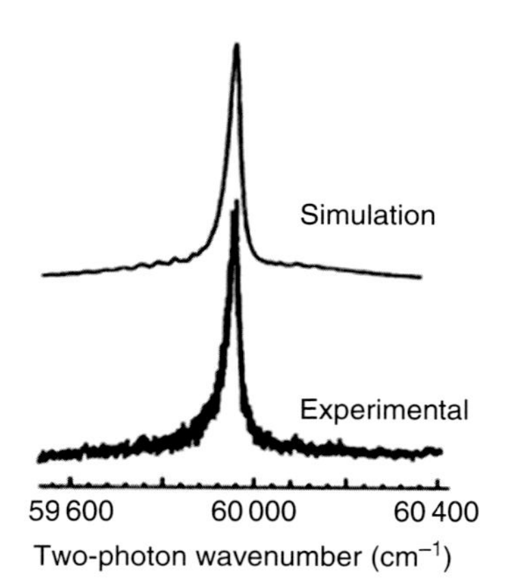

Figure 3.12 Simulated and experimental 2 + 1 REMPI spectra of CH_3 formed in the catalytic decomposition of a 1% CH_4 in H_2 gas mixture. The Ta (tantalum) wire temperature was 2475 K, and the position from the wire was 4 mm. The rotational temperature was assumed to be 1150 K in the simulation. Source: Smith et al. 2001 [71]. Reprinted with permission of Elsevier.

detect radicals typical in CVD processes by REMPI. More information can be found in a report by M.N.R. Ashfold et al. [77]. Figure 3.12 shows a REMPI spectrum of CH_3 ($3p_z$ $^2A''_2$- $\tilde{X}$ $^2A''_2$) obtained by J.A. Smith et al. [71]. As has been mentioned in Section 3.4, the Doppler width measurements of REMPI signals of H atoms at 243.1 nm is useful to determine the translational temperature because of the absence of fine structures [70, 78]. Figure 3.13 illustrates the REMPI spectra of H atoms at two radial distances from a hot wire. It is clear that the translational temperature decreases with the increase in the distance [70].

3.6.2 Cavity Ringdown Spectroscopy

In CRDS, the time profile of the laser intensity leaking from an optical cavity is measured. The optical cavity is consisted of two highly reflective concave mirrors. Both pulsed and CW lasers can be used. When CW lasers are used, the profiles are recorded just after the termination of the laser. In the presence of

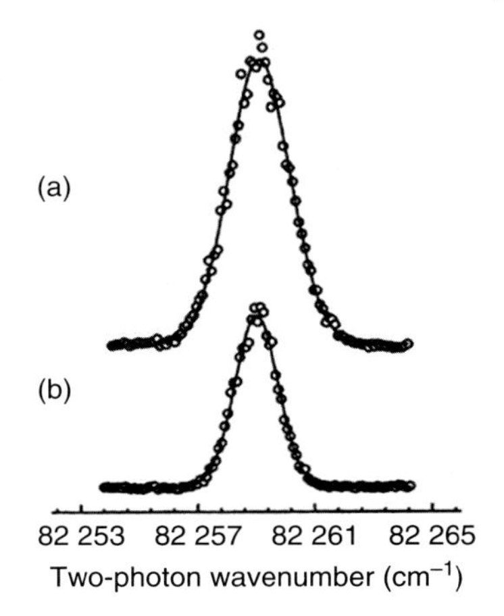

Figure 3.13 Doppler-broadened REMPI spectra of H atoms recorded in 2.7 kPa of pure H_2 with the laser focus at 0.5 mm (upper) and 10 mm (lower) from the bottom of the hot coiled Ta wire held at 2375 K. The solid curve in each case is a least squares fit to a Gaussian function, the width of which yields a measure of the local temperature of 1750 and 840 K, respectively. Source: Smith et al. 2000 [70]. Reprinted with permission of Elsevier.

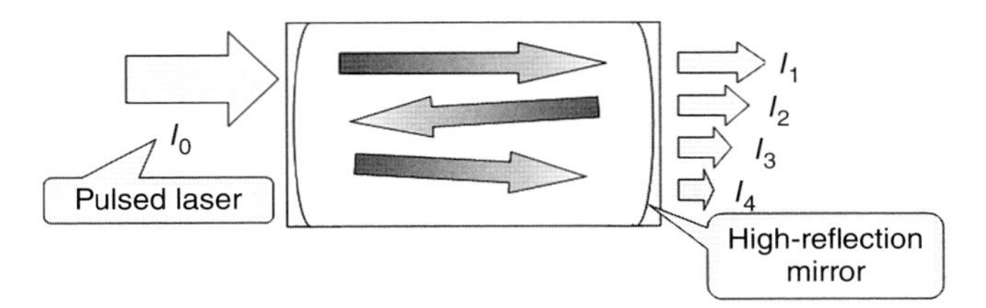

Figure 3.14 Principle of cavity ringdown spectroscopy. After a single path, $t = \ell/c$ and $I_1(\ell/c) = I_0(1 - R)^2\exp(-\sigma N\ell)$, after 1.5 round trip, $t = 3\ell/c$ and $I_2(3\ell/c) = I_0(1 - R)^2R^2\exp(-3\sigma N\ell)$, and at time t, $I(t) = I_0(1 - R)^2R^{(ct/\ell-1)}\exp(-\sigma Nct) = I_0(1 - R)^2R^{-1}\exp\{(-\sigma N + \ln R/\ell)ct\}$.

some absorbing species, the decay is faster than in their absence because of the additional losses caused by photoabsorption, and it is possible to evaluate the density by measuring the decay time, *ringdown time*. The measuring principle is illustrated in Figure 3.14.

The time dependence of the transmitted light intensity, $I(t)$, is given by [79, 80]

$$I(t) = I_0(1 - R)^2R^{-1}\exp\left\{\left(-\sigma N + \frac{\ln R}{l}\right)ct\right\} \tag{3.19}$$

Here, I_0 is the incident light intensity, R is the reflectivity of the mirrors, l is the path length, and c is the speed of light. N and σ are the density and the absorption cross section of the species to be detected, respectively. The product of N and σ is equivalent to the absorption coefficient, k. Just for simplicity, we consider a system where the Beer–Lambert law holds. As weak lasers are used, stimulated emission processes can be ignored. When R is 99.99% and l is 50 cm, the ringdown time in the absence of absorbing species is calculated to be 17 μs and the effective path length is 5.0 km. It is easy to derive the following relationship between the ringdown time in the presence of absorbing species, τ, and that in its absence, τ_0:

$$1/\tau - 1/\tau_0 = \sigma Nc \tag{3.20}$$

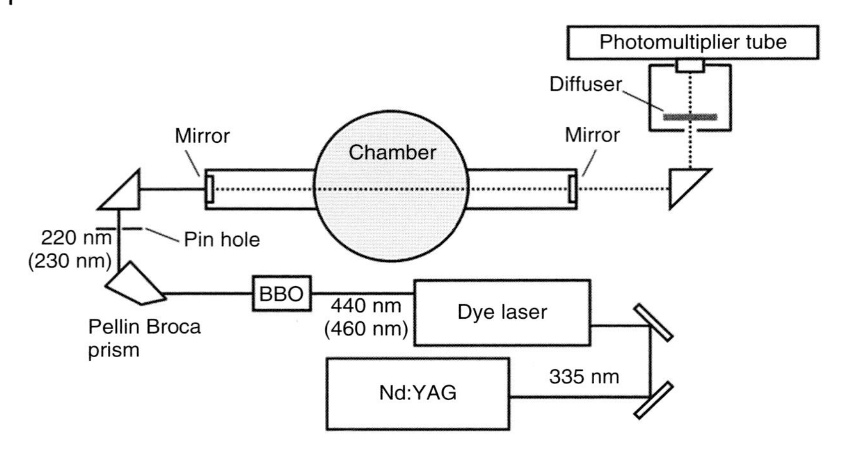

Figure 3.15 Schematic diagram of the apparatus for cavity ringdown spectroscopy to detect SiH_3.

Then, it is possible to evaluate the density of absorbing species, N, if the absorption cross section, σ, is known. It should be noted that the ringdown time does not depend on the laser intensity. In other words, the fluctuation in laser intensity cannot be a problem. The problem is that wide-band high-reflection mirrors are not available, especially in the UV region, and it is impossible to scan the wavelength widely.

A schematic view of the experimental equipment is shown in Figure 3.15. The insertion of a diffuser in front of the PMT is effective to ensure the uniform distribution of the laser beam on the light receiving surface. To protect the mirrors from deposition, it is desired to flow rare gas near the mirrors, but, in this case, the density of absorbing species may not be uniform and the effective path length may change.

In Cat-CVD processes, this technique has mainly been applied to the detection of CH_3 [81, 82] and SiH_3 [83, 84]. Typical ringdown decay forms in the absence and in the presence of SiH_3 produced in a SiH_4/NH_3 system are illustrated in Figure 3.16 [84]. SiH_3 has also been identified by CRDS in plasma processes

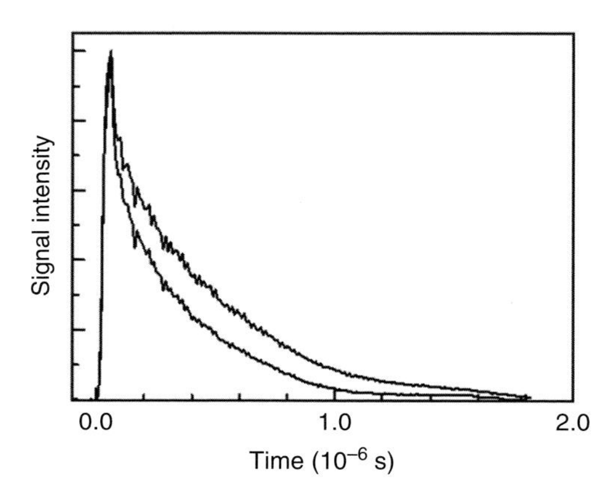

Figure 3.16 Ringdown decay forms observed at 230 nm in the presence (lower) and in the absence (upper) of 10 sccm of SiH_4 flow. The decay forms were measured in the presence of a 500 sccm of NH_3 flow and the total pressure was 20 Pa when the catalyzer was not heated. The W wire temperature was 2.30×10^3 K. Source: Umemoto et al. 2003 [84]. Reprinted with permission of Elsevier.

Table 3.5 Wavelengths to detect radicals by CRDS.

Radical	Transition	Wavelength (nm)	References
CH	$A\,^2\Delta - X\,^2\Pi$	431[a)]	[87]
CH_3	$B\,^2A'_1 - \tilde{X}\,^2A''_2$	214, 217[b)]	[82]
NH	$A\,^3\Pi - X\,^3\Sigma^-$	336[a)]	[88]
NH_2	$\tilde{A}\,^2A_1 - \tilde{X}\,^2B_1$	597[b)]	[88]
Si	$3s^23p4s\,^3P_J - 3s^23p^2\,^3P_J$	250.7–251.9	[89]
SiH	$A\,^2\Delta - X\,^2\Pi$	413[a)]	[89]
SiH_2	$\tilde{A}\,^1B_1 - \tilde{X}\,^1A_1$	582[b)]	[90]
SiH_3	$\tilde{A}\,^2A'_1 - \tilde{X}\,^2A_1$	215–230[b)]	[83–86]
SH	$A\,^2\Sigma^+ - X\,^2\Pi$	326[a)]	[91]
CS	$A\,^1\Pi - X\,^1\Sigma^+$	258[a)]	[91]

a) Band origin of the (0,0) band.
b) Typical wavelength(s) employed.

[85, 86]. The wavelengths employed are given in Table 3.5, together with those for other species typical in CVD processes. In this technique, care must be paid to the contribution of Rayleigh and Mie scattering caused by small particles [86].

3.6.3 Tunable Diode Laser Absorption Spectroscopy

TDLAS is effective for the detection of polyatomic molecules with more than four atoms, which are hard to be detected by LIF [13]. Most polyatomic radicals are infrared active and can be detected by photoabsorption in the mid-infrared region, 2.5–25 μm or 400–4000 cm^{-1}, which corresponds to the fundamental vibrational excitation. CH_3 radicals produced by catalytic decomposition of CH_4 have been identified by TDLAS in the near 606 cm^{-1} region [92, 93]. SiH_3 can also be detected at around 720 cm^{-1} [94, 95], although TDLAS has not been applied to the catalytic decomposition systems.

As the absorption cross sections of molecular species are much smaller than those of atoms, multipath absorption is essential. The bandwidth of single-mode diode lasers used in TDLAS is less than the Doppler width at room temperature. Then, the simple Beer–Lambert law can be applied and the absolute column density can be easily determined if the absorption cross section is known. It is also possible to determine the translational temperature from the Doppler width measurements. The rotational temperature can also be evaluated from rotationally resolved spectra.

3.7 Mass Spectrometric Techniques

Mass spectrometry has widely been employed to identify radical species in the gas phase. There are three stages in the mass spectrometric analysis: ionization, mass selection, and ion detection. The first step should be paid the most attention when radical species are detected because the densities of source material (and possibly end product) molecules are much more abundant than those of radicals, and these stable molecules may be decomposed to produce fragment ions. In order to avoid this difficulty, several ionization techniques have been proposed: photoionization, threshold ionization, and ion attachment. All of these techniques are less molecular specific compared to the laser techniques, which are more selective.

3.7.1 Photoionization Mass Spectrometry

The photoionization technique is further classified into single-photon and multiphoton ones. The REMPI technique, described in Section 3.6.1, can be combined with a mass spectrometric technique to improve the selectivity and sensitivity. In single-photon ionization (SPI), the ninth harmonic of a Nd^{3+}:YAG laser at 118 nm, generated by tripling the third harmonic by Xe (xenon), is widely used because this energy, 10.5 eV, is sufficient to ionize many molecular radicals but is not enough to cause dissociative ionization, thereby avoiding the appearance of daughter ions [96–100]. The produced ions are usually mass selected by a time-of-flight mass spectrometer. Figure 3.17 shows a comparison of SPI and electron impact mass spectra of silacyclobutane at room temperature [99]. Silacyclobutane can be ionized, but the fragmentation is minor. A conventional light source has also been utilized to ionize the catalytic decomposition products of SiH_4 [101]. One of the merits in this single-photon technique compared to REMPI is that multiple species can be detected at one time. On the other hand, as the ionization process is not resonant, information on the internal state distributions as well as the translational energies cannot be obtained.

The VUV laser SPI technique coupled with time-of-flight mass spectrometry has been extensively applied to the catalytic CVD processes of organosilicon compounds by Y.J. Shi and coworkers [99, 100, 102–105]. They mainly investigated the compounds that can be a single-source precursor of SiC films. They have found that the decomposition products are different from those of thermal decomposition, showing that the decomposition is not thermal but catalytic [102, 104]. Besides the identification of the direct decomposition products on hot wire surfaces, they made clear many succeeding reactive processes in the gas phase. For example, they confirmed the insertion of SiH_2, $SiH(CH_3)$, and $Si(CH_3)_2$ into Si—H bonds besides radical recombination and addition reactions [99, 104, 105].

It should also be noted that this combination of laser spectroscopic and mass spectrometric techniques can be applied to the detection of organic radicals, which play key roles in initiated CVD (*i*CVD). The details of *i*CVD will be presented in Chapter 6.

3.7.2 Threshold Ionization Mass Spectrometry

In the threshold ionization technique, also called appearance potential technique, the energy of the impact electron is reduced so as not to produce fragment ions. Of course, the electron energy must be higher than the ionization potentials of

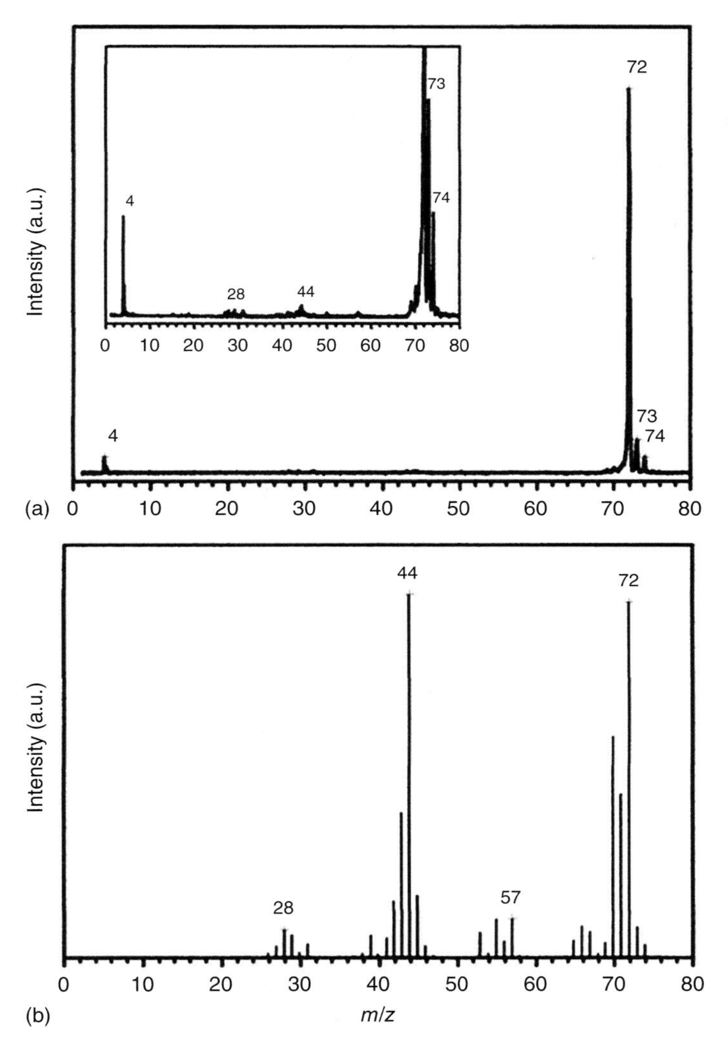

Figure 3.17 Mass spectra of silacyclobutane at room temperature. (a) 10.5 eV single-photon ionization, (b) 70 eV electron impact. Source: Shi et al. 2007 [99]. Reprinted with permission of John Wiley & Sons.

radicals to be detected. For example, when the electron energy is chosen between 9.8 and 12.6 eV, it is possible to ionize CH_3, but not to ionize CH_4. The dissociative ionization to produce $CH_3^+ + H$ from CH_4 does not take place, either. This is a very useful technique, and many papers have been published concerning the catalytic decomposition both for carbon [69, 106] and silicon [107–109] systems. With this technique, it can be clearly shown that the direct production of SiH, SiH_2, and SiH_3 is minor in the catalytic decomposition of SiH_4 [108]. One of the problems in this technique is the presence of false radical signals caused by the spread in the electron energy distribution. It is still difficult to narrow the energy range of bombarding electrons within 0.5 eV [108].

3.7.3 Ion Attachment Mass Spectrometry

Ion attachment is one of the softest ionization methods, and no fragmentation is expected [110]. Li^+ ions emitted from small mineral beads fused to a filament are attached to the sample molecules. This attachment reaction is exothermic, but the excess energy is generally much smaller than the energies of chemical bonds, and no bond dissociation is expected. In order to reduce the detachment of Li^+, the attached ions must be stabilized by collisions with third bodies to remove the excess energies. The attached ions can be mass selected by a conventional quadrupole mass spectrometer. In this ion attachment method, the signal intensity depends on the attaching probability of Li^+ ions. Then, the sensitivity correction is required in the quantitative evaluation. The Li^+ ion affinity evaluated by *ab initio* calculations can be used as a measure of the attaching probability. This technique has been used to identify the catalytic decomposition products of $(CH_3)_3SiNHSi(CH_3)_3$ (HMDS) as will be mentioned in Section 4.5.9. Typical mass spectra of HMDS and its decomposed products shall be shown in Figure 4.16.

3.8 Determination of Gas-Phase Composition of Stable Molecules

Mass spectrometry can also be used to determine the gas-phase composition of stable molecules. When the catalyzers are not heated, the compositions can simply be calculated from the flow rate fractions. It is not so simple when the catalyzers are heated and gaseous end products are produced. Mass spectrometry is useful in such systems. For example, N_2 and H_2 are produced as end products in the catalytic decomposition of NH_3. The decomposition efficiency of NH_3 as well as the production efficiencies of N_2 and H_2 can be determined by using a conventional mass spectrometer using 70 eV electron bombardment. The mass spectrometer can be attached to the chamber through a sampling hole, ~0.1 mm in diameter, and is differentially pumped, typically down to 5×10^{-4} Pa. The absolute sensitivities can be easily calibrated by flowing a mixture of NH_3, N_2, and H_2 of a known composition. Figure 3.18 shows the dependence of NH_3, N_2,

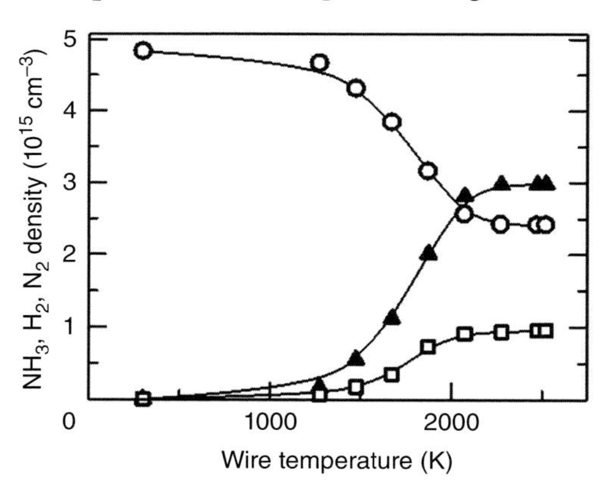

Figure 3.18 Dependence of NH_3 (circle), N_2 (square), and H_2 (triangle) densities on the W wire temperature measured mass spectrometrically. The NH_3 flow rate was 500 sccm. The pressure was 20 Pa when the wire was not heated. Source: Umemoto et al. 2003 [41]. Copyright 2003: The Japan Society of Applied Physics.

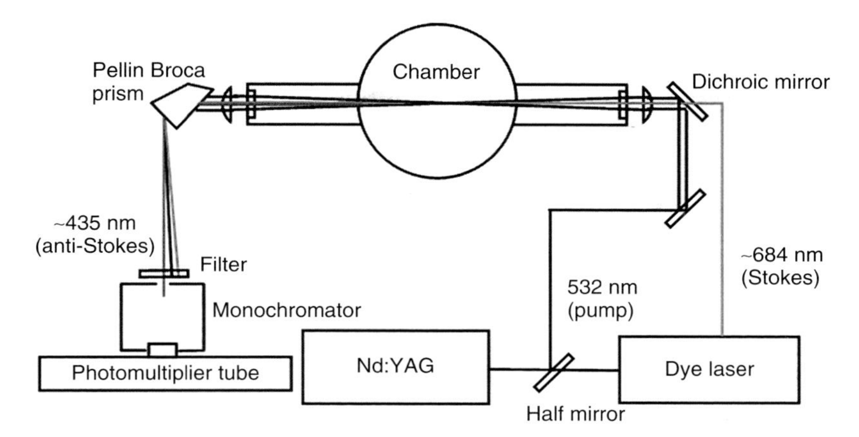

Figure 3.19 Schematic diagram of the apparatus for coherent anti-Stokes Raman scattering to detect H_2. A folded BoxCARS geometry, which has better spatial resolution than collinear CARS, is employed in this diagram.

and H_2 densities on the W wire temperature measured mass spectrometrically [41]. The NH_3 density levels off over 2.0×10^3 K. This can be explained by the presence of reproduction processes of NH_3 from radical species, as will be discussed in Section 4.5.4.

It is also possible to detect H_2, N_2, and other di- and polyatomic molecules by coherent anti-Stokes Raman scattering (CARS) [5, 43, 69]. CARS is a third-order nonlinear optical process, and a schematic diagram of the experimental apparatus is shown in Figure 3.19. In the detection, a pump beam excites the ground-state molecules to a virtual state. A Stokes beam dumps this state to a vibrationally excited state. This state is excited to another virtual state with a pump beam again. Finally, an anti-Stokes beam is emitted from this virtual state. As not only energy but also momentum must be conserved, the anti-Stoles beam is also directional. With this technique, it is possible to determine the rotational and vibrational state distributions with a spatial resolution. The weak point is its low sensitivity, especially at low pressures, because the CARS signal is proportional to the square of the population difference between the two vibrational levels.

A tunable diode laser absorption technique, introduced in Section 3.6.3, can also be employed to identify the stable products in the catalytic decomposition [93]. Gas chromatographic analyses of the end products may also be informative, but, as far as the present authors know, no such systematic work has been carried out.

3.A Term Symbols Used in Atomic and Molecular Spectroscopy

Many term symbols are used in atomic and molecular spectroscopy. In this appendix, the minimum explanations are given. For more details, please refer to Refs. [51, 111, 112].

H(1s $^2S_{1/2}$) means that the electronic spin state of atomic hydrogen with a 1s electron is doublet and that this state has no orbital angular momentum. The total angular momentum is the same as the electronic spin angular momentum and is $\sqrt{(1/2) \times (3/2)}\hbar$, where $\hbar$ is the Planck constant divided by 2π. H(2p $^2P_{1/2,3/2}$) means that atomic hydrogen with a 2p electron is also spin doublet and that this state has an orbital angular momentum of $\sqrt{1 \times 2}\hbar$. The total angular momentum is either $\sqrt{(1/2) \times (3/2)}\hbar$ or $\sqrt{(3/2) \times (5/2)}\hbar$. In atomic silicon represented by $3s^2 3p^2\,^3P_1$, the total spin angular momentum, the total orbital angular momentum, and the total angular momentum are all $\sqrt{1 \times 2}\hbar$.

In the photoabsorption and emission, there are selection rules. The spin multiplicity may not change, although this rule is not strict for heavy elements. This can also be mentioned that the total spin angular momentum quantum number, S, may not change; $\Delta S = 0$. The difference in the total orbital angular momentum quantum numbers, L, must be either 0 or ± 1; $\Delta L = 0, \pm 1$. The difference in the total angular momentum quantum number, J, must also be either 0 or ± 1. In addition, the transition for a state with $J = 0$ to another state with $J = 0$ is forbidden; $\Delta J = 0, \pm 1$ (not $J = 0$ to $J = 0$). Besides these, the difference in the orbital angular momentum quantum numbers of the transiting electron, l, must be ± 1; $\Delta l = \pm 1$. For example, the transition between H(2p $^2P_{1/2}$) and H(1s $^2S_{1/2}$) is allowed but that between H(2s $^2S_{1/2}$) and H(1s $^2S_{1/2}$) is forbidden because $\Delta l = 0$. The transition between Si($3s^2 3p4s\,^3P_2$) and Si($3s^2 3p^2\,^3P_1$) is allowed but that between Si($3s^2 3p4s\,^3P_2$) and Si($3s^2 3p^2\,^3P_0$) is forbidden because $\Delta J = 2$.

The situation is similar in diatomic molecules. BH(X $^1\Sigma^+$) means that the electronic ground-state of BH is a spin singlet state and the component of the electronic orbital angular momentum along the internuclear axis is zero, whereas BH(A $^1\Pi$) means that the component of the electronic orbital angular momentum along the internuclear axis of the first electronically excited singlet state of BH is $\hbar$. Here, X and A represent the ground state and the first electronically excited state with the same spin multiplicity of the ground state, respectively. The general selection rule for diatomic molecules is $\Delta S = 0$ and $\Delta\Lambda = 0, \pm 1$, where Λ is the component of the electronic orbital angular momentum divided by $\hbar$. In addition, there are finer selection rules; for example, in this $^1\Pi$–$^1\Sigma^+$ transition, the difference in the total angular momentum quantum number, ΔJ, must be 0 or ± 1.

In polyatomic molecules with more than three atoms, symbols in the group theory are used. NH_2($\tilde{X}\,^2B_1$) means that the ground state of NH_2 is spin doublet and has a C_{2v} symmetry, which has one twofold axis and two planes of symmetry through the axis. The eigenvalues of the electronic wave function for the twofold rotation and one of the reflections are −1. NH_2($\tilde{A}\,^2A_1$) means that the first excited state of NH_2 with the same spin multiplicity of the ground state has also a C_{2v} symmetry and all the eigenvalues for the symmetry operations are +1. As used above, symbols with tilde, such as $\tilde{X}$ and $\tilde{A}$, are used in polyatomic molecules.

References

1 Arthur, N.L. and Miles, L.A. (1997). Arrhenius parameters for the reaction of H atoms with SiH_4. *J. Chem. Soc., Faraday Trans.* 93: 4259–4264.
2 Wu, S.Y., Raghunath, P., Wu, J.S., and Lin, M.C. (2010). Ab initio chemical kinetic study for reactions of H atoms with SiH_4 and Si_2H_6: comparison of theory and experiment. *J. Phys. Chem. A* 114: 633–639.
3 Gates, S.M. (1996). Surface chemistry in the chemical vapor deposition of electronic materials. *Chem. Rev.* 96: 1519–1532.
4 Matsuda, A. (2004). Thin-film silicon –growth process and solar cell application–. *Jpn. J. Appl. Phys.* 43: 7909–7920.
5 Umemoto, H. (2010). Production and detection of H atoms and vibrationally excited H_2 molecules in CVD processes. *Chem. Vap. Deposition* 16: 275–290.
6 Umemoto, H. (2015). Gas-phase diagnoses in catalytic chemical vapor deposition (hot-wire CVD) processes. *Thin Solid Films* 575: 3–8.
7 Ashfold, M.N.R., May, P.W., Petherbridge, J.R. et al. (2001). Unravelling aspects of the gas phase chemistry involved in diamond chemical vapor deposition. *Phys. Chem. Chem. Phys.* 3: 3471–3485.
8 Duan, H.L., Zaharias, G.A., and Bent, S.F. (2002). Detecting reactive species in hot wire chemical vapor deposition. *Curr. Opin. Solid State Mater. Sci.* 6: 471–477.
9 Döbele, H.F., Czarnetzki, U., and Goehlich, A. (2000). Diagnostics of atoms by laser spectroscopic methods in plasmas and plasma-wall interaction studies (vacuum ultraviolet and two-photon techniques). *Plasma Sources Sci. Technol.* 9: 477–491.
10 Amorim, J., Baravian, G., and Jolly, J. (2000). Laser-induced resonance fluorescence as a diagnostic technique in non-thermal equilibrium plasmas. *J. Phys. D: Appl. Phys.* 33: R51–R65.
11 Tachibana, K. (2002). VUV to UV laser spectroscopy of atomic species in processing plasmas. *Plasma Sources Sci. Technol.* 11: A166–A172.
12 Döbele, H.F., Mosbach, T., Niemi, K., and Schulz-von der Gathen, V. (2005). Laser-induced fluorescence measurements of absolute atomic densities: concepts and limitations. *Plasma Sources Sci. Technol.* 14: S31–S41.
13 Röpcke, J., Lombardi, G., Rousseau, A., and Davies, P.B. (2006). Application of mid-infrared tuneable diode laser absorption spectroscopy to plasma diagnostics: a review. *Plasma Sources Sci. Technol.* 15: S148–S168.
14 Toyoda, H., Childs, M.A., Menningen, K.L. et al. (1994). Ultraviolet spectroscopy of gaseous species in a hot filament diamond deposition system when C_2H_2 and H_2 are the input gases. *J. Appl. Phys.* 75: 3142–3150.
15 Abe, K., Ida, M., Izumi, A. et al. (2009). Estimation of hydrogen radical density generated from various kinds of catalysts. *Thin Solid Films* 517: 3449–3451.

16 Childs, M.A., Menningen, K.L., Anderson, L.W., and Lawler, J.E. (1996). Atomic and radical densities in a hot filament diamond deposition system. *J. Chem. Phys.* 104: 9111–9119.
17 Mitchell, A.C.G. and Zemansky, M.W. (1934). *Resonance Radiation of Excited Atoms*. London: Cambridge University Press.
18 Huber, K.P. and Herzberg, G. (1979). *Molecular Spectra and Molecular Structure IV. Constants of Diatomic Molecules*. New York: Van Nostrand Reinhold.
19 Halpern, J.B., Hancock, G., Lenzi, M., and Welge, K.H. (1975). Laser induced fluorescence from NH_2 ($^2A_1$). State selected radiative lifetimes and collisional de-excitation rates. *J. Chem. Phys.* 63: 4808–4816.
20 Fukushima, M., Mayama, S., and Obi, K. (1992). Jet spectroscopy and excited state dynamics of SiH_2 and SiD_2. *J. Chem. Phys.* 96: 44–52.
21 Kono, A., Koike, N., Okuda, K., and Goto, T. (1993). Laser-induced-fluorescence detection of SiH_2 radicals in a radio-frequency silane plasmas. *Jpn. J. Appl. Phys.* 32: L543–L546.
22 Umemoto, H., Nishihara, Y., Ishikawa, T., and Yamamoto, S. (2012). Catalytic decomposition of PH_3 on heated tungsten wire surfaces. *Jpn. J. Appl. Phys.* 51: 086501/1–086501/9.
23 Hertl, M. and Jolly, J. (2000). Laser-induced fluorescence detection and kinetics of SiH_2 radicals in $Ar/H_2/SiH_4$ RF discharges. *J. Phys. D: Appl. Phys.* 33: 381–388.
24 https://www.nist.gov/pml/atomic-spectra-database.
25 Hirabayashi, A., Nambu, Y., and Fujimoto, T. (1986). Excitation anisotropy in laser-induced-fluorescence spectroscopy—high-intensity, broad-line excitation. *Jpn. J. Appl. Phys.* 25: 1563–1568.
26 Umemoto, H. (2004). Chapter 12, Section 2, Elementary chemical reactions. In: *Kagakubenran*, 5e (ed. Y. Iwasawa, K. Ogura, M. Kitamura, K. Suzuki and K. Yamanouchi). Tokyo: Maruzen [in Japanese].
27 Nemoto, M., Suzuki, A., Nakamura, H. et al. (1989). Electronic quenching and chemical reactions of SiH radicals in the gas phase. *Chem. Phys. Lett.* 162: 467–471.
28 Kenner, R.D., Pfannerberg, S., and Stuhl, F. (1989). Collisional quenching of PH(A $^3\Pi_i$, $v = 0$) at 296 and 415 K. *Chem. Phys. Lett.* 156: 305–311.
29 Holstein, T. (1951). Imprisonment of resonance radiation in gases. II. *Phys. Rev.* 83: 1159–1168.
30 Umemoto, H. and Moridera, M. (2008). Production and detection of reducing and oxidizing radicals in the catalytic decomposition of H_2/O_2 mixtures on heated tungsten surfaces. *J. Appl. Phys.* 103: 034905/1–034905/6.
31 Petsalakis, I.D. and Theodorakopoulos, G. (2007). Theoretical study of nonadiabatic interactions, radiative lifetimes and predissociation lifetimes of excited states of BH. *Mol. Phys.* 105: 333–342.
32 Brzozowski, J., Bunker, P., Elander, N., and Erman, P. (1976). Predissociation effects in the A, B, and C states of CH and the interstellar formation rate of CH via inverse predissociation. *Astrophys. J. Part 1* 207: 414–424.

33 Patel-Misra, D., Parlant, G., Sauder, D.G. et al. (1991). Radiative and nonradiative decay of the NH(ND) A $^3\Pi$ electronic state: predissociation induced by the $^5\Sigma^-$ state. *J. Chem. Phys.* 94: 1913–1922.
34 Dimpfl, W.L. and Kinsey, J.L. (1979). Radiative lifetimes of OH(A $^2\Sigma$) and Einstein coefficients for the A–X system of OH and OD. *J. Quant. Spectrosc. Radiat. Transfer* 21: 233–241.
35 Larsson, M. (1987). Ab initio calculations of transition probabilities and potential curves of SiH. *J. Chem. Phys.* 86: 5018–5026.
36 Fitzpatrick, J.A.J., Chekhlov, O.V., Morgan, D.R. et al. (2002). Predissociation dynamics in the A $^3\Pi$ state of PH: an experimental and ab initio investigation. *Phys. Chem. Chem. Phys.* 4: 1114–1122.
37 Kono, A., Hirose, S., Kinoshita, K., and Goto, T. (1998). Translational temperature measurement for SiH_2 in RF silane plasma using CW laser induced fluorescence spectroscopy. *Jpn. J. Appl. Phys.* 37: 4588–4589.
38 Nozaki, Y., Kongo, K., Miyazaki, T. et al. (2000). Identification of Si and SiH in catalytic chemical vapor deposition of SiH_4 by laser induced fluorescence spectroscopy. *J. Appl. Phys.* 88: 5437–5443.
39 Umemoto, H., Kanemitsu, T., and Tanaka, A. (2014). Production of B atoms and BH radicals from B_2H_6/He/H_2 mixtures activated on heated W wires. *J. Phys. Chem. A* 118: 5156–5163.
40 Umemoto, H., Ohara, K., Morita, D. et al. (2002). Direct detection of H atoms in the catalytic chemical vapor deposition of the SiH_4/H_2 system. *J. Appl. Phys.* 91: 1650–1656.
41 Umemoto, H., Ohara, K., Morita, D. et al. (2003). Radical species formed by the catalytic decomposition of NH_3 on heated W surfaces. *Jpn. J. Appl. Phys.* 42: 5315–5321.
42 Umemoto, H. (2010). A clean source of ground-state N atoms: decomposition of N_2 on heated tungsten. *Appl. Phys. Express* 3: 076701/1–076701/3.
43 Umemoto, H., Funae, T., and Mankelevich, Y.A. (2011). Activation and decomposition of N_2 on heated tungsten filament surfaces. *J. Phys. Chem. C* 115: 6748–6756.
44 Umemoto, H. and Kusanagi, H. (2008). Catalytic decomposition of O_2, NO, N_2O and NO_2 on a heated Ir filament to produce atomic oxygen. *J. Phys. D: Appl. Phys.* 41: 225505/1–225505/5.
45 Fernando, W.T.M.L. and Bernath, P.F. (1991). Fourier transform spectroscopy of the A $^1\Pi$-X $^1\Sigma^+$ transition of BH and BD. *J. Mol. Spectrosc.* 145: 392–402.
46 Bernath, P.F., Brazier, C.R., Olsen, T. et al. (1991). Spectroscopy of the CH free radical. *J. Mol. Spectrosc.* 147: 16–26.
47 Brazier, C.R., Ram, R.S., and Bernath, P.F. (1986). Fourier transform spectroscopy of the A $^3\Pi$-X $^3\Sigma^-$ transition of NH. *J. Mol. Spectrosc.* 120: 381–402.
48 Dieke, G.H. and Crosswhite, H.M. (1962). The ultraviolet bands of OH. *J. Quant. Spectrosc. Radiat. Transfer* 2: 97–199.
49 Klynning, L. and Lindgren, B. (1966). The spectra of silicon hydride and silicon deuteride. *Ark. Fys.* 33: 73–91.
50 Pearse, R.W.B. (1930). The λ 3400 band of phosphorus hydride. *Proc. R. Soc. London, Ser. A* 129: 328–354.

51 Herzberg, G. (1950). *Molecular Spectra and Molecular Structure I. Spectra of Diatomic Molecules*. New York: Van Nostrand Reinhold.
52 Umemoto, H., Morimoto, T., Yamawaki, M. et al. (2004). Catalytic decomposition of HCN on heated W surfaces to produce CN radicals. *J. Non-Cryst. Solids* 338–340: 65–69.
53 Luh, W.-T. and Stwalley, W.C. (1983). The X $^1\Sigma^+$, A $^1\Pi$, and B $^1\Sigma^+$ potential energy curves and spectroscopy of BH. *J. Mol. Spectrosc.* 102: 212–223.
54 Childs, D.R. (1964). Vibrational wave functions and Franck–Condon factors of various band systems. *J. Quant. Spectrosc. Radiat. Transfer* 4: 283–290.
55 Fairchild, P.W., Smith, G.P., Crosley, D.R., and Jeffries, J.B. (1984). Lifetimes and transition probabilities for NH(A $^3\Pi_i$-X $^3\Sigma^-$). *Chem. Phys. Lett.* 107: 181–186.
56 Luque, J. and Crosley, D.R. (1998). Transition probabilities in the A $^2\Sigma^+$–X $^2\Pi_i$ electronic system of OH. *J. Chem. Phys.* 109: 439–448.
57 Smith, W.H. and Liszt, H.S. (1971). Franck–Condon factors and absolute oscillator strengths for NH, SiH, S_2 and SO. *J. Quant. Spectrosc. Radiat. Transfer* 11: 45–54.
58 Rostas, J., Cossart, D., and Bastien, J.R. (1974). Rotational analysis of the PH and PD A $^3\Pi_i$–X $^3\Sigma^-$ band systems. *Can. J. Phys.* 52: 1274–1287.
59 Niemi, K., Mosbach, T., and Döbele, H.F. (2003). Is the flow tube reactor with NO_2 titration a reliable absolute source for atomic hydrogen? *Chem. Phys. Lett.* 367: 549–555.
60 Niemi, K., Schultz-von der Gathen, V., and Döbele, H.F. (2001). Absolute calibration of atomic density measurements by laser-induced fluorescence spectroscopy with two-photon excitation. *J. Phys. D: Appl. Phys.* 34: 2330–2335.
61 Jolly, J. and Booth, J.-P. (2005). Atomic hydrogen densities in capacitively coupled very high-frequency plasmas in H_2: effect of excitation frequency. *J. Appl. Phys.* 97: 103305/1–103305/6.
62 Aldén, M., Hertz, H.M., Svanberg, S., and Wallin, S. (1984). Imaging laser-induced fluorescence of oxygen atoms in a flame. *Appl. Opt.* 23: 3255–3257.
63 Adams, S.F. and Miller, T.A. (1998). Two-photon absorption laser-induced fluorescence of atomic nitrogen by an alternative excitation scheme. *Chem. Phys. Lett.* 295: 305–311.
64 Das, P., Ondrey, G., van Veen, N., and Bersohn, R. (1983). Two photon laser induced fluorescence of carbon atoms. *J. Chem. Phys.* 79: 724–726.
65 Preppernau, B.L., Pearce, K., Tserepi, A. et al. (1995). Angular momentum state mixing and quenching of $n = 3$ atomic hydrogen fluorescence. *Chem. Phys.* 196: 371–381.
66 Tonokura, K., Inoue, K., and Koshi, M. (2002). Chemical kinetics for film growth in silicon HWCVD. *J. Non-Cryst. Solids* 299-302: 25–29.
67 Comerford, D.W., Cheesman, A., Carpenter, T.P.F. et al. (2006). Experimental and modeling studies of B atom number density distributions in hot filament activated B_2H_6/H_2 and $B_2H_6/CH_4/H_2$ gas mixtures. *J. Phys. Chem. A* 110: 2868–2875.

68 Corat, E.J. and Goodwin, D.G. (1993). Temperature dependence of species concentrations near the substrate during diamond chemical vapor deposition. *J. Appl. Phys.* 74: 2021–2029.

69 Zumbach, V., Schäfer, J., Tobai, J. et al. (1997). Experimental investigation and computational modeling of hot filament diamond chemical vapor deposition. *J. Chem. Phys.* 107: 5918–5928.

70 Smith, J.A., Cook, M.A., Langford, S.R. et al. (2000). Resonance enhanced multiphoton ionization probing of H atoms and CH_3 radicals in a hot filament chemical vapour deposition reactor. *Thin Solid Films* 368: 169–175.

71 Smith, J.A., Cameron, E., Ashfold, M.N.R. et al. (2001). On the mechanism of CH_3 radical formation in hot filament activated CH_4/H_2 and C_2H_2/H_2 gas mixtures. *Diamond Relat. Mater.* 10: 358–363.

72 Johnson, R.D. III, Tsai, B.P., and Hudgens, J.W. (1989). Multiphoton ionization of SiH_3 and SiD_3 radicals: electronic spectra, vibrational analyses of the ground and Rydberg states, and ionization potentials. *J. Chem. Phys.* 91: 3340–3359.

73 Irikura, K.K. and Hudgens, J.W. (1992). Detection of methylene ($\tilde{X}\ ^3B_1$) radicals by 3 + 1 resonance-enhanced multiphoton ionization spectroscopy. *J. Phys. Chem.* 96: 518–519.

74 Irikura, K.K., Johnson, R.D. III, and Hudgens, J.W. (1992). Two new electronic states of methylene. *J. Phys. Chem.* 96: 6131–6133.

75 Chen, P., Pallix, J.B., Chupka, W.A., and Colson, S.D. (1987). Resonant multiphoton ionization spectrum and electronic structure of CH radical. New states and assignments above 50000 cm^{-1}. *J. Chem. Phys.* 86: 516–520.

76 Johnson, R.D. III, and Hudgens, J.W. (1989). New electronic state of silylidyne and silylidyne-d radicals observed by resonance-enhance multiphoton ionization spectroscopy. *J. Phys. Chem.* 93: 6268–6270.

77 Ashfold, M.N.R., Clement, S.G., Howe, J.D., and Western, C.M. (1993). Multiphoton ionisation spectroscopy of free radical species. *J. Chem. Soc., Faraday Trans.* 89: 1153–1172.

78 Redman, S.A., Chung, C., Rosser, K.N., and Ashfold, M.N.R. (1999). Resonance enhanced multiphoton ionisation probing of H atoms in a hot filament chemical vapour deposition reactor. *Phys. Chem. Chem. Phys.* 1: 1415–1424.

79 Scherer, J.J., Paul, J.B., O'Keefe, A., and Saykally, R.J. (1997). Cavity ringdown laser absorption spectroscopy: history, development, and application to pulsed molecular beams. *Chem. Rev.* 97: 25–51.

80 Wheeler, M.D., Newman, S.M., Orr-Ewing, A.J., and Ashfold, M.N.R. (1998). Cavity ring-down spectroscopy. *J. Chem. Soc., Faraday Trans.* 94: 337–351.

81 Wahl, E.H., Owano, T.G., Kruger, C.H. et al. (1996). Measurement of absolute CH_3 concentration in a hot-filament reactor using cavity ring-down spectroscopy. *Diamond Relat. Mater.* 5: 373–377.

82 Wahl, E.H., Owano, T.G., Kruger, C.H. et al. (1997). Spatially resolved measurements of absolute CH_3 concentration in a hot-filament reactor. *Diamond Relat. Mater.* 6: 476–480.

83 Nozaki, Y., Kitazoe, M., Horii, K. et al. (2001). Identification and gas phase kinetics of radical species in Cat-CVD processes of SiH_4. *Thin Solid Films* 395: 47–50.
84 Umemoto, H., Morimoto, T., Yamawaki, M. et al. (2003). Deposition chemistry in the Cat-CVD processes of SiH_4/NH_3 mixture. *Thin Solid Films* 430: 24–27.
85 Kessels, W.M.M., Leroux, A., Boogaarts, M.G.H. et al. (2001). Cavity ring down detection of SiH_3 in a remote SiH_4 plasma and comparison with model calculations and mass spectrometry. *J. Vac. Sci. Technol., A* 19: 467–476.
86 Nagai, T., Smets, A.H.M., and Kondo, M. (2008). Formation of SiH_3 radicals and nanoparticles in SiH_4-H_2 plasmas observed by time-resolved cavity ringdown spectroscopy. *Jpn. J. Appl. Phys.* 47: 7032–7043.
87 Lommatzsch, U., Wahl, E.H., Aderhold, D. et al. (2001). Cavity ring-down spectroscopy of CH and CD radicals in a diamond thin film chemical vapor deposition reactor. *Appl. Phys. A* 73: 27–33.
88 van den Oever, P.J., van Helden, J.H., Lamers, C.C.H. et al. (2005). Density and production of NH and NH_2 in an Ar-NH_3 expanding plasma jet. *J. Appl. Phys.* 98: 093301/1–093301/10.
89 Kessels, W.M.M., Hoefnagels, J.P.M., Boogaarts, M.G.H. et al. (2001). Cavity ring down study of the densities and kinetics of Si and SiH in a remote Ar-H_2-SiH_4 plasma. *J. Appl. Phys.* 89: 2065–2073.
90 Friedrichs, G., Fikri, M., Guo, Y., and Temps, F. (2008). Time-resolved cavity ringdown measurements and kinetic modeling of the pressure dependences of the recombination reactions of SiH_2 with the alkenes C_2H_4, C_3H_6, and t-C_4H_8. *J. Phys. Chem. A* 112: 5636–5646.
91 Buzaianu, M.D., Makarov, V.I., Morell, G., and Weiner, B.R. (2008). Detection of SH and CS radicals by cavity ringdown spectroscopy in a hot filament chemical vapor deposition environment. *Chem. Phys. Lett.* 455: 26–31.
92 Celii, F.G., Pehrsson, P.E., Wang, H.-t., and Butler, J.E. (1988). Infrared detection of gaseous species during the filament-assisted growth of diamond. *Appl. Phys. Lett.* 52: 2043–2045.
93 Hirmke, J., Hempel, F., Stancu, G.D. et al. (2006). Gas-phase characterization in diamond hot-filament CVD by infrared tunable diode laser absorption spectroscopy. *Vacuum* 80: 967–976.
94 Itabashi, N., Nishiwaki, N., Magane, M. et al. (1990). Spatial distribution of SiH_3 radicals in RF silane plasma. *Jpn. J. Appl. Phys.* 29: L505–L507.
95 Loh, S.K. and Jasinski, J.M. (1991). Direct kinetic studies of $SiH_3 + SiH_3$, H, CCl_4, SiD_4, Si_2H_6, and C_3H_6 by tunable infrared diode laser spectroscopy. *J. Chem. Phys.* 95: 4914–4926.
96 Duan, H.L., Zaharias, G.A., and Bent, S.F. (2001). Probing radicals in hot wire decomposition of silane using single photon ionization. *Appl. Phys. Lett.* 78: 1784–1786.
97 Zaharias, G.A., Duan, H.L., and Bent, S.F. (2006). Detecting free radicals during the hot wire chemical vapor deposition of amorphous silicon carbide films using single-source precursors. *J. Vac. Sci. Technol., A* 24: 542–549.

98 Nakamura, S. and Koshi, M. (2006). Elementary processes in silicon hot wire CVD. *Thin Solid Films* 501: 26–30.

99 Shi, Y.J., Lo, B., Tong, L. et al. (2007). In situ diagnostics of the decomposition of silacyclobutane on a hot filament by vacuum ultraviolet laser ionization mass spectrometry. *J. Mass Spectrom.* 42: 575–583.

100 Shi, Y. (2015). Hot wire chemical vapor deposition chemistry in the gas phase and on the catalyst surface with organosilicon compounds. *Acc. Chem. Res.* 48: 163–173.

101 Tange, S., Inoue, K., Tonokura, K., and Koshi, M. (2001). Catalytic decomposition of SiH_4 on a hot filament. *Thin Solid Films* 395: 42–46.

102 Badran, I., Forster, T.D., Roesler, R., and Shi, Y.J. (2012). Competition of silene/silylene chemistry with free radical chain reactions using 1-methylsilacyclobutane in the hot-wire chemical vapor deposition process. *J. Phys. Chem. A* 116: 10054–10062.

103 Toukabri, R., Alkadhi, N., and Shi, Y.J. (2013). Formation of methyl radicals from decomposition of methyl-substituted silanes over tungsten and tantalum filament surfaces. *J. Phys. Chem. A* 117: 7697–7704.

104 Toukabri, R. and Shi, Y.J. (2014). Dominance of silylene chemistry in the decomposition of monomethylsilane in the presence of a heated metal filament. *J. Phys. Chem. A* 118: 3866–3874.

105 Shi, Y. (2017). Role of free-radical chain reactions and silylene chemistry in using methyl-substituted silane molecules in hot-wire chemical vapor deposition. *Thin Solid Films* 635: 42–47.

106 McMaster, M.C., Hsu, W.L., Coltrin, M.E., and Dandy, D.S. (1994). Experimental measurements and numerical simulations of the gas composition in a hot-filament-assisted diamond chemical-vapor-deposition reactor. *J. Appl. Phys.* 76: 7567–7577.

107 Doyle, J., Robertson, R., Lin, G.H. et al. (1988). Production of high quality amorphous silicon films by evaporative silane surface decomposition. *J. Appl. Phys.* 64: 3215–3223.

108 Holt, J.K., Swiatek, M., Goodwin, D.G., and Atwater, H.A. (2002). The aging of tungsten filaments and its effect on wire surface kinetics in hot-wire chemical vapor deposition. *J. Appl. Phys.* 92: 4803–4808.

109 Zheng, W. and Gallagher, A. (2008). Radical species involved in hotwire (catalytic) deposition of hydrogenated amorphous silicon. *Thin Solid Films* 516: 929–939.

110 Morimoto, T., Ansari, S.G., Yoneyama, K. et al. (2006). Mass-spectrometric studies of catalytic chemical vapor deposition processes of organic silicon compounds containing nitrogen. *Jpn. J. Appl. Phys.* 45: 961–966.

111 Herzberg, G. (1944). *Atomic Spectra and Atomic Structure*. New York: Dover Publications.

112 Herzberg, G. (1967). *Molecular Spectra and Molecular Structure III. Electronic Spectra and Electronic Structure of Polyatomic Molecules*. Princeton: D. Van Nostrand.

4

Physics and Chemistry of Cat-CVD

In this chapter, the dynamics of gas molecules introduced into a catalytic chemical vapor deposition (Cat-CVD) chamber are mentioned in detail to reveal the deposition mechanism, along with chemical reactions on catalyzer surfaces and in gas phase. The features of Cat-CVD and the difference from plasma-enhanced chemical vapor deposition (PECVD), explained roughly in Chapter 2, are more clearly illustrated.

4.1 Kinetics of Molecules in Cat-CVD Chamber

4.1.1 Molecules in Cat-CVD Chamber

A typical Cat-CVD apparatus is again illustrated in Figure 4.1. In this apparatus, a sample is set on the sample holder and source gases are introduced into the chamber through a showerhead. The system is sometimes called "deposition down system" or "sample face up system." Here, the chamber has a cylindrical shape with an inner diameter of D_{ch} and an inner height of H_{ch}. The sample holder or the substrate holder has also a shape of cylinder with a diameter of D_{sub}. The exhaust gas is pumped out from the bottom of the chamber. A catalyzing wire is spanned in the area of $A_{cat} \times B_{cat}$, keeping it parallel to the substrate holder. The catalyzer is usually heated by applying electric currents directly on it.

All molecules introduced through the showerhead into the chamber are moving around inside the chamber with thermal velocity v_{th}, described by Eq. (2.2), and colliding with molecules in space and also colliding with solid surfaces following Eqs. (2.6) and (2.7).

Here, as an example, let us imagine the dynamics of gas molecules in a chamber when amorphous silicon (a-Si) films are deposited, under the following deposition parameters: (i) source gas molecules are only SiH_4, (ii) gas pressure P_g is 1 Pa, (iii) the chamber has a cylindrical shape, and both D_{ch} and H_{ch} are 50 cm, and thus, the volume of the chamber V_{ch} is approximately 1×10^5 cm^3 as already shown as an example in Section 2.1.3, (iv) the temperature of chamber walls is approximated to be 27 °C (300 K) and the catalyzer temperature, T_{cat}, is 1800 °C, (v) a catalyzing wire with a diameter of 0.05 cm and a length of 300 cm is set in the spanning area of $A_{cat} = 24$ cm $\times$ $B_{cat} = 29$ cm, and thus, the total surface area of

Catalytic Chemical Vapor Deposition: Technology and Applications of Cat-CVD,
First Edition. Hideki Matsumura, Hironobu Umemoto, Karen K. Gleason, and Ruud E.I. Schropp.

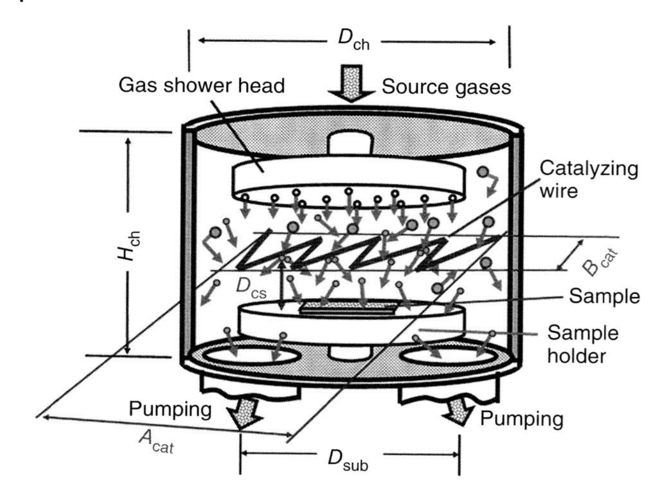

Figure 4.1 Schematic view of molecules moving around inside a chamber.

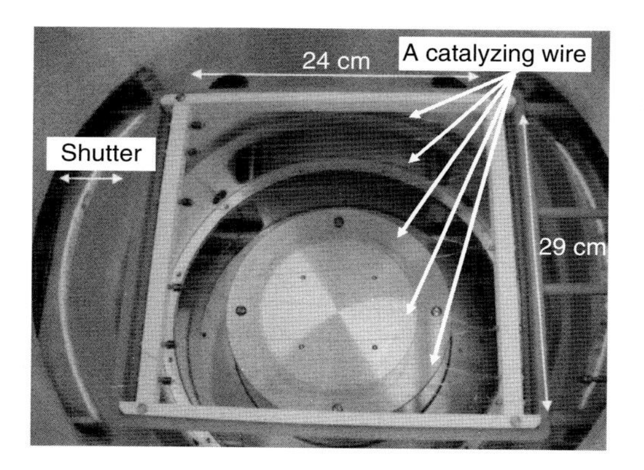

Figure 4.2 A photograph of the catalyzing wire and inside the chamber.

the catalyzing wire, S_{cat}, is about 47 cm^2, and (vi) the flow rate of SiH_4 gas without any dilution, FR(SiH_4), is 50 sccm.

All these parameters are summarized in Table 4.1, as one of the typical values for a-Si deposition by Cat-CVD. The deposition rate of about 1.5 nm/s is expected in keeping device quality of a-Si for this example, when the distance between the catalyzer and the substrate, D_{cs}, is about 10 cm.

Figure 4.2 shows a photograph of the catalyzing wire with S_{cat} = 47 cm^2, which is installed in the chamber described in Table 4.1. A circle mark seen at a center of the substrate holder in the photograph shows the trace of a crystalline silicon (c-Si) wafer of 20 cm in diameter. The diameter of the substrate holder itself is 30 cm as also summarized in Table 4.1. As seen in the photograph, although the catalyzing wire is not spanned in dense but rather sparse, still, S_{cat} of 47 cm^2 can be obtained easily. This means that the influence of thermal radiation from the heated catalyzer is limited. The influence of thermal radiation in Cat-CVD will be mentioned in Chapter 7.

Table 4.1 Parameters for deposition of Cat-CVD a-Si, used in the present explanation.

Parameters	Abbreviations	Conditions
1. Shape of the chamber and its size		
Shape		Cylindrical
Inner diameter	D_{ch}	50 cm
Height	H_{ch}	50 cm
Inner volume	V_{ch}	Approximately 1×10^5 cm^3
2. Catalyzing materials		W, Ta or Ta alloys
3. Size of catalyzing wire		
Spanning area	$A_{cat} \times B_{cat}$	24 cm × 29 cm
Diameter		0.05 cm
Length		300 cm
4. Total surface of area of such a catalyzer	S_{cat}	47 cm^2
5. Diameter of a cylindrical substrate holder	D_{sub}	30 cm
6. Distance between the catalyzing wire and the substrate	D_{cs}	10 cm
7. Temperature of the catalyzer	T_{cat}	1800 °C
8. Gas pressure	P_g	1 Pa
9. Flow rate of SiH_4	FR(SiH_4)	50 sccm
10. Temperature of the substrate	T_s	200–300 °C
11. Uniform deposition area		20 cm in diameter
12. Deposition rate of a-Si	D_R	About 1.5 nm/s

The number of SiH_4 molecules colliding with a solid surface with an area of 1 cm^2 within one second is evaluated as 1.02×10^{18} cm^{-2}/s (for 1800 °C) to 2.69×10^{18} cm^{-2}/s (for 27 °C) as shown in Table 2.6, when P_g is 1 Pa. Thus, the number of SiH_4 molecules colliding with the surface of the catalyzing wire of S_{cat} = 47 cm^2 in unit time is 4.79×10^{19}/s (for 1800 °C) to 1.26×10^{20}/s (for 27 °C). In the chamber with a volume of $V_{ch} = 1 \times 10^5$ cm^3, SiH_4 molecules of 3.50×10^{18} (for 1800 °C) to 2.42×10^{19} (for 27 °C) are present in total as evaluated from Table 2.1. Then, a single SiH_4 molecule should collide with the catalyzer by 5 (=$1.26 \times 10^{20}/2.42 \times 10^{19}$) times (for 27 °C) to 14 (=$4.79 \times 10^{19}/3.50 \times 10^{18}$) times (for 1800 °C) within one second in average. When the flow rate of SiH_4 gas, FR(SiH_4), is 50 sccm, the residence time t_{res} of a SiH_4 molecule in the chamber is evaluated to be 1.2 seconds from Eq. (2.8). Then, a SiH_4 molecule introduced into the chamber should collide with the catalyzer for 6 (=5 × 1.2) times (for 27 °C) to 17 (=14 × 1.2) times (for 1800 °C) during the residence time, 1.2 seconds. The temperature of SiH_4 molecules before colliding with the catalyzer is expected to be higher than 27 °C, as the chamber walls as well as the substrate are heated by the hot catalyzer. As the collision number in unit time increases with the gas temperature, the above collision number at 27 °C, six times, should be the minimum.

The dissociation probability of SiH_4 on the metal catalyzer, α, is a function of T_{cat}. That varies from 0.1 to 0.3 as T_{cat} increases from 1700 to 2000 °C, and it is about 0.15 for W, Ta, and TaC catalyzers of T_{cat} = 1800 °C, as shown in Figure 7.27. When α is 0.15, $1-(1-0.15)^6 = 0.62$ = 62% of SiH_4 molecules are expected to be decomposed before evacuation when the gas temperature T_g is 27 °C. This efficiency increases if T_g is higher than 27 °C and the number of collisions with the catalyzer is larger than six times. For instance, when T_g is 250 °C, the typical substrate temperature, the number of collisions of SiH_4 with the heated catalyzer before evacuation becomes eight times, and thus, 73% of SiH_4 molecules are decomposed. If we consider the case of T_{cat} = 2000 °C and $\alpha = 0.3$, 88% of SiH_4 molecules are decomposed before evacuation even for T_g = 27 °C.

4.1.2 Comparison with PECVD for Decomposition

As already mentioned in Chapter 2, the efficiency of gas use for formation of films in Cat-CVD is much higher than that in PECVD. S. Osono et al. at ULVAC Inc., Japan, developed a large size Cat-CVD apparatus and deposited device quality a-Si films with a size of about 1 m × 1.5 m [1]. The apparatus was a vertical type, in which catalyzing wires were hung vertically, and substrates were placed vertically on both sides of catalyzing wires hung at the center. The image of the apparatus is the same as that shown in Figure 2.12, and the photograph is shown in Figure 7.41. FR(SiH_4) for depositing a-Si films with the deposition rate larger than 1 nm/s was 100–500 sccm for T_{cat} = 1700 °C and P_g = 0.1–10 Pa. A similar size chamber was prepared for PECVD, and the results were compared. According to their experience, to obtain the same quality a-Si films with the same deposition rate to PECVD, FR(SiH_4) in Cat-CVD could be reduced to 1/5 to 1/10 of PECVD, although the value has a fluctuation depending on the design of the apparatus.

In the case of PECVD, molecules of source gases are decomposed by the collisions with energetic electrons as explained in Chapter 2. The number of collisions between energetic electrons and molecules are explained in Section 2.3.5 and Eq. (2.14). As an example, the SiH_4 decomposition is mentioned for the case of P_g = 1 Pa and gas temperature of 27 °C (300 K) in the same section. Density of collisions that can decompose a SiH_4 molecule is estimated as 3.63×10^{14} cm^{-3}/s.

In PECVD, the decomposition rate, the number of decomposed species in unit time, is proportional to the effective volume. One point to be noted is that only plasma space, which is much smaller than the chamber volume, has meaning for such decomposition. If anode and cathode electrodes have the same diameter of 30 cm and the distance between them, D_{AC}, is 10 cm, the plasma space is approximately 7.1×10^3 cm^3. Therefore, only 2.58×10^{18} dissociative collisions are expected per second in the plasma chamber.

In Cat-CVD, on the other hand, the decomposition rate is independent of the chamber volume. When S_{cat} is 47 cm^2, SiH_4 molecules of 4.79×10^{19}–1.26×10^{20} are colliding with the catalyzer per second depending on the gas temperature assumed, as mentioned in Section 4.1.1. Even if the decomposition probability α is as low as 0.1, still 4.79×10^{18}–1.26×10^{19} SiH_4 molecules are decomposed. These numbers are larger by several times than that in PECVD.

The experimental results showing the high utilization efficiency of source gas molecules are explained by these simple model calculations.

4.1.3 Influence of Surface Area of Catalyzer

Figure 4.3 shows the reported results on the deposition rate of a-Si films vs. reciprocal of T_{cat} for a small size chamber. The data described in two Refs. [2, 3] are summarized in the figure. Gas pressure P_g was 1.3 and 13 Pa, and the substrate temperature, T_s, was 200 and 300 °C. Detailed deposition parameters, including the chamber size, are not described in their references. The net SiH_4 flow rate, FR(SiH_4), for P_g = 13 Pa samples was 4 sccm and that for P_g = 1.3 Pa samples appears a few sccm. The chamber size appears quite small, considering the size of the catalyzing wire. Data were taken for S_{cat} = 1, 1.5, and 3 cm^2. The saturation of deposition rates can be observed at catalyzer temperatures over 1700 –1800 °C.

Looking at many other data, the saturation of deposition rate is observed when SiH_4 flow rate is small. When the flow rate is reduced with keeping the pressure constant, the residence time is increased. In consequence, the number of collisions that a SiH_4 molecule experiences before evacuation increases. This brings about the depletion of the source gas molecules. This should be the main cause of the saturation, although other factors, such as the change in the wall temperature, may contribute.

Figure 4.3 shows that the deposition rate is likely to increase as S_{cat} increases when other deposition conditions are the same. If we carefully look at the data, the deposition rate appears to be almost proportional to S_{cat} when the deposition rate is not saturated. The deposition rate prepared with S_{cat}=3 cm^2 is almost three times larger than that with S_{cat} = 1 cm^2 below 1900 K. This result is reasonable in the absence of depletion, as the number of collisions must be proportional to the surface area. When the deposition rate saturates, the area dependence is less remarkable, probably because of the depletion of SiH_4 molecules near the catalyzer.

The figure tells us many things about fundamental features of Cat-CVD.

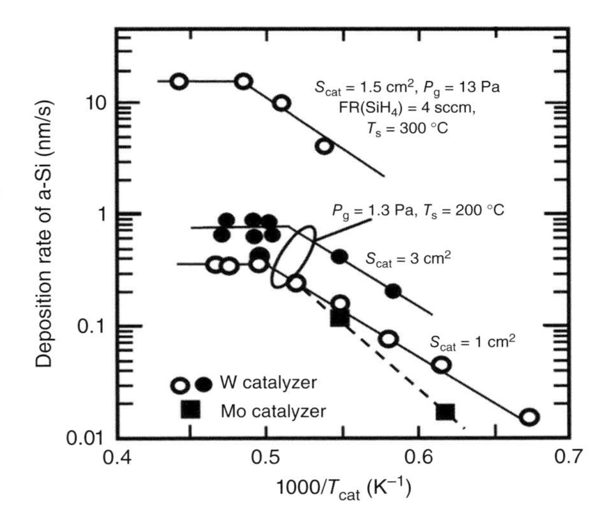

Figure 4.3 Deposition rates of a-Si films as a function of reciprocal of catalyzer temperature T_{cat}. Surface area of catalyzer S_{cat} is taken as a parameter. Source: From Horbach et al. 1989 [2] and Tsuji et al. 1996 [3].

4.2 What Happens on Catalyzer Surfaces – Catalytic Reactions

Next, we demonstrate the possible reactions happening on the catalyzer surfaces for decomposition of molecules. Before we start to explain individual reactions, here, we briefly mention an image showing what happens on the catalyzer. As an example, we just introduce the case of SiH_4 decomposition on a W (tungsten) catalyzer.

A.G. Sault and D.W. Goodman studied the dissociative adsorption processes of SiH_4 on clean W surfaces [4]. An ultrahigh vacuum chamber pumped with an ion pump was used. W was heated at about 2300 K in vacuum below 10^{-6} Pa to make the surfaces clean by removing carbon and oxygen. After that, SiH_4 was introduced at 120 or 350 K to form an adsorbed layer on W surfaces. The surfaces were observed by Auger electron spectroscopy (AES) and low-energy electron diffraction (LEED). The desorbed species were monitored after temperature-programmed desorption (TPD). It was revealed that at 120 K, SiH_4 is absorbed by forming SiH_3 and H, but at 350 K, SiH_4 undergoes complete dissociation to form Si + 4H. They also confirmed that Si atoms are ejected at 2300 K. They concluded that SiH_4 molecules are dissociatively adsorbed on clean W surfaces even at temperatures lower than room temperature (RT) and that, when the temperature is elevated, SiH_4 on W surfaces is decomposed completely and Si atoms are released.

The above conclusion is consistent with the results of J. Doyle et al. who showed that Si and H atoms are released mainly when SiH_4 is decomposed on W wires heated over 1800 K [5]. Later, this decomposition process of SiH_4 to atomic species, Si + 4H, has been confirmed by many investigators, as will be mentioned in Section 4.5.3. These results are in contrast to those of thermal or plasma decomposition. For instance, M. Koshi et al. reported that the major products of thermal decomposition of SiH_4 in the gas phase are SiH_2 and H_2 [6]. Plasma decomposition is much less selective, and Si, SiH, SiH_2, and SiH_3 are produced, although the steady-state density of SiH_3 is much higher than that of others because of its low reactivity [7].

From these experiments, we can speculate that one of the primary reactions on W surfaces is a dissociative adsorption reaction to produce SiH_3 + H, which may proceed even at temperatures lower than RT. By elevation of W temperatures, SiH_3 is further decomposed to SiH_2 + H. SiH_4 may also be decomposed to SiH_2 + 2H on W surfaces. This is a well-known thermal CVD regime. When W temperatures are elevated to over 1000 °C, SiH_2 cannot stay any more, and SiH_4 is decomposed to Si + 4H immediately after touching with heated W. On W surfaces at over 1000 °C, any other decomposed forms of SiH_4 cannot exist. A set of five empty sites is necessary to decompose SiH_4 molecules on W surfaces. If such a set cannot be provided, SiH_4 molecules are repelled out from W surfaces.

This situation is schematically drawn in Figure 4.4, (a) for the temperature of a W catalyzer, T_{cat}, less than RT, (b) T_{cat} at about RT to 1000 °C, and (c) T_{cat} over 1000 °C [8]. In Cat-CVD, there is an additional very important point, that is, to keep W surfaces clear, these Si + 4H have to be released in the space. For such

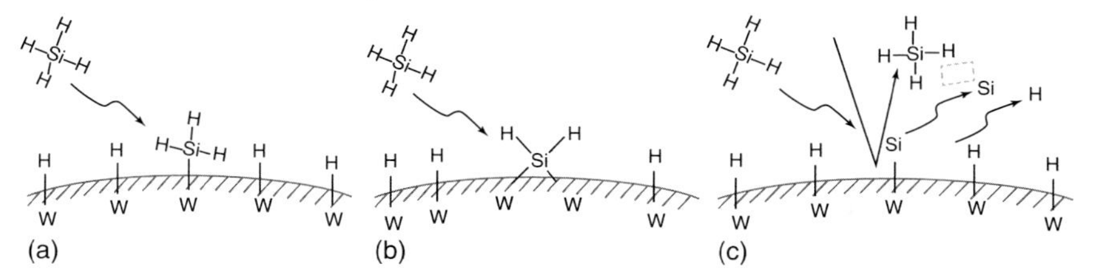

Figure 4.4 Decomposition modes of SiH_4 molecules on a W catalyzer. (a) for $T_{cat} < RT$, (b) $T_{cat} = RT$ to 1000 °C, and (c) $T_{cat} > 1000$ °C. Source: Matsumura et al. 2004 [8] Reprinted with permission of Elsevier.

clearance, even higher temperature of W catalyzers is required as A.G. Sault and D.W. Goodman have already confirmed in their report. H atoms are easily desorbed when T_{cat} of W is kept over 1000 °C; however, as the bond between Si and W is not so week, desorption of Si atoms is not easy. If such adsorbed Si atoms start to accumulate at a certain particular W site and if the size of Si—W bonded sites exceeds over a critical radius, they form W-silicide, and W surfaces cannot be returned to pure W any more. To avoid this, the temperature of W catalyzers has to be elevated more than 1800 °C.

In Cat-CVD, avoidance of this silicide formation is one of the most important issues to make Cat-CVD in practical use. For reducing the maximum temperature of catalyzing wires, many groups are likely to use Ta (tantalum) instead of W for SiH_4 decomposition, as silicide formation can be avoided when T_{cat} is over 1700 °C, slightly lower than the case of W. The selection of catalyzing materials is a very important issue in Cat-CVD. This topic will be discussed in detail in Chapter 7.

Through these explanations, you may understand what is happening on the catalyzer surfaces. In this example, it is clearly demonstrated that SiH_4 molecules are not decomposed simply pyrolytically but that they are decomposed through the reactions on the catalyzer surfaces. The elemental reactions are dissociative adsorption and the release of such dissociative species from the catalyzer surfaces. The heat is used to enhance desorption of decomposed species from the catalyzer surfaces and to keep the surface clean. That is, the decomposition happens based on catalytic cracking reactions.

4.3 Poisoning of Surface Decomposition Processes

As an even clear evidence of surface catalytic reactions on wire surfaces, next, we show the existence of poisoning. Poisoning, suppression of catalytic activities by either competitive adsorption onto active sites or alloy formation, is one of the well-known phenomena in catalytic reactions.

As an example, we show the poisoning of NH_3 decomposition by SiH_4. N_2 and H_2 are produced as end products in the catalytic decomposition of NH_3, although this is not a single-step reaction as will be explained in Section 4.5.4. The decomposition efficiency of NH_3 as well as the production efficiencies of N_2 and H_2

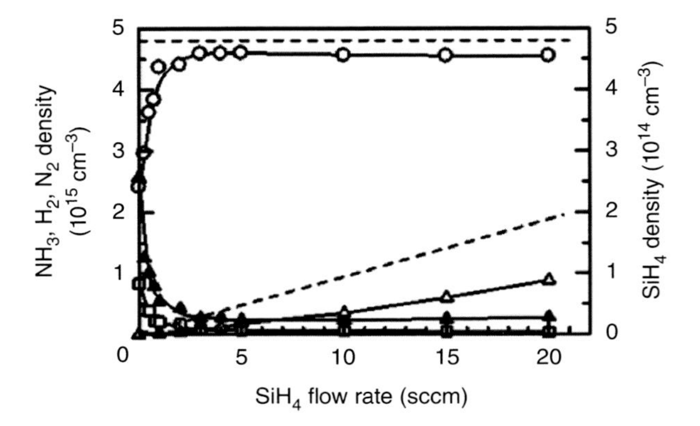

Figure 4.5 NH_3 (open circle), N_2 (open square), H_2 (closed triangle), and SiH_4 (open triangle) densities as a function of the flow rate of SiH_4. The NH_3 flow rate was fixed at 500 sccm. The W wire temperature was 2.30×10^3 K. Dashed lines represent the NH_3 and SiH_4 densities when the wire is not heated. Source: Umemoto et al. 2003 [9]. Reprinted with permission of Elsevier.

can be determined by using a conventional mass spectrometer, as has been mentioned in Section 3.8. The decomposition efficiency increases with the wire temperature and saturates at about 50% over 2.0×10^3 K, as is shown in Figure 3.18. When a tiny amount of SiH_4 is introduced into the chamber, the situation changes drastically. Figure 4.5 shows the NH_3, N_2, H_2, and SiH_4 densities as a function of SiH_4 flow rate when the NH_3 flow rate was 500 sccm, and the W wire temperature was 2.30×10^3 K. The decomposition efficiency of NH_3 decreases from about 50% to 5% by the introduction of only 3 sccm of SiH_4 and levels off over 3 sccm [9]. This decrease can be explained by the temporal poisoning of the catalyzer surfaces by SiH_4. It has also been shown that the decomposition efficiency of NH_3 recovers by the addition of H_2 when the NH_3 pressure is low [10]. H atoms produced from H_2 may reactivate the catalyzer surfaces poisoned by SiH_4.

As we showed in the previous section, SiH_4 is decomposed to Si + 4H by using five sites on W surfaces. The desorption of H atoms is easy, but the desorption of Si requires a large activation energy and Si atoms are likely to remain on the W surfaces. On the other hand, according to the study on NH_3 decomposition, mentioned later in Section 4.5.4, contrary to SiH_4 decomposition, NH_3 is not fully decomposed on W surfaces to N + 3H, which is decomposed to NH_2 + H by using two sites on W surfaces, and NH_2 and H are released to space. When SiH_4 is introduced, the most bonding sites on W surfaces are immediately occupied by Si + 4H or actually by Si, and thus, it is hard for NH_3 to find the remaining two sites for its decomposition. By this way, the decomposition of NH_3 is drastically suppressed. The incomplete suppression of the decomposition of NH_3 can be explained by considering that, although less efficient, NH_3 can be decomposed on silicide surfaces [10].

When the temperature of W is elevated to 2.50×10^3 K, the suppression of NH_3 decomposition is more gradual, as is shown in Figure 4.6. The desorption of Si atoms bonded to W must become easier and more free sites may come to appear.

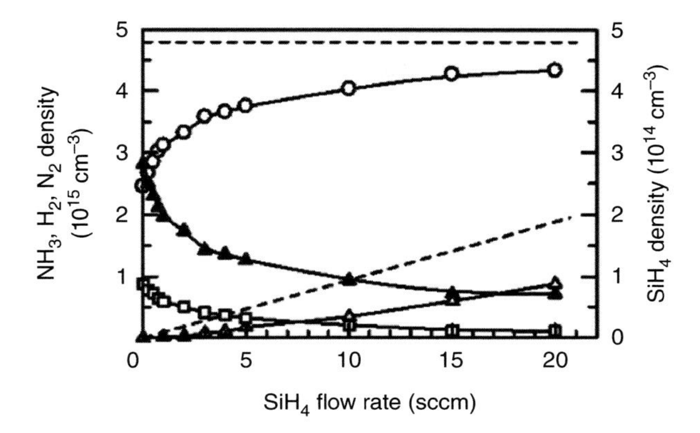

Figure 4.6 NH_3 (open circle), N_2 (open square), H_2 (closed triangle), and SiH_4 (open triangle) densities as a function of the flow rate of SiH_4. The NH_3 flow rate was fixed at 500 sccm. The W wire temperature was 2.50×10^3 K. Dashed lines represent the NH_3 and SiH_4 densities when the wire is not heated.

These results clearly demonstrate the importance of surface catalytic reactions in the decomposition mechanism.

Further evidence for catalytic decomposition can be found in the wire material dependence of the decomposition efficiencies [11, 12] as well as the decrease in the deposition rate after aging [13].

4.4 Gas Temperature Distribution in Cat-CVD Chambers

Information on the temperature distribution is important for modeling because the reaction rate constants depend on the surrounding temperature. As has been discussed in Section 3.3.7, the temperature is around 350 K when the distance from the wire is around 10 cm and the pressure is several pascals, depending little on the wire temperature. Then, in the deposition of amorphous and polycrystalline Si films, the temperature distribution has been paid less attention.

In the deposition of carbon-related films, such as diamond, on the other hand, the distance between the wire and the substrate must be shorter, in the order of 1 cm, while the pressure is higher, several kilopascals, compared with silicon-related systems. In such cases, the gas temperature strongly depends on the wire temperature as well as the distance from the wire. Much attention has been paid to the temperature distribution in this short-distance region. In Section 3.6.1, we have shown that the resonance-enhanced multiphoton ionization (REMPI) spectral profiles of H atoms depend on the radial distance [14, 15]. Coherent anti-Stokes Raman scattering (CARS) [16–18] and cavity ringdown spectroscopy (CRDS) [19] have also been used to determine the rotational temperature distributions of molecular species. The gas temperature decreases smoothly with the radial distance, as expected. Such distance dependences of the temperature can be reproduced quantitatively by a formula presented by Y.A. Mankelevich and coworkers, who solved a set of conservation equations for mass, momentum, energy, and species concentration numerically [15, 20]. In the course of such studies, the presence of temperature discontinuity between the wire surfaces and the gas phase was confirmed [14–19]. The discontinuity is as large as several hundred kelvins. The translational temperature of H atoms just after the formation from SiH_4 was also measured to be $\sim 1.0 \times 10^3$ K lower than that of the wire [21]. Although less remarkable, another temperature discontinuity is present between the substrate and the gas phase [16–18].

Hot radicals are relaxed translationally and rotationally by collisions with surrounding gas molecules. Translational relaxation is the most effective when the masses of the two colliding species are equal. Imagine that the translational temperature of Si atoms formed from SiH_4 on hot wire surfaces is 1.3×10^3 K. The thermal velocity should be 9.9×10^2 m/s. This velocity should be reduced to 6.2×10^2 m/s by just one elastic head-on collision with a H_2 molecule at 300 K, although the mass of Si is 14 times larger than that of H_2. The velocity of Si after one elastic collision corresponds to 510 K. The translational temperature of H_2 may be higher than 300 K near wires and most collisions may not be head-on. It can be concluded, nevertheless, that radical species produced on hot wire surfaces may well be thermalized when the pressure is more than a few pascals and the wire–substrate distance is more than a few centimeters.

4.5 Decomposition Mechanisms on Metal Wire Surfaces and Gas-Phase Kinetics

Gas-phase diagnoses have revealed that selective production of radicals is possible in the catalytic decomposition, in contrast to plasma processes. The decomposition efficiency of N_2 is low, but ground-state N atoms can be produced selectively, without coproducing metastable excited species [22]. The major decomposition products of SiH_4 are Si and H atoms [5, 11, 21, 23]. Similarly, P and H atoms are mainly produced in the decomposition of PH_3 [24]. In contrast, the major products are NH_2 and H for NH_3 [25]. The causes of these behaviors are discussed in this section.

4.5.1 Catalytic Decomposition of Diatomic Molecules: H_2, N_2, and O_2

H atoms can be produced efficiently from H_2 on heated catalyzer surfaces. Table 4.2 summarizes the typical results of the absolute density measurements of H atoms [26]. This efficient production can be explained quantitatively by a model calculation with adjustable rate parameters for the dissociative adsorption and subsequent desorption [27, 28]:

$$H_2 + S^* \leftrightarrow H(ads) + H,$$

$$H(ads) \leftrightarrow H + S^*.$$

Here, H(ads) represents adsorbed atomic hydrogen and S* stands for a vacant site on the metal catalyzer. This model is, in principle, the same as that presented in Section 4.2 for the catalytic decomposition of SiH_4. With this model, the linear

Table 4.2 Absolute densities of H atoms observed in catalytic decomposition of H_2 [26].[a)]

Material gas	Detection technique	T_{cat} (K)	$[H_2]$ (Pa)	Maximum [H] observed (cm^{-3})
H_2	RA[b)]	1.9×10^3	1.9	3×10^{11}
H_2	VUV laser absorption[c)] Two-photon LIF[d)]	2.2×10^3	7.5	1.8×10^{14}
H_2	VUV laser absorption Two-photon LIF VUV LIF[e)]	2.2×10^3	17	1.2×10^{13}
H_2	SR absorption[f)]	2.8×10^3	2.7×10^3	4×10^{15}
H_2	Two-photon LIF	2.6×10^3[j)]	3.0×10^3	4.8×10^{15}
$CH_4(2\%)/H_2$	Two-photon LIF	3.0×10^3[k)]	4.0×10^3	1.2×10^{17}
$CH_4(0.5\%)/Ar(0.5\%)/H_2$	TIMS[g)] REMPI-TOF[h)]	2.4×10^3[l)]	3.0×10^3	1.2×10^{15}
$CH_4(0.5\%)/H_2$	THG[i)]	2.4×10^3	1.3×10^4	4×10^{15}
$CH_4(1\%)/Ar(7\%)/H_2$	TIMS	2.6×10^3	2.7×10^3	4×10^{14}

a) The catalyzer is W otherwise stated.
b) Resonance absorption.
c) Vacuum ultraviolet laser absorption.
d) Two-photon laser-induced fluorescence at 205 nm.
e) Vacuum ultraviolet laser-induced fluorescence.
f) Synchrotron radiation absorption.
g) Threshold ionization mass spectrometry.
h) (2 + 1) Resonance-enhanced multiphoton ionization at 243 nm with time-of-flight mass spectrometry.
i) Third harmonic generation.
j) Ta catalyzer.
k) TaC catalyzer.
l) Catalyzer unknown.

relationship between the logarithm of the H atom density and the reciprocal of the catalyzer temperature, as well as the saturation of the H atom density against the H_2 pressure, can be explained.

Similar mechanisms can also be applied to the decomposition processes of N_2 and O_2 [22, 28]. The decomposition efficiencies of N_2 are much lower than those of H_2, and the N atom densities obtained by catalytic decomposition are much lower than those in plasma processes [22]. The low decomposition efficiencies of N_2 compared to H_2 can be explained by its large bond energy. The N≡N triple bond, 945 kJ/mol, is too strong to be broken easily on catalyzer surfaces. As catalytic decomposition is a mild process, the bond energy dependence must appear more remarkably than plasma processes. The bond energy of the O=O double bond is 498 kJ/mol and a little larger than that of the H—H single bond, 436 kJ/mol. The production of O atoms from O_2 on heated Ir (iridium) surfaces can be observed, and the production efficiency is comparable to that

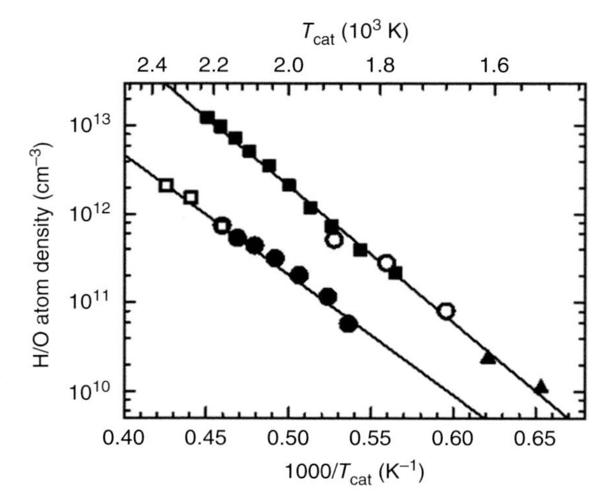

Figure 4.7 H and O atom densities as a function of the reciprocal of the Ir catalyzer temperature measured in pure H_2 and pure O_2 systems. Closed square, H atoms measured by two-photon laser-induced fluorescence (LIF); open circle, H atoms measured by vacuum ultraviolet (VUV) laser absorption; closed triangle, H atoms measured by VUV LIF; open square, O atoms measured by VUV laser absorption; closed circle, O atoms measured by VUV LIF. The H_2 flow rate was 100 sccm and the pressure was 17 Pa, while the O_2 flow rate was 1.00 sccm and the pressure was 0.8 Pa. Source: From Umemoto and Kusanagi 2008 [29] and Umemoto et al. 2009 [30].

for H atoms [29, 30]. Figure 4.7 compares the Ir wire temperature dependences of H and O atom densities obtained by using the same chamber. The apparent activation energies, the gradients of the plots multiplied by the gas constant, are much smaller than the bond dissociation energies. The activation energy for H_2 is 295 kJ/mol, whereas that for O_2 is 260 kJ/mol. This means that the decomposition involves more than two steps, such as dissociative adsorption followed by desorption from metal surfaces [27, 28]. The difference in the absolute densities of H and O atoms can almost be attributed to the differences in thermal velocities and pressures. It should be remembered that the atomic density saturates against the material gas pressure. The maximum H atom density in Figure 4.7, $1.2 \times 10^{13}/cm^3$ at 17 Pa and 2.20×10^3 K, is one order less than that reported in Ref. [31], $1.8 \times 10^{14}/cm^3$ at 7.5 Pa and 2.20×10^3 K. This difference should be ascribed to the difference in the chamber size. The chamber diameter was 10 cm in the former case, whereas it was 45 cm in the latter. The H atom recombination loss on chamber walls must be larger when a small chamber is used. Besides the effect of volume/surface ratio, the recombination probability increases with the wall temperature [32], which is higher in small chambers. The H atom density measurements using a small chamber with a W wire shows that the effect of the difference in wire materials is rather minor [33].

Finally, it should be noted that catalytic decomposition can be clean sources of ground-state $N(2s^22p^3\,{}^4S)$ and $O(2s^22p^4\,{}^3P_J)$ atoms. No metastable excited species, such as $N(2s^22p^3\,{}^2D)$, $O(2s^22p^4\,{}^1D_2)$, $N_2(A^3\Sigma_u{}^+)$, and $O(a^1\Delta_g)$, are produced.

4.5.2 Catalytic Decomposition of H_2O

Most high melting point metals, including W, Ta, and Mo (molybdenum), can easily be oxidized when exposed to oxidizing agents, including H_2O, at elevated temperatures. Ir is an exception and can be used even in the presence of pure O_2 [29, 30]. Production of O, H, and OH was confirmed in the catalytic decomposition of H_2O on a heated Ir wire [34]. The radical densities showed non-Arrhenius temperature dependences, in contrast to the results on the decomposition of H_2 and O_2. Especially, the OH density decreased with the increase in the wire temperature over 2.10×10^3 K. The exit channel must change from the production of H + OH to that of 2H + O with the increase in the wire temperature.

4.5.3 Catalytic Decomposition of SiH_4 and SiH_4/H_2 and the Succeeding Gas-Phase Reactions

The main products in the decomposition of SiH_4 are Si and H atoms when the wire temperature is high enough [5, 11, 21, 23, 35–37]. Mass spectrometric measurements carried out under collision-free conditions show that the direct production of SiH_x ($1 \leq x \leq 3$) is minor [5, 11, 21, 23, 36, 37]. The release of molecular H_2 is also minor [5, 11]. In the decomposition of many molecular species, a linear relationship can be observed between the logarithm of the radical density and the reciprocal of the catalyzer temperature, such as shown in Figure 4.7. On the other hand, such plots for the production of Si and H atoms from SiH_4 are nonlinear, and the apparent activation energies are temperature dependent. The activation energies are large at low wire temperatures, whereas they are small at high temperatures as is shown in Figure 4.8. These results suggest that the production of Si and H atoms is controlled by surface reactions on wire surfaces at low temperatures, while the high-temperature regime is dominated by mass transport limitations [12, 23]. The apparent activation energy obtained at low temperatures depends on the wire materials, showing that the decomposition

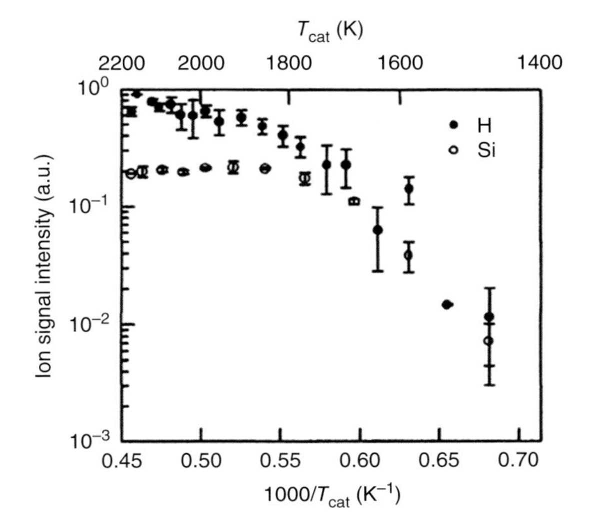

Figure 4.8 Logarithm of the REMPI signal intensities of H (closed circle) and Si (open circle) atoms measured in the presence of pure SiH_4 under collision-free conditions as a function of the reciprocal of the W wire temperature. Source: Tonokura et al. 2002 [36]. Reprinted with permission of Elsevier.

is not thermal but catalytic [11, 12]. The saturation of radical densities at high temperatures is consistent with the results of the saturation of deposition rates as shown in Figure 4.3.

Si atoms produced on wire surfaces may react with SiH_4. The rate constant has been measured to be large [38]. Theoretical calculations suggest the spin-forbidden production of dibridged singlet-state $Si(H_2)Si$, with two H atoms in bond-centered positions, through $HSiSiH_3$ [39, 40]. Si_2H_2, possibly $Si(H_2)Si$, has actually been detected mass spectrometrically in the absence of a H_2 flow [11]. $Si(H_2)Si$ thus produced may contribute to the deposition of amorphous Si films, besides SiH_3 produced in the $SiH_4 + H \rightarrow SiH_3 + H_2$ reaction [41]. This point shall be discussed at the last part of this chapter, Section 4.6. The direct deposition of Si atoms should be minor, at least under practical deposition conditions, because of the rapid reaction with SiH_4.

In the presence of an excess amount of H_2, the efficient production of SiH_3 has been confirmed [42]. SiH_3 must be produced in the gas-phase reaction between H and SiH_4 [41], and SiH_3 thus produced can be a good precursor of polycrystalline Si films because of its high surface mobility [7, 43]. When the SiH_3 density is too high, SiH_2 as well as higher silanes may be produced in self-reactions, such as $SiH_3 + SiH_3 \rightarrow SiH_2 + SiH_4$ and $SiH_2 + SiH_4 + M \rightarrow Si_2H_6 + M$ [44, 45], but the contribution of such species may be minor under typical low-pressure deposition conditions. It has also been confirmed that SiH_4 is reproduced in reactions of H atoms with the Si compounds deposited [35].

4.5.4 Catalytic Decomposition of NH_3 and the Succeeding Gas-Phase Reactions

N_2 and H_2 are two of the major end products in the decomposition of NH_3, but these are not direct products on wire surfaces, as has been mentioned in Section 3.8. Radical species are produced at the first stage. The decomposition efficiency of NH_3 is around 50% when the W wire temperature is over 2.0×10^3 K, as Figure 3.18 shows. On the other hand, the densities of radical species, H, NH, and NH_2, still increase over 2.0×10^3 K [25]. Figure 4.9 shows the relation between the NH_2 density and the wire temperature. The saturation of the decomposition efficiency

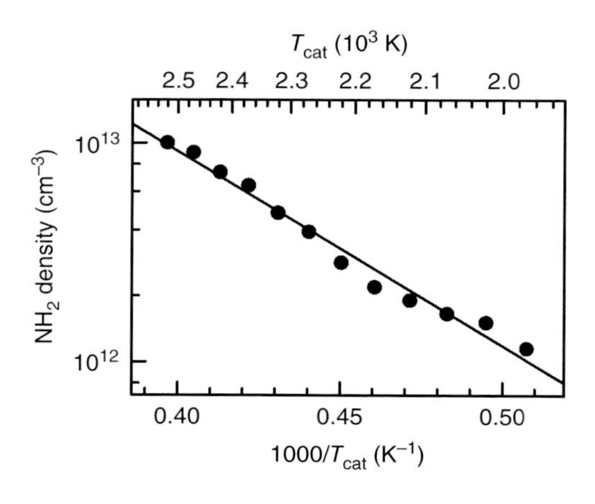

Figure 4.9 NH_2 density as a function of the reciprocal of the W wire temperature. The NH_3 flow rate was 500 sccm. The pressure was 20 Pa when the catalyzer was not heated. Source: Umemoto et al. 2003 [25]. Copyright 2003: The Japan Society of Applied Physics.

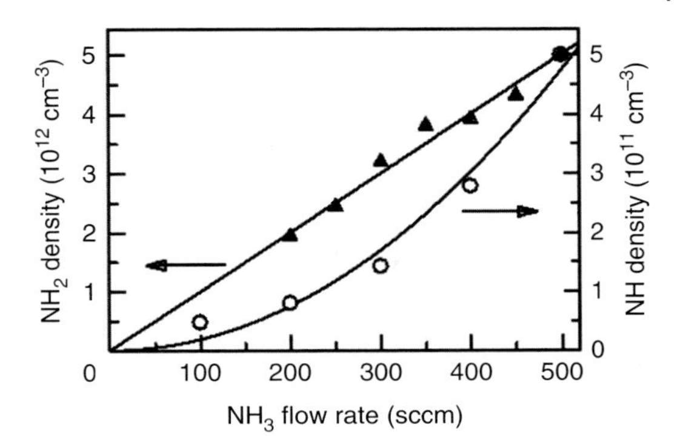

Figure 4.10 NH_3 flow rate dependence of the NH (circle) and NH_2 (triangle) densities. The flow rate of 500 sccm corresponds to the NH_3 pressure of 20 Pa when the W catalyzer was not heated. The W catalyzer temperature was 2.30×10^3 K. Source: Umemoto et al. 2003 [25]. Copyright 2003: The Japan Society of Applied Physics.

of NH_3 should be ascribed to the presence of reproduction processes of NH_3 from radical species. H atoms must be produced not only from NH_3 but also from one of the end products, H_2. It should be noted that the steady-state density of H_2 is higher than that of NH_3 at high temperatures. H atoms thus produced may react with NH_2 to reproduce NH_3. By the balance of decomposition and formation reactions of NH_3, the density may depend little on the wire temperature over 2.0×10^3 K.

Figure 4.10 shows the NH_3 flow rate dependence of NH_2 and NH densities [25]. It is clear that the NH_2 density is more than one order of magnitude larger and increases linearly against the flow rate, whereas the NH density increases quadratically. It has also been confirmed that the H atom density increases linearly with the increase in the NH_3 flow rate. Then, it can be concluded that NH_2 and H are the direct products on wire surfaces, while NH is formed in the succeeding gas-phase $NH_2 + H \rightarrow NH + H_2$ reaction. In such a case, the NH density may increase in proportion to the product of the densities of NH_2 and H, and, in consequence, should be proportional to the square of the NH_3 flow rate. The $NH_2 + H \rightarrow NH + H_2$ reaction has a fairly large activation energy but may proceed near a filament [46]. The steady-state density of N atoms was found to be low. N atoms must be removed rapidly by the fast $N + NH \rightarrow N_2 + H$ and $N + NH_2 \rightarrow N_2 + H + H$ reactions [47, 48], if produced in the $NH + H \rightarrow N + H_2$ reaction.

NH_3 cannot be fully decomposed to atomic species, although SiH_4 can be. This difference should be attributed to the difference in bond energies. The $H_3Si{-}H$ bond energy is 384 kJ/mol, whereas the $H_2N{-}H$ bond energy is 453 kJ/mol. Although the difference is just 69 kJ/mol, this difference must be critical in catalytic decomposition. As will be discussed in Section 4.5.6, PH_3 is directly decomposed to $P + 3H$. This can also be explained by the weak $H_2P{-}H$ bond energy, 351 kJ/mol.

4.5.5 Catalytic Decomposition of CH_4 and CH_4/H_2 and the Succeeding Gas-Phase Reactions

In the CH_4/H_2 systems, the major carbon-containing radical is CH_3 [15, 49]. The CH_3 density was found to peak at the location of several millimeters radially away from the wire surface [50]. This off-filament peak has been explained by considering the gas phase $H + CH_4 \rightarrow H_2 + CH_3$ reaction. The efficient accommodation of carbon species into wires (carburization of wires) may also be the cause of this unique behavior. The rate constant for the $H + CH_4$ reaction is extremely small at room temperature because of its large activation energy, 58 kJ/mol, but that is larger than 3×10^{-13} cm^3/s over 1.0×10^3 K [51, 52].

There are few studies for the decomposition of CH_4 in the absence of a H_2 flow. K.L. Menningen et al. succeeded in the observation of CH_3 in a CH_4/He system, although the density was one order of magnitude smaller than that observed in a CH_4/H_2 system [53]. The inefficient direct production of CH_3 is not surprising, as the H_3C—H bond energy, 439 kJ/mol, is much larger than that of H_3Si—H, 384 kJ/mol. The absence of lone-paired electrons may also be one of the causes.

4.5.6 Catalytic Decomposition of PH_3 and PH_3/H_2 and the Succeeding Gas-Phase Reactions

P, PH, PH_2, and H can be observed in the catalytic decomposition of PH_3 both in the presence and in the absence of a H_2 flow [24, 54]. In PH_3/He systems without a H_2 flow, P and H atom densities increased linearly against the PH_3/He flow rate, while the PH and PH_2 densities increased nonlinearly; the PH_2 density increased in proportion to the square of the flow rate, whereas the PH density increased in proportion to the cube, as shown in Figures 4.11 and 4.12. The absolute density of P atoms was the highest, whereas the PH and PH_2 densities were much lesser than that of P atoms. The P atom density changed little when H_2 was introduced, but PH and PH_2 densities increased with the H_2 flow rate. These results show that the major products on the wire surfaces are P and H atoms, while PH_2 and PH are produced in succeeding reactions with H atoms in the gas phase, $PH_x + H \rightarrow PH_{x-1} + H_2$ $(2 \leq x \leq 3)$. As PH and PH_2 densities are much lower than that of P atoms unless H_2 is introduced, catalytic decomposition of pure PH_3 can be a clean source of P atoms, together with the catalytic decomposition of neat P_4 vapor [54].

The decomposition efficiency of PH_3 in the absence of a H_2 flow increased with the wire temperature up to 2.0×10^3 K and then showed saturation [24]. This is

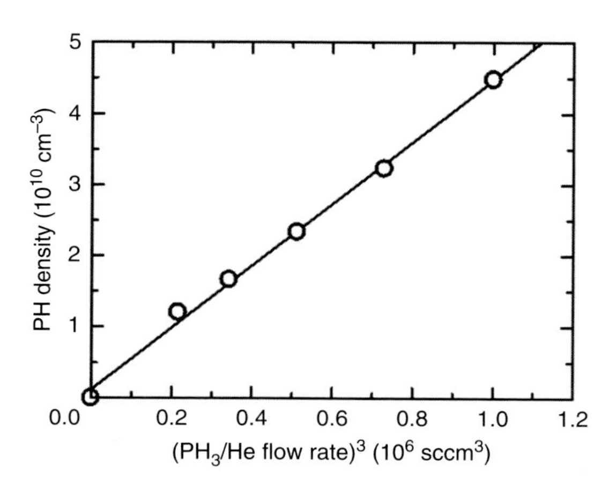

Figure 4.11 PH density as a function of the cube of the PH_3/He (2.0% dilution) flow rate. The W catalyzer temperature was 2.38×10^3 K and the total pressure at 100 sccm was 12 Pa. Source: Umemoto et al. 2012 [24]. Copyright 2012: The Japan Society of Applied Physics.

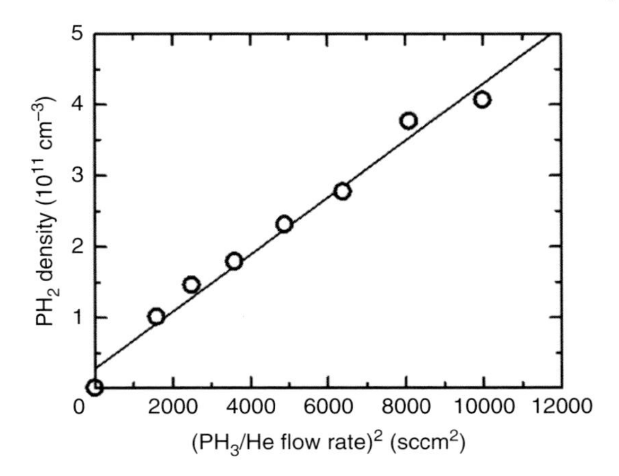

Figure 4.12 PH_2 density as a function of the square of the PH_3/He (2.0% dilution) flow rate. The W catalyzer temperature was 2.58×10^3 K and the total pressure at 100 sccm was 12 Pa. Source: Umemoto et al. 2012 [24]. Copyright 2012: The Japan Society of Applied Physics.

similar to the case for NH_3 and suggests the presence of reproduction processes of PH_3 from radical species. In the presence of an excess amount of H_2, the P, PH, and PH_2 densities were comparable, and the apparent decomposition efficiency of PH_3 was small. This small efficiency can be explained by rapid cyclic reactions including decomposition, deposition, and etching of deposited P films to reproduce PH_3 [24].

4.5.7 Catalytic Decomposition of B_2H_6 and B_2H_6/H_2 and the Succeeding Gas-Phase Reactions

B and BH could be identified in the B_2H_6/H_2 and B_2H_6/He/H_2 systems [55, 56]. The densities in the absence of a H_2 flow were very small, suggesting that the direct production of B and BH on wire surfaces is minor. Mass spectrometric measurements, on the other hand, show that the decomposition efficiency of B_2H_6 is high even in the absence of a H_2 flow, 88% when the B_2H_6/He (2.0% dilution) flow rate is 5 sccm and 67% at 20 sccm [57]. The direct decomposition product of B_2H_6 is considered to be BH_3. B and BH, and possibly BH_2, must be produced in the H atom shifting reactions, $BH_x + H \rightarrow BH_{x-1} + H_2$ $(1 \leq x \leq 3)$, in the gas phase. The inefficient direct production of B and BH on wire surfaces may be ascribed to the large bond energy of H_2B—H, 421 kJ/mol. The dimerization energy of BH_3 is much smaller, 134 kJ/mol. Figure 4.13 shows the W wire temperature dependence of the B and BH densities. The saturation over 2.1×10^3 K may be explained by a transition in the rate-limiting step from surface reactions to mass transport, similar to the SiH_4 systems. The accommodation of B atoms into the wire may also be the cause of this non-Arrhenius behavior. The apparent activation energies for the production of B and BH below 2.05×10^3 K are 462 and 245 kJ/mol, respectively. The difference, 217 kJ/mol, almost agrees with the activation energy to produce H atoms in this system and consistent with a model that B is formed in the reaction of BH + H. It was also found that metal wires can easily be boronized and the boronized wires can be a source of B atoms when heated in the presence of pure H_2 [56]. In contrast to Si and P, etching of deposited B compounds by H atoms was not observed.

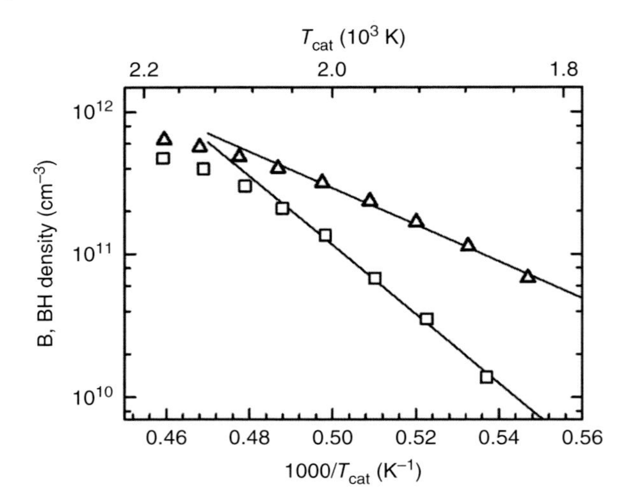

Figure 4.13 B (squares) and BH (triangles) densities as a function of the reciprocal of the W wire temperature. The flow rates of B_2H_6/He (2.0% dilution) and H_2 were 10 and 20 sccm, respectively. The total pressure was 3.9 Pa. Source: From Umemoto et al. 2014 [56].

The detection of BH_2 and BH_3 has not been reported in the catalytic decomposition, but there are some reports in plasma processes. BH_2 has been detected by an intracavity laser spectroscopic technique in the 646 and 735 nm regions, which correspond to the $\tilde{A}^2B_1-\tilde{X}^2A_1$ transition [58]. It is difficult to detect BH_3 in the visible or UV regions but is possible by tunable diode laser absorption spectroscopy (TDLAS), at around 2596 cm^{-1} [59].

4.5.8 Catalytic Decomposition of H_3NBH_3 and Release of B Atoms from Boronized Wires

H_3NBH_3 (borazane) is not explosive or toxic and can be a safe and low-cost source of B atoms. H_3NBH_3 can easily be decomposed to NH_3 and BH_3, from which B atoms can be produced by reactions with H atoms [60]. One of the concerns is the contamination by N atoms. This concern can be avoided by using boronized wires. W wires can easily be boronized not only by B_2H_6/H_2 but also by H_3NBH_3/H_2, and the boronized wires can be a clean and safe source of B atoms [61, 62]. Sustained release of B atoms, sufficient for *surface doping*, can be achieved when W wires boronized by H_3NBH_3/H_2 for 60 minutes are heated in the presence of H_2, as is shown in Figure 4.14. As for *surface doping*, an explanation shall be given in Chapter 10. As W wires are not nitrided, contamination by N atoms can be avoided.

4.5.9 Catalytic Decomposition of Methyl-Substituted Silanes and Hexamethyldisilazane (HMDS)

Organosilicon compounds are less explosive compared to inorganic silicon hydrides and can be much safer sources of SiC and SiCN films. Y.J. Shi and coworkers have studied the decomposition processes of methyl-substituted silanes. It was found that the amount of CH_3 decreases with the increasing number of methyl substitution, while the activation energy to produce CH_3

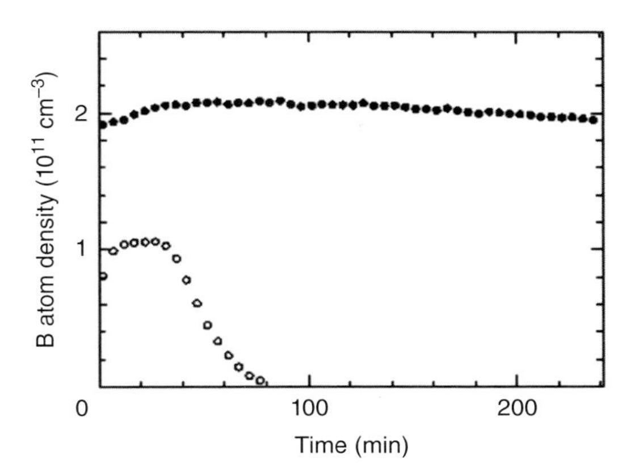

Figure 4.14 Time dependences of B atom densities in the presence of a H_2 flow of 20 sccm measured after boronization. Boronization was carried out for 60 minutes (closed circle) or 60 seconds (open circle) in the presence of ~0.01 Pa of H_3NBH_3 and 2.1 Pa of H_2, whereas the B atom density was measured in the presence of 2.1 Pa of neat H_2. The W wire temperature was kept at 2.29×10^3 K during both the boronization and the B atom density measurement. Source: Umemoto and Miyata 2016 [61]. Reprinted with permission of The Chemical Society of Japan.

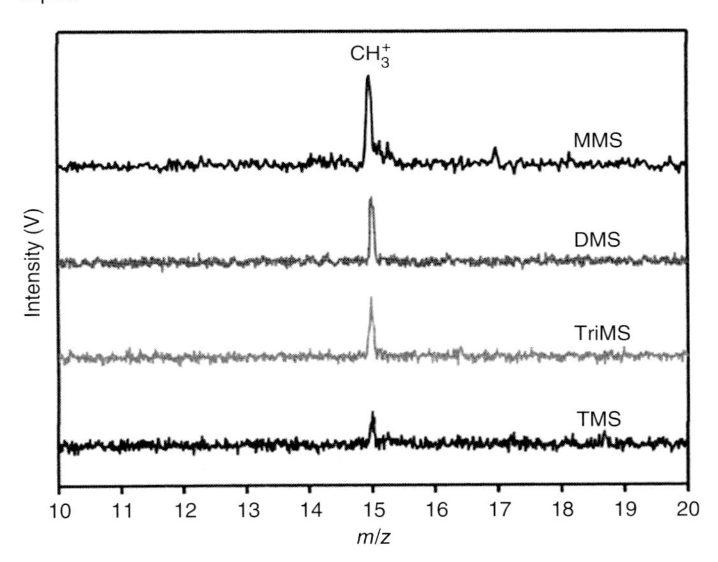

Figure 4.15 Single-photon ionization mass spectra (10.5 eV) of CH_3SiH_3 (MMS), $(CH_3)_2SiH_2$ (DMS), $(CH_3)_3SiH$ (TriMS), and $(CH_3)_4Si$ (TMS) obtained under collision-free conditions. The W wire temperature was 2100 °C. Source: Shi 2015 [64].

increases [63, 64]. Figure 4.15 shows the single-photon ionization (SPI) mass signals of CH_3 radicals produced. The same trend has also been reported for the production of Si atoms [65]. These results suggest that the decomposition is initiated by the Si—H bond breakage and followed by Si—CH_3 bond cleavage to form CH_3 radicals, except for tetramethylsilane, which shows the highest activation energy.

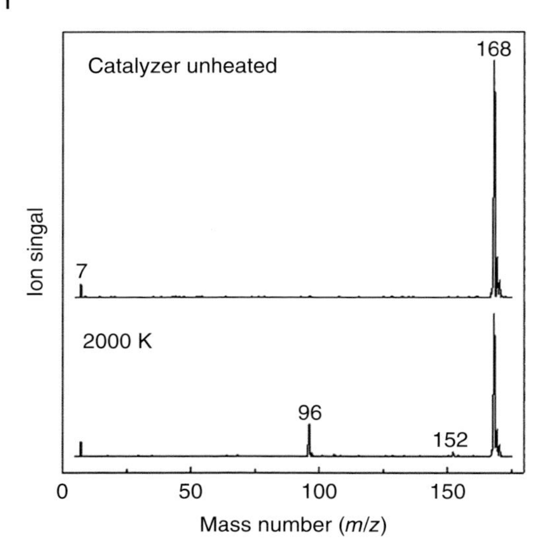

Figure 4.16 Ion attachment mass spectra of $(CH_3)_3SiNHSi(CH_3)_3$ (HMDS) with and without heating the W catalyzer. The flow rate was 1.5 sccm while the pressure in the decomposition chamber was 1.0 Pa. Source: Morimoto et al. 2006 [67]. Copyright 2006: The Japan Society of Applied Physics.

Hexamethyldisilazane ($(CH_3)_3SiNHSi(CH_3)_3$, HMDS) is also expected as one of the safe and low-cost source materials to fabricate SiCN films [66]. T. Morimoto et al. recorded Li^+ ion attachment mass spectra of HMDS with and without heating a W catalyzer [67]. When the catalyzer was not heated, as is shown in Figure 4.16, there were just two peaks assignable to Li^+ and HMDS·Li^+, HMDS attached by Li^+ (mass-to-charge ratio of 168). When the wire was heated up to 2.00×10^3 K, peaks for $(CH_3)_3SiNH_2 \cdot Li^+$ and $(CH_3)_3SiN{=}Si(CH_3)_2 \cdot Li^+$ appeared, while the parent peak decreased. The appearance of the $(CH_3)_3SiNH_2 \cdot Li^+$ signal suggests the breakage of Si—N bonds in HMDS. $(CH_3)_3SiNH$ radicals produced on the catalyzer surfaces must abstract or recombine with H atoms to produce $(CH_3)_3SiNH_2$. Neither the pair product, $Si(CH_3)_3$, nor its derivative, $HSi(CH_3)_3$, was observed, but this lack of a $HSi(CH_3)_3 \cdot Li^+$ signal can be attributed to the small Li^+ ion affinities. On the other hand, the lack of $(CH_3)_3SiNHSi(CH_3)_2H \cdot Li^+$ cannot be attributed to the difference in the Li^+ ion affinities and suggests the absence of Si—C bond scission. The lone-paired electrons of N atoms may interact with the catalyzer to break the Si—N bonds selectively. A steric hindrance must be present to break Si—C bonds.

4.5.10 Summary of Catalytic Decomposition of Various Molecules on Metal Wires

The typical species generated on heated metal wires and the active species in CVD chambers are summarized in Table 4.3. The metal material is W otherwise stated.

4.6 Si Film Formation Mechanisms in Cat-CVD

Finally, we would like to present models of Si-film formation by Cat-CVD. It should be noted, at first, that there is still a controversy regarding the film precursors. Then, we would like to integrate the common points as much as possible.

Table 4.3 Major nascent decomposition products and active species in Cat-CVD processes.

Material gas	Major nascent decomposition product(s)	Major active species	References
H_2	H	H	[26]
O_2 [a)]	O	O	[29]
$O_2 + H_2$ [b)]		O, H, OH	[30, 33]
N_2	N	N	[22]
P_4	P	P	[54]
H_2O [a)]	H, O, OH	H, O, OH	[34]
SiH_4 [c)]	Si, H	SiH_3, $Si(H_2)Si$, H	[11, 23, 40]
$SiH_4 + H_2$		SiH_3, H	[42]
NH_3	NH_2, H	NH_2, H	[25]
$SiH_4 + NH_3$		SiH_3, NH_2, H	[9]
CH_4	CH_3, (H)[d)]	CH_3, (H)[d)]	[53]
$CH_4 + H_2$ [c)]		CH_3, H	[15, 49]
PH_3	P, H	P, H	[24]
$PH_3 + H_2$		P, PH, PH_2, H	[24]
B_2H_6	(BH_3)[d)]	(BH_3)[d)]	[56]
$B_2H_6 + H_2$ [c)]		B, BH, (BH_2)[d)], (BH_3)[d)], H	[55, 56]

a) Ir wire.
b) W and Ir wires.
c) W and Ta wires.
d) Species in parentheses have not been identified experimentally.

In the fabrication of a-Si films, SiH_4 is used with or without H_2 dilution. The typical deposition conditions are listed in Table 4.1. We just mention the nondiluted case for simplicity. As has been described in Section 4.5.3, the direct decomposition products of SiH_4 on wire surfaces are Si and H atoms. Then, when the total pressure is low and the mean free path is larger than the chamber size, Si atoms should be the only precursor of Si films. However, under such conditions, high-quality films cannot be obtained, suggesting the importance of gas-phase reactions [68]. The typical pressure is around 1 Pa and the mean free path is much shorter than the distance between the catalyzer and the substrate. Under such practical deposition conditions, Si and H atoms released from the wire may soon react with SiH_4.

H atoms may react with the parent SiH_4 molecules to produce SiH_3 radicals; $SiH_4 + H \rightarrow SiH_3 + H_2$. SiH_3 thus produced may be one of the deposition precursors. Although the rate constant for the above reaction is not so large, 3×10^{-13} cm^3/s at room temperature [41], the production of SiH_3 has been confirmed experimentally in the presence of 1.0 Pa of pure SiH_4 [42]. Of course, not all the H atoms may be converted to SiH_3. Some of the H atoms may recombine to produce H_2 on chamber walls. Some may react with Si compounds deposited on chamber walls to reproduce SiH_4. In any case, SiH_3 should be produced and must be long-lived in the gas phase because of its unreactive nature. The $SiH_3 + H_2 \rightarrow SiH_4 + H$ reaction is too endothermic. In the $SiH_3 + SiH_4$ reaction, SiH_3 should be reproduced again. Then, SiH_3 has a high probability to survive and deposit on the substrate surfaces.

Si atoms should react with SiH_4 in the gas phase. The rate constant is temperature independent and is as large as 2×10^{-10} cm^3/s [38]. This large rate constant means that the reaction occurs almost in every collision. In other words, almost all Si atoms may be lost in gas-phase reactions under practical deposition conditions. The products of the $Si + SiH_4$ reaction have not been fully identified, although the production of dibridged $Si(H_2)Si$ and H_2 has been proposed based on density functional theoretical calculations [39, 40]. W. Zheng and A. Gallagher have actually identified Si_2H_2 species as one of the main products mass spectroscopically [11]. The production processes of $2SiH_2$, $SiH_3 + SiH$, as well as $H_2 + SiSiH_2$, are all endothermic and may be ignored. The large and temperature-independent rate constant for the $Si + SiH_4$ reaction suggests that this reaction is not a simple abstractive one to produce $SiH + SiH_3$ and that the direct product must be triplet-state $HSiSiH_3$ [39, 40]. This triplet $HSiSiH_3$ may undergo intersystem crossing to singlet $HSiSiH_3$, which may easily be converted to $Si(H_2)Si + H_2$ via H_2SiSiH_2. Unless collisionally stabilized, the lifetimes of $HSiSiH_3$ and H_2SiSiH_2 are too short to contribute to deposition [39]. $Si(H_2)Si$ is unreactive to nonradical species in the gas phase [69] and may have a large probability to reach the substrate surfaces. Density functional theoretical calculations show that $Si(H_2)Si$ may adsorb not only on Si dangling bonds but also on H-terminated Si surfaces [40]. In conclusion, a-Si films should start to grow from the products of gas-phase reactions, possibly SiH_3 and $Si(H_2)Si$. This growth model will be mentioned in Chapter 5 when we explain the formation of Cat-CVD a-Si.

The mechanism for the formation of polycrystalline silicon films is rather simple. In the fabrication of polycrystalline silicon, highly H_2-diluted SiH_4 is used as a material gas. The typical deposition conditions are given in Table 5.3. As H_2 can easily be decomposed on heated metal surfaces, a large amount of H atoms are present. These H atoms may react with SiH_4 to produce SiH_3 in the gas phase. SiH_3 radicals thus produced should be the only major precursors of polycrystalline Si films. The deposition processes shall also be presented in Chapter 5.

When the radical density is high, reactions between two radical species may take place. For example, two SiH_3 radicals may react to produce $SiH_2 + SiH_4$ or $HSiSiH_3 + H_2$ [44, 45, 70]. Higher silanes, such as Si_2H_6 and Si_3H_8, may be produced from these species. However, according to the simulation by S. Nakamura et al., contribution of such species to the film growth is rather minor under practical deposition conditions [40].

References

1 Osono, S., Kitazoe, M., Tsuboi, H. et al. (2006). Development of catalytic chemical vapor deposition apparatus for large size substrates. *Thin Solid Films* 501: 61–64.

2 Horbach, C., Beyer, W., and Wagner, H. (1989). Deposition of a-Si:H by high temperature thermal decomposition of silane. *J. Non-Cryst. Solids* 114: 187–189.

3 Tsuji, N., Akiyama, T., and Komiyama, H. (1996). Characteristics of the hot-wire CVD reactor on a-Si:H deposition. *J. Non-Cryst. Solids* 198–200: 1054–1057.

4 Sault, A.G. and Goodman, D.W. (1990). Reactions of silane with W(110) surface. *Surf. Sci.* 235: 28–46.

5 Doyle, J., Robertson, R., Lin, G.H. et al. (1988). Production of high-quality amorphous silicon films by evaporative silane surface decomposition. *J. Appl. Phys.* 64: 3215–3223.

6 Koshi, M., Kato, S., and Matsui, H. (1991). Unimolecular decomposition of SiH_4, SiH_3F, and SiH_2F_2 at high temperatures. *J. Phys. Chem.* 95: 1223–1227.

7 Matsuda, A. (2004). Thin-film silicon – growth process and solar cell application-. *Jpn. J. Appl. Phys.* 43: 7909–7920.

8 Matsumura, H., Umemoto, H., and Masuda, A. (2004). Cat-CVD (hot-wire CVD): how different from PECVD in preparing amorphous silicon. *J. Non-Cryst. Solids* 338–340: 19–26.

9 Umemoto, H., Morimoto, T., Yamawaki, M. et al. (2003). Deposition chemistry in the Cat-CVD processes of SiH_4/NH_3 mixture. *Thin Solid Films* 430: 24–27.

10 Ansari, S.G., Umemoto, H., Morimoto, T. et al. (2006). H_2 dilution effect in the Cat-CVD processes of the SiH_4/NH_3 system. *Thin Solid Films* 501: 31–34.

11 Zheng, W. and Gallagher, A. (2008). Radical species involved in hotwire (catalytic) deposition of hydrogenated amorphous silicon. *Thin Solid Films* 516: 929–939.

12 Duan, H.L. and Bent, S.F. (2005). The influence of filament material on radical production in hot wire chemical vapor deposition of a-Si:H. *Thin Solid Films* 485: 126–134.

13 Frigeri, P.A., Nos, O., and Bertomeu, J. (2015). Degradation of thin tungsten filaments at high temperature in HWCVD. *Thin Solid Films* 575: 34–37.

14 Redman, S.A., Chung, C., Rosser, K.N., and Ashfold, M.N.R. (1999). Resonance enhanced multiphoton ionisation probing of H atoms in a hot filament chemical vapor deposition reactor. *Phys. Chem. Chem. Phys.* 1: 1415–1424.

15 Ashfold, M.N.R., May, P.W., Petherbridge, J.R. et al. (2001). Unravelling aspects of the gas phase chemistry involved in diamond chemical vapor deposition. *Phys. Chem. Chem. Phys.* 3: 3471–3485.

16 Chen, K.-H., Chuang, M.-C., Penney, C.M., and Banholzer, W.F. (1992). Temperature and concentration distribution of H_2 and H atoms in hot-filament chemical-vapor deposition of diamond. *J. Appl. Phys.* 71: 1485–1493.
17 Connell, L.L., Fleming, J.W., Chu, H.-N. et al. (1995). Spatially resolved atomic hydrogen concentrations and molecular hydrogen temperature profiles in the chemical-vapor deposition of diamond. *J. Appl. Phys.* 78: 3622–3634.
18 Zumbach, V., Schäfer, J., Tobai, J. et al. (1997). Experimental investigation and computational modeling of hot filament diamond chemical vapor deposition. *J. Chem. Phys.* 107: 5918–5928.
19 Lommatzsch, U., Wahl, E.H., Aderhold, D. et al. (2001). Cavity ring-down spectroscopy of CH and CD radicals in a diamond thin film chemical vapor deposition reactor. *Appl. Phys. A* 73: 27–33.
20 Mankelevich, Y.A., Rakhimov, A.T., and Suetin, N.V. (1996). Two-dimensional simulation of a hot-filament chemical vapor deposition reactor. *Diamond Relat. Mater.* 5: 888–894.
21 Tange, S., Inoue, K., Tonokura, K., and Koshi, M. (2001). Catalytic decomposition of SiH_4 on a hot filament. *Thin Solid Films* 395: 42–46.
22 Umemoto, H., Funae, T., and Mankelevich, Y.A. (2011). Activation and decomposition of N_2 on heated tungsten filament surfaces. *J. Phys. Chem. C* 115: 6748–6756.
23 Duan, H.L., Zaharias, G.A., and Bent, S.F. (2002). Detecting reactive species in hot wire chemical vapor deposition. *Curr. Opin. Solid State Mater. Sci.* 6: 471–477.
24 Umemoto, H., Nishihara, Y., Ishikawa, T., and Yamamoto, S. (2012). Catalytic decomposition of PH_3 on heated tungsten wire surfaces. *Jpn. J. Appl. Phys.* 51: 086501/1–086501/9.
25 Umemoto, H., Ohara, K., Morita, D. et al. (2003). Radical species formed by the catalytic decomposition of NH_3 on heated W surfaces. *Jpn. J. Appl. Phys.* 42: 5315–5321.
26 Umemoto, H. (2010). Production and detection of H atoms and vibrationally excited H_2 molecules in CVD processes. *Chem. Vap. Deposition* 16: 275–290.
27 Comerford, D.W., Smith, J.A., Ashfold, M.N.R., and Mankelevich, Y.A. (2009). On the mechanism of H atom production in hot filament activated H_2 and CH_4/H_2 gas mixtures. *J. Chem. Phys.* 131: 044326/1–044326/12.
28 Mankelevich, Y.A., Ashfold, M.N.R., and Umemoto, H. (2014). Molecular dissociation and vibrational excitation on a metal hot filament surface. *J. Phys. D: Appl. Phys.* 47: 025503/1–025503/12, 069601/1.
29 Umemoto, H. and Kusanagi, H. (2008). Catalytic decomposition of O_2, NO, N_2O and NO_2 on a heated Ir filament to produce atomic oxygen. *J. Phys. D: Appl. Phys.* 41: 225505/1–225505/5.
30 Umemoto, H., Kusanagi, H., Nishimura, K., and Ushijima, M. (2009). Detection of radical species produced by catalytic decomposition of H_2, O_2 and their mixtures on heated Ir surfaces. *Thin Solid Films* 517: 3446–3448.
31 Umemoto, H., Ohara, K., Morita, D. et al. (2002). Direct detection of H atoms in the catalytic chemical vapor deposition of the SiH_4/H_2 system. *J. Appl. Phys.* 91: 1650–1656.

32 Rousseau, A., Granier, A., Gousset, G., and Leprince, P. (1994). Microwave discharge in H_2: influence of H-atom density on the power balance. *J. Phys. D: Appl. Phys.* 27: 1412–1422.
33 Umemoto, H. and Moridera, M. (2008). Production and detection of reducing and oxidizing radicals in the catalytic decomposition of H_2/O_2 mixtures on heated tungsten surfaces. *J. Appl. Phys.* 103: 034905/1–034905/6.
34 Umemoto, H. and Kusanagi, H. (2009). Catalytic decomposition of $H_2O(D_2O)$ on a heated Ir filament to produce O and OH(OD) radicals. *Open Chem. Phys. J.* 2: 32–36.
35 Nozaki, Y., Kongo, K., Miyazaki, T. et al. (2000). Identification of Si and SiH in catalytic chemical vapor deposition of SiH_4 by laser induced fluorescence spectroscopy. *J. Appl. Phys.* 88: 5437–5443.
36 Tonokura, K., Inoue, K., and Koshi, M. (2002). Chemical kinetics for film growth in silicon HWCVD. *J. Non-Cryst. Solids* 299–302: 25–29.
37 Holt, J.K., Swiatek, M., Goodwin, D.G., and Atwater, H.A. (2002). The aging of tungsten filaments and its effect on wire surface kinetics in hot-wire chemical vapor deposition. *J. Appl. Phys.* 92: 4803–4808.
38 Koi, M., Tonokura, K., Tezaki, A., and Koshi, M. (2003). Kinetic study for the reactions of Si atoms with SiH_4. *J. Phys. Chem. A* 107: 4838–4842.
39 Holt, J.K., Swiatek, M., Goodwin, D.G. et al. (2001). Gas phase and surface kinetic processes in polycrystalline silicon hot-wire chemical vapor deposition. *Thin Solid Films* 395: 29–35.
40 Nakamura, S., Matsumoto, K., Susa, A., and Koshi, M. (2006). Reaction mechanism of silicon Cat-CVD. *J. Non-Cryst. Solids* 352: 919–924.
41 Arthur, N.L. and Miles, L.A. (1997). Arrhenius parameters for the reaction of H atoms with SiH_4. *J. Chem. Soc., Faraday Trans.* 93: 4259–4264.
42 Nozaki, Y., Kitazoe, M., Horii, K. et al. (2001). Identification and gas phase kinetics of radical species in Cat-CVD processes of SiH_4. *Thin Solid Films* 395: 47–50.
43 Perrin, J., Shiratani, M., Kae-Nune, P. et al. (1998). Surface reaction probabilities and kinetics of H, SiH_3, Si_2H_5, CH_3, and C_2H_5 during deposition of *a*-Si:H and *a*-C:H from H_2, SiH_4, and CH_4 discharges. *J. Vac. Sci. Technol., A* 16: 278–289.
44 Matsumoto, K., Koshi, M., Okawa, K., and Matsui, H. (1996). Mechanism and product branching ratios of the $SiH_3 + SiH_3$ reaction. *J. Phys. Chem.* 100: 8796–8801.
45 Nakamura, S. and Koshi, M. (2006). Elementary processes in silicon hot wire CVD. *Thin Solid Films* 501: 26–30.
46 Röhrig, M. and Wagner, H.G. (1994). The reactions of NH(X $^3\Sigma^-$) with the water gas components CO_2, H_2O, and H_2. *Symp. (Int.) Combust.* 25: 975–981.
47 Caridade, P.J.S.B., Rodrigues, S.P.J., Sousa, F., and Varandas, A.J.C. (2005). Unimolecular and bimolecular calculations for HN_2. *J. Phys. Chem. A* 109: 2356–2363.
48 Whyte, A.R. and Phillips, L.F. (1984). Products of reaction of nitrogen atoms with amidogen. *J. Phys. Chem.* 88: 5670–5673.

49 Childs, M.A., Menningen, K.L., Anderson, L.W., and Lawler, J.E. (1996). Atomic and radical densities in a hot filament diamond deposition system. *J. Chem. Phys.* 104: 9111–9119.

50 Wahl, E.H., Owano, T.G., Kruger, C.H. et al. (1997). Spatially resolved measurements of absolute CH_3 concentration in a hot-filament reactor. *Diamond Relat. Mater.* 6: 476–480.

51 Sutherland, J.W., Su, M.-C., and Michael, J.V. (2001). Rate constants for H + CH_4, CH_3 + H_2, and CH_4 dissociation at high temperature. *Int. J. Chem. Kinet.* 33: 669–684.

52 Bryukov, M.G., Slagle, I.R., and Knyazev, V.D. (2001). Kinetics of reactions of H atoms with methane and chlorinated methanes. *J. Phys. Chem. A* 105: 3107–3122.

53 Menningen, K.L., Childs, M.A., Chevako, P. et al. (1993). Methyl radical production in a hot filament CVD system. *Chem. Phys. Lett.* 204: 573–577.

54 Umemoto, H., Kanemitsu, T., and Kuroda, Y. (2014). Catalytic decomposition of phosphorus compounds to produce phosphorus atoms. *Jpn. J. Appl. Phys.* 53: 05FM02/1–05FM02/4.

55 Comerford, D.W., Cheesman, A., Carpenter, T.P.F. et al. (2006). Experimental and modeling studies of B atom number density distributions in hot filament activated B_2H_6/H_2 and $B_2H_6/CH_4/H_2$ gas mixtures. *J. Phys. Chem. A* 110: 2868–2875.

56 Umemoto, H., Kanemitsu, T., and Tanaka, A. (2014). Production of B atoms and BH radicals from B_2H_6/He/H_2 mixtures activated on heated W wires. *J. Phys. Chem. A* 118: 5156–5163.

57 Umemoto, H. and Miyata, A. (2015). Decomposition processes of diborane and borazane (ammonia-borane complex) on hot wire surfaces. *Thin Solid Films* 595: 231–234.

58 Miller, D.C., O'Brien, J.J., and Atkinson, G.H. (1989). *In situ* detection of BH_2 and atomic boron by intracavity laser spectroscopy in the plasma dissociation of gaseous B_2H_6. *J. Appl. Phys.* 65: 2645–2651.

59 Lavrov, B.P., Osiac, M., Pipa, A.V., and Röpcke, J. (2003). On the spectroscopic detection of neutral species in a low-pressure plasma containing boron and hydrogen. *Plasma Sources Sci. Technol.* 12: 576–589.

60 Umemoto, H., Miyata, A., and Nojima, T. (2015). Decomposition processes of H_3NBH_3 (borazane), $(BH)_3(NH)_3$ (borazine), and $B(CH_3)_3$ (trimethylboron) on heated W wire surfaces. *Chem. Phys. Lett.* 639: 7–10.

61 Umemoto, H. and Miyata, A. (2016). A clean source of B atoms without using explosive boron compounds. *Bull. Chem. Soc. Jpn.* 89: 899–901.

62 Umemoto, H. and Miyata, A. (2017). Hot metal wires as sinks and sources of B atoms. *Thin Solid Films* 635: 78–81.

63 Toukabri, R., Alkadhi, N., and Shi, Y.J. (2013). Formation of methyl radicals from decomposition of methyl-substituted silanes over tungsten and tantalum filament surfaces. *J. Phys. Chem. A* 117: 7697–7704.

64 Shi, Y. (2015). Hot wire chemical vapor deposition chemistry in the gas phase and on the catalyst surface with organosilicon compounds. *Acc. Chem. Res.* 48: 163–173.

65 Zaharias, G.A., Duan, H.L., and Bent, S.F. (2006). Detecting free radicals during the hot wire chemical vapor deposition of amorphous silicon carbide films using single-source precursors. *J. Vac. Sci. Technol., A* 24: 542–549.
66 Harada, T., Nakanishi, H., Ogata, T. et al. (2011). Evaluation of corrosion resistance of SiCN-coated metals deposited on an NH_3-radical-treated substrate. *Thin Solid Films* 519: 4487–4490.
67 Morimoto, T., Ansari, S.G., Yoneyama, K. et al. (2006). Mass-spectrometric studies of catalytic chemical vapor deposition processes of organic silicon compounds containing nitrogen. *Jpn. J. Appl. Phys.* 45: 961–966.
68 Molenbroek, E.C., Mahan, A.H., Johnson, E.J., and Gallagher, A.C. (1996). Film quality in relation to deposition conditions of *a*-Si:H films deposited by the "hot wire" method using highly diluted silane. *J. Appl. Phys.* 79: 7278–7292.
69 Nakajima, Y., Tonokura, K., Sugimoto, K., and Koshi, M. (2001). Kinetics of Si_2H_2 produced by the 193 nm photolysis of disilane. *Int. J. Chem. Kinet.* 33: 136–141.
70 Koshi, M., Miyoshi, A., and Matsui, H. (1991). Rate constant and mechanism of the SiH_3 + SiH_3 reaction. *Chem. Phys. Lett.* 184: 442–447.

5

Properties of Inorganic Films Prepared by Cat-CVD

In the previous chapter, the reaction process of source gases is explained in detail. The features of catalytic chemical vapor deposition (Cat-CVD) films are all determined by the reaction process of the source gases, as briefly mentioned in the final section of the previous chapter. Here, the properties of individual films prepared by Cat-CVD are mentioned in detail. The reader will understand that the features of Cat-CVD films are strongly affected by the reaction processes.

5.1 Properties of Amorphous Silicon (a-Si) Prepared by Cat-CVD

5.1.1 Fundamentals of Amorphous Silicon (a-Si)

5.1.1.1 Birth of Device Quality Amorphous Silicon (a-Si)

Amorphous silicon (a-Si) is a relatively new material used in semiconductor industries. As already mentioned in Chapter 2, the first a-Si prepared by plasma deposition method was obtained in 1969 by Chittick et al. [1]. Later, in 1975, Spear and LeComber discovered that the valency control of plasma-deposited a-Si is possible by phosphorus (P) and boron (B) doping [2]. They discovered that the plasma deposited a-Si, i.e. plasma enhanced chemical vapor deposition (PECVD) a-Si, might be used in electronic devices by making n-type or p-type a-Si, similar to crystalline silicon (c-Si) p–n junction devices. Since then, many researchers have been developing this new frontier of material science and device application. Since the time was just after the oil crisis of 1973, primarily solar cell applications collected strong attention, as the a-Si thin film solar cells could be made by simple processes with low cost.

The a-Si thin film solar cells were put in market in 1980 as an energy source of pocket calculators and watches. Initially, a-Si solar cells were a major player as low-cost solar cells, until the cost of c-Si solar cells enormously dropped in the 2000s. However, a-Si is still used in c-Si solar cells as a key material of a-Si/c-Si heterojunction solar cells [3, 4]. These heterojunction solar cells are expected to realize top-class energy conversion efficiency within a c-Si-based solar cell technology.

Just after the research movement of a-Si solar cells, in 1979, the usefulness of application to thin film transistors (TFTs) that would be used in the pixel control

Catalytic Chemical Vapor Deposition: Technology and Applications of Cat-CVD,
First Edition. Hideki Matsumura, Hironobu Umemoto, Karen K. Gleason, and Ruud E.I. Schropp.

of liquid crystal display (LCD) was claimed by Le Comber and Spear again [5]. The a-Si TFT is suitable for controlling the brightness of all pixels individually, which construct the pictures of flat panels. This kindled the development of the new industrial area of the LCD industry. If you watch a flat panel display television or the screen of a smartphone, you may appreciate the fruit of much research in this area. Here, we start with an explanation of the quality of a-Si made by Cat-CVD as a useful application of Cat-CVD technology.

5.1.1.2 Band Structure of Amorphous Materials

As explained above, a-Si is one of the most important materials. However, it is quite different from c-Si, which is the primary material used in semiconductor industries. The fundamental physics such as that of the band structure is different. Thus, we start this section with the explanation of the fundamentals of amorphous materials.

Contrary to crystalline solids, in amorphous materials, the periodic atomic arrays are not formed, although it is known that the short-range order, which is defined as a structure with atomic configuration similar to that of a crystalline material over a short distance, can be seen even in a-Si [6, 7]. As the a-Si thin films with a quality suitable for electronic devices are easily obtained at temperatures lower than 300 °C by PECVD and Cat-CVD, the usefulness of a-Si in modern industry is widely recognized, as mentioned above.

At first, we explain briefly the energy band structure of amorphous materials. Figure 5.1 shows a schematic illustration of the energy levels of electron orbitals when an isolated atom forms a molecule with another identical isolated atom. When two identical atoms are at a distance within which the orbitals of the independent atoms begin to overlap, each orbital of the two atoms is modified in order to facilitate the two atoms to make a molecule. In this case, the energy level of the new orbital splits into two levels. The upper level is called the antibonding

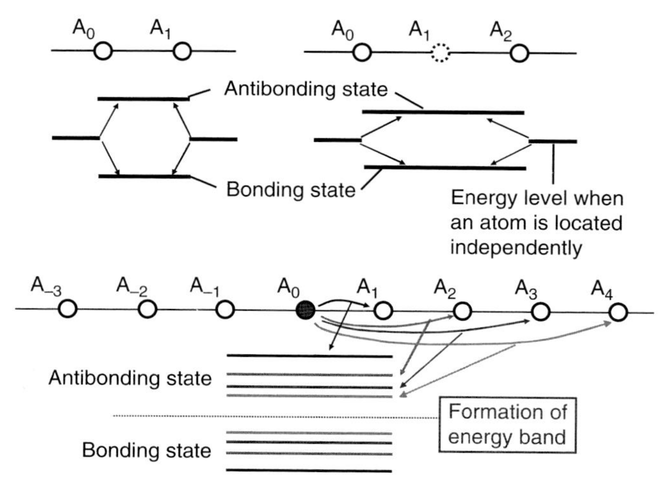

Figure 5.1 Construction of a molecule from two identical atoms and the energy levels of a new molecular orbital. When many atoms are assembled, such energy levels are converted to an energy band.

state and the lower level is the bonding state. The gap between antibonding and bonding states is likely to narrow as the distance of two atoms increases.

Here, let us consider a solid consisting of many identical atoms and let us also focus on a single atom A_0. The energy gap between the antibonding and bonding states for the bond of atom A_0 with its nearest-neighboring atom A_1 is the largest because the distance of the two atoms is the shortest. The energy gap between the antibonding and bonding states for the bond of atom A_0 with the second nearest atom A_2 is smaller than the energy gap for the bond of atom A_0 with atom A_1. Thus, if we consider the bonds of atom A_0 with many other atoms, many energy levels are created in the energy gap between antibonding and bonding states of A_0–A_1 bond, and the energy gap is filled with these various energy levels, thus forming an energy band.

This is the story for an outer orbital of an atom. If two atoms are approaching nearer, an inner orbital begins to overlap with the inner orbital of other atoms. Thus, an energy band is also formed from the inner orbitals. This situation is illustrated in Figure 5.2. At the distance of the nearest-neighbor atoms a_0 in the figure, the energy gap in which electrons cannot exist occurs at energies between the bottom of the energy band formed by the outer orbitals and the top of the band formed by the inner orbitals. In this way, the energy band structure in solids is constructed. The energy gap between two energy bands is essentially formed because of the existence of discrete energy levels in each atom, not because of the periodic arrays of the atoms. The periodicity of an atomic array is not a substantial requirement for the formation of a band structure. Here, if the Fermi level is located within the band gap as shown in Figure 5.2, the upper band becomes the conduction band and the lower band becomes the valence band.

When we consider the materials with a diamond crystalline structure such as diamond carbon (C), Si, and germanium (Ge), the above explanation can be

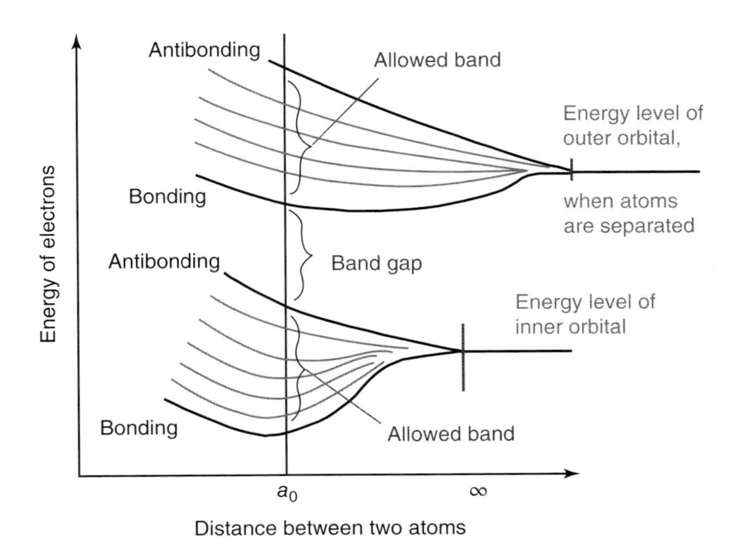

Figure 5.2 Birth of band structure and band gap.

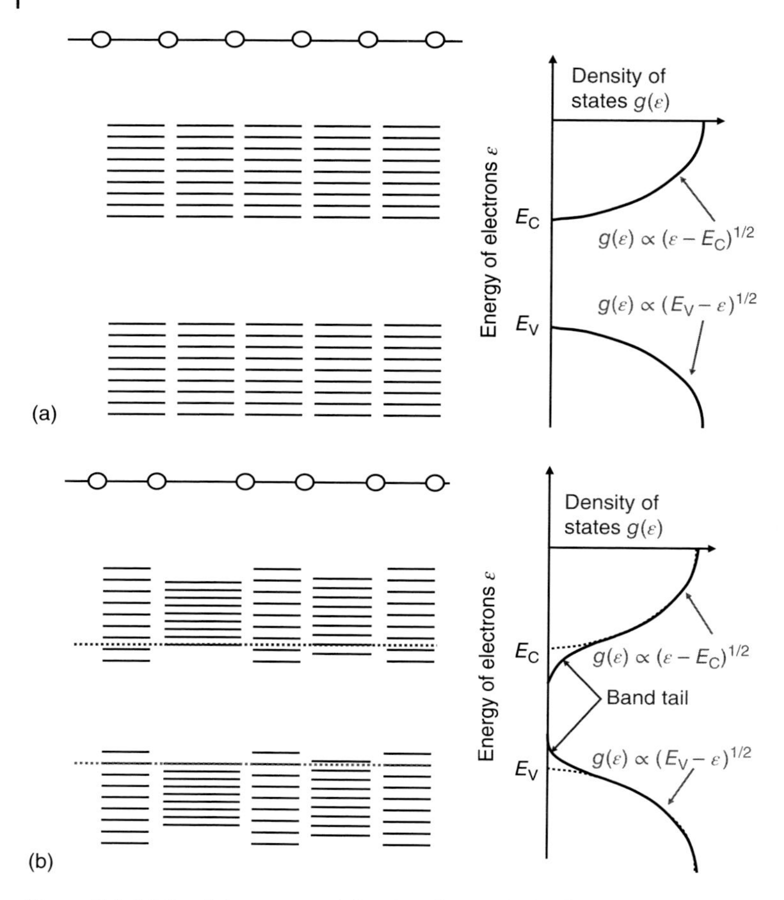

Figure 5.3 (a) Band structure and density of states (DOS) for a solid with periodic atomic array and (b) band structure and DOS for amorphous solid.

further detailed. When two atoms approach within a certain distance, the outer orbital, the p-orbital, and the inner orbital, the s-orbital, form a new hybrid orbital, called a sp^3 hybrid orbital. That is, the approach of two atoms not only gives rise to a band structure but also forms new orbitals. However, the concept of forming a band structure can still be applied in this case.

Here, if the atoms are arranged periodically, the energy band at any location should be the same. Thus, all energy levels inside the energy band should be the same, too. Thus, electrons can travel freely within such a band. In this case, it is known that the density of states (DOS) for electrons in the solid is proportional to the square root of the electron energy. This is demonstrated in Figure 5.3a. Here, the DOS is defined as the number of electronic states per unit volume and per unit energy range. When $g(\varepsilon)\mathrm{d}\varepsilon$ is defined as the DOS in an energy range between ε and $\varepsilon + \mathrm{d}\varepsilon$, it is known that $g(\varepsilon)$ is proportional to $(\varepsilon)^{1/2}$ when electrons can move freely [8].

If atoms are not periodically arranged, the width of the energy band fluctuates at each atomic position, depending on the nearest-neighbor distance as shown in Figure 5.3b. The major part of the band is still the same as that of the periodic structure. However, the bottom and the top of the energy bands fluctuate. The DOS for this case is also drawn in the figure. The fluctuating parts at the bottom or top of the bands are called "band tails." The shape of the band tail cannot be defined by a straightforward mathematical expression. From Figure 5.3, it can be imagined that a smaller deviation from a periodic structure causes a smaller fluctuation of the band edges, that is, less extensive band tails.

Of course, the present explanation is too much simplified to explain all the phenomena. Above we used the fluctuation in atomic distances in a linear array of bonded atoms. However, in real a-Si, three-dimensional structures should be considered. It is known that the variation of bond length is not the largest relative fluctuation. The fluctuation of bond angles is larger and is more often considered as the dominant origin of nonperiodic structures [6]. Even so, a different overlapping of orbitals induces the band structures as explained above, and the concept is still adoptable.

5.1.1.3 General Properties of a-Si

a-Si films prepared by chemical vapor deposition (CVD) contain many hydrogen (H) atoms, and these H atoms play an important role in terminating the Si dangling bonds that are induced in a-Si during film growth. The presently usable a-Si is mainly hydrogenated a-Si (a-Si:H), although a-Si films, in which other atoms such as fluorine (F) terminate Si dangling bonds, have been reported [9–11]. The fluorinated a-Si (a-Si:F) or hydrogenated and fluorinated a-Si (a-Si:F:H) have collected much attention because of its stable properties [11]. However, they have not become major industrial materials. The a-Si:H is the most prominent amorphous semiconductor in the industry.

The main properties of a-Si are summarized in Table 5.1, although some of them are dependent on specific deposition methods and parameters. In the table, the typical properties of a-Si prepared by PECVD and Cat-CVD are described, to show a rough outline. Detailed data will be mentioned later.

In crystalline semiconductors, there are two types of light absorption or optical transition modes of electrons: direct and indirect transitions [12]. In the direct transition, electrons transit directly from the valence band to the conduction band. However, in the indirect transition, electrons can only transit to the bottom of the conduction band from the top of the valence band when they receive the required momentum from the thermal vibration of crystalline atomic arrays because the conservation of both energy and momentum has to be satisfied during the transition process of electrons.

In other words, in the direct transition, the momentum of the top of the valence band is the same as that of the bottom of the conduction band in the energy (ε)–the momentum (k) space of crystalline semiconductors [12]. However, in the indirect transition, the momentum of the bottom of the conduction band is different from that of the valence band in the ε–k space. This difference arises from the crystalline structure in which the thermal vibration of the atomic arrays has a special effect on the electron transition, and the ε–k space is formed when the atoms in the crystalline semiconductor make periodic arrays. This rule of electron transition in the crystalline semiconductors with periodic atomic arrays is sometimes called as the "k-selection rule."

Table 5.1 Typical properties of hydrogenated amorphous silicon (a-Si:H) prepared by PECVD and Cat-CVD using SiH_4 source gas.

	Preparation method of i-a-Si	
	PECVD	Cat-CVD
T_s for obtaining device quality films (°C)	100–300	100–400
Typical optical band gap (eV)	1.70–1.85	1.65–1.80
Dark conductivity (S/cm)	10^{-12} to 10^{-10}	10^{-13} to 10^{-10}
Photoconductivity in AM-1, 100 mW/cm^2 light[a)] (S/cm)	10^{-5} to 10^{-4}	10^{-5} to 10^{-4}
Photosensitivity in AM-1, 100 mW/cm^2 light[a)] = Photoconductivity/dark conductivity	10^5 to 10^6	10^5 to 10^6
Mobility of electrons (cm^2/Vs)	0.5–1.5	0.5–1.5
ESR spin density[b)] (cm^{-3})	1×10^{15} to 2×10^{16}	1×10^{15} to 2×10^{16}
Hydrogen (H) content in device quality a-Si (at.%)	10–20	1–10
Characteristic energy of Urbach tail, E_u (meV)	50–60	45–55

a) The spectrum of AM-1 (air mass 1), 100 mW/cm^2, corresponds to sun light with normal incidence at the equator.
b) Spin density measured by electron spin resonance (ESR) experiments, which is equivalent to the defect density due to Si dangling bonds.

It is known that c-Si is an indirect gap semiconductor. However, as the k-selection rule for optical transitions cannot hold in a-Si because of the lack of periodicity, a-Si behaves like a direct gap semiconductor. For this reason, although a thickness of more than 100 μm is required for the absorption of all sunlight in c-Si, only a few micrometer thick a-Si is enough to absorb most of the sunlight. This raised the expectation that a-Si solar cells could save material costs with respect to c-Si, although a-Si solar cells are not mainstream cells because of the lower energy conversion efficiency compared to c-Si solar cells at the moment.

As the band gap of a-Si is much wider than that of c-Si, the conductivity of intrinsic a-Si (i-a-Si) in the dark is usually much lower than that of c-Si. For instance, the optical band gap of a-Si is 1.65–1.85 eV dependent on deposition conditions, whereas it is 1.12 eV in c-Si at 300 K. Thus, the conductivity of i-a-Si in the dark is 10^{-13} to 10^{-10} S/cm, whereas that of intrinsic c-Si is 10^{-6} to 10^3 S/cm as realistic values because of both the narrower band gap and the presence of residual impurities in c-Si. Although the conductivity of i-a-Si is very low, as it has photoconductive property, the actual conductivity of a-Si used in solar cells is high during light exposure. The photosensitivity, defined as the ratio of photoconductivity σ_p to dark conductivity σ_d, is often used as a measure of a-Si quality. A large photoconductivity is desired as this implies that the photoexcited carriers can travel without being trapped by defects and easily reach the electrodes and thus that the defect density of a-Si can be expected to be low.

As will be mentioned later, the hydrogen (H) content in a-Si, C_H, in a Cat-CVD a-Si film is usually less than that of PECVD a-Si. In Cat-CVD, as already explained in the previous chapter, as the atomic H density in the gas phase in the chamber is higher than that in PECVD, H atoms act to pull out other H atoms from the growing surface. Thus, the density of residual H atoms in Cat-CVD a-Si films becomes smaller than that in PECVD films.

Another optical property often measured is the Urbach tail energy E_u, as this reflects an important property of the band tail. Thus, here, we briefly explain the optical properties of amorphous materials, based on the knowledge of the band structure mentioned above.

As already mentioned, the DOS in the conduction band $g_C(\varepsilon')$ is expressed by $g_C(\varepsilon') = A_C\ (\varepsilon')^{1/2}$, where the energy ε' is measured from the energy at the bottom of the conduction band E_C. Similarly, the DOS in the valence band $g_V(\varepsilon'')$ is expressed by $g_V(\varepsilon'') = A_V(\varepsilon'')^{1/2}$, where the energy ε'' is measured toward lower direction from the energy of the top of the valence band E_V. On the other hand, as the DOS in the conduction band tail $g_{C\ tail}(\varepsilon''')$ is not known, we assume that it can be expressed by $g_{C\ tail}(\varepsilon''') = A_{C\ tail}\ \exp(-\varepsilon'''/E'_u)$, where the energy ε''' is measured toward lower direction from E_C. In the above explanation, A_C, $A_{C\,tail}$, and A_V all express proportionality constants, and E'_u refers to the characteristic energy showing the band tail steepness.

Figure 5.4 shows a schematic illustration of electron transitions from the energy ε at the valence band to the conduction band by receiving the photon energy $h\nu_2$ and also to the conduction tail states by receiving the photon energy $h\nu_1$. When we consider the transition of an electron, due to a photon of the energy $h\nu$, from an energy level in the valence band to a level in the conduction band, where ν refers to the frequency of the photon or exciting light, the optical absorption constant α has the relation shown in Eq. (5.1). The derivation of Eq. (5.1) is too complicated to explain in this book, and if you like to know the derivation, please read a relevant book, such as Ref. [6].

$$\alpha(h\nu) = A_1 \int \frac{g_V(\varepsilon)g_C(\varepsilon + h\nu)}{h\nu}\, d\varepsilon \tag{5.1}$$

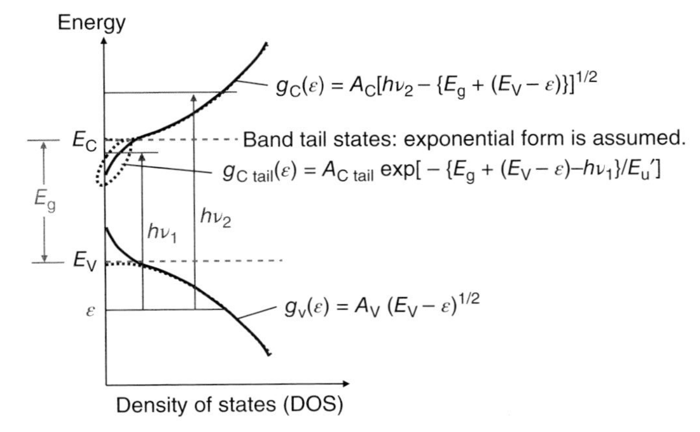

Figure 5.4 Schematic illustration of electron transition from the energy ε at the valence band to the conduction band with the photon energy of $h\nu_2$ or to the conduction band tail with $h\nu_1$.

When we consider the transition from the energy ε at the valence band to the conduction band by the photon energy $h\nu_2$, as $g_C(\varepsilon)$ and $g_V(\varepsilon)$ are expressed by

$$g_C(\varepsilon) = A_C[h\nu_2 - \{E_g + (E_V - \varepsilon)\}]^{1/2}$$
$$g_V(\varepsilon) = A_V(E_V - \varepsilon)^{1/2}$$

the following equations hold:

$$\alpha(h\nu_2) = A_2 \int_0^{E_V} \frac{[(h\nu_2 - E_g) - (E_V - \varepsilon)\}]^{1/2}(E_V - \varepsilon)^{1/2}}{h\nu_2} d\varepsilon$$
$$= \frac{A_2(h\nu_2 - E_g)}{h\nu_2} \int_0^{E_V} \left(1 - \frac{E_V - \varepsilon}{h\nu_2 - E_g}\right)^{1/2} \left(\frac{E_V - \varepsilon}{h\nu_2 - E_g}\right)^{1/2} d\varepsilon \quad (5.2)$$

$y \equiv \frac{E_V - \varepsilon}{h\nu_2 - E_g}$ and then, $d\varepsilon = -(h\nu_2 - E_g)dy$

$$\alpha(h\nu_2) = \frac{A_2(h\nu_2 - E_g)^2}{h\nu_2} \int_{E_V/(h\nu_2 - E_g)}^{0} (1 - y)^{1/2} y^{1/2}\, dy = A_3 \frac{(h\nu_2 - E_g)^2}{h\nu_2} \quad (5.3)$$

$$\sqrt{\alpha(h\nu_2)h\nu_2} = B(h\nu_2 - E_g) \quad (5.4)$$

where A_1, A_2, A_3, and B are constants. The relation described by Eq. (5.4) is called the Tauc relation.

Figure 5.5 demonstrates the Tauc plots for Cat-CVD a-Si [13] and PECVD a-Si [14]. The electron transition for the Tauc relation is illustrated in the inset of

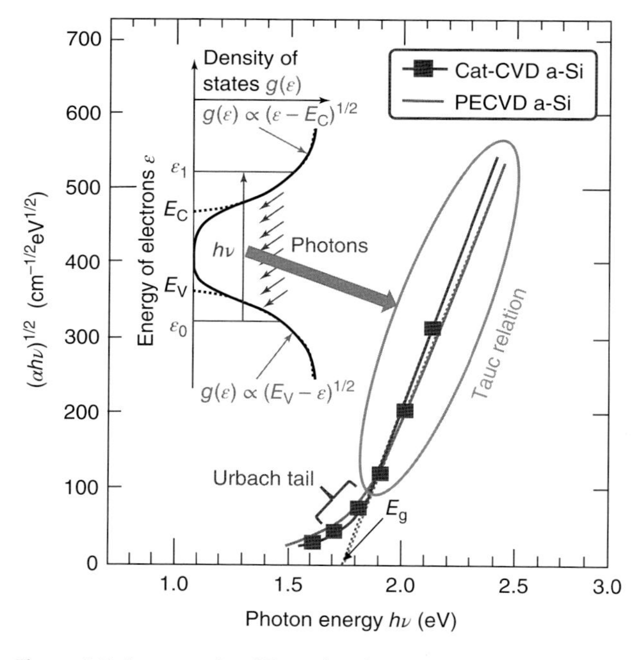

Figure 5.5 An example of Tauc plots for Cat-CVD a-Si and PECVD a-Si. In the inset, the transition of electrons causing Tauc relation also explained schematically. Source: Data from Matsumura 1985 [13] and Tsai and Fritzsche 1979 [14].

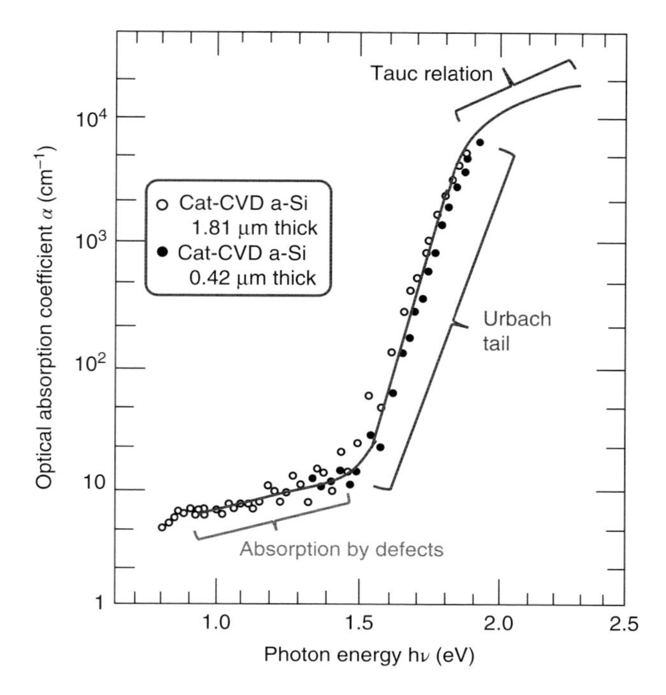

Figure 5.6 Optical absorption constant a as a function of photon energy $h\nu$ for Cat-CVD a-Si. Source: Data from Yamaguchi et al. 1990 [15].

the figure. The thickness of Cat-CVD a-Si is about 1 μm and that for PECVD is roughly the same. From the figure, the band gap E_g is estimated from the intercept of the Tauc linear approximations with the horizontal axis, which is 1.75 eV for both cases.

For the energy range lower than the range for which the Tauc relation applies, the electron transitions begin to include those involving band tail states. If electrons in the valence band are excited to the band tail of the conduction band or if electrons at the band tail of the valence band are excited to the conduction band, the Tauc relation does not hold any more. The optical absorption in this energy range just below the energy range of the Tauc relation is called the Urbach tail.

If we plot the logarithm of the optical absorption coefficient α as a function of photon energy $h\nu$, it is found that there is a linear relation with the photon energy as shown in Figure 5.6. In the figure, the results of two Cat-CVD a-Si films are demonstrated as examples [15]. The optical absorption is measured by the photodeflection spectroscopy (PDS), which is a powerful tool to measure the small optical absorption coefficients down to 1–10 cm^{-1} [16]. Two Cat-CVD a-Si films are prepared at $T_s = 250\,°C$, but the thickness of films is 1.81 and 0.42 μm. The linear relation is seen in the figure in the range from 1.55 to 1.75 eV. A similar relation is observed for any amorphous material. Because of the linear relation of the Urbach tail in a logarithmic plot, α can be expressed as a function proportional to $\exp(h\nu/E_u)$. This E_u is called as the characteristic energy of the Urbach tail, as shown earlier in Table 5.1. E_u is estimated to be about 50 meV in this case.

When we consider the transition from the valence band to the conduction band tail, as, as mentioned above, the following relation is assumed:

$$g_{\mathrm{C\,tail}}(\varepsilon''') = A_{\mathrm{C\,tail}}\ \exp(-\varepsilon'''/E_{\mathrm{u}}')$$

the following Eq. (5.5) can be derived, where E_{u}' is a constant characteristic energy expressing the shape of the band tail,

$$\alpha(h\nu_1) = A_4 \int_0^{E_{\mathrm{V}}} \frac{\exp[-\{E_{\mathrm{g}} + (E_{\mathrm{V}} - \varepsilon) - h\nu_1\}/E_{\mathrm{u}}'](E_{\mathrm{V}} - \varepsilon)^{1/2}}{h\nu_1} \mathrm{d}\varepsilon \tag{5.5}$$

where A_4 is a constant.

And then, as $h\nu_1$ is constant for ε variation, the following equation is derived, where A_5 is a constant.

$$\alpha(h\nu_1) = \frac{A_5}{h\nu_1} \exp\{-(E_{\mathrm{g}} - h\nu_1)/E_{\mathrm{u}}'\} \tag{5.6}$$

Compared with the exponential component in the equation, the effect of the term $(1/h\nu_1)$ is small, and therefore, this equation demonstrates that the α is approximately proportional to $\exp\{-(E_{\mathrm{g}}-h\nu_1)/E_{\mathrm{u}}'\}$. That is, the characteristic energy of Urbach tail E_{u}, derived from the slope of Figure 5.6, is exactly the same as the characteristic energy of band tail E_{u}'. This also means that the DOS at the band tail has an exponential form with the characteristic energy derived from the experiment of optical absorption. The degree of order in the structure of a-Si can be derived from the data of the Urbach tail measurement.

In this explanation, the transition from the valence band to the conduction band tail is treated. However, the transition from the valence band tail to the conduction band also shows the same characteristics. In actual a-Si, it is believed that the major origin of Urbach tail is due to the transition from the valence band tail to the conduction band, as the band tail of the valence band is larger than that of the conduction band.

It is reported that the width of the valence band of a-Si is about 12 eV and that of the conduction band is about 7 eV [17]. The characteristic energy of Urbach tail of Cat-CVD a-Si is about 50 meV or less as shown in Table 5.1 and Figure 5.6. Thus, the fluctuation of band edge at the valence band in Cat-CVD a-Si is only about 50 meV/12 eV = 0.004 of the total band width.

Although the experimentally obtained information is mainly concerned with the valence band tail, we can speculate that the similar order of band fluctuation exists in the conduction band. Since most of the electrons move at the bottom of the conduction energy band, the mobility is strongly affected by this band fluctuation. The electron mobility in c-Si with a low doping concentration is about 1500 $\mathrm{cm^2Vs}$; however, the mobility is about 1 $\mathrm{cm^2/Vs}$ for the best a-Si and only 0.1–0.5 $\mathrm{cm^2/Vs}$ in slightly lower grade a-Si.

As shown in Table 5.1 and briefly explained already, the photoconductivity is also affected by this band tail in addition to the existence of the gap states due to defects. The photoconductivity σ_{p} is proportional to $\mu\Delta n$, where μ and Δn refer to the electron mobility and the steady-state density of photogenerated electrons during photoexcitation, respectively. The mobility is influenced by the magnitude of the band fluctuation and Δn is related to the density of the defects in the band gap. If the defect density is small enough, the photoexcited electrons can survive longer and Δn becomes larger. A photosensitivity σ_p/σ_d for Cat-CVD a-Si is 10^5 to 10^6, which is good enough as shown in Figure 5.13, later. This also demonstrates again that the density of defect energy levels in the band gap and the band tail are both sufficiently small for Cat-CVD a-Si films and that Cat-CVD a-Si appears to have a simple and more ordered structure.

5.1.2 Fundamentals of Preparation of a-Si by Cat-CVD

5.1.2.1 Deposition Parameters

Regarding the preparation of a-Si films by Cat-CVD, there are some key parameters. In Cat-CVD, selection of materials of catalyzing metals is the first key issue. In case of a-Si deposition, as mentioned in the previous chapter, desorption of Si atoms requires high catalyzer temperatures over 1700 °C in order to avoid silicide formation. Thus, usually, metals with high-melting point and low vapor pressure are required. Low vapor pressure is required to avoid contamination from catalyzing metals. As realistic candidates, usually tungsten (W), tantalum (Ta), and their alloys, are used because the melting points of W and Ta are 3382–3422 °C and 2985–3017 °C, respectively, and the vapor pressure at 1900 °C is about 2×10^{-8} Pa for W and about 3×10^{-7} Pa for Ta [18].

The second issue is a proper setting of the temperature of catalyzer, T_{cat}. The T_{cat} is decided from the compromise between avoiding silicide formation and avoiding metal contamination from the catalyzing wire. The surface area of catalyzing metal wires, S_{cat}, is another issue concerned with the catalyzer. The S_{cat} is determined by the total length of the catalyzing metal wire and the diameter of the wire. The value is also a compromise between the suppression of thermal radiation and deposition rates along with the desired uniformity of films. The effect of thermal radiation, contamination from the catalyzer, and film uniformity are all discussed in Chapter 7.

The distance between the catalyzer and the substrate, D_{cs}, is a factor that decides the deposition rate and the thermal radiation from the catalyzer. In many experiments, D_{cs} is shorter than 10 cm; however, in practical machines used for large area mass production, it is longer than 15 cm.

The gas pressure, P_g, and flow rate of silane (SiH_4) as a source gas, denoted by FR(SiH_4), are keys to decide the deposition rates of a-Si films. In addition, the substrate temperatures, T_s, are a parameter that decides the film quality.

Typical data, concerned with these parameters, are summarized in Table 5.2, based on various reports [19–23]. The data of our JAIST group are almost the same as those shown in Table 4.1. Reported parameters shown in the table are all for a chamber with a small size, although Cat-CVD mass production machines have been in the market with a deposition area is $100 \times 100\ cm^2$ to $150 \times 170\ cm^2$. In addition, in mass production, tantalum (Ta) or Ta alloy is used instead of W catalyzers, to extend the lifetime of the catalyzer. The lifetime of catalyzing wires will be discussed in detail in Chapter 7.

5.1.2.2 Structural Studies on Cat-CVD a-Si: Infrared Absorption

Here, we explain the measurement of infrared (IR) absorption spectra because it is a simple method to resolve the structural properties of Cat-CVD a-Si films.

Table 5.2 Deposition parameters for a-Si by Cat-CVD reported in various references.

Parameters	Setting values of parameters			
	A.H. Mahan et al. [19]	R.E.I. Schropp [20]	M. Heinze et al. [21]	Our group [22, 23]
Material of catalyzer	W	Ta	W	W
Deposition area (cm^2)	<5 × 5[a)]	5 × 5	<5 × 5[a)]	24 × 29
Catalyzer-substrate distance, D_{cs} (cm)	<5[a)]	5	1–4	4–20
Surface area of catalyzing, S_{cat} (cm^2)	About 2[a)]	2.5–3	About 0.3	44—50
Temperature of catalyzer, T_{cat} (°C)	1900	1700	1650	1750–1800
Temperature of substrate, T_s (°C)	40–600	250	400	270–300
Gas pressure, P_g (Pa)	0.13–1.3	2	1–5	1.1
Flow rate of SiH_4, FR(SiH_4) (sccm)	20	90		10–50
Flow rate of H_2, FR(H_2) (sccm)	0	0		0–100
Typical deposition rate, D_R (nm/s)	1	1	1–5	1—3
D_R of PECVD (comparison) (nm/s)	0.15–0.25	0.1–0.2		

a) Data are not clearly mentioned in their reports. Values written here are estimated ones from other parameters described in the reports from these groups.

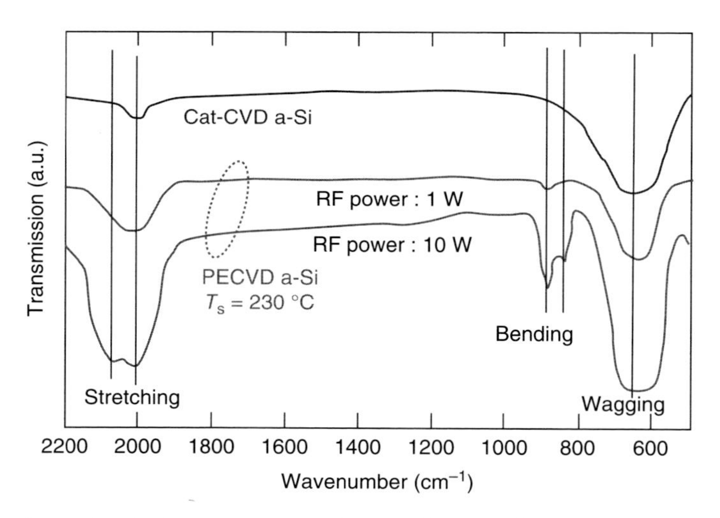

Figure 5.7 IR absorption spectra for Cat-CVD a-Si and PECVD a-Si films. Source: Data from Lucovsky et al. 1979 [24].

There are a number of typical IR absorption peaks that express what types of chemical bonds exist in the films.

Figure 5.7 shows the IR absorption spectra for one Cat-CVD a-Si film and two PECVD a-Si films. The Cat-CVD a-Si film is prepared at T_{cat} = 1700 °C, T_s = 300 °C, and P_g = 1.3 Pa, whereas PECVD a-Si films are prepared by radio frequency (RF) power of 1 and 10 W at T_s = 230 °C [24]. In the samples made with

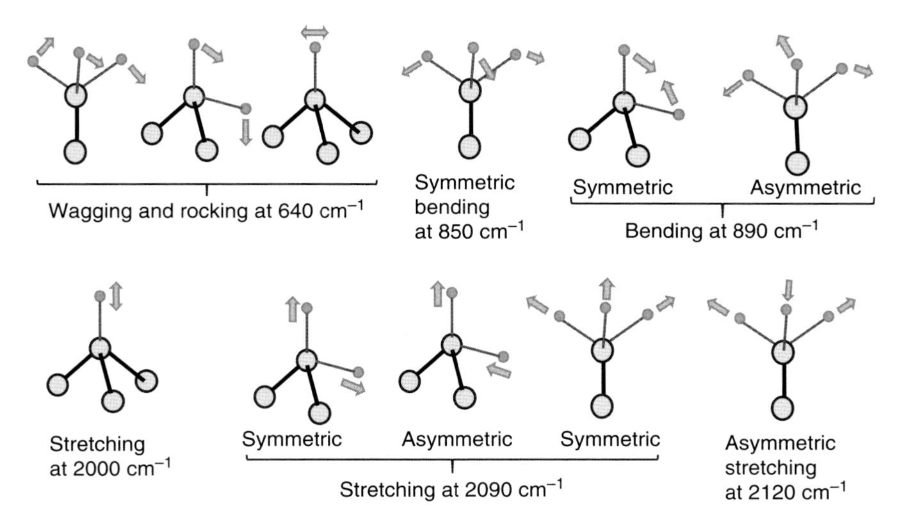

Figure 5.8 Schematic illustration of each vibration mode of Si—H-related bonds. Large spheres denote Si atoms and small-spheres H atoms.

higher RF power, several additional absorption peaks can be observed, whereas the number absorption peaks seen in Cat-CVD a-Si appears limited.

M.H. Brodsky et al. assigned typical IR absorption peaks to silicon–hydrogen bonding configurations in PECVD a-Si films [25]. The wagging and rocking vibration peaks for Si—H, $Si—H_2$, and $Si—H_3$ bonds are observed at about 640 cm^{-1}; the symmetric bending vibration of $Si—H_3$ bonds at 850 cm^{-1}; the symmetric bending vibration of $Si—H_2$ bonds and the asymmetric bending vibration of $Si—H_3$ bonds both at 890 cm^{-1}; the stretching vibration of Si—H bonds is at 2000 cm^{-1}; the stretching vibration of Si—H bonds at the surfaces of voids; the asymmetric stretching vibration of $Si—H_2$ bonds, the symmetric stretching vibration of $Si—H_3$ bonds are at 2090 cm^{-1}, and the asymmetric stretching vibration of $Si—H_3$ bonds is at 2120 cm^{-1}. These vibration modes are schematically illustrated in Figure 5.8.

They also provided the evaluation method of the density of H atoms in a-Si from Si-H, N_H, from the stretching vibration at 2000 cm^{-1} and the wagging vibration at 640 cm^{-1} as described in Eq. (5.7).

$$N_H = A_s \int \frac{\alpha(h\nu)}{h\nu}\, d(h\nu) \tag{5.7}$$

Here, for stretching vibration, $A_s = 1.4 \times 10^{20}$ cm^{-2}, and for wagging vibration, $A_s = 1.6 \times 10^{19}$ cm^{-2}, although other values have been reported after their reports. For instance, $A_s = 2.1 \times 10^{19}$ cm^{-2} for the wagging vibration is often used based on the report by A.A. Langford et al. [26]. From the area of IR absorption peaks, the density of H atoms is easily evaluated.

5.1.3 General Properties of Cat-CVD a-Si

This section deals with the properties of Cat-CVD a-Si. There are many reports on the quality of a-Si films prepared by Cat-CVD using different deposition systems.

When we change the size of the deposition chamber or change the deposition system, the deposition parameters to obtain the best quality a-Si films should be adjusted. In most cases, if we keep the residence time of source gas molecules the same, we can obtain equivalent properties of a-Si films in different deposition systems. Therefore, we show the typical properties of a-Si based on various reports in this section. We confirmed that we were always able to obtain similar properties after optimizing the deposition parameters.

A.H. Mahan et al. at the National Renewable Energy Laboratory (NREL, at the time the report was issued, it was called SERI (Solar Energy Research Institute), USA, reported the properties of a-Si made by Cat-CVD in comparison to those of PECVD a-Si [19]. Their deposition parameters are summarized in Table 5.2, although some of the deposition parameters were not given explicitly in their reports, and thus, we presumed them based on private communications.

The first unique feature of Cat-CVD a-Si that stands out is the low hydrogen (H) content C_H in device quality a-Si films. For PECVD a-Si, if we reduce C_H by elevating T_s, for instance, the quality of the films is enormously degraded. However, in Cat-CVD a-Si, we found that a-Si keeps its quality, even at high T_s.

Figure 5.9 shows the reported data by R.S. Crandall et al. of NREL group for C_H, evaluated from IR spectra, of both Cat-CVD a-Si and PECVD a-Si films and by our group for Cat-CVD a-Si, as a function of substrate temperature T_s [27, 28].

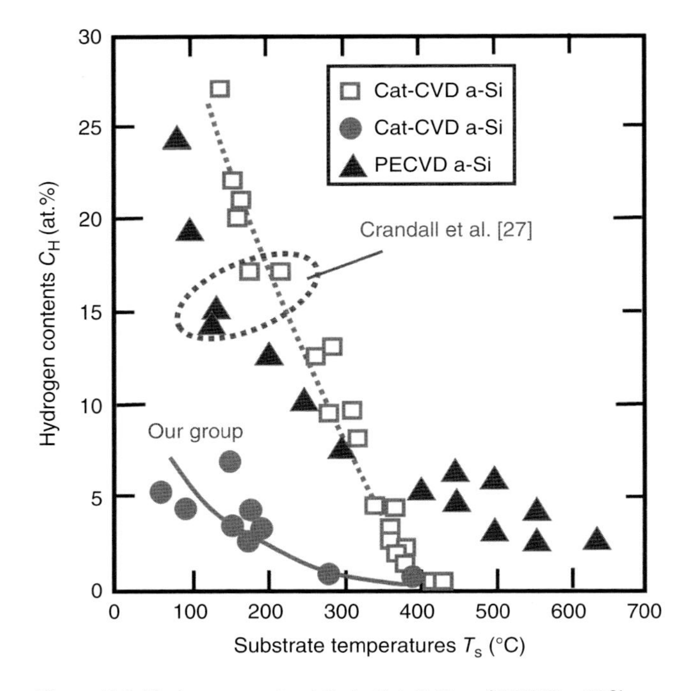

Figure 5.9 Hydrogen content C_H in Cat-CVD and PECVD a-Si films as a function of substrate temperature T_s during film formation. Source: From Crandall et al. 1992 [27] and Matsumura et al. 1999 [28].

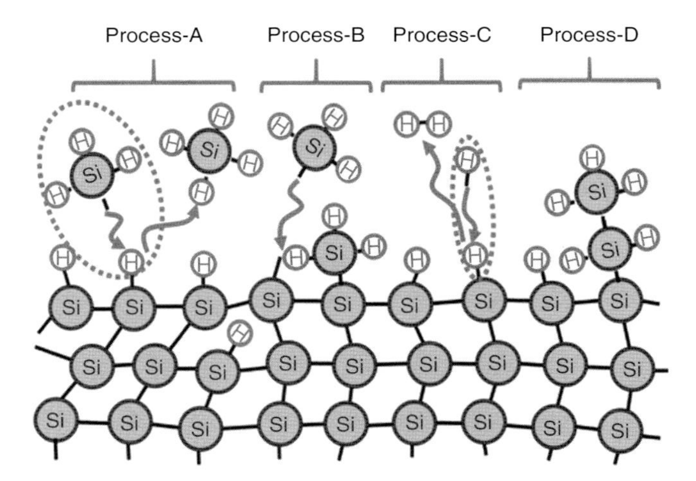

Figure 5.10 A model of the growing surface of a-Si films. Removal of H atoms covering the surface is a key for proper growth of the films.

In the case of Cat-CVD films, at about T_s from 300 to 400 °C, we obtain a-Si with C_H less than a few atomic % (at.%).

Here, the rough image of reactions happening on the growing a-Si surface is shown in Figure 5.10. It is believed that the growing surface of a-Si films is covered with H atoms [29, 30]. These H atoms are removed through the reaction of $\equiv$Si—H + SiH_3 → $\equiv$Si— + SiH_4 or $\equiv$Si—H + H → $\equiv$Si— + H_2, leaving a Si dangling bond on the surface (Processes A or C in the figure). As the next step, another SiH_3 arrives at the surface and makes a bond with such a Si dangling bond to form a $\equiv$Si—SiH_3 configuration (Process B). It is believed that the a-Si layer is constructed in this way. However, this is an idealized model, and often the removal of H atoms is not fast enough, and H atoms remain within the growing a-Si films. Sometimes $\equiv$Si—SiH_2—SiH_2—SiH_3 chains remain inside, as illustrated in Process D. If the removal of H atoms is not sufficient, many H atoms remain inside the a-Si film and the C_H becomes large.

As already explained in the previous chapter, the amount of H atoms that exist in the gas phase in a Cat-CVD chamber is usually much larger than that in a PECVD chamber. Thus, the H atoms are also active in pulling away residual H atoms from the growing surface, as illustrated in Process C in the figure. Thus, the C_H of films prepared by Cat-CVD is commonly lower than that in PECVD films.

This model is supported by a-Si films prepared by a mix of SiH_4 and deuterium (D_2) gas instead of H_2 gas. D has the same chemical properties as H, and it can easily be resolved by IR absorption spectra as the absorption peak of Si—D bonds is located at a different position than that of Si—H bonds. Figure 5.11 shows the H and D densities of Cat-CVD a-Si, prepared with the SiH_4 and D_2 gas mixture. The H and D densities are plotted as a function of D_2 flow rate, FR(D_2), at a fixed SiH_4 flow rate of 25 sccm. When FR(D_2) is 0 sccm, only H atoms are detected, and the C_H is about 5 at.%, whereas the Si atomic density is about 5×10^{22} cm^{-3}

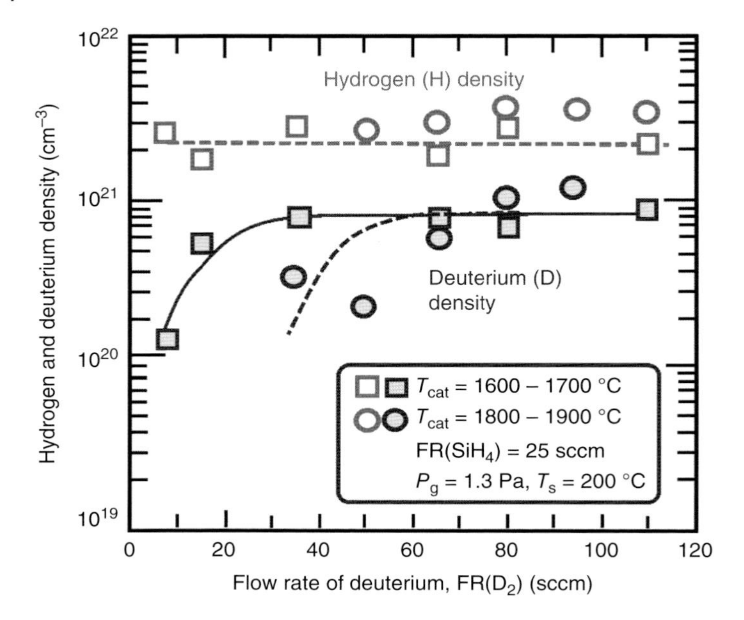

Figure 5.11 Hydrogen (H) and deuterium (D) density in Cat-CVD a-Si films prepared with SiH_4 and D_2 gas mixture.

or slightly less. At increasing FR(D_2), the C_H does not change much. That is, H atoms incorporated in a-Si films are mainly provided by SiH_4, probably through SiH_3 reactions following the mechanism mentioned in the previous chapter, in spite of the existence of many H and D atoms in the gas phase. The D content is likely to increase as FR(D_2) increases. However, the concentration of D atoms incorporated in a-Si films is always limited and much lower than that of H atoms provided from SiH_4. The results appear to support the model in which residual H atoms in a-Si mainly come from SiH_4.

As mentioned above in Figure 5.9, C_H in our Cat-CVD a-Si is usually lower than that of PECVD a-Si. By this reduction of C_H, in the first instance, we can adjust the optical band gap of a-Si, E_{gopt} and narrow it to less than 1.65 eV. As for a-Si solar cells, the E_{gopt} is about 1.8 eV and a little bit wider than the optimum value for collecting sun light efficiently, because a narrower E_{gopt} is desired. From that point of view, Cat-CVD films are collecting attention. Figure 5.12 demonstrates E_{gopt} vs. C_H for Cat-CVD a-Si films prepared by the NREL group and our group [31]. In the figure, the results for PECVD a-Si films and a-Si made by a chemical annealing process using reactive hydrogen [32] are also demonstrated for comparison.

In PECVD, it is hard to sustain the quality for the films with C_H less than 10%. Thus, actually, it is hard to obtain device quality PECVD a-Si films with an optical band gap less than 1.7 eV. On the other hand, in Cat-CVD, we can obtain a-Si with such a narrower band gap.

We summarize other properties of Cat-CVD a-Si films as a function of hydrogen content C_H, by using data reported by A.H. Mahan et al. [19] in Figure 5.13.

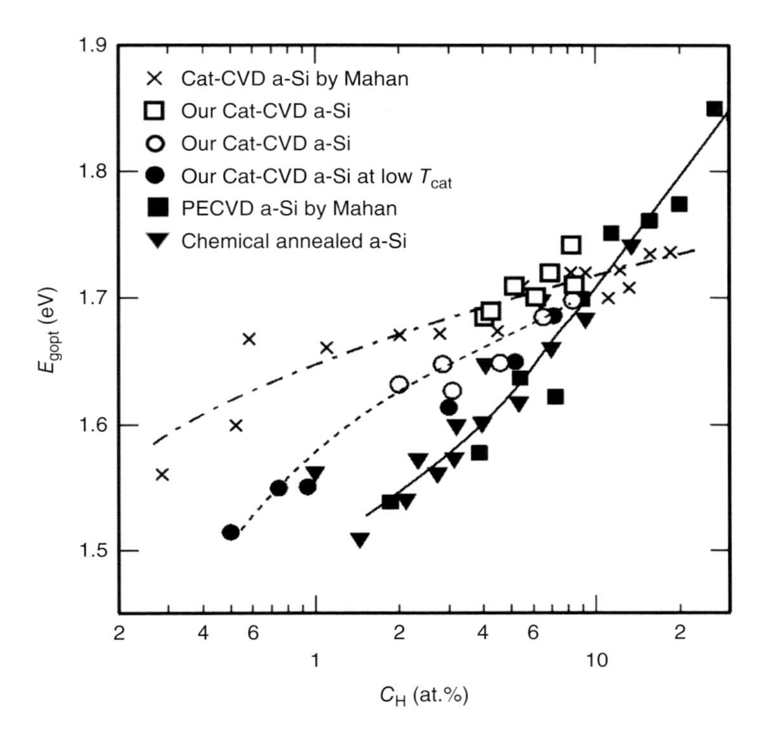

Figure 5.12 Optical band gap, E_{gopt}, as a function of hydrogen content C_H. Source: Reprinted with permission from Ref. [31] after translating original figure in Japanese to English. Copyright (1991): The Japan Society of Applied Physics.

In the figure, optical band gap E_{gopt}, characteristic energy of Urbach tail E_u, ambipolar diffusion coefficient L_D, and ratio of photoconductivity σ_p to dark conductivity σ_d, σ_p/σ_d, are all shown, in comparison to the same properties for PECVD a-Si films.

The σ_p was measured using the illumination of ELH lamp with 100 mW/cm^2. As the PECVD technique shown in the figure, RF-PECVD (13.56 MHz) was used. The unique properties of Cat-CVD a-Si are demonstrated clearly. In the case of PECVD a-Si, C_H more than 10% seems to be necessary to maintain the quality of the films. At C_H less than 10%, E_u begins to increase and σ_p/σ_d begins to decrease, showing the degradation of film quality. Contrary to this, in Cat-CVD films, these physical parameters of films are kept or even improved at low C_H down to 0.5%. H atoms in a-Si are useful because they can terminate Si dangling bonds. However, 10% C_H as found in PECVD fa-Si films seems to be too much, and the excess H atoms might form clusters or other undesirable structures, causing instability of a-Si properties. In Cat-CVD films, the role of H atoms appears mainly for terminating Si dangling bonds in bulk a-Si films.

E_u of Cat-CVD films appears slightly smaller than that of the PECVD ones, and correlated with this, L_D is slightly larger than that of PECVD. The value of σ_p/σ_d is a quite convenient measure to characterize the optoelectronic quality of

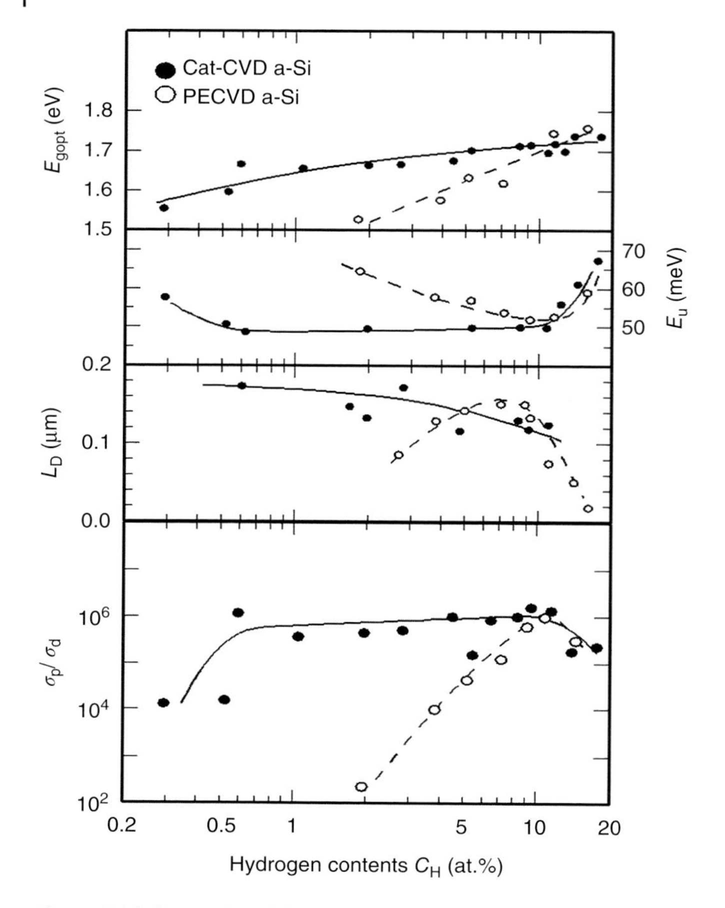

Figure 5.13 Properties of Cat-CVD a-Si films as a function of hydrogen content C_H in the films. The results are compared with those of films made with the conventional RF-PECVD a-Si films. Source: Reprinted with permission from Ref. [33]. Copyright (1998): The Japan Society of Applied Physics.

a-Si films because the measurement of conductivity is not so difficult. The value, 10^5–10^6, is a state-of-the art value.

Figure 5.14 shows the dark conductivity σ_d, the photoconductivity σ_p, and the photosensitivity, σ_p/σ_d, of Cat-CVD a-Si films as a function of deposition rate D_R. The data for Cat-CVD a-Si are taken from the report by B.P. Nelson et al. [34]. Cat-CVD a-Si films were prepared at T_s = 300 °C but with various SiH_4 flow rates FR(SiH_4) from 10 to 100 sccm and gas pressures P_g from 1.6 to 20 Pa. The deposition rates were adjusted by changing FR(SiH_4) and P_g. The photoconductivity was measured under AM-1.5 light with 100 mW/cm^2. The AM-1.5 light is equivalent to the sun-light observed in Europe, China, Korea, Japan, United States, and the other countries of similar latitude. In the figure, the photosensitivities of other a-Si films prepared by RF-PECVD and μ-wave-excited PECVD are shown for comparison in the figure. The RF-PECVD a-Si films were prepared at

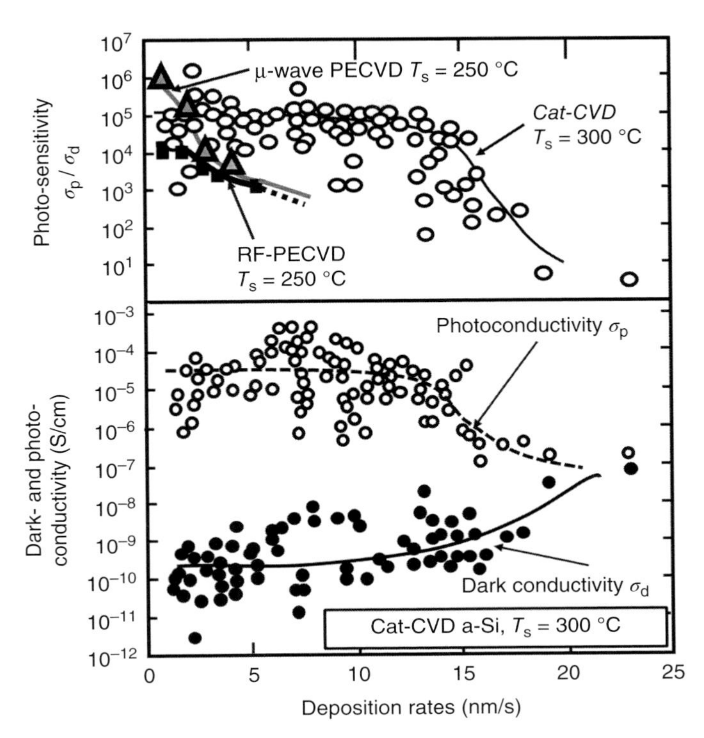

Figure 5.14 The ratio of photoconductivity (σ_p) and dark conductivity (σ_d), σ_p/σ_d, of Cat-CVD a-Si as a function of deposition rate D_R. Similar results for RF-PECVD and micro (μ)-wave-PECVD a-Si are also demonstrated for comparison. Source: Data in Refs. [34–36] are summarized to draw this figure.

$T_s = 250\,°C$, and the deposition rates were changed by changing RF powers from 2 to 200 W for the electrode with a diameter of 10 cm [35]. The photoconductivity was measured under the illumination of monochromatic light of 600 nm wavelength. The μ-wave PECVD a-Si films were prepared at $T_s = 250\,°C$ by using a 2.45 GHz microwave [36]. The photoconductivity was measured in AM-1.5 light with 100 mW/cm^2.

The results are surprising. The Cat-CVD films keep their quality up to D_R of 15 nm/s. In PECVD a-Si films, for both RF- and μ-wave-excited PECVD, the quality of a-Si degrades when D_R exceeds 2 nm/s. The high rate deposition keeping their quality is a key for low-cost mass production. The superiority of Cat-CVD a-Si to PECVD a-Si is apparent.

It is well known that the optoelectronic properties of a-Si films are likely to degrade after prolonged light exposure. This phenomenon is called the "Staebler–Wronski effect" [37], which has been the cause of headache for many a-Si researchers. It is also known that the degradation is usually smaller for a-Si films with low C_H. The results shown in Figure 5.13 induce expectation for it.

Figure 5.15 shows the defect density of silicon dangling bonds (DBs) for Cat-CVD and RF-PECVD a-Si films before and after exposure to light from a

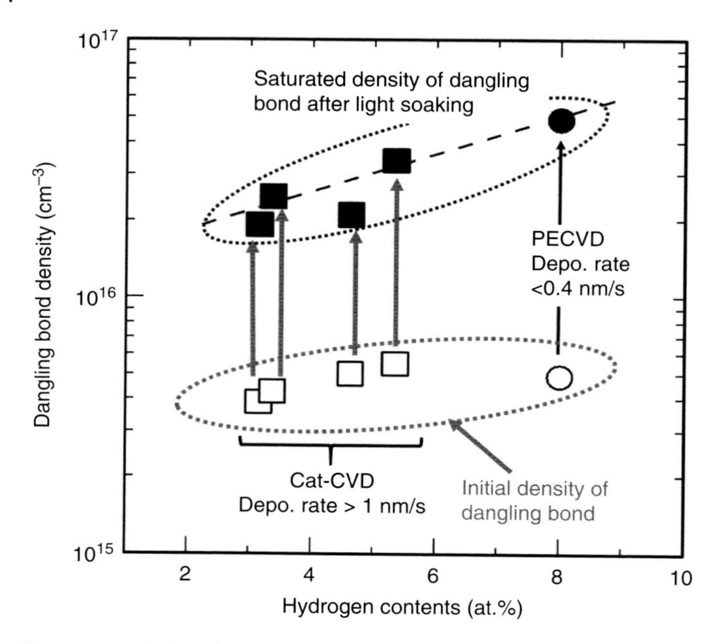

Figure 5.15 Defect density due to dangling bonds (DBs) in Cat-CVD and RF-PECVD a-Si films before and after exposure to a xenon (Xe) lamp with infrared (IR) cut filter and with light intensity of 340 mW/cm^2. Source: Matsumura 2001 [38]. Reprinted with permission of The Institute of Electronics, Information and Communication Engineers.

xenon (Xe) lamp. The IR component of Xe light is cut by IR cutoff filters in the experiments to avoid heating of the samples during light soaking. The intensity of the light is set at 340 mW/cm^2, which is much stronger than sun light, which has a light intensity of about 100 mW/cm^2. The DB density is measured by electron spin resonance (ESR). The exposure time for light soaking is long enough to reach the saturated state of the DB density. It is shown that the initial DB density for Cat-CVD and RF-PECVD a-Si films appears almost the same. However, after exposure to light, the saturated DB density of Cat-CVD a-Si with a lower C_H appears smaller than that of RF-PECVD a-Si. The results reveal that (i) small C_H induces smaller degradation and (ii) Cat-CVD a-Si is more stable than RF-PECVD films as lower C_H a-Si films can be made by Cat-CVD. The deposition rates for PECVD a-Si films are about 0.4 nm/s and those for Cat-CVD films are larger than 1 nm/s. In spite of the faster deposition, Cat-CVD a-Si films are more stable than the PECVD ones. The superiority of Cat-CVD over PECVD is again demonstrated.

We show another evidence of better stability of Cat-CVD a-Si compared to PECVD a-Si. Here, we show the results of the degradation of a-Si due to light soaking by summarizing two papers reported by A.H. Mahan et al. [19, 39]. Figure 5.16 shows such summarized data for the ratio of the carrier diffusion length in a-Si before light exposure, L_D(Initial), to that after saturation because of light soaking, L_D(Saturation), and also for the ratio of defect density of a-Si before light exposure, N_{defect}(Initial), to that after light exposure, N_{defect}(Saturated). The data are

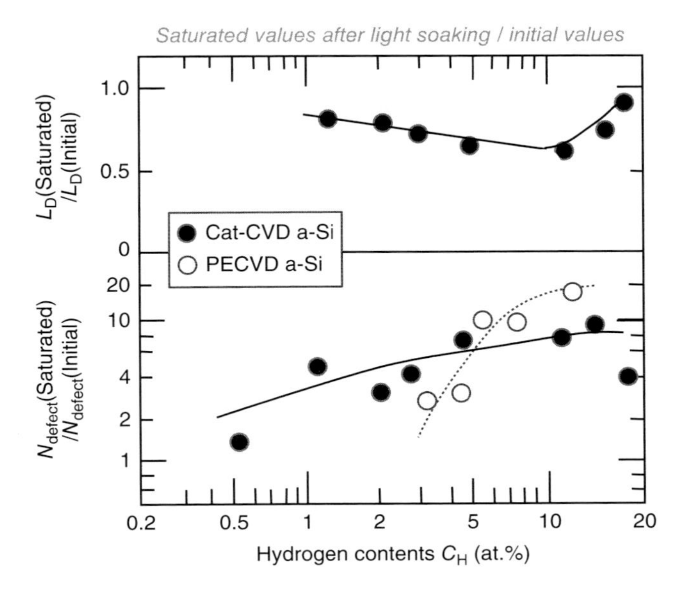

Figure 5.16 Degradation of a-Si films by light soaking with AM-1.5, 100 mW/cm^2 for both Cat-CVD and PECVD a-Si films. Degradation is finally saturated and saturated physical values are compared with the initial ones. Source: From Mahan et al. 1991 [19, 39].

plotted as a function of hydrogen contents C_H in the a-Si films. In the figure, the results on Cat-CVD a-Si are compared with those on PECVD a-Si. In device quality PECVD a-Si films in which usually C_H is larger than 10%, the saturated defect density is almost one order of magnitude larger than the initial value. However, in Cat-CVD a-Si films, it can be suppressed to less than several times the initial value, although even in Cat-CVD a-Si, the degradation is still observed. Nevertheless, the superiority of Cat-CVD a-Si against RF-PECVD a-Si is clear.

5.1.4 Deposition Mechanism of a-Si in Cat-CVD Process – Growth Model

There have been many reports about the deposition mechanism of a-Si in PECVD systems. Among various reports, the image presented by A. Matsuda appears to be most widely supported [30, 40]. In PECVD, energetic electrons collide with SiH_4 molecules to decompose them. The cross section for producing SiH_3 appears the highest, as stripping only one H atom from a SiH_4 molecule requires the lowest amount of energy compared to other more complicated decompositions. This mechanism can be thought of as the excitation of a SiH_4 molecule by the collision with an energetic electron and the subsequent decomposition of the excited SiH_4 to various fragments.

Here, we consider a model for the formation of a-Si films based on the SiH_3 precursor model by A. Matsuda et al. combined with the growth model of the diamond structure [41].

The lifetime of SiH_3 is long as the number of reactions in which SiH_3 is involved is limited. Thus, it is expected that the most abundant decomposed species in the PECVD chamber will be SiH_3. This SiH_3 reaches the growing surfaces as a precursor for the formation of a-Si films. When SiH_3 reaches the surface of a-Si, terminated with H atoms, it is weakly adsorbed by physical adsorption by Van der Waals forces. Because of this weak physical bond, SiH_3 can migrate on the surface over a long distance and finally find a Si dangling bond to make a chemical bond. Sometimes, however, it will desorb from the surface into the gas phase. This image is drawn in Figure 5.17a. One idea for the formation of such a Si dangling bond is that H atoms are pulled away from the H-terminated surface by the reaction: $SiH_3 + H{-}Si \rightarrow SiH_4 + {-}Si$. This is drawn in Figure 5.17b. The Si dangling bond may also be formed by H-extraction by another H atom as shown in Figure 5.17c. However, although SiH_3 has a dipole, and it is easy to have Van der Waals physical bonding with H-terminated a-Si surface, H does not have such a dipole, and long distance migration cannot be expected in this case. The relatively inactive property of SiH_3 allows a large surface diffusion length on the a-Si surface. Thus, Si dangling bonds are efficiently terminated with only few bonding errors, leading to device quality low-defect a-Si films. This is drawn in Figure 5.17d,e.

This is the story of the first adsorption of Si atoms on a H-terminated a-Si surface, which is partially mentioned already in Figure 5.10. The first adsorbed Si atom may be an obstacle to the surface migration of the second SiH_3. However, because of this obstacle, the possibility of stay of the second SiH_3 at the position near to the first fixed Si atom may increase, and thus, a new Si dangling bond may be created near to the first Si atom. Then, the second Si atom may adsorb at the neighbor to the first Si atom and thus the film is grown. The image of this story is simply illustrated in Figure 5.17g–i. However, in consideration of these a-Si film formation processes, we should consider the three-dimensional structure of newly coming SiH_3 species and the former SiH_3 fixed at a-Si surfaces.

Figure 5.18 shows a schematic view of two SiH_3 species when they make a cross-linking reaction. It is known that the activation energy for this cross-linking reaction is relatively large and estimated to be 94 kcal/mol (4.1 eV/bond) to 102 kcal/mol (4.4 eV/bond), although the estimation is carried out for molecular states [42]. As the cross-linking reaction has a large activation energy, there may be a possibility of more complicated intermediate reaction paths. In addition, the geometrical relation for establishing a cross-link of two SiH_3 species appears quite limited. The position of base Si atoms is decided, and the cross-linking reaction starts only when two SiH_3 species have proper bonds with base Si atoms located exactly at a position that fits. Therefore, it can be imagined that this cross-linking reaction can be started when the proper direction of ordered Si atoms is prepared as a base.

The first principle calculation will be a good tool to reveal this process for future discussion. However, as there are no concrete discussions about the real growth mechanism of a-Si, here, we make a brave speculation on an image of the growth of a-Si layers, considering the analogy of recent study on the growth model of diamond by CVD [41].

In growth of a-Si, Si atoms may be positioned under some rules, as the bond length and bond angle are not far different from those of the c-Si. It is believed for a-Si that Si atoms, in a short range, are located in an atomic array structure almost similar to that of c-Si. That is, when a-Si growth is considered, we can

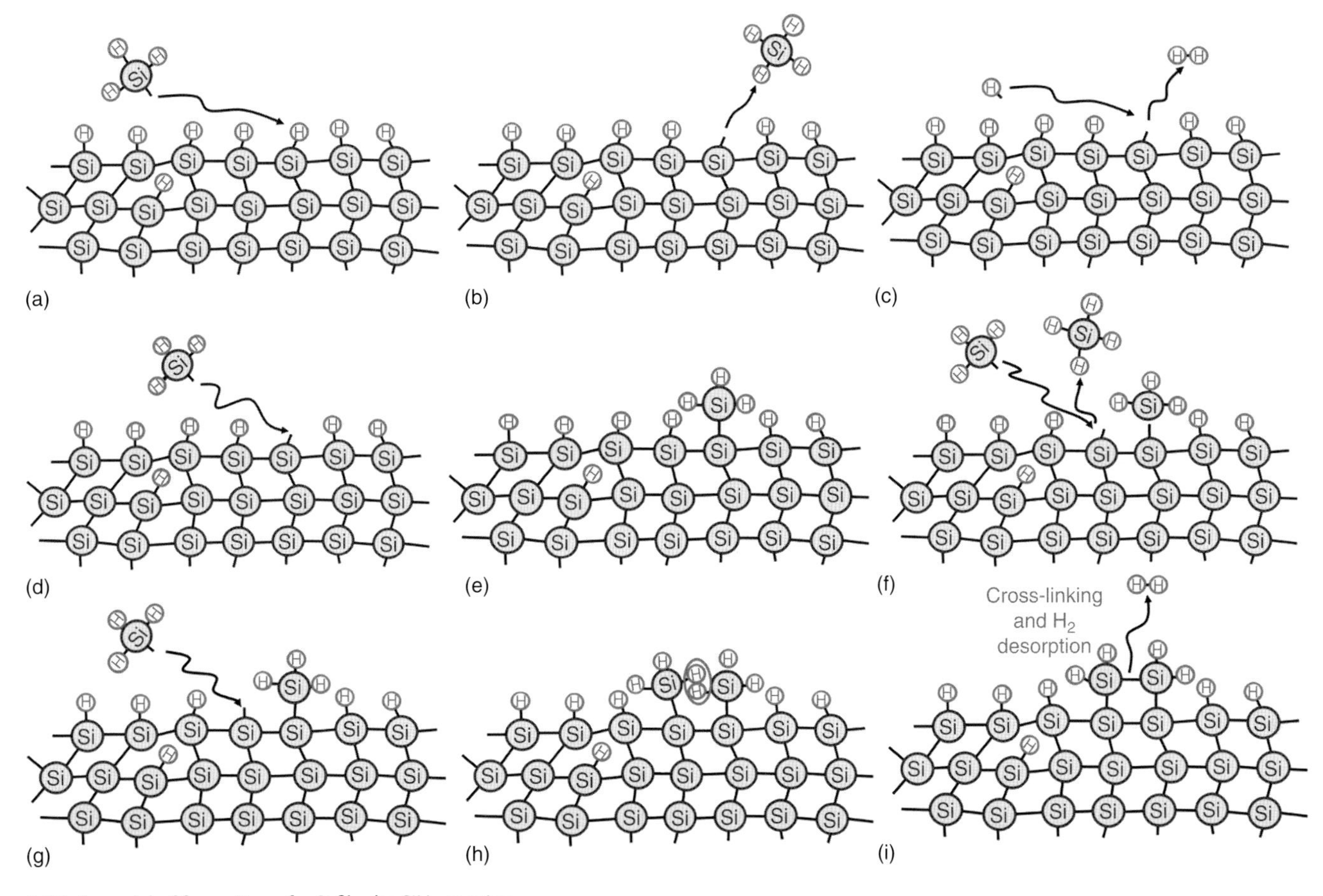

Figure 5.17 A model of formation of a-Si film by SiH_3 species.

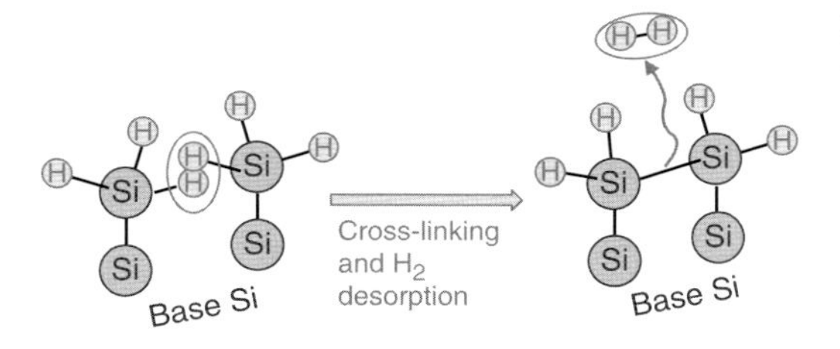

Figure 5.18 A model of cross-linking of two SiH_3 species.

adopt the growth of c-Si as an approximate image of the growth of a-Si. This helps to consider the growth of Si atoms in a-Si.

Figure 5.19 shows the image for the growth of Si atomic layers on a (111) plane of c-Si from SiH_3 species. Figure 5.19a shows the image of the H-terminated Si layer with a dangling bond, which may be created by the removal of a H atom by SiH_3 (a black dot in the figure expresses a Si dangling bond) and Figure 5.19b shows the attachment of the first SiH_3 at such a dangling bond. Because the existence of this first attached SiH_3 disturbs the free surface migration of other SiH_3, the second Si dangling bond is made at the neighbor of the first attached SiH_3 by removing the H atom by an impinging SiH_3. This is shown in Figure 5.19c. Figure 5.19d shows the attachment of the second SiH_3 at the second dangling bond. If the spatial distance between H atoms of the first SiH_3 and that of the second attached SiH_3 is small enough to facilitate the cross-linking reaction (–Si–H + H–Si– → –Si–Si– + H_2), these two SiH_3 can make a next Si layer. However, as shown in the figure, the distance between the two H atoms of two SiH_3 is not so short in three-dimensional space. Thus, for making the Si—Si bond by the cross-linking of two H atoms of two SiH_3 species, the attachment of the third SiH_3 at a Si dangling bond created at a SiH_3 branch, shown in Figure 5.19e, is required to form a new Si layer as shown in Figure 5.19f. Figure 5.19f demonstrates that three SiH_3 species are necessary to form the Si ring structure fitting to the growth on the (111) plane. That is, if we consider the three-dimensional structure, satisfying the requirements for cross-linking reaction is getting complicated and more SiH_3 becomes necessary to form the next Si layer.

Figure 5.20 demonstrates similar growth steps for the Si growth on the (110) plane. Figure 5.20a shows H-covered Si surface with a dangling bond. The first SiH_3 attaches at such a Si dangling bond as shown in Figure 5.20b. Figure 5.20c shows the creation of another dangling bond in the attached first SiH_3. In this case, if the second SiH_3 makes a bond with this dangling bond, the distance between H atoms of the second SiH_3 and other H atoms at Si surface becomes quite short, as two SiH_3 are located on a single plane and finding another H atom on the same plane from the Si surface is also possible. By this way, as shown in Figure 5.20d, a Si ring is formed by using only two SiH_3, if we consider the growth on the (110) plane.

Figure 5.21 shows the growth steps on a (100) plane of the Si layer. If we consider similarly, we can easily understand the Si growth on the (100) plane. That is, as shown in Figure 5.21a–f, three SiH_3 species are necessary to form a Si ring, just similar to the growth on the (111) plane.

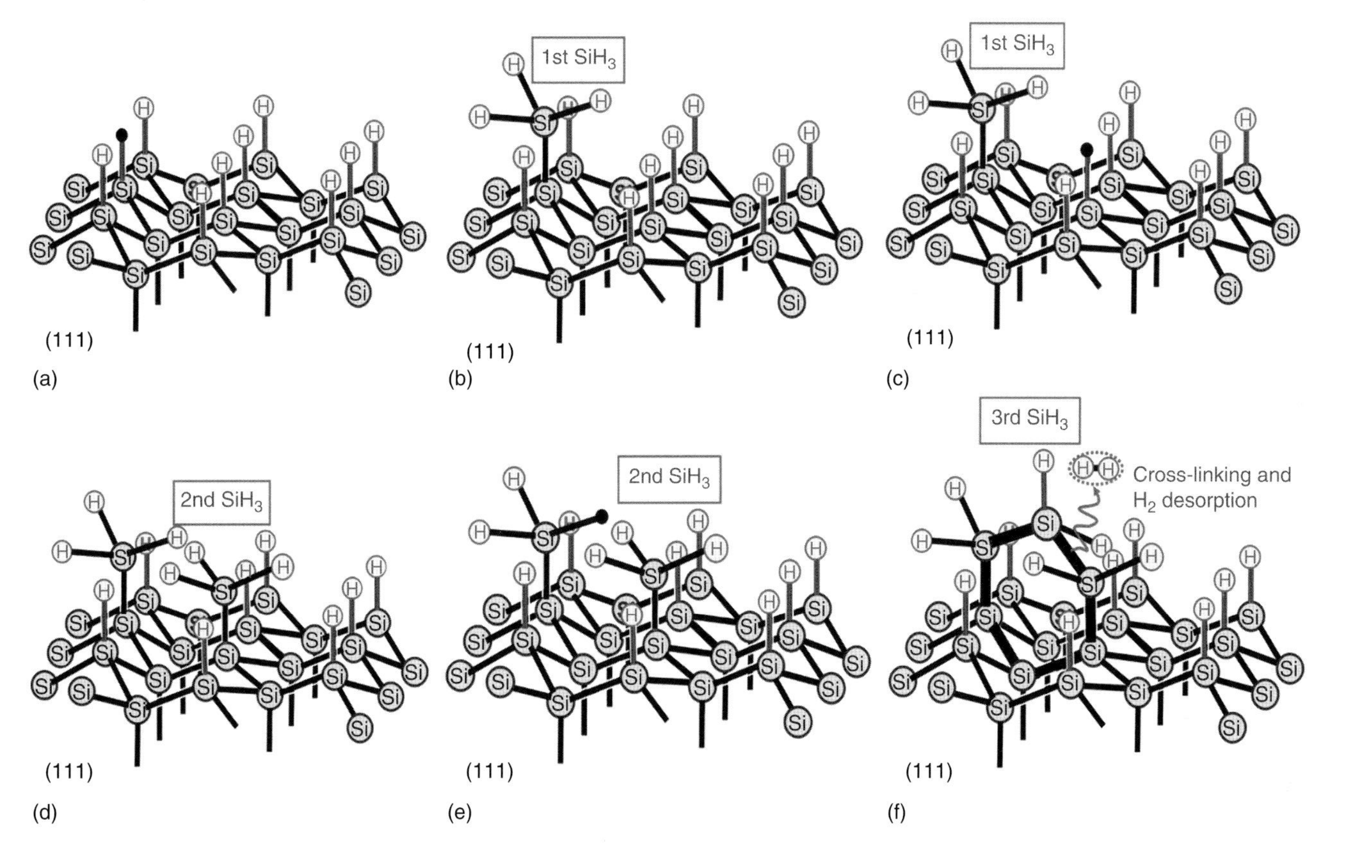

Figure 5.19 A model of Si film growth on (111) plane along ⟨111⟩ direction from SiH_3 species. Three SiH_3 species are required for film growth. A black dot indicates the Si dangling bond. (a) Shows H-terminated Si surface with a Si-dangling bond, (b) the first SiH_3 makes a bond with such a dangling bond, (c) a second Si dangling bond is created at the surface nearest to the first Si—SiH_3 bond, (d) the second SiH_3 makes another bond with the second dangling bond but the distance between H atoms in the first SiH_3 and H atoms in the second SiH_3 is too far to make cross-linking, (e) Si dangling bond is created in the first SiH_3, and (f) the third SiH_3 can make a bond with the Si dangling bond created at the first SiH_3, in this case, the distance between H atoms in the third SiH_3 and H atoms in the second SiH_3 becomes short enough to make a cross-linking, and so that, the second Si layer can be successfully constructed using three SiH_3.

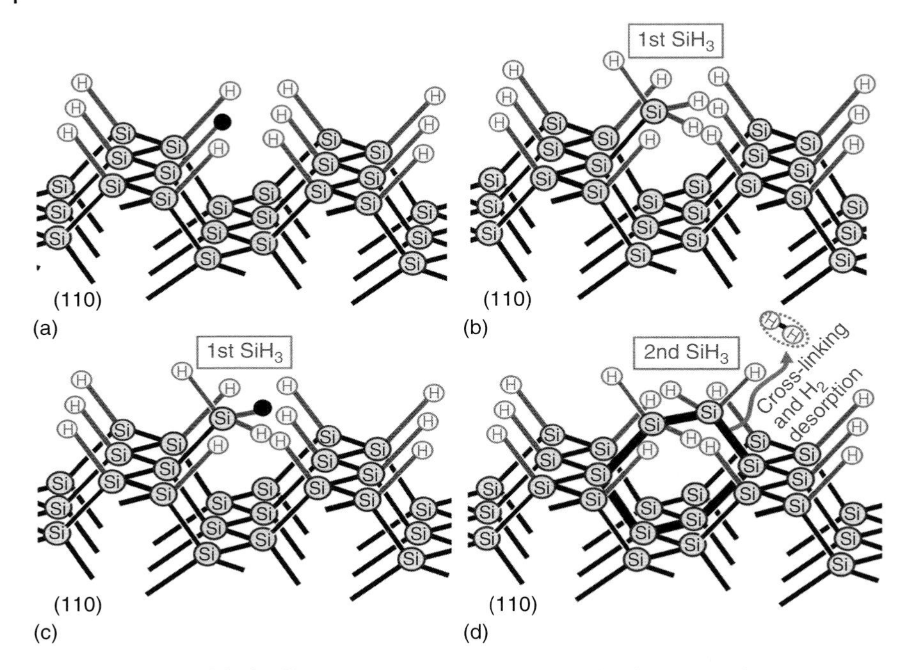

Figure 5.20 A model of Si film growth on (110) plane along ⟨110⟩ direction from SiH_3 species. Two SiH_3 species are required for film growth. A black dot indicates the Si dangling bond. (a) shows H-terminated Si surface with a Si-dangling bond, (b) the first SiH_3 makes a bond with such a dangling bond, (c) Si dangling bond is created in the first SiH_3, and (d) the second SiH_3 can make a bond with the Si dangling bond created at the first SiH_3. Since the distance between H atoms in the second SiH_3 and H atoms on the surface is short enough to make cross-linking, the second Si layer can be successfully constructed using only two SiH_3 in this case.

From these three figures, considering the Si growth by the attachment of SiH_3 species coming from gas phase, the growth on the (110) plane appears the most probable. It is known that when a Si crystal is grown by liquid-phase epitaxy (LPE) or by molecular beam epitaxy (MBE), the crystalline growth on the (100) plane is the most probable as the number of bonds on the (100) plane is the smallest. However, when we consider the Si growth from SiH_3 species, the (110) plane becomes the preferential plane of crystal growth. Related to this, we will later discuss the preferential orientation in Si (micro-)crystalline growth.

Although Figures 5.19–5.21 show the growth of c-Si from SiH_3 species, as mentioned firstly, this phenomena can be applied approximately to the growth of a-Si because the short-range order exists also in a-Si layers. Thus, we can make an image of a-Si film growth by using SiH_3 species.

Of course, the explanation is based on speculation. At the moment, there is no experimental evidence to directly support the model. However, as mentioned later, the fact that polycrystalline silicon (poly-Si) growth by using SiH_3 species is preferentially oriented on the (110) plane appears to support the above-mentioned three-dimensional growth model.

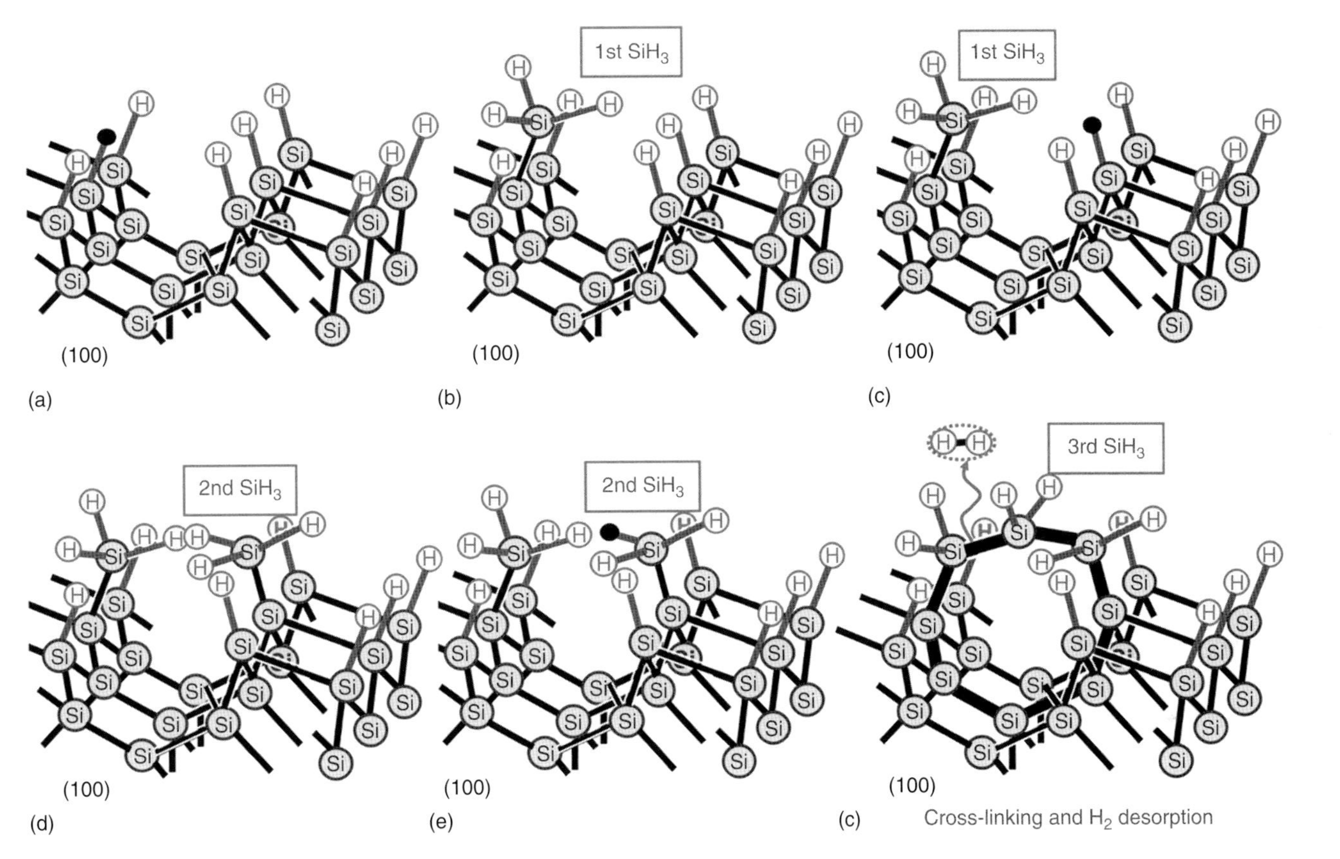

Figure 5.21 A model of Si film growth on the (100) plane along ⟨100⟩ direction from SiH_3 species. Three SiH_3 species are required for film growth. A black dot indicates the Si dangling bond. (a) Shows H-terminated Si surface with a Si-dangling bond, (b) the first SiH_3 makes a bond with such a dangling bond, (c) another second Si dangling bond is created at the surface nearest to the first Si—SiH_3 bond, (d) the second SiH_3 makes another bond with the second dangling bond but the distance between H atoms in the first SiH_3 and H atoms in the second SiH_3 is too far to make cross-linking, (e) Si dangling bond is created in the second SiH_3, and (f) the third SiH_3 can make a bond with the Si dangling bond created at the second SiH_3, in this case, the distance between H atoms in the third SiH_3 and H atoms in the first SiH_3 becomes short enough to make a cross-linking and the second Si layer can be successfully constructed using three SiH_3.

5.2 Crystallization of Silicon Films and Microcrystalline Silicon (μc-Si)

5.2.1 Growth of Crystalline Si Film

Apart from a-Si, its crystallized counterpart, poly-Si, and the films including crystallites are collecting much attention. The films including small size Si crystallites embedded in the films are often called microcrystalline silicon (μc-Si) or nanocrystalline silicon (nc-Si). The interest in poly-Si started in view of application in Si-integrated circuits and also solar cells. In most of the cases, it was made at high temperatures. However, poly-Si, μc-Si, and nc-Si films prepared by low-temperature deposition collected much attention for the fabrication of low-cost solar cells and TFTs, which are both expected to obtain better performance than a-Si films.

There had been some discussions on the crystallization effect of a-Si, particularly after the report by S. Usui and M. Kikuchi in 1979 [43] in which they demonstrate the discovery of highly doped low-resistivity Si film prepared by PECVD. As far as the authors know, this paper is the first one showing the deposition of poly-Si or μc-Si at low-substrate temperatures by PECVD, although many works about the low-temperature μc-Si formation by PECVD have been carried out since then. On the other hand, in 1991, H. Matsumura reported that poly-Si or μc-Si films could be obtained at low-substrate temperatures around 300 °C on glass substrates even by Cat-CVD [44]. At that time, the apparatus was the same one used for the preparation of a-Si, but SiH_4 gas was diluted by H_2 at a relatively low gas pressure. Observations by X-ray diffraction (XRD) analysis showed that the grain size was derived to be100 nm, based on the well-known Scherrer formula [45].

Typical deposition parameters for obtaining μc-Si or poly-Si films by Cat-CVD are summarized in Table 5.3. Except for the gas flow ratio of SiH_4 and H_2 and the relatively low gas pressure, most of the deposition parameters are the same as those for a-Si deposition.

Figure 5.22 shows the result of XRD analysis for one of such Si films. The X-rays were generated by electron bombardment of a copper (Cu) target, and thus, two peaks were observed: Cu Kα and Cu Kβ X-rays. A sharp diffraction peak from

Table 5.3 Typical deposition parameters for obtaining μc-Si or poly-Si films.

Parameters	Setting values
Catalyzing material	Tungsten (W) wire
Temperature of catalyzer, T_{cat} (°C)	1700–1900
Temperature of substrate, T_s (°C)	200–400
Gas pressure during deposition, P_g (Pa)	0.06–0.13
Flow rate of SiH_4, $FR(SiH_4)$ (sccm)	0.1–1.5
Flow rate of H_2, $FR(H_2)$ (sccm)	10–100
Distance between catalyzer and substrate, D_{cs} (cm)	4–5

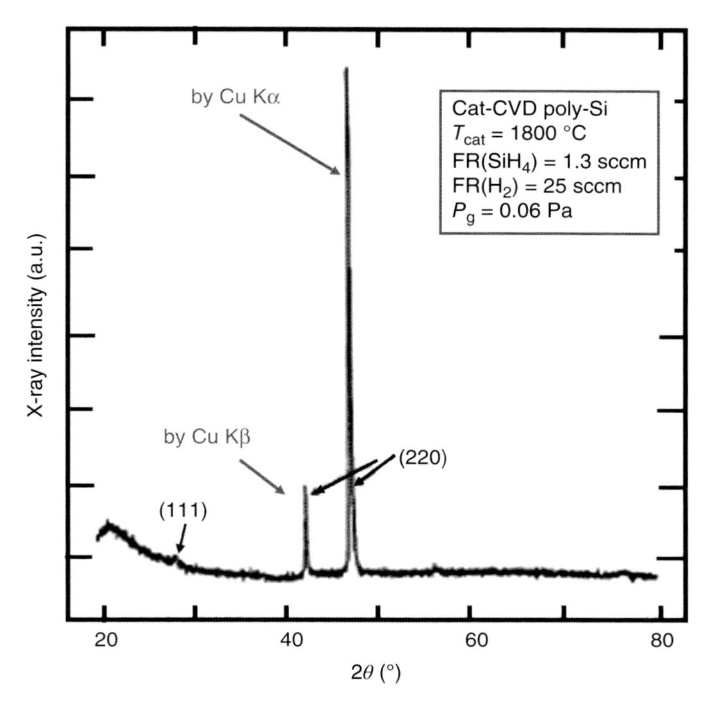

Figure 5.22 X-ray diffraction spectrum of a Si film prepared by Cat-CVD under the condition of high H_2 dilution of SiH_4.

(220) direction of c-Si can be observed. The signal from the (111) plane appears quite week in this case. From the figure, it is apparent that the Si film including crystallites is made at substrate temperatures lower than 400 °C.

In the observation of XRD, although the (110) plane of c-Si cannot be seen because of the structure factor rule [46], the (220) plane is observed, and thus, the information on the (110) plane can be obtained from the data of (220) planes. The preferential orientation of c-Si growth on the (110) plane has been well explained by the model illustrated in Figures 5.19–5.21 for the growth of a-Si layer. Of course, by changing the deposition conditions considerably, we can make poly-Si showing a (111) peak, apart from a (220) peak. When the density of SiH_3 species arriving at the growing surface is high enough to increase the growth rate on other planes, or when the substrate temperatures are high enough to increase the rate of the film formation, the limitations imposed on the preferential growth are mitigated and the poly-Si films with other orientations apart from (220) can grow. However, in any case, (220) is still the major orientation plane when poly-Si is formed by deposition at low temperatures.

It is well known that in the case of crystalline growth of Si films by LPE or by MBE, the prudential growth direction is usually on the (100) plane, as the number of bonds to obtain growth on the (100) plane is smallest. However, the growth by SiH_3 species appears different, probably because of the necessity of cross-linking and H elimination reactions during the growth as mentioned already in Section 5.1.4.

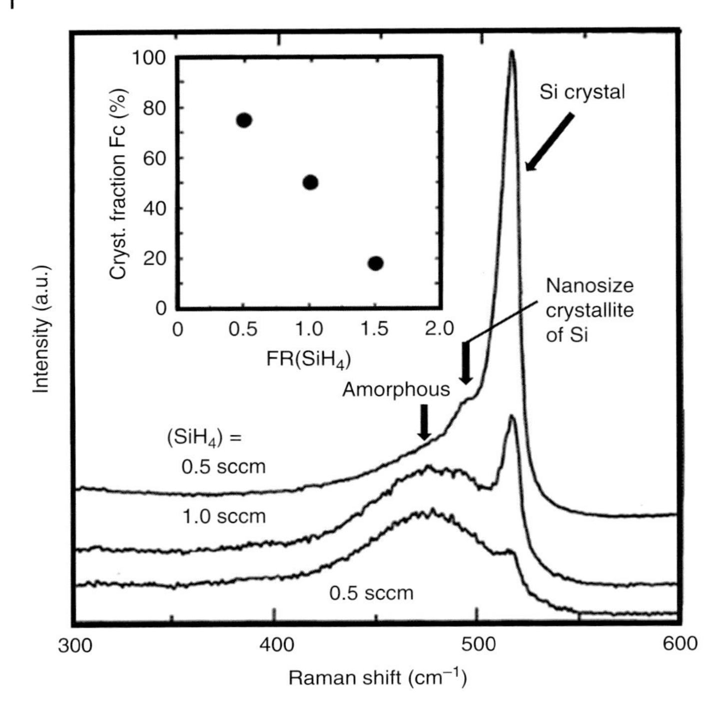

Figure 5.23 Various Raman spectra of Cat-CVD Si films, taking FR(SiH_4) as a parameter. Source: Matsumura 1998 [33]. Reprinted with permission of Japanese Journal of Applied Physics

By Raman spectroscopy, we can also confirm the crystallization of Si films. In Figure 5.23, the Raman spectra are demonstrated taking SiH_4 flow rates, FR(SiH_4), as a parameter, for a fixed H_2 flow rate of 30 sccm. In the Raman spectrum, it is known that the transverse optical (TO) phonon peak at 520 cm^{-1} is attributed to the signal from the c-Si and that at 480 cm^{-1} is from amorphous Si. Sometimes, an additional peak is seen at about 495–510 cm^{-1}, and it is assigned to nanosize c-Si-grains [47, 48].

The crystalline fraction F_c, which is defined as the percentage of the volume of the crystalline phase in the total solid volume, is a parameter that represents to what extent crystallization has proceeded. That is, if the volume of the amorphous phase and that of the crystalline phase are represented by X_a and X_c, respectively, the crystalline fraction F_c is defined as $X_c/(X_a+X_c)$. F_c is usually simply estimated from the peak area ratio of the crystalline phase to the amorphous phase. This crystalline fraction is shown in the inset of the figure. The figure demonstrates that F_c increases as the dilution ratio by H_2 increases.

From X-ray analysis and Raman measurement, it is again clear that by diluting SiH_4 with H_2, the crystalline phase is easily formed in the Cat-CVD process.

5.2.2 Structure of Cat-CVD Poly-Si

From various studies on the growth of poly-Si films, it is known that crystalline growth is not uniform along the growth direction. When poly-Si films are grown

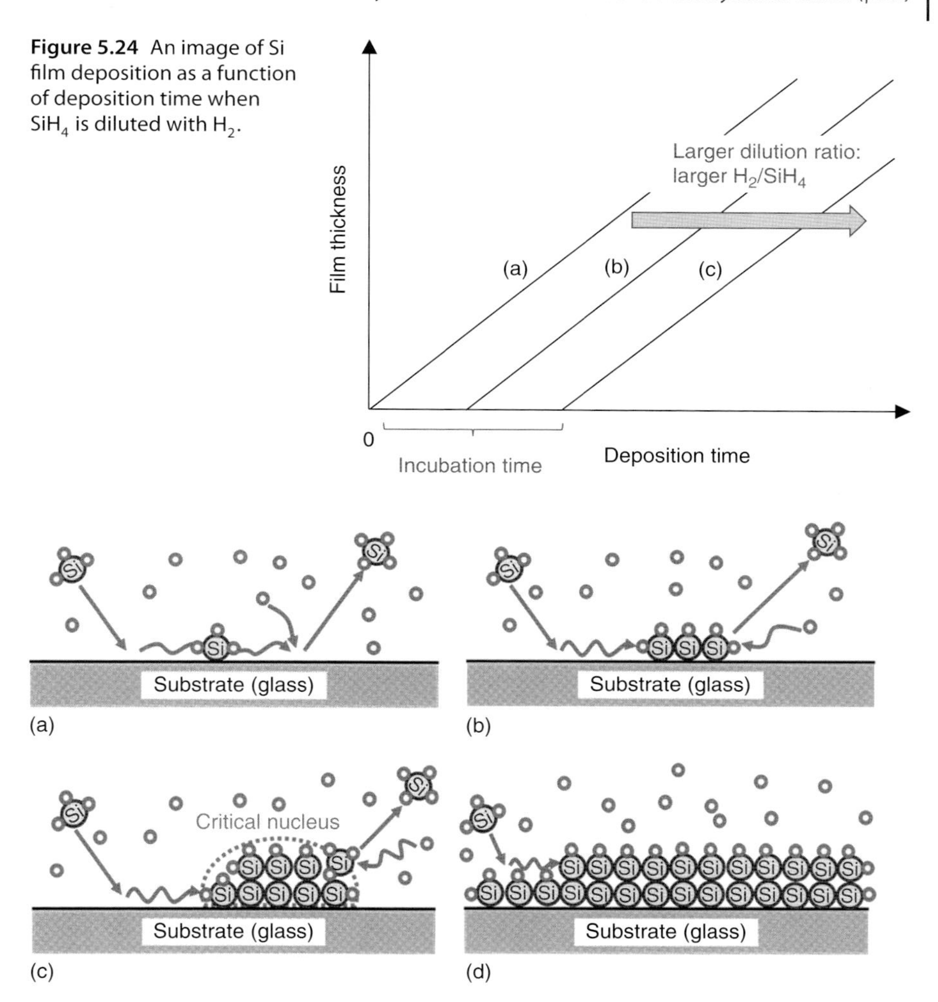

Figure 5.24 An image of Si film deposition as a function of deposition time when SiH_4 is diluted with H_2.

Figure 5.25 Growth model of Si films from SiH_3 precursor. Large circles express Si atoms and small circles express H atoms. (a) Shows a view of the motion of Si atoms on the substrate surface, adsorption, and desorption of Si atoms are making a balance, (b) some of Si atoms start to accumulate by some reasons, but some Si atoms are still desorbing from the surface, (c) the number of Si atoms accumulating on the substrate surface is exceeding that of critical nucleus, then, due to the attractive forces from the nucleus the desorption is suppressed, and (d) Si films start to grow.

on glass substrates, sometimes, the deposition does not start until a certain time has passed, as illustrated in Figure 5.24. When the dilution ratio of H_2 to SiH_4 is large, that is, when the flow rate ratio of $FR(H_2)/FR(SiH_4)$ is larger than several 10s, Si films can begin to grow after several 10 seconds or sometimes after a few minutes. This time duration without deposition is often called "*incubation time.*"

In film growth models, the image shown in Figure 5.25 is usually considered. Here, we consider that the precursor for Si film deposition is SiH_3.

When many SiH_3 species arrive at the surface of a substrate such as a glass plate, some of them are reflected from the substrate immediately, but some of them make a weak physical bond based on Van der Waals forces and begin to migrate on the substrate surface. If such SiH_3 do not meet other SiH_3 species that are also migrating on the surface, it would return from the surface to gas phase after a certain time, as shown in Figure 5.25a. However, if SiH_3 meets other SiH_3 and produces a Si—Si bond, the release to the gas phase is suppressed by the existence of a Si—Si bond. If the number of SiH_3 assembling on the surface is not large, the probability for returning to the gas phase is still larger than that for adding new SiH_3 to the assembled body, as there are many H atoms that make the Si—H bonds and finally form SiH_4 to pull the Si atoms out from the assembled body. This is shown in Figure 5.25b. Here, if the number of Si atoms assembling at a point on the surface exceeds at a certain size, the energy for condensing the body by keeping many Si—Si bonds inside exceeds the energy for releasing Si atoms to the gas phase. Thus, Si films start to grow. This size of the assembled body is called "critical nucleus" as explained in Figure 5.25c, and when the size of the assembled body exceeds a critical nucleus size, Si films are automatically grown, as shown in Figure 5.25d. The incubation times for poly-Si growth can be explained by the time required to create many critical nuclei.

For large dilution, as H atoms prevent to grow ($Si-H_2$) chain by removing H atoms, and then prevent residual H atoms inside the Si films, Si films are likely to be crystallized and such crystallization will occur on the (110) plane preferentially, as already explained by the model. When H dilution is larger, because of the etching effect or the removal of Si atoms by making SiH_4, the incubation time becomes longer.

Even a-Si films whose H contents are small enough (around a few atomic percent) can be polycrystallized, when many H atoms act on the films. Si films are etched, but at the same time, the a-Si films are crystallized [49]. This means that when H atoms that disturb the crystallization of a-Si inside the film are removed, Si films are easily crystallized, probably because the formation of a crystal requires lower energy than the formation of a complicated nonperiodic structure. These phenomena are also observed in the crystalline growth of Si films by MBE. When a low amount of H atoms is introduced in the vacuum of the MBE chamber, the crystalline growth is disturbed through the amorphization of surface layer by H atoms [50].

When Si films are prepared on glass substrates with pure SiH_4 or relatively low H dilution of SiH_4, the film starts to grow without the incubation time as showed in Figure 5.24a. However, in this case, in spite of H dilution, the crystalline phase sometimes cannot be observed until the thickness of the Si film exceeds a certain value. In this case, if we observe the film structure by the cross-sectional view of a transmission electron microscope (TEM), for instance, an initial layer on glass substrates is amorphous, but when the thickness exceeds above a certain value, the crystalline phase appears to grow. The cross-sectional TEM image is demonstrated in Figure 5.26a. Before the films including the crystalline phase start to grow, amorphous incubation layer exists.

As the dilution ratio increases, the incubation time is likely to increase as already shown in Figure 5.24b,c. If too much H atoms are supplied, the films cannot grow any more as the etching by H atoms is too strong to form Si films. When enough incubation time is detected, the crystalline phase is found from

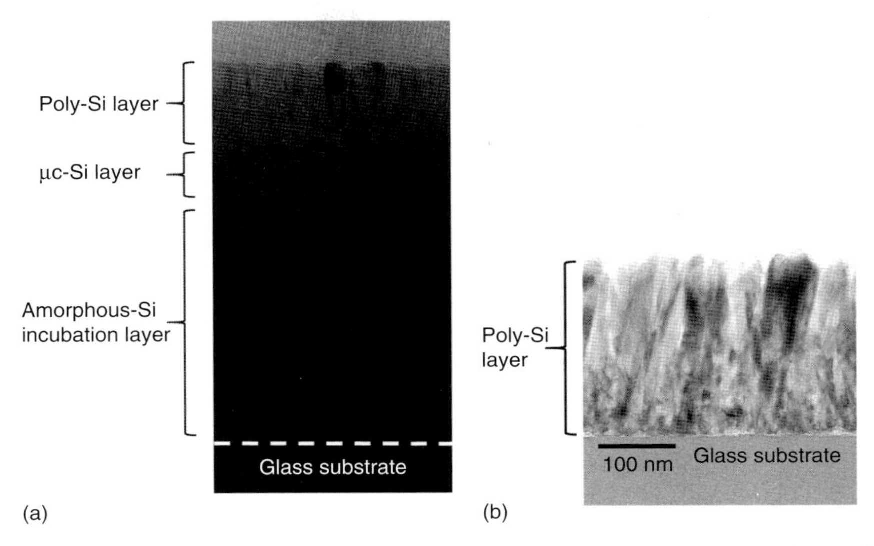

Figure 5.26 Cross-sectional images of TEM, for the samples (a) with medium H_2 dilution of SiH_4 ($H_2/SiH_4 = 10$) and (b) with high H_2 dilution of SiH_4 ($H_2/SiH_4 = 100$).

the boundary between Si films and the glass substrate. The cross-sectional TEM image is demonstrated in Figure 5.26b. As seen in the figure, crystallites were grown from the glass substrates directly, although the substrate temperature was as low as around 300 °C in this case.

The formation of crystals by Cat-CVD was successful. One problem was found, however, in these poly-Si films the films were easily oxidized. The composition of the Si films has also been measured by secondary ion mass spectroscopy (SIMS). Figure 5.27 shows the SIMS profiles together with schematic images of the Si films. Figure 5.27a expresses the result for the sample prepared with a medium dilution ratio, say, $H_2/SiH_4 = 10$. Figure 5.27b shows the results for high dilution ratio, say, $H_2/SiH_4 = 100$. For high dilution ratio, poly-Si is grown in columnar structures directly from the bottom at the glass substrate as indicated in Fgure 5.26b, but oxygen atoms can easily penetrate down to the bottom after air break.

In Figure 5.28, the phenomena are summarized again by using the schematic images of structures inside polycrystallized Si films. The profiles of oxygen content, C_O, along the depth, measured by SIMS, are also schematically shown in the figure. When the dilution ratio is small, there is an incubation layer of a-Si just before the film with crystalline phase starts to grow. The amorphous layer and the initial crystallized layer with small size grains do not contain many oxygen atoms, probably as they are densely packed. When the dilution ratio becomes large, poly-Si layer starts to grow from the substrate surface, but it contains many oxygen atoms, which are believed to be incorporated from air. When poly-Si grains become large, complete packing becomes difficult, and a lot of empty spaces are created at the grain boundary. Through this empty space, the oxygen atoms are easily penetrating and forming oxidized layer at the boundary surface.

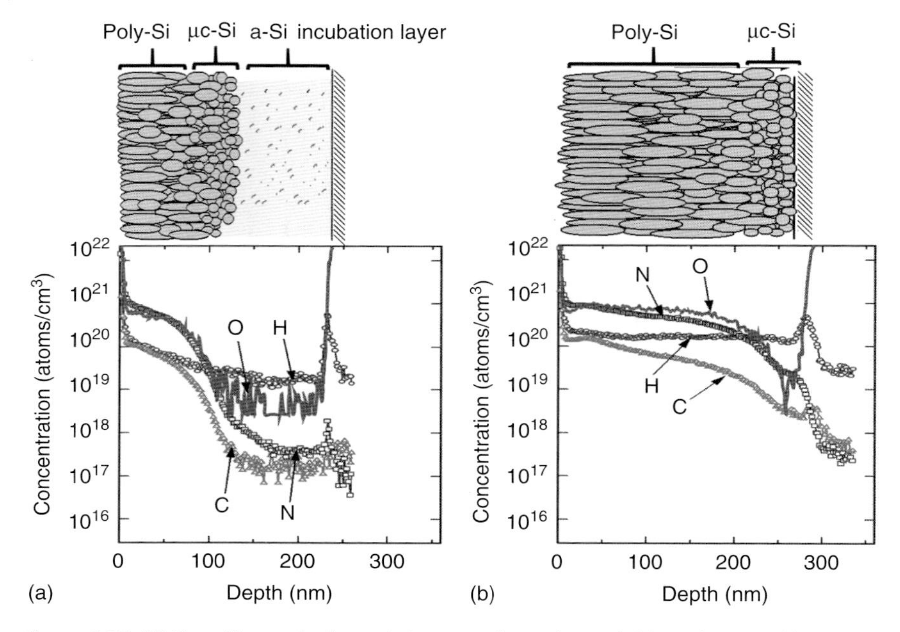

Figure 5.27 SIMS profiles and schematic images of samples with (a) medium H_2 dilution and (b) high dilution of H_2.

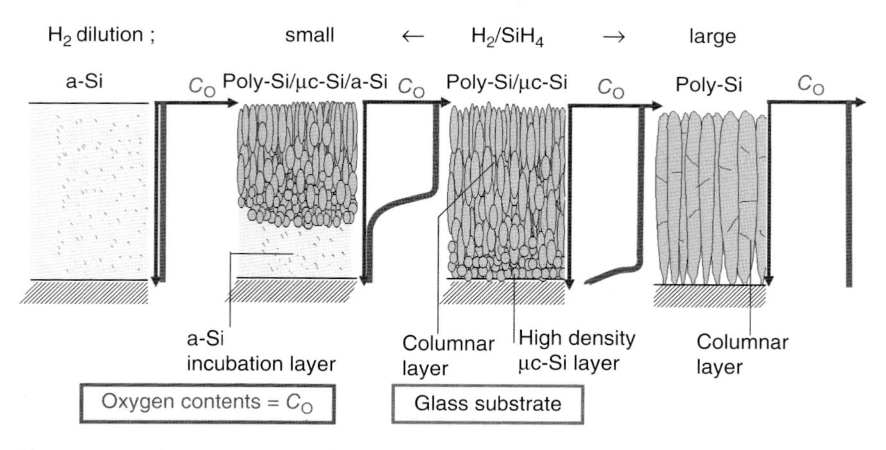

Figure 5.28 Schematic images of structures of poly-Si films prepared with various H_2 dilutions of SiH_4. The profiles of oxygen content (C_O) are also demonstrated. Source: From H. Matsumura et al. 2003 [51].

5.2.3 Properties of Cat-CVD Poly-Si Films

As poly-Si films made by Cat-CVD have structures varying along the growth direction, the properties of films would be dependent on the depth of the films. To know the electrical properties such as mobility of poly-Si films, H. Kasai et al. at SONY Corp. made a TFT with the structure shown in Figure 5.29 [51, 52]. Poly-Si films are deposited on the glass substrates, in this case, quartz substrates.

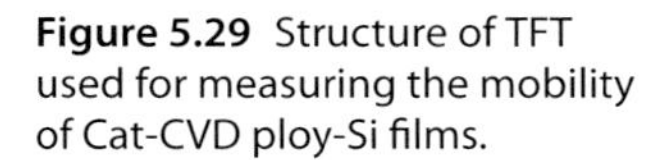
Figure 5.29 Structure of TFT used for measuring the mobility of Cat-CVD ploy-Si films.

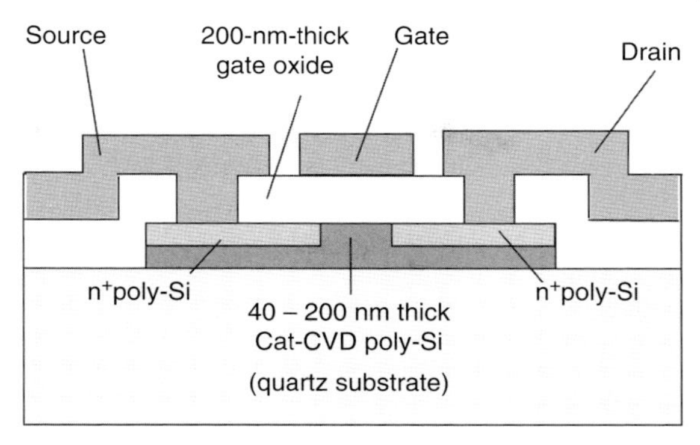

The poly-Si films are etched chemically to make the film with a certain desired thickness. After that, n^+-poly-Si layer are formed by phosphorus ion implantation and successive rapid thermal annealing for two seconds. The gate oxide is deposited by PECVD. After forming metal contacts, the TFT was completed. Poly-Si is made with medium dilution.

By etching of Cat-CVD poly-Si films with original thickness of about 220 nm, various thicknesses of poly-Si films can be obtained. Figure 5.30 shows the crystalline fraction, the oxygen concentration, and the mobility of the TFTs as a function of residual poly-Si thickness. The crystalline fraction does not change much by etching. On the other hand, the oxygen concentration at the surface of poly-Si is high, and it is likely to decrease with the depth of the film. If you look at the mobility, it shows the maximum value at the position where the crystalline fraction is high enough and oxygen concentration is relatively low. The figure shows that the mobility of Cat-CVD poly-Si can reach 40 cm^2/Vs, in spite of the low temperature deposition. The value is a good one for the application of TFT, which will be discussed later in Chapter 8.

The properties of Cat-CVD poly-Si films have also been studied by measuring the electrical properties by forming Van der Pauw patterns [53] or Hall patterns on the Cat-CVD poly-Si films as illustrated in Figure 5.31. In this case, the depth dependence cannot be known. However, as the current can flow in the easiest path, automatically, we can measure the properties of the region of the layer with the highest mobility or the lowest resistivity.

For the carrier density n of the film, it is known that the following relationship, Eq. (5.8), holds for the variation of measuring temperature T [54, 55]. In addition, for the carrier mobility of the film μ, Eq. (5.9) holds,

$$n^2 = AT^3 \exp(-E_g/kT) \tag{5.8}$$

$$\mu = B(L/\sqrt{T}) \exp(-\phi_B/kT) \tag{5.9}$$

where E_g, k, L, and ϕ_B refer to the band gap energy at $T = 0$ K, Boltzmann's constant, grain size, and barrier height at the grain boundary, respectively [55, 56]. Also, A and B refer to the proportional constants.

Figure 5.32 shows n^2/T^3 as a function of the reciprocal T. The gradient of semilogarithmic plots of this relation is equivalent to E_g, and it is derived from the figure to be 1.21 eV. This is exactly the band gap value of c-Si at $T = 0$ K [54].

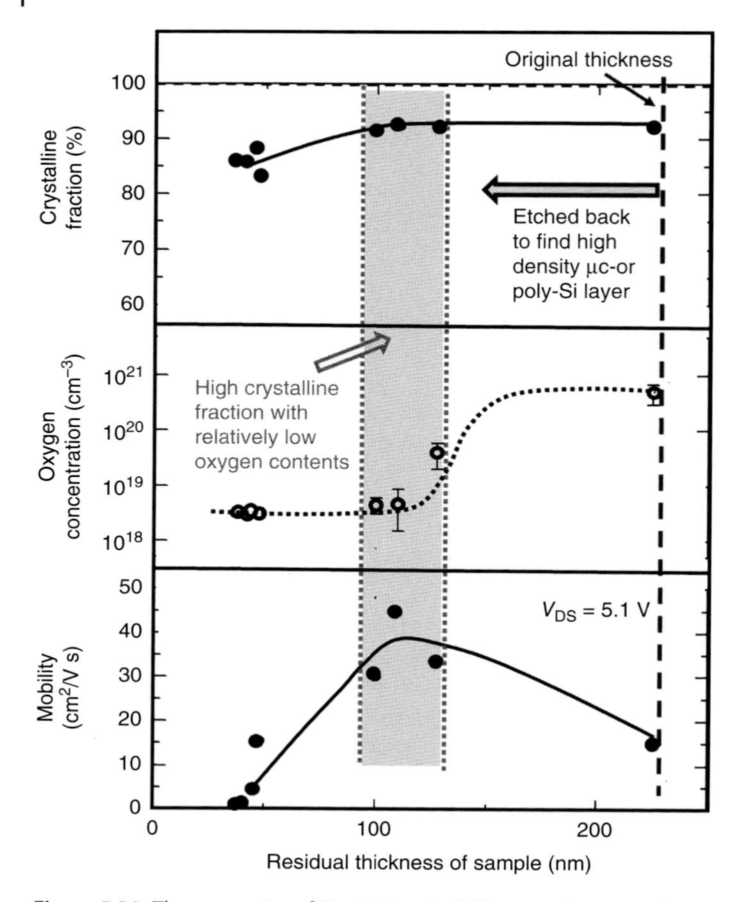

Figure 5.30 The properties of Cat-CVD poly-Si films as a function of thickness or depth.

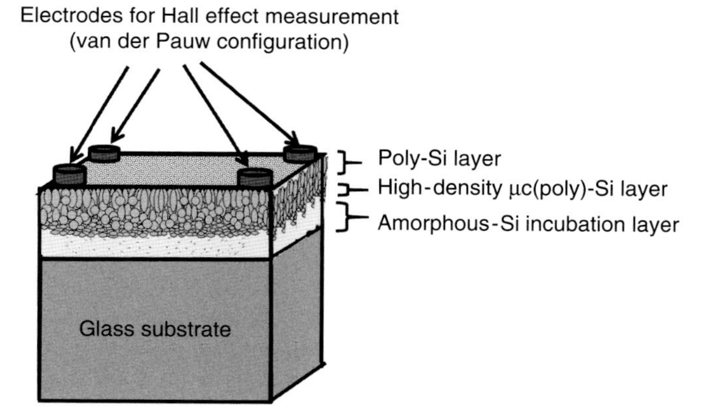

Figure 5.31 Structure of samples for measuring the electrical properties of Cat-CVD poly-Si films.

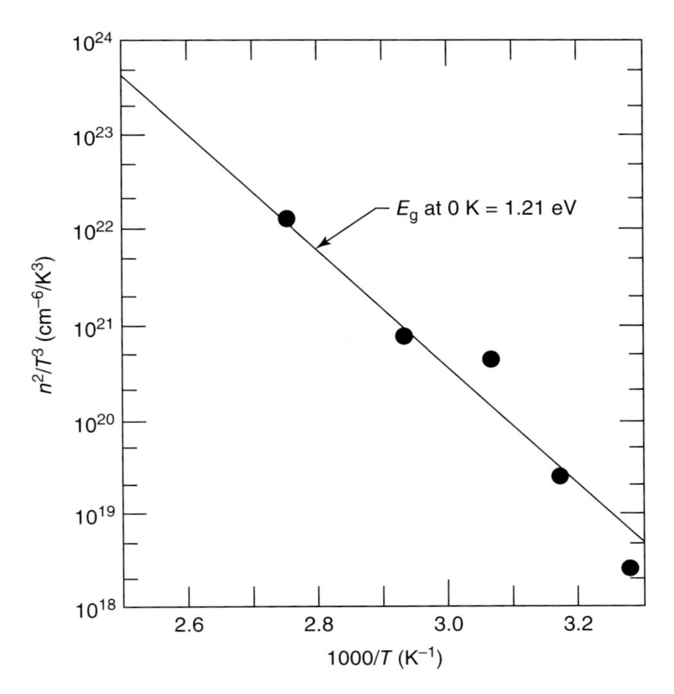

Figure 5.32 Relation between carrier density n and measuring temperature T. Source: Matsumura et al. 1994 [55]. Reprinted with permission of Japanese Journal of Applied Physics.

That is, it can be confirmed that we could definitely measure the property of the film consisting of c-Si grains.

The relationship shown in Eq. (5.9) is derived from the Seto model, in which carriers are transferred through grain boundaries with a certain barrier height ϕ_B [56]. From the gradient of $\mu\sqrt{T}$ vs. $1/T$, ϕ_B can be evaluated as shown in Figure 5.33.

All data concerned with Cat-CVD poly-Si films are summarized in Figure 5.34. In the figure, the grain size derived from the XRD experiment, the barrier height at the grain boundaries, and the mobility are plotted for fixed flow rate of SiH_4, FR(SiH_4), at 1 sccm as a function of H_2 flow rate, FR(H_2). In the preparation of poly-Si films, $T_{cat} = 1800\,°C$, $T_s = 300\,°C$, and $P_g = 0.67$ Pa [33].

5.2.4 Si Crystal Growth on Crystalline Si

We know that poly-Si films are obtained on the glass substrate with or without incubation times or incubation a-Si layers. So, if such poly-Si films are deposited on c-Si, what happens? This was attempted by many groups. Here, our results are demonstrated because as far as the authors know, it may be the first report on epitaxial growth by Cat-CVD [57].

Figure 5.35 shows a TEM cross-sectional view of a Si film prepared with a large H_2 dilution ratio on a Si wafer [57]. This low-temperature Si epigrowth is one of the features of Cat-CVD technology. This is probably due to the high H atomic

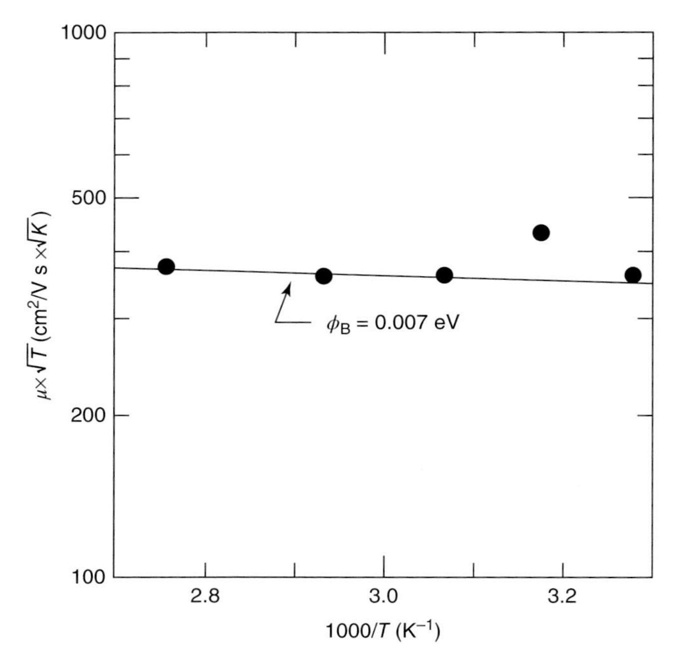

Figure 5.33 Barrier height of Cat-CVD poly-Si grains. Source: Matsumura et al. 1994 [55]. Reprinted with permission of Japanese Journal of Applied Physics.

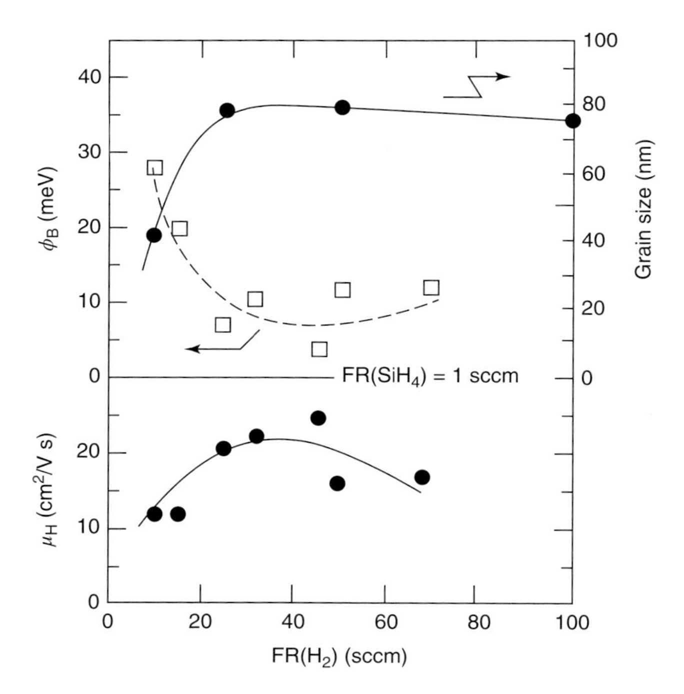

Figure 5.34 Summarized data for Cat-CVD poly-Si for various dilution ratios, measured by the Van der Pauw method at the surface of samples. Source: Matsumura 1998 [33]. Reprinted with permission of Japanese Journal of Applied Physics.

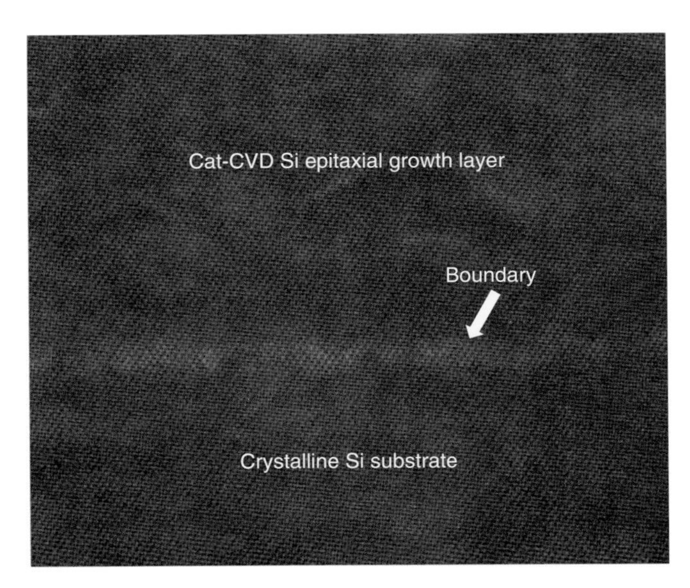

Figure 5.35 Si epitaxial growth on crystalline Si at substrate temperatures around 100 °C. Source: Data from Yamoto et al. 1999 [57].

density in the gas phase of the chamber. This may be applied for the fabrication of integrated circuits by making a new type of transistor with elevated source and drain.

5.3 Properties of Silicon Nitride (SiN_x)

5.3.1 Usefulness of Silicon Nitride (SiN_x) Films

The reason why Si became a dominant material is that Si can be covered with thermally grown high-quality insulating silicon dioxide (SiO_2) films. When Si wafers are exposed to oxygen atmosphere at high temperatures over 800–1000 °C, the surface of Si is converted to SiO_2, and this SiO_2 is an excellent insulator to avoid leakage of electric currents through the surface of Si. In electronic devices, surface coating is one of the most important processes for stable operation of such devices. For Si devices, thermally grown SiO_2 is widely used. However, as the formation of it requires temperatures higher than several hundred degree Celsius at least, it cannot be used in some processes or some semiconductors that cannot be used at high temperatures. In such a case, the insulating films have to be made at temperatures lower than 400 °C, or sometimes lower than 100 °C. For such requirements, the deposition of high-quality insulating films at low temperatures is expected to be the most feasible way. For such a purpose, deposition of SiO_2 is of course important; however, if we deviate from thermally grown processes, a wide variety of insulators is usable. Silicon nitride (Si_3N_4 or SiN_x) is one of the major viable insulators, as high-quality films can be relatively easily obtained, and it is suitable for coating on various semiconductors such as gallium arsenide (GaAs).

SiN_x is a useful material as gate insulator in a-Si TFT. The SiN_x films are also useful for coating plastic substrate or organic materials as gas barrier films. As the application of SiN_x films is so wide, we explain below the quality of SiN_x films prepared by Cat-CVD as typical insulating films.

5.3.2 Fundamentals for the Preparation of SiN_x

Silicon nitride (described by "Si_3N_4" for stoichiometric material, however, in this book, just described as "SiN_x") films are one of the most useful materials for electronic devices, as mentioned above. The SiN_x films have been prepared with various methods such as thermal CVD using various sources and PECVD. SiN_x preparation by Cat-CVD first succeeded in 1987, and this was a relatively new method compared with other known methods. SiN_x films can be obtained by Cat-CVD, using the mixture of pure ammonia (NH_3) and silane (SiH_4) and also the mixture of NH_3, SiH_4 diluted with H_2. In most cases, the former simple NH_3 and SiH_4 mixture is widely used; however, the latter NH_3, SiH_4, and H_2 mixture gives powerful conditions for obtaining good conformal step coverage. As a method using safe source gases, there are reports to prepare Cat-CVD SiN_x films using a mixture of hexamethyldisilazane (HMDS), NH_3, and H_2. They are explained later in this book. Contrary to these methods, in the case of PECVD, sometimes, N atoms are provided from N_2 gas instead of NH_3. As N_2 cracking is not easy on heated W wires, in most cases, NH_3 is used as the N source in Cat-CVD.

Typical parameters for SiN_x deposition by Cat-CVD are summarized in Table 5.4. When H_2 is part of the mixture, the flow rate of NH_3 gas is usually much larger than that of SiH_4, and this situation can also be observed in PECVD. However, when SiH_4 and NH_3 mixture is highly diluted by H_2, the mixing ratio of NH_3 to SiH_4 becomes almost unity. This case will be explained later.

5.3.3 SiN_x Preparation from NH_3 and SiH_4 Mixture

The most success in the initial stage of development of Cat-CVD technology was the deposition of silicon nitride (SiN_x). When we attempted to make SiN_x films, it was not easy to obtain it, as the gas mixing ratio of ammonia (NH_3) gas to silane (SiH_4) gas was so large for the gas pressure used at the first experiment. The relationship among the various flow rates of SiH_4, FR(SiH_4), for the fixed flow rate of NH_3, the deposition rate to obtain stoichiometric SiN_x with the refractive index of about 2.0 and gas pressure during deposition are demonstrated in Figure 5.36 [38].

The results shown in Figure 5.36 suggests that (i) the deposition rate depends mainly on the flow rate of SiH_4 for a fixed NH_3 flow rate and (ii) the mixing ratio of NH_3 to SiH_4 to obtain stoichiometric SiN_x with the refractive index of 2.0 is likely to reduce as a pressure during the deposition increases. That is, when the gas pressure is low, we have to introduce a large amount of NH_3 gas to keep the large mixing ratio. As the gas pressure increases, the deposition rate increases and the NH_3 mixing ratio can be reduced to obtain films with the refractive index of 2.0. To obtain SiN_x by Cat-CVD, this is the fundamental guiding principle.

Table 5.4 Examples of deposition parameters of Cat-CVD SiN_x films.

	Setting values of parameters			
Parameters	**ULVAC**	**Our group-1**	**Our group-2**	**Utrecht group [58]**
Material of catalyzer	W wire	W wire	W wire	Ta wire
Deposition area	<20 cmØ	$<10\times10\,cm^2$	$20\times20\,cm^2$	$5\times5\,cm^2$
Catalyzer–substrate distance, D_{cs} (cm)	<5	3–5	10–15	4
Diameter of wire (mmØ)	0.5	0.5	0.5	0.5
Length of wire (cm)	300	200	300	60
Surface area of catalyzing, S_{cat} (cm^2)	50	33	50	18.8
Temperature of catalyzer, T_{cat} (°C)	1900	1750	1750	2100
Temperature of substrate, T_s (°C)	40–600	200–400	200–400	450
Gas pressure, P_g (Pa)	0.13–1.3	1–6	10–20	20
Flow rate of SiH_4, FR(SiH_4) (sccm)	7	1	6	22
Flow rate of NH_3, FR(NH_3) (sccm)	10	30–100	300	300
Flow rate of H_2, FR(H_2) (sccm)	30	0	0	0
Typical deposition rate, D_R (nm/s)	1	0.2–0.5	0.5–1.5	3.1
Typical refractive index	—	1.95–2.05	2.00–2.05	1.96
Coverage at steps	Conformal	—	—	Conformal on textured structure

Ø refers to a diameter.

The number of molecules colliding with the catalyzer surface depends on the gas pressure P_g or the density of molecules during deposition and their thermal velocity, as explained in Chapter 2.

The thermal velocity is dependent on the mass of molecules: it is inversely proportional to the root of the mass number. Comparing the dependence on the gas pressure of the number of NH_3 molecules colliding with that of SiH_4, the number of NH_3 is larger than that of SiH_4 by about 1.4 times, as the mass number of NH_3 is 17 and that of SiH_4 is 32. This means that an increase of gas pressure is more effective in increasing the number of decomposed species from NH_3 than that of SiH_4. As explained in Chapter 4, five sites are required for the decomposition of SiH_4 on the catalyzer surface, but only two sites are enough for NH_3 decomposition. Thus, even if the number of SiH_4 molecules attacking on the catalyzer surface increases by increasing P_g, the decomposition efficiency would not increase proportionally to it, although NH_3 decomposition would be easier than SiH_4 decomposition, even if the number of NH_3 molecules attacking on the catalyzer surface is 1.4 times larger than that of SiH_4 molecules. Probably for these reasons, the mixing ratio is likely to be reduced as the gas pressure increases.

The structural properties, such as H contents in Cat-CVD SiN_x films, can be easily determined by the measurement of IR absorption spectrum.

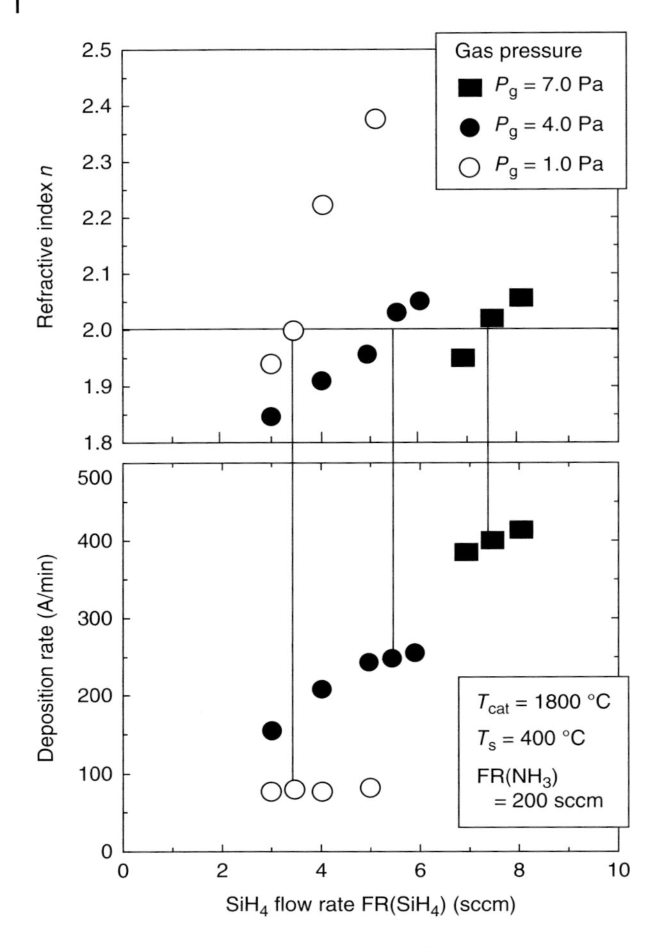

Figure 5.36 Refractive index of deposited SiN_x and deposition rates are plotted as a function of the flow rate of SiH_4, FR(SiH_4), taking gas pressure P_g as a parameter. Source: Matsumura 2001 [38]. Reprinted with permission of The Institute of Electronics, Information and Communication Engineers.

Figure 5.37 shows the typical IR spectrum of a Cat-CVD SiN_x film, which is prepared at T_{cat} = 1750 °C, T_s = 280 °C, P_g = 5 Pa, and the flow rates of SiH_4, FR(SiH_4) = 1 sccm, and that of NH_3, FR(NH_3) = 100 sccm. The spectrum is compared with SiN_x films prepared by PECVD for T_s = 250 and 320 °C [59]. The IR absorption by the stretching vibration of Si—N bonds is observed at about 880 cm^{-1}, and the IR absorptions by the stretching vibration of Si—H bonds and N—H bonds are observed at 2180 and 3320 cm^{-1}, respectively, and the N–H bending vibration is at 1180 cm^{-1} [60]. W.A. Lanford and M.J. Rand studied on the way to estimate H contents from the area of the IR absorption peaks [61], based on similar thinking way to Section 5.1.2.2.

The absorbance is expressed with arbitrary units and for all samples normalized to the peak height of the Si–N stretching vibration at 830 cm^{-1}, just for the comparison of relative peak heights or relative contents of Si–H, N–H stretching

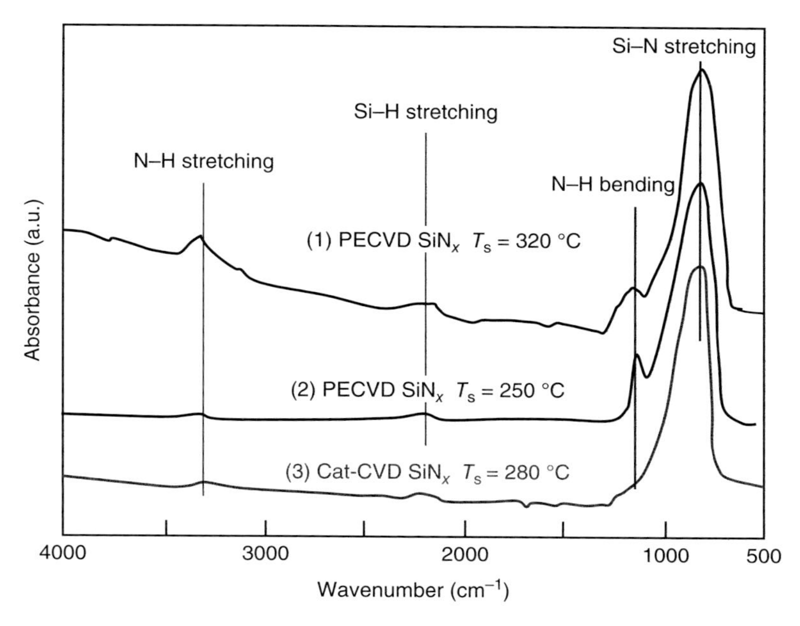

Figure 5.37 Infrared (IR) absorption spectra of typical Cat-CVD SiN_x and PECVD SiN_x films. PECVD SiN_x spectra are cited from two different reports. Source: Parsons et al. 1991 [59]. Reproduced with permission of Japanese Journal of Applied Physics.

vibrations, and N–H bending vibrations. The refractive index of all films is around 2.0, and the atomic ratio of N to Si is about 1.3 or slightly larger, that is, near to stoichiometric Si_3N_4. Two PECVD samples in the figure are particularly prepared by SiH_4, N_2, and He, to reduce the H content in the SiN_x films [59]. The authors of Ref. [59] claimed that they could make SiN_x films with low C_H, by PECVD. However, comparing the Si–H and N–H stretching and N–H bending peak heights among various samples, N–H of PECVD SiN_x appears still larger than that of the Cat-CVD film. That is, C_H in Cat-CVD SiN_x is definitely smaller than that in PECVD SiN_x.

The atomic ratio of N to Si atoms is also evaluated from X-ray photoelectron spectroscopy (XPS) or Rutherford backscattering (RBS). As an example, our XPS result is demonstrated in Figure 5.38. In this case, Cat-CVD SiN_x films are deposited at $T_{cat} = 1800\,°C$, $T_s = 300\,°C$, $P_g = 1$ Pa, and the flow rates of SiH_4, FR(SiH_4) = 1.1 sccm and that of NH_3, FR(NH_3) = 60 sccm. The refractive index n of the films is about 2.0. The refractive index of device quality Cat-CVD SiN_x is usually 1.95–2.05, depending on the H content. When the H content is low, say, for instance, lower than 5 at.%, n is around 2.00–2.05. In the figure, O atoms observed at the surface region is due to the influence of adsorbed O atoms on the surface, and the spectra only after 400 seconds etch time have meaning.

The figure demonstrates that the film is made uniform along the depth. The atomic ratio of N/Si is about 1.33. In the case of Cat-CVD SiN_x with low H content, the value is likely to be near to stoichiometry, 1.33. If you compare the similar properties with PECVD SiN_x films prepared at the same T_s, the H content is apparently smaller than that of PECVD films.

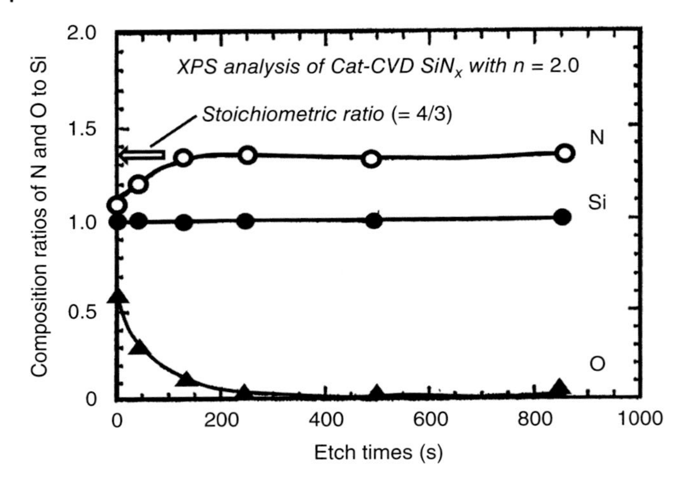

Figure 5.38 Depth profile of N, Si, and O contents in Cat-CVD SiN_x, measured by X-ray photoemission spectroscopy (XPS). Etching time corresponds to the depth. Hundred seconds etching time is almost equivalent to 100 nm in depth.

Figure 5.39 shows the densities of N—H bonds and Si—H bonds in Cat-CVD SiN_x films to evaluate C_H in the films as a function of the refractive index of SiN_x films, n. The Cat-CVD films are prepared at $T_{cat} = 1800\,°C$, $T_s = 350\,°C$, $P_g = 4$ Pa, the flow rate of SiH_4, FR(SiH_4) = 3–7 sccm, and that of NH_3, FR(NH_3) = 200 sccm. The refractive index was changed by changing SiH_4 flow rates. When n is lower than 2.04, N—H bonds are dominant to keep H atoms inside SiN_x films. When n exceeds over 2.04, both the densities of N–H and Si—H bonds are low; the H content is only a few atomic percent.

In the figure, similar results of PECVD SiN_x films, which are reported in Ref. [59], are also included for comparison. The authors of the report made efforts to obtain low C_H SiN_x, and indeed, we can say that the result shows the property of relatively low C_H SiN_x. However, in spite of the efforts, as shown in the figure, C_H of PECVD SiN_x is still larger than that of Cat-CVD SiN_x, if we compare the results with the stoichiometric Cat-CVD SiN_x films with $n = 2.0$.

The dense films are also effective to achieve chemical resistivity against etching by buffered hydrofluoric acid (BHF) solution. Figure 5.40 shows the etching rates in 5% buffered HF solution as a function of n of Cat-CVD SiN_x. In the figure, typical etching rates of PECVD SiN_x in the same etchant are indicated for comparison, although even in PECVD SiN_x, by reducing C_H, the etching rate can be reduced [61]. The results for general PECVD films and those for specially modified PECVD SiN_x are shown by a hashed bar. It is apparent that Cat-CVD SiN_x films are resistive against chemical etching because of large atomic density of Si and N atoms.

The reason why hydrogen contents in Cat-CVD SiN_x can be low and the films become dense is probably the same as for the Cat-CVD a-Si films. As high-density H atoms exist in the gas phase, they can abstract other H atoms from the growing film surface and reduce the number of H atoms left inside the films. This happened even in SiN_x films prepared at low T_s. Figure 5.41 shows the atomic density

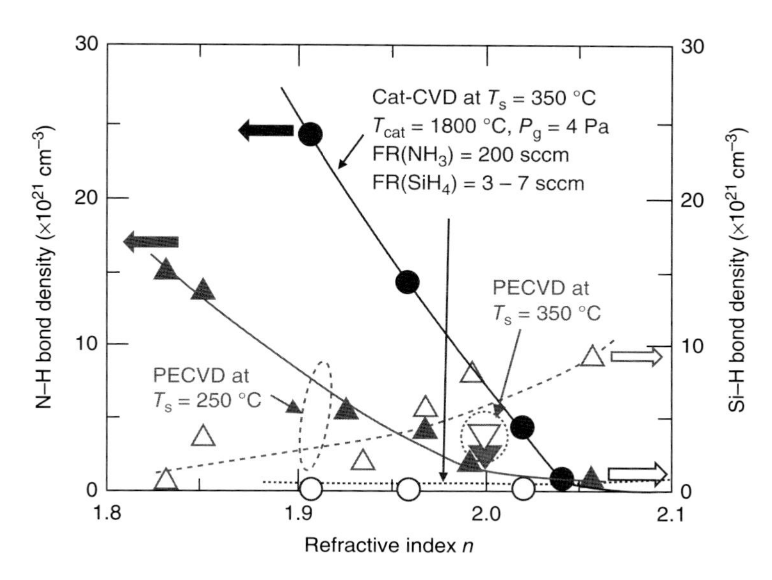

Figure 5.39 Hydrogen content estimated from the densities of N—H bonds and Si—H bonds in the SiN_x films. Results for PECVD and Cat-CVD films are demonstrated. Cat-CVD SiN_x film is deposited at $T_s = 350\,°C$ only, whereas PECVD ones are deposited at $T_s = 250$ and $350\,°C$. Source: Parsons et al. 1991 [59]. Reproduced with permission of Japanese Journal of Applied Physics.

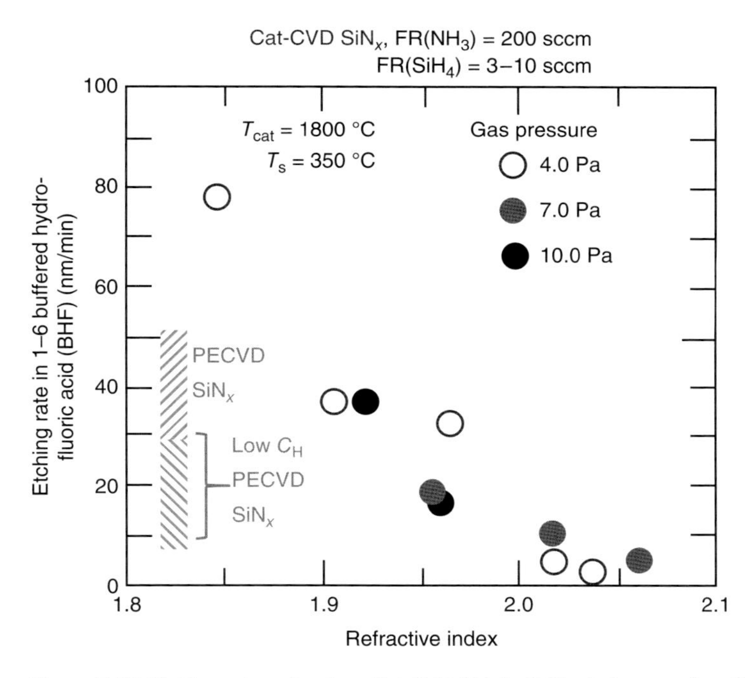

Figure 5.40 Etching rates of various Cat-CVD SiN_x in BHF solution as a function of refractive index of the films.

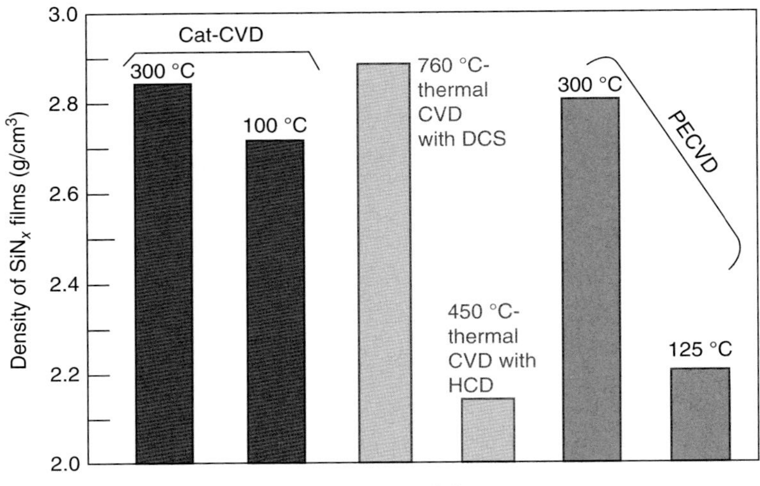

Figure 5.41 Density of SiN_x films prepared by various methods with various substrate temperatures described in the figure.

of Cat-CVD SiN_x deposited at T_s = 300 and 100 °C. Similar atomic density of thermal CVD SiN_x films prepared by using dichlorosilane (DCS) at T_s = 760 °C and hexachlorodisilane (HCD) at T_s = 450 °C, and PECVD SiN_x films prepared at T_s = 300 °C and 125 are shown for comparison. It is apparent that Cat-CVD SiN_x films are quite dense and that the atomic density is equivalent to that of high-temperature thermal CVD films.

The high density of the films leads to their possible use as gas barrier films. Figure 5.42 shows IR absorption spectra of PECVD SiN_x deposited at T_s = 320 °C and Cat-CVD SiN_x at T_s = 280 °C before and after a pressure cooker test (PCT). The data before PCT are the same as those shown in Figure 5.37. The PCT is a test to know the resistivity of samples to moisture penetration or moisture-induced modification of the film itself. In this case, the samples are left for 96 hours at 120 °C in H_2O vapor at a pressure of 2 bar. Although the test is hard, many electronic devices guaranteeing their lifetimes over 100 years should pass this test. In the figure, it is demonstrated that almost no change is found in Cat-CVD SiN_x film after PCT, but that the PECVD SiN_x film is oxidized or oxygen has penetrated through the SiN_x films to oxidize the c-Si substrate on which the film is deposited. Figure 5.37 demonstrates that Cat-CVD films have high gas barrier ability.

This gas barrier property is also observed in Cat-CVD SiN_x films deposited at T_s lower than 100 °C. However, for a reduced T_s, other recipes become necessary.

5.3.4 SiN_x Preparation from Mixture of NH_3, SiH_4, and a Large Amount of H_2

When T_s is lowered, the activity of removing hydrogen from the growing surface becomes low, and thus, the hydrogen content in the film becomes large. That is,

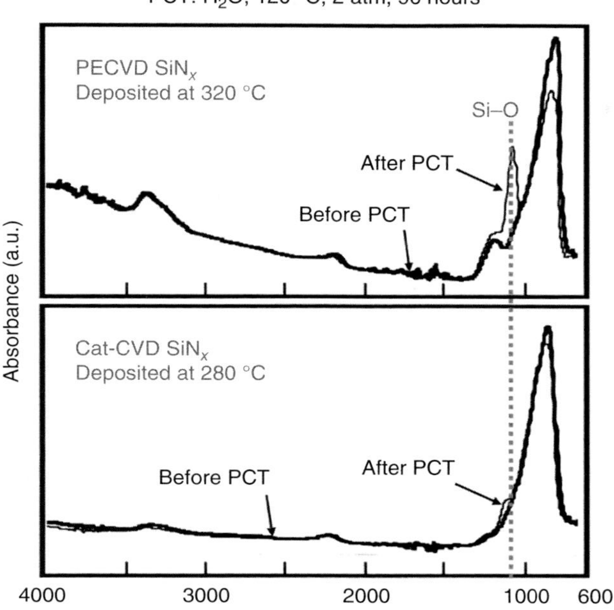

Figure 5.42 Infrared absorption spectra of PECVD and Cat-CVD SiN_x films before and after pressure cooker test (PCT). PCT is a test keeping samples in 100% H_2O vapor at 120 °C for 96 hours.

when T_s is lowered, the atomic density of such films becomes low and they lose gas barrier ability. To avoid this, when T_s is lowered, a large amount of H_2 gas is added into the mixture of SiH_4 and NH_3 for obtaining proper SiN_x films. These deposition conditions are also useful to obtain films with conformal step coverage as described later. Typical deposition parameters for this low-temperature deposition are summarized in Table 5.5 [62]. The Cat-CVD chamber is the same as that of "our group" shown in Table 5.2 or Table 4.1, although some of the parameters are different.

As shown in Table 5.5, even if S_{cat} increases and D_{cs} decreases, the substrate temperatures can be still kept lower than 100 °C under strong thermal radiation. The substrate temperatures can be kept low by cooling the substrate holder. This control of substrate temperatures is mentioned in detail in Chapter 7.

As shown in Figure 5.36, when the mixing ratio of SiH_4 to NH_3 increases, the deposition rate is likely to increase, and the gas pressure for obtaining the films with refractive index of 2.0, which is a measure of stoichiometric SiN_x, is likely to increase. This tendency is still surviving even for highly H_2-diluted conditions. In Table 5.5, $P_g = 35$ Pa is the best condition to obtain $n = 2.0$ films in the conditions of apparatus, but at $P_g = 50$ Pa, the flow rate of SiH_4, 30 sccm, may not be enough to obtain stoichiometric films.

The 24-hour PCT results of low-temperature deposited Cat-CVD SiN_x films are demonstrated in Figure 5.43. Cat-CVD SiN_x films in the figure are prepared

Table 5.5 Deposition parameters of low-temperature deposited Cat-CVD SiN_x films.

	Condition-1	Condition-2	Condition-3
Catalyzing materials (5 mmØ wire)	W	W	W
Total surface area of catalyzer, S_{cat} (cm^2)	44	64	64
Distance between catalyzer and substrate, D_{cs} (cm)	20	5	5
Temperature of catalyzer, T_{cat} (°C)	1750–1800	1750–1800	1750–1800
Gas pressure, P_g (Pa)	20	35	50
Flow rate of SiH_4, FR(SiH_4) (sccm)	10	30	30
Flow rate of NH_3, FR(NH_3) (sccm)	20	20	20
Flow rate of H_2, FR(H_2) (sccm)	400	400	400
Substrate temperature, T_s (°C)	<100	<100	<100
Refractive index, n	2.0	2.0	1.95
Deposition rate (nm/min)	10	110	85

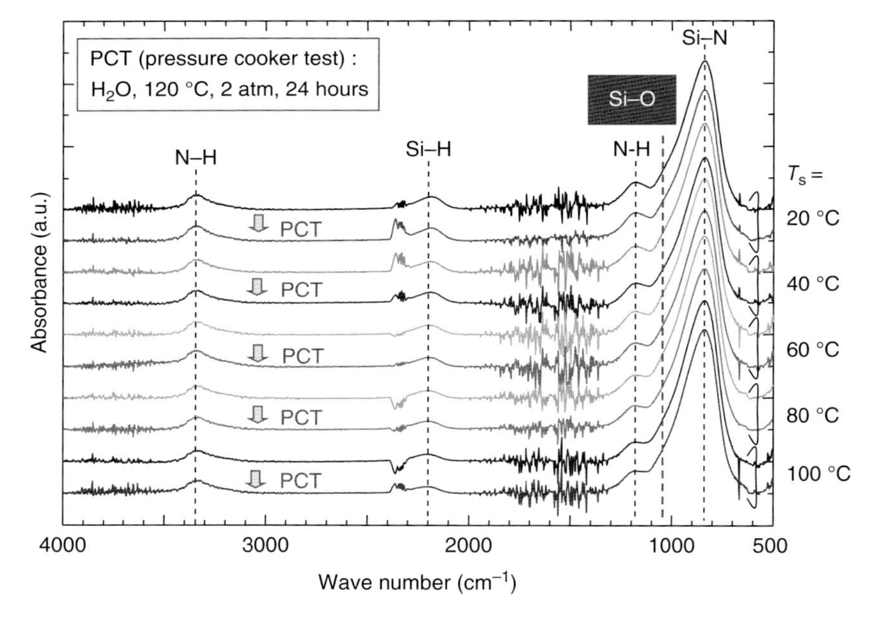

Figure 5.43 Infrared absorption spectra of Cat-CVD SiN_x films deposited at various substrate temperatures, before and after 24 hours PCT.

by condition-1, that is, at T_{cat} = 1800 °C, P_g = 20 Pa, FR(SiH_4) = 10 sccm, FR(NH_3) = 20 sccm, and FR(H_2) = 400 sccm. In this case, PCT is carried out for 24 hours at 120 °C in H_2O vapor at a pressure of 2 bar.

The stress of Cat-CVD SiN_x films has also been studied by various groups. Figure 5.44 shows the inner stress of the samples shown in Figure 5.43, prepared at low temperatures. Inner stress includes the intrinsic stress and thermal stress

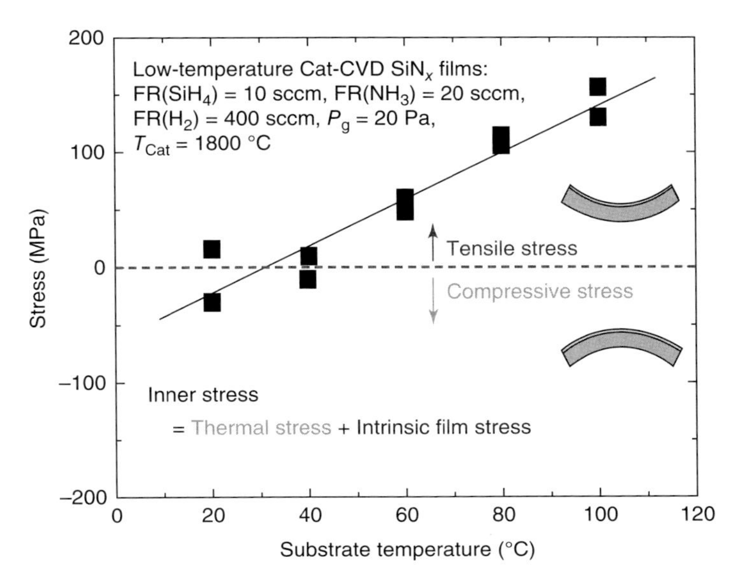

Figure 5.44 Stress of Cat-CVD SiN_x films deposited on crystalline silicon at low substrate temperatures.

because of the difference of thermal expansion between the films and substrates. In this case, Cat-CVD SiN_x films are deposited on c-Si substrates. In the figure, the inner stress is shown as a function of substrate temperature during deposition. As shown in the figure, it is clear that the film stress is relatively low and it can be controlled by T_s. The value of stress can also be adjusted by changing the mixing ratio of H_2.

Figure 5.45 also shows the results of PCT for Cat-CVD SiN_x films prepared with relatively high deposition rates such as the condition-2 or condition-3 in Table 5.5. However, only P_g is varied from 10 to 50 Pa. The films made with the deposition rates larger than 100 nm/min can still have the ability of a gas barrier. In the figure, it can be seen that when P_g is lower than 20 Pa, the gas barrier ability is lost. When P_g is less than 35 Pa, the films become Si rich, far from stoichiometric ones, and lose their gas barrier ability. From this, keeping stoichiometric value of $n = 2.0$ appears essential to obtain films with gas barrier ability.

The figure demonstrates the possibility of Cat-CVD SiN_x films as gas barrier films used in organic devices or for other packaging purposes. This application is mentioned in Chapter 8.

5.3.5 Conformal Step Coverage of SiN_x Prepared from the Mixture of NH_3, SiH_4, and a Large Amount of H_2

The recipe using a large amount of H_2 induces another advantage of Cat-CVD SiN_x. In 2004, Q. Wang et al. discovered that when SiN_x films were deposited on substrates with fine structures by using highly H_2-diluted mixtures of SiH_4 and NH_3, such fine-structured substrates are covered with conformal Cat-CVD SiN_x films [63]. The experiments have been checked by many groups.

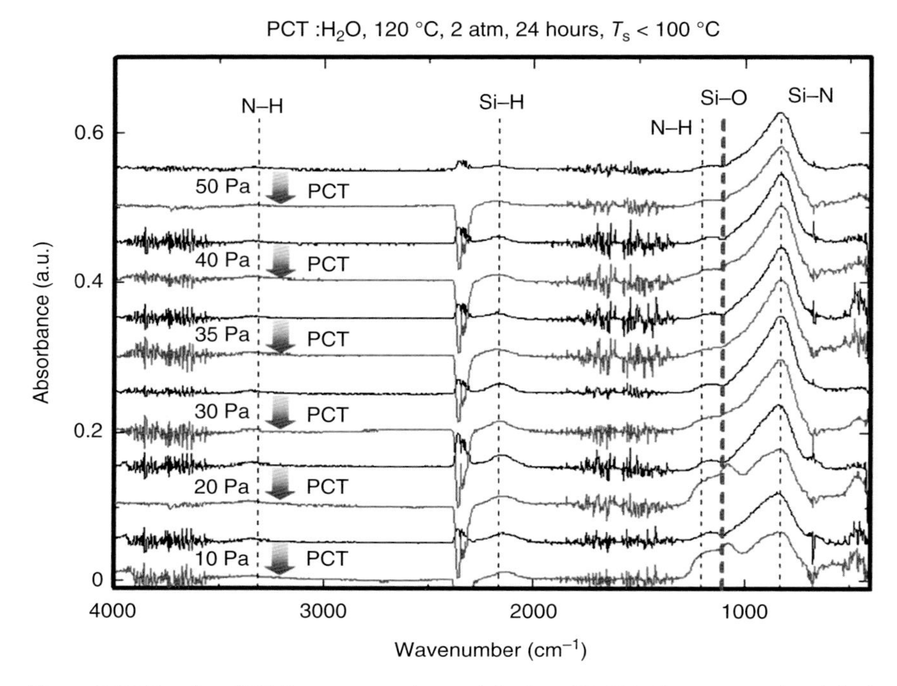

Figure 5.45 Results of PCT (pressure cooker test) for Cat-CVD SiN_x films prepared with high deposition rates over 100 nm/min for various gas pressures P_gs.

Figure 5.46 shows the photographs of cross-sectional images of c-Si samples covered with Cat-CVD SiN_x, taken by scanning electron microscope (SEM). The ratios of film thickness deposited on the top, bottom, and side are described in the figure. (The photographs are provided from ULVAC Co. Ltd. Japan, based on their report [64].) The figure demonstrates that by using highly H_2-diluted conditions, conformal step coverage is possible with Cat-CVD SiN_x films.

The reason why H_2 dilution realizes such conformal coverage is not clear. As mentioned in Chapter 4, on a W catalyzer surface, NH_3 is decomposed to $NH_2 + H$, and also high-density SiH_3 exists in the gas phase in a Cat-CVD chamber. These NH_2 and SiH_3 are adsorbed on the surface of the substrate. Of course, it may be safe to say that NH_x and SiH_y are absorbed on the substrate surface. Such adsorbed species are converted to N or Si or more chemically active forms by attack of H atoms. As most of the SiN_x forming reactions are carried out on the substrate surface with the help from atomic H, the coverage becomes conformal.

Finally, the comparison of typical properties of Cat-CVD SiN_x films with other SiN_x films prepared at normal substrate temperatures of 250–400 °C for PECVD and Cat-CVD and at 700–900 °C for thermal CVD is summarized in Table 5.6 [65]. The superiority of Cat-CVD SiN_x to other SiN_x films is clearly demonstrated.

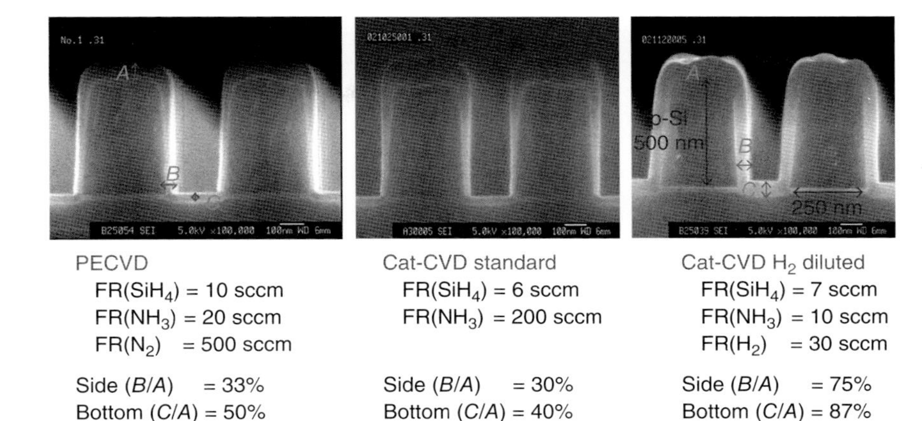

Figure 5.46 SEM cross-sectional images for Cat-CVD SiN_x films deposited on substrates with fine structures of 500 nm height and 250 nm width. The thicknesses at the top, side, and bottom are symbolized by *A*, *B*, and *C*, respectively. Source: Courtesy of ULVAC [64].

Table 5.6 Typical properties of stoichiometric SiN_x films prepared by various methods. BHF refers to buffered hydro-fluoric acid.

	Thermal CVD (700–900 °C)	PECVD (250–400 °C)	Cat-CVD (250–400 °C)
Refractive index, n	2.0–2.1	1.9–2.0	1.9–2.0
Deposition rate (nm/min)	30	30–300	5–180 [65]
Breakdown field (MV/cm)	10	5–10	5–10
H content, C_H (at.%)	2–3	10–20	<3
Stress (MPa)	1000–2000	400	70–400
Etching rate by BHF (nm/min)	2	20–50	2–5

5.3.6 Cat-CVD SiN_x Prepared from HMDS

In the fabrication of SiN_x films, SiH_4 gas is used in the previous sections. However, SiH_4 is not a safe gas to handle, and thus, the use of SiN_x films is not widely spread in various fields of industries, apart from semiconductor industry. Thus, if the films are obtained by using safe sources, the use of SiN_x films will spread even more widely. As a solution to it, A. Izumi and K. Oda succeeded in obtaining high-quality SiN_x films by Cat-CVD using HMDS (structural formula: $(CH_3)_3$–Si–NH–Si–$(CH_3)_3$) and NH_3 [66]. The quality of these SiN_x films is lower than that of SiN_x using SiH_4, but the results appear promising. In their first report, they did not control the flow rate of HMDS, probably because HMDS was vaporized by pumping, and such a vaporized HMDS gas was introduced directly into the deposition chamber.

Table 5.7 Various deposition conditions for preparing Cat-CVD SiN_x films from HMDS.

	First experiment	Avoiding W carburation
Flow rate of HMDS (sccm)	1	0.5
Flow rate of NH_3, FR(NH_3) (sccm)	—	50
Flow rate of H_2, FR(H_2) (sccm)	—	40
Catalyzer temperatures, T_{cat} (°C)	1900	1900
Gas pressure, P_g (Pa)	1	50
Distance between catalyzer and substrate (cm)	8	8
Temperature of cooling liquid of substrate holder	Water cooling	Water cooling

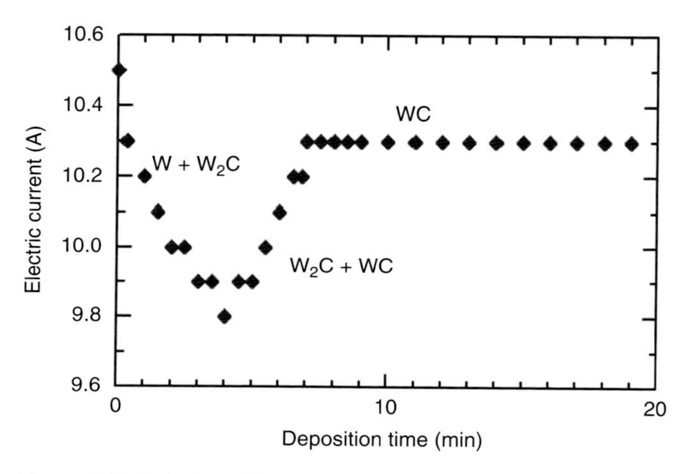

Figure 5.47 Variation of heating currents of W catalyzer in HMDS atmosphere.

Table 5.7 shows the deposition conditions used in the experiments of our group, using the same chamber as shown in Figure 4.2 and Table 4.1.

When HMDS gas was introduced into the Cat-CVD chamber using a W catalyzing wire, the W surface was immediately carburized (in this book, the term "carburized" or "carburization" is used instead of using "carbided" or "carbidation.") and the deposition became unstable. Figure 5.47 shows the variation of electric currents that are fed through the W wire to keep the wire temperature constant, as a function of deposition time. The current is very unstable. Figure 5.48 also shows XRD spectra of a W catalyzer, which was used for five minutes in the deposition process. In the figure, the surface materials formed after the exposure of HMDS gas are indicated. It is clear that the instability of the current to keep the wire temperature constant is caused by the change in the nature of the surface of the W catalyzer due to carburization. When a W catalyzer is carburized, it is easily broken.

To avoid this carburization, T_{cat}, P_g, FR(NH_3), and FR(H_2) varied in a wide range, and finally, it was discovered that an increase of P_g and FR(NH_3) are effective to avoid carburization of the W catalyzer. This condition is also shown in

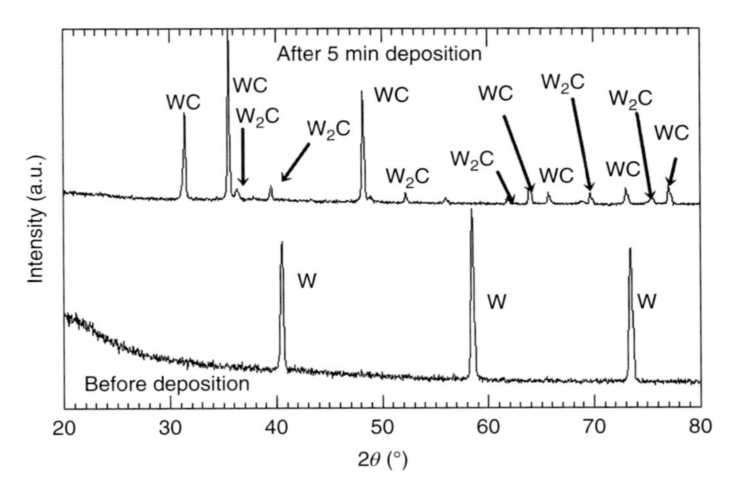

Figure 5.48 XRD spectra of W catalyzing wire before use and after five minutes deposition of SiN_x films using HMDS.

Table 5.7. The IR absorption spectra of SiN_x films, which are actually SiN_xC_y films since carbon (C) atoms are included, are demonstrated in Figure 5.49. In the figure, the spectra are shown by taking P_g as a parameter. When P_g is less than 30 Pa, the W catalyzer is likely to be carburized, but the carburization is suppressed when P_g becomes 50 Pa. From the figure, Si—(CH_3) bonding peaks are likely to decrease when the film is made using the conditions to suppress the carburization. In addition, the films made at P_g lower than 30 Pa appear to be oxidized easily after deposition.

5.4 Properties of Silicon Oxynitride (SiO_xN_y)

5.4.1 SiO_xN_y Films Prepared by SiH_4, NH_3, H_2, and O_2 Mixtures

Cat-CVD SiN_x is a dense and hard material, and thus, its properties are useful for application as coatings for semiconductors. However, sometimes, more flexible and soft films are required to cover soft substrates. There is also a requirement to obtain films with even lower refractive index than that of SiN_x. In addition, there is another requirement to obtain stress-controllable films. For satisfying such requirements, silicon oxynitride (SiO_xN_y) films have been investigated for a long time, using PECVD or other methods. Therefore, formation of good-quality SiO_xN_y films by Cat-CVD is also expected to be possible.

Table 5.8 shows an example of deposition parameters for SiO_xN_y films by Cat-CVD using a mixture of SiH_4, NH_3, H_2, and helium (He)-diluted O_2. Introduction of a large amount of oxygen (O_2) oxidizes the surface of W catalyzer, and as the melting temperature of WO_x is only about 1500 °C, the films are easily contaminated by W because of the evaporation of WO_x. However, the O_2 introduction does not damage the W catalyzer when reducing gases are the majority in the mixture.

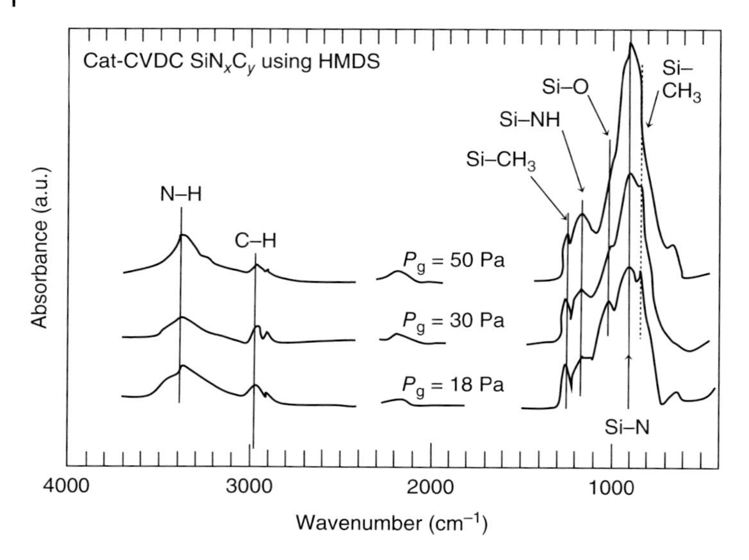

Figure 5.49 Infrared absorption spectra for Cat-CVD SiN_xC_y films prepared with a mixture of HMDS of FR(HMDS) = 0.5 sccm, $FR(NH_3)$ = 50 sccm and $FR(H_2)$ = 40 sccm, T_s<80 °C, and P_g is changed as a parameter.

Table 5.8 Deposition parameters of SiO_xN_y films by Cat-CVD.

Parameters	Conditions
Catalyzing material	W wire
Catalyzer temperature, T_{cat} (°C)	1750–1800
Total surface area of catalyzer, S_{cat}(cm^2)	44
Distance between catalyzer and substrate, D_{cs}(cm)	20
Gas pressure, P_g(Pa)	20
Flow rate of SiH_4, $FR(SiH_4)$ (sccm)	10
Flow rate of NH_3, $FR(NH_3)$ (sccm)	20
Flow rate of H_2, $FR(H_2)$ (sccm)	400
Net flow rate of O_2, $FR(O_2)$, 2% in He (sccm)	0–10
Substrate temperature, T_s(°C)	80

Figure 5.50 shows the IR absorption spectra of SiO_xN_y films made by Cat-CVD using a mixture of SiH_4, NH_3, H_2, and O_2, taking the flow rate of mixing O_2 gas as a parameter. The O_2 gas is actually diluted to 2% in He gas. The flow rates of O_2 are expressed by net values (the flow of pure O_2). It can be confirmed that the IR absorption peaks due to Si—O bonds increases as the O_2 mixing ratio increases, whereas the peaks due to Si—N bonds decrease.

Figure 5.51 also shows the refractive index of films of similar Cat-CVD SiO_xN_y films as a function of flow rate of mixing O_2 gas. The figure shows that a small

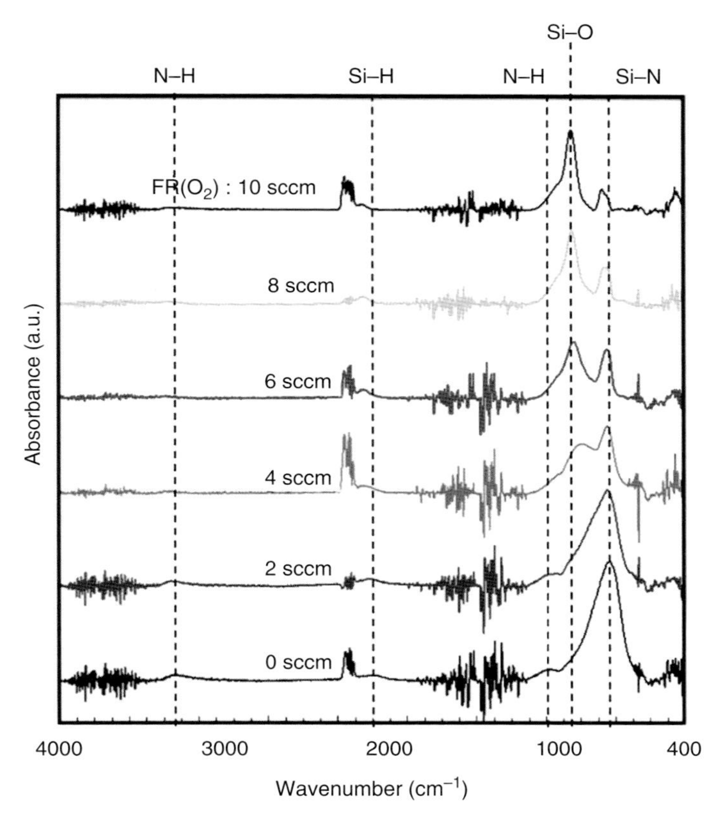

Figure 5.50 Infrared absorption spectra of Cat-CVD SiO_xN_y films. Source: Ogawa et al. 2008 [67]. Reprinted with permission of Elsevier.

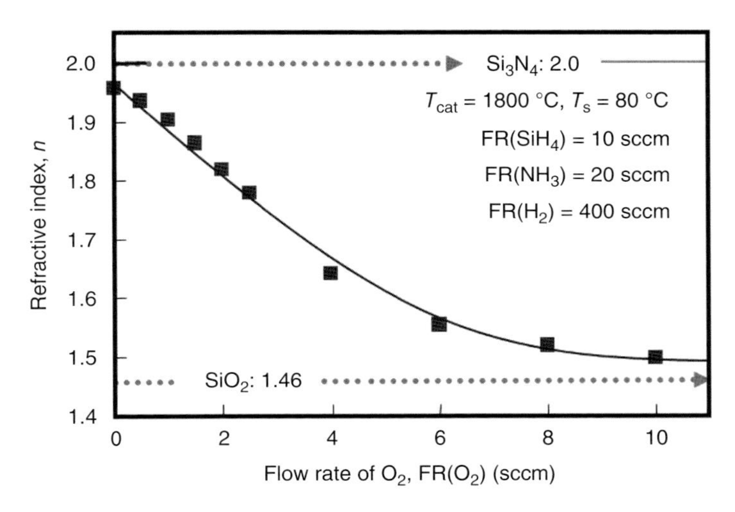

Figure 5.51 Refractive index of Cat-CVD SiN_xO_y films as a function of flow rate of mixing O_2 gas. Source: Ogawa et al. 2008 [67]. Reprinted with permission of Elsevier.

Table 5.9 Contents of elements in Cat-CVD SiO_xN_y films, measured by Rutherford backscattering (RBS) method.

O_2 flow rate, $FR(O_2)$ (sccm)	*x*(O)	*y*(N)	O/(Si + O + N) (at.%)
0	0	1.3	0
1	0.29	1.2	12
2	0.65	1.2	23
4	1.1	1.3	32
6	1.3	0.70	43
8	1.7	0.41	55
10	1.8	0.18	60
Si_3N_4	0	1.33	0
Si_2ON_2	0.50	1.00	20
SiO_2	2.0	0	67

amount of O_2 gas causes a large change of refractive index. That can be controlled from about 2.0 to almost 1.5 by changing the mixing ratio of O_2 gas [67]. In Table 5.9, the atomic ratios of Si, N, and O atoms, which are evaluated from the data obtained by RBS experiments using 2 MeV He ions, are also listed.

By mixing O_2 gas, we can also control the stress of films by adjusting the mixing ratio of O_2 gas. Figure 5.52 shows the inner stress of SiO_xN_y films prepared at about 80 °C as a function of the flow rate of O_2 gas. By selecting a proper mixing ratio of O_2 gas, we can obtain stress-free films. In this case, at about 1.2 sccm of O_2

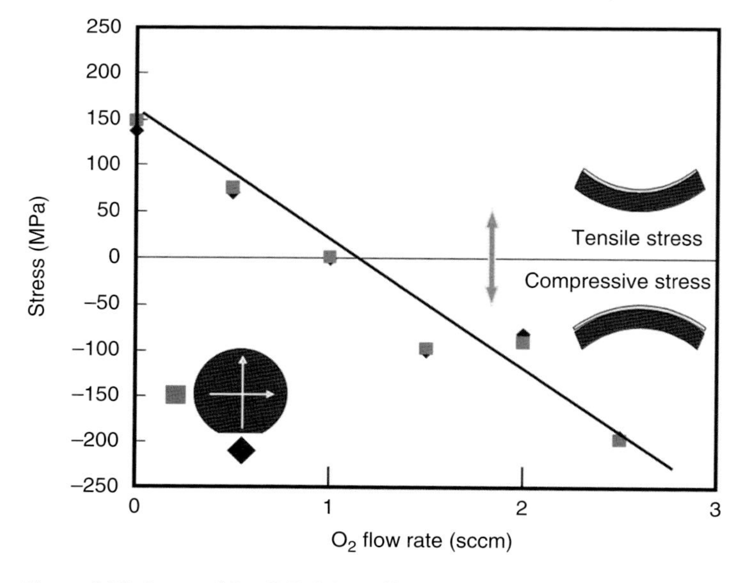

Figure 5.52 Stress of Cat-CVD SiO_xN_y films as a function of flow rate of mixing O_2 gas.

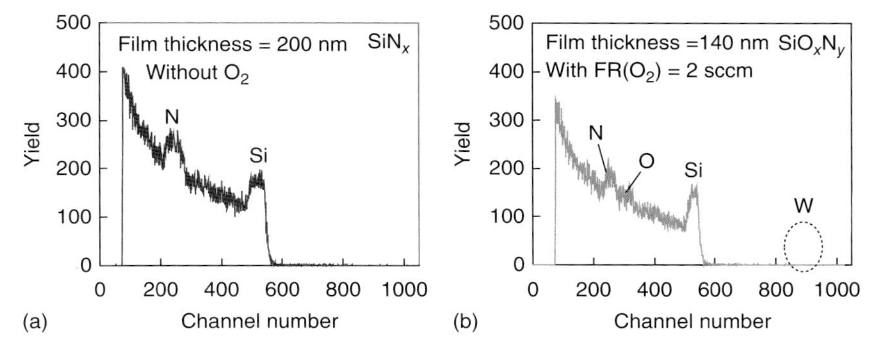

Figure 5.53 RBS spectra of (a) Cat-CVD SiN_x and (b) Cat-CVD SiO_xN_y.

gas, stress-free films are obtained. The required flow of O_2 gas to make the films stress free may change when gas pressure or substrate temperatures are changed.

One may worry about W contamination of the films caused by WO_x, which easily evaporates from the wire. Figure 5.53 shows the RBS spectra for pure SiN_x (Figure 5.53a) and SiO_xN_y prepared with 2 sccm of O_2 gas. The RBS spectrum is drawn by expressing the number of backscattered He ions, described as yield, as a function of the energy of backscattered He ions. The energy of He ions is also expressed by the channel number, in which the larger numbers correspond to higher energies. In this measurement, as the scattering cross section increases as the mass number of elements increases, the detection sensitivity becomes very high for W and the detection limit for W in the SiO_xN_y films is quite small, usually less than 1 ppm. The figure demonstrates that W contamination in SiO_xN_y is negligible even if O_2 is included in the gas mixture. As discussed later in Chapter 9, when reducing gas is introduced as the major constituent of the gas mixture, W surface is not oxidized even if source gases contain O atoms.

This SiO_xN_y film is an important material for making SiN_x/SiO_xN_y stacked gas barrier films, antireflection coatings of solar cells, and other applications, which will be demonstrated later in Chapter 8.

5.4.2 SiO_xN_y Films Prepared by HMDS, NH_3, H_2, and O_2 Mixtures

Similar to SiN_x, SiO_xN_y films can also be made by using HMDS. This experiment was carried out by using a small size chamber [68]. Deposition parameters are summarized in Table 5.10 along with the information on the chamber. O_2 gas was diluted to 2% in He gas, but the flow rate is expressed by the net value.

As is the case for SiO_xN_y prepared with a SiH_4, NH_3, and O_2 mixture, SiO_xN_y films can be made easily by mixing a small amount of O_2 gas. Figure 5.54 shows the IR spectra of $SiO_xN_yC_z$ films prepared with a HMDS, NH_3, H_2, and O_2 mixture, taking O_2 flow rate as a parameter. In the deposition without O_2 flow, the absorption peaks of Si—N, N—H, and Si—NH bonds are dominant. However, by mixing a very small amount of O_2, the IR absorption starts to shift from the Si—N bond to the Si—O bond. The films easily become O-containing films. The figure also demonstrates that the films include carbon (C) atoms. Thus, the film is not

Table 5.10 Deposition parameters for the preparation of SiO_xN_y films from HMDS.

Parameters	Setting values
Inner volume of chamber, V_{ch}	3000 cm^3
Material of catalyzer	Tungsten (W)
Spanning area of catalyzing wire	7 cm × 7 cm
Deposition area	5 cm × 5 cm
Surface area of catalyzer, S_{cat} (cm^2)	4.4
Temperature of catalyzer, T_{cat} (°C)	1800
Flow rate of HMDS, FR(HMDS) (sccm)	0.5
Flow rate of NH_3, FR(NH_3) (sccm)	200
Flow rate of H_2, FR(H_2) (sccm)	100
Net flow rate of O_2, FR(O_2), in He (sccm)	0–6
Substrate temperature, T_s(°C)	<100
Gas pressure, P_g	2.3 Torr, 300 Pa

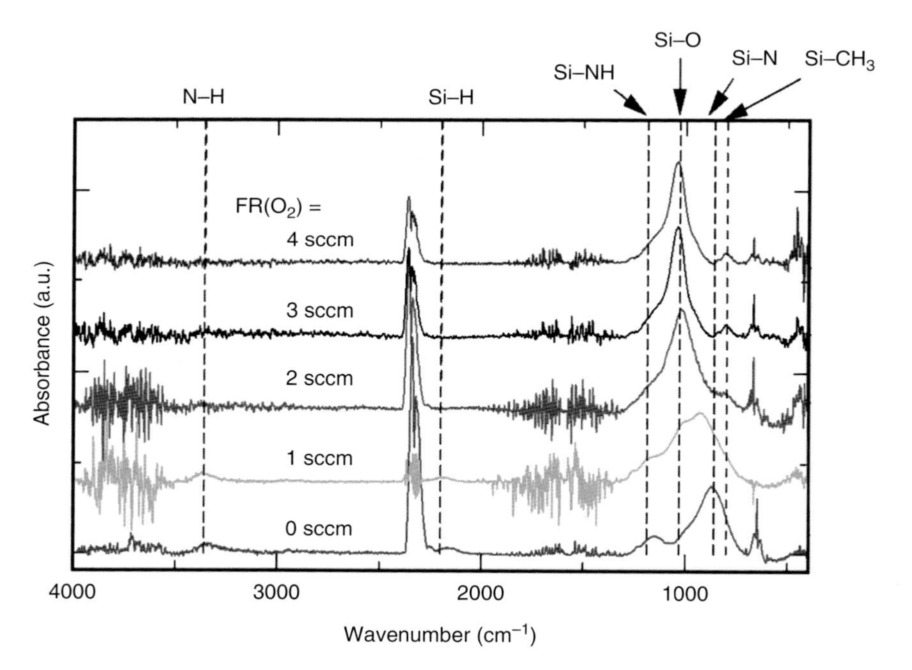

Figure 5.54 IR spectra of $SiO_xN_yC_z$ films, taking mixing O_2 flow rates as a parameter. Source: Oyaidu et al. 2008 [68]. Reprinted with permission of Elsevier.

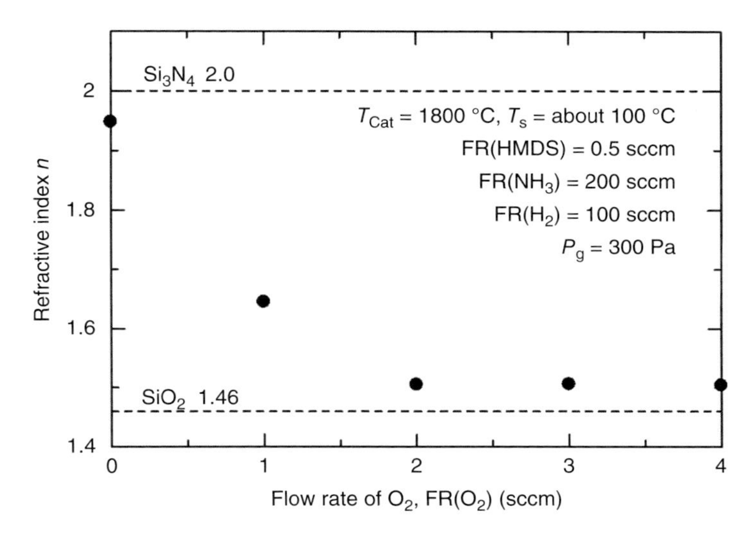

Figure 5.55 Refractive index of $SiO_xN_yC_z$ films as a function of mixing O_2 flow rate. Source: Oyaidu et al. 2008 [68]. Reprinted with permission of Elsevier.

simply SiO_xN_y, but actually $SiO_xN_yC_z$. Following this change of bond configuration, the refractive index of the film also changes from 1.95–2.0 to 1.48, reaching the value of SiO_2, by this mixing of O_2 gas. The results are shown in Figure 5.55.

The $SiO_xN_yC_z$ films show better gas barrier properties than the SiO_xN_y films prepared with SiH_4, NH_3, H_2, and O_2 mixtures. Figure 5.56 demonstrates the water vapor transmission rate (WVTR) of $SiO_xN_yC_z$ films prepared with a

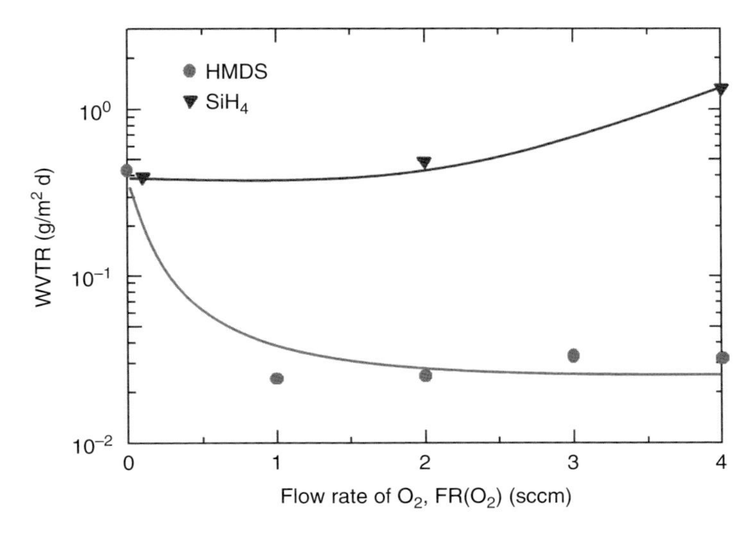

Figure 5.56 WVTRs of SiO_xN_y prepared with a SiH_4, NH_3, H_2, and O_2 mixture and $SiO_xN_yC_z$ prepared with a HMDS, NH_3, H_2, and O_2 mixture, as a function of mixing O_2 flow rate. Source: Oyaidu et al. 2008 [68]. Reprinted with permission of Elsevier.

HMDS, NH_3, and H_2 mixture, as a function of flow rate of mixing O_2 gas. The WVTR is a measure showing gas barrier ability and defined as the total weight of water vapor transmitting through a film of an area of 1 m^2/d. In the figure, WVTR values for SiO_xN_y films prepared with SiH_4, NH_3, H_2, and O_2 gas mixtures, already mentioned above, are also shown for comparison. All films are deposited on polyethylene terephthalate (PET) substrates, and the thickness of the deposited films is about 50–80 nm. The initial WVTR of PET substrates before film deposition is about 5.75 g/m^2 d. That is, by the deposition of SiN_x or SiN_xC_y films on the PET substrates, which is the film without O_2, the WVTR is likely to decrease by one order of magnitude. By mixing O_2, it drops by two orders of magnitude for HMDS films. The better gas barrier ability of HMDS-based films is demonstrated. The application of Cat-CVD for forming gas barrier films will be discussed later in Chapter 8.

5.5 Properties of Silicon Oxide (SiO_2) Films Prepared by Cat-CVD

Silicon dioxide (SiO_2) is one of the most important materials in electronic devices. It can be grown on c-Si substrates by a thermal oxidation process. However, in the silicon process, there are many processes requiring that SiO_2 is formed by deposition at temperatures lower than 400 °C. For such low temperature deposition, PECVD is a widely used technology. If SiO_2 can be made by Cat-CVD, any concerns about plasma damages or plasma-induced problems can be removed.

However, as already mentioned, the formation of oxide films by using catalytic cracking reactions with metal wires does not appear easy, as many metals with high melting temperatures are easily oxidized, and in most of the cases, the melting temperatures of such oxidized metals drastically decrease compared with those of nonoxidized pure metals. Even so, if the oxidant is highly diluted by reducing gas such as H_2 gas, metals such as W and Ta stay free from oxidation. This property is utilized in silicon oxidization, described later in Chapter 9.

The other approach to realize the preparation of oxidized films is to use metals that are particularly resistive to oxidization. Among such metals, iridium (Ir) is known, although it is not a low-cost material.

K. Saito et al. reported on the preparation of silicon dioxide (SiO_2) films by using Cat-CVD [69]. At first, they checked the usability of a range of metals as a catalyzing metal wire in a Cat-CVD apparatus. W, Ta, Ir, and Pt were heated in the vacuum chamber under the introduction of tetraethyl orthosilicate silane (or tetraethoxysilane; TEOS) or O_2. It was found that the melting point of Pt is too low, whereas metal oxides are easily formed for W and Ta. Then, they attempted to deposit SiO_2 by using Ir. A SiH_4 and N_2O mixture or a TEOS and N_2O mixture were used as source gases. Their deposition parameters are summarized in Table 5.11.

They measured the depth profiles of components in deposited SiO_2 by Auger electron spectroscopy (AES) and confirmed that stoichiometric SiO_2 could be formed by Cat-CVD at a substrate temperature of 400 °C.

Table 5.11 Deposition parameters of SiO_2 films by Cat-CVD using a mixture of SiH_4 and N_2O or a mixture of TEOS and N_2O.

Parameters	SiO_2 using SiH_4	SiO_2 using TEOS
Materials of catalyzing wire	W, Ta, Ir, Pt	W, Ta, Ir, Pt
Catalyzer temperature, T_{cat} (°C)	1800	1800
Substrate temperature, T_s (°C)	400	400
Catalyzer–substrate distance, D_{cs} (cm)	4.5	2.5
Flow rate of SiH_4, FR(SiH_4) (sccm)	5	—
Flow rate of TEOS, FR(TEOS) (sccm)	—	40
Flow rate of N_2O, FR(N_2O) (sccm)	500	600
Gas pressure, P_g (Pa)	50	50

After that, they made metal/SiO_2/c-Si structures to measure the leakage currents of Cat-CVD-formed SiO_2 films. The current densities through the SiO_2 layers are plotted as a function of applied electric field upon the films. The results are shown in Figure 5.57. In the figure, the results of SiO_2 made by PECVD using a mixture of TEOS and N_2O are plotted for comparison. It is demonstrated that Cat-CVD SiO_2 prepared with SiH_4 shows slightly lower leakage currents than PECVD SiO_2. The breakdown voltage of Cat-CVD SiO_2 is also high enough for use as insulator and is about 10 MV/cm. Cat-CVD films are again equivalent to PECVD at least or even slightly better than it, although SiO_2 made by TEOS requires some more effort.

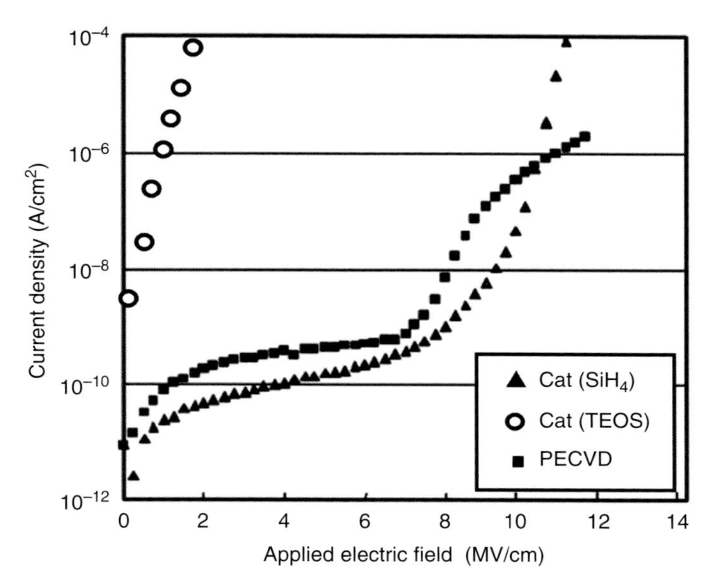

Figure 5.57 Current density vs. applied electric field for metal/SiO_2/Si structures. Source: Saito et al. 2003 [69]. Reprinted with permission of Elsevier.

5.6 Preparation of Aluminum Oxide (Al_2O_3) Films by Cat-CVD

Aluminum oxide (Al_2O_3) is another important material that is widely used as a passivation film or as an insulator in electronic devices. The preparation of Al_2O_3 by Cat-CVD was firstly attempted by Y.-I. Ogita and T. Tomita [70]. They tried to obtain Al_2O_3 by simply mixing trimethyl aluminum (TMA) and O_2. TMA is a liquid at atmospheric pressure and room temperature. However, it is easily vaporized by vacuum pumping. Their Cat-CVD apparatus is schematically illustrated in Figure 5.58. In the figure, MFC, QMS, TMP, and RP refer to a mass flow controller, a quadrupole mass spectrometer, a turbo molecular pump, and a rotary pump, respectively. In their system, TMA vapor is carried by nitrogen (N_2) gas after bubbling and mixed with O_2 in the deposition chamber. The O_2 gas is provided directly to just near the substrate by introducing a tube inside the chamber. They attempted to obtain Al_2O_3 by simply using W or Ir as a catalyzing wire; however, as W is easily oxidized, the major data were obtained by using an Ir catalyzer.

Figure 5.59 demonstrates the results of QMS measurement when an Ir wire is used as a catalyzer. The diameter of the Ir wire is 0.254 mmØ, and the length is 25 cm. As the temperature of the catalyzer, T_{Cat}, increased over 600 °C, CH_4, CH_3, and C_2H_4 signals began to be detected when only TMA gas was introduced into the chamber. However, for this elevation of T_{Cat}, O_2 signals did not respond at all when only O_2 gas was introduced. This means that O_2 cannot be decomposed on an Ir catalyzer at temperatures less than 1000 °C, although TMA is decomposed over 600 °C. However, when TMA is mixed with O_2, the signal of O_2 starts to decrease; that is, O_2 starts to decompose by reactions with TMA fragments. From this point onward, Al_2O_3 films start to be deposited. Although they did not show the composition of the films, 17-nm-thick films show excellent insulating property.

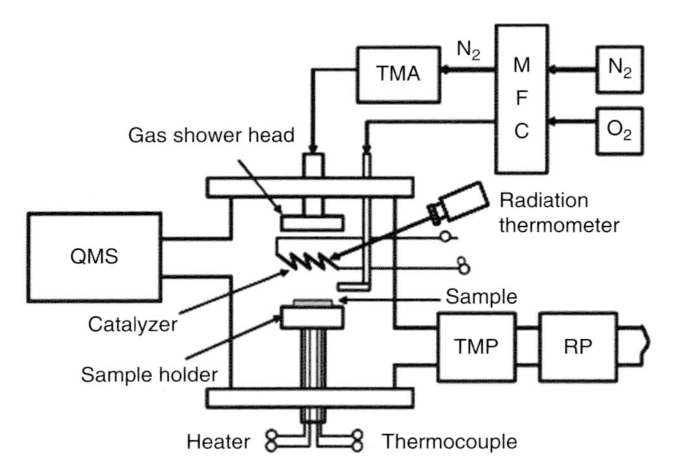

Figure 5.58 Schematic diagram of Cat-CVD apparatus used for Al_2O_3 deposition. Source: Ogita and Tomita 2006 [70]. Reprinted with permission of Elsevier.

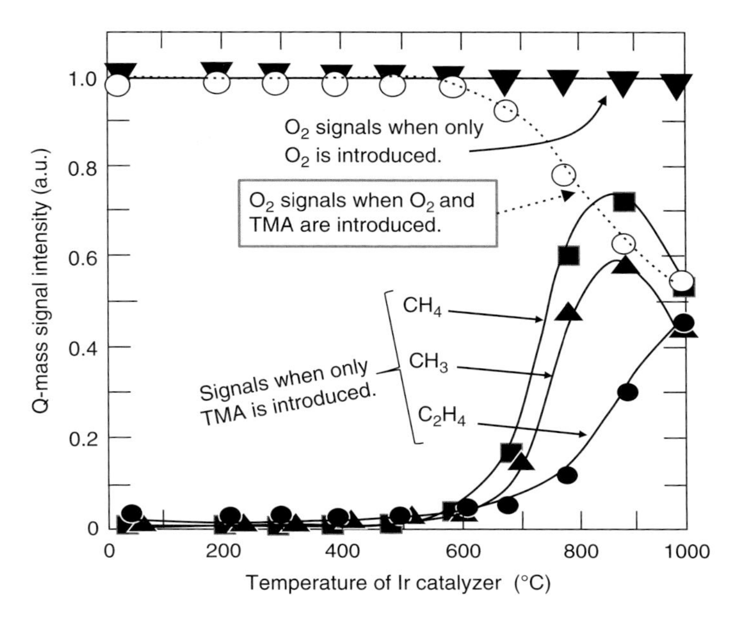

Figure 5.59 QMS signals of O_2 and TMA decomposed species as a function of Ir catalyzer temperature. Source: Data from Ogita and Tomita 2006 [70].

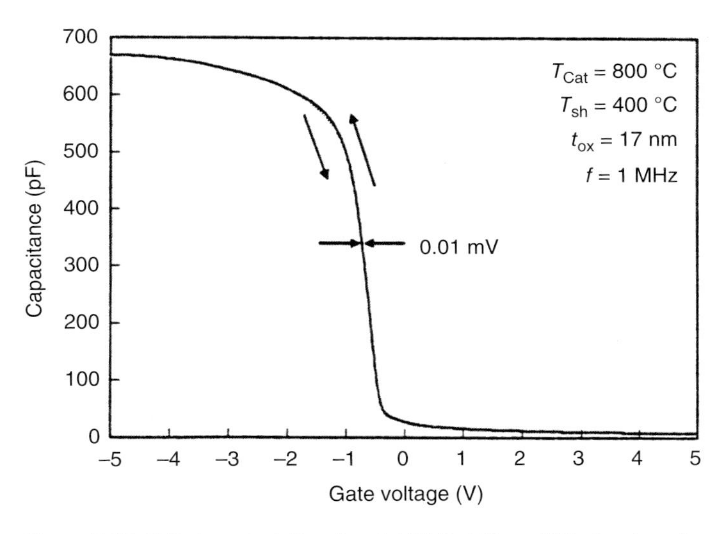

Figure 5.60 *C–V* characteristics of metal/Al_2O_3/silicon, MIS, structure. Source: Ogita and Tomita 2006 [70]. Reprinted with permission of Elsevier.

Figure 5.60 shows the capacitance–voltage (*C–V*) characteristics of a sample with a metal/17-nm-thick Cat-CVD Al_2O_3 film/c-Si (MIS) structure. The figure indicates that the insulating quality is excellent for use as a gate insulator of the MIS structure. The deposition conditions used for the preparation of Al_2O_3 films are summarized in Table 5.12.

Table 5.12 Typical Cat-CVD parameters for the preparation of Al_2O_3 films.

Parameters	Setting values for Al_2O_3 preparation
Size of a cylindrical chamber	20 cmØ diameter, 20 cm height
Catalyzing materials (size)	Ir, (0.254 mmØ diameter and 25 cm length)
Source gas (flow rate)	TMA (1 sccm)
	O_2 (1.5 sccm)
Catalyzer temperature, T_{cat} (°C)	RT–1000
Catalyzer–substrate distance, D_{cs} (cm)	4
Substrate temperature, T_s(°C)	400
Gas pressure, P_g (Pa)	100

As Ir is a very expensive material, and as it is known that an organic source gas can sometimes be decomposed by using a metal catalyzer containing nickel (Ni), as explained in Figure 6.3, they tried to use other metal wires for the decomposition of TMA. They discovered that TMA could be decomposed at catalyzer temperatures from 200 to 600 °C when Chromel, which is an alloy of 90% Ni and 10% chromium (Cr), is used as a catalyzer. TMA could also be decomposed at temperatures from 100 to 600 °C when stainless steel (SUS) 304 was used, which is an alloy of 10% Ni, 20% Cr, and about 70% iron (Fe).

The research of Cat-CVD deposition of Al_2O_3 does not appear completed. Further progress will be made in the future.

5.7 Preparation of Aluminum Nitride (AlN) by Cat-CVD

In the case of Al_2O_3 deposition by Cat-CVD, care has to be taken to prevent oxidation of catalyzing metal wires, and this appears to limit further development of Cat-CVD Al_2O_3 application. Instead of Al_2O_3, aluminum nitride (AlN) is easy to obtain as we need not use an oxidant during film formation. Source gas materials are only TMA and NH_3. AlN is a good insulator with a band gap of 6.3 eV and has large heat conductivity. Because of this property, AlN is often applied as a potential barrier layer in gallium nitride (GaN) blue emission diodes. Application of AlN as passivation on compound semiconductors appears promising. AlN is also very resistive to acid and alkaline chemicals. Thus, it is a very useful coating material. The heat conductivity of AlN is quite large, and thus, AlN is often used as a heat-radiating plate.

The melting temperature is near to 2200–2400 °C, and the refractive index is about 1.9–2.2. AlN films are also used as optical components and may be useful as antireflection coating.

When AlN films are used for passivation on compound semiconductors, as the surface of compound semiconductors is usually more delicate than that of c-Si, the plasma-less deposition using Cat-CVD may give some advantages.

The first work on AlN preparation by Cat-CVD was reported by J.L. Dupuie and E. Gulari in 1991 [71], and later, detailed data were reported by them [72].

Table 5.13 Deposition parameters of AlN by Cat-CVD.

Parameters	Setting values for AlN
Catalyzing materials (size)	W (diameter; 0.25 mm, length: 93 cm)
Source gas (flow rate)	TMA (1–4 sccm)
	NH_3 (61.9 sccm)
Catalyzer temperature, T_{cat} (°C)	1750
Catalyzer–substrate distance, D_{cs} (cm)	4
Substrate temperature, T_s (°C)	310–460
Gas pressure, P_g (Pa)	66.7
Deposition rates (nm/min)	50–200

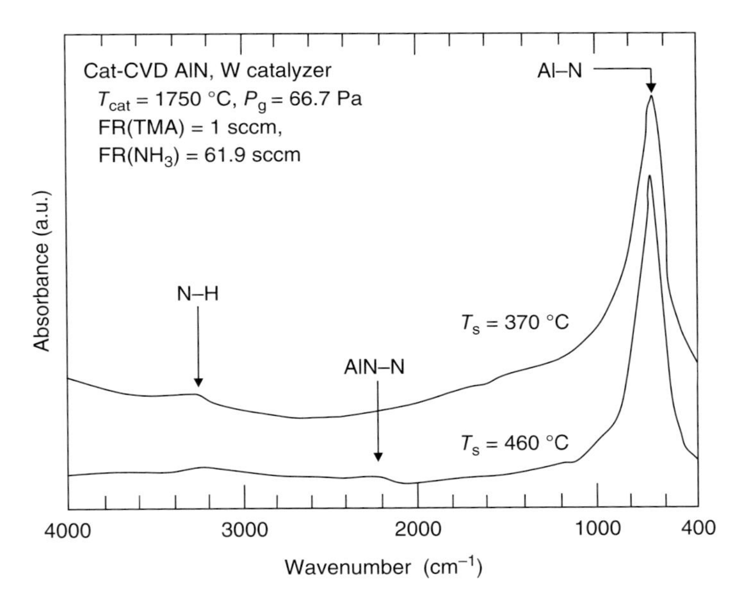

Figure 5.61 Infrared absorption spectra of two Cat-CVD AlN films. One is deposited at 370 °C and the other at 460 °C. Source: Data from Dupuie and Gulari 1992 [72].

Their deposition conditions are summarized in Table 5.13. They studied the structure of their AlN by IR absorption and the atomic composition of the films by XPS.

The IR spectra of Cat-CVD AlN films deposited at T_s = 370 °C and 460 °C are shown in Figure 5.61. The absorption peaks of Al—N bonds are clearly observed in both films. The deposition conditions are also described in the figure. The absorption peak of AlN films deposited at T_s = 460 °C appears sharper than that of AlN deposited at 370 °C. Generally speaking, the width of the IR absorption peak is related to the circumstances surrounding the bonds. If there are no restrictions for vibration, the width of absorption peak is likely to be wider. From the figure, it is concluded that the bond vibration is more stressed in the 460 °C

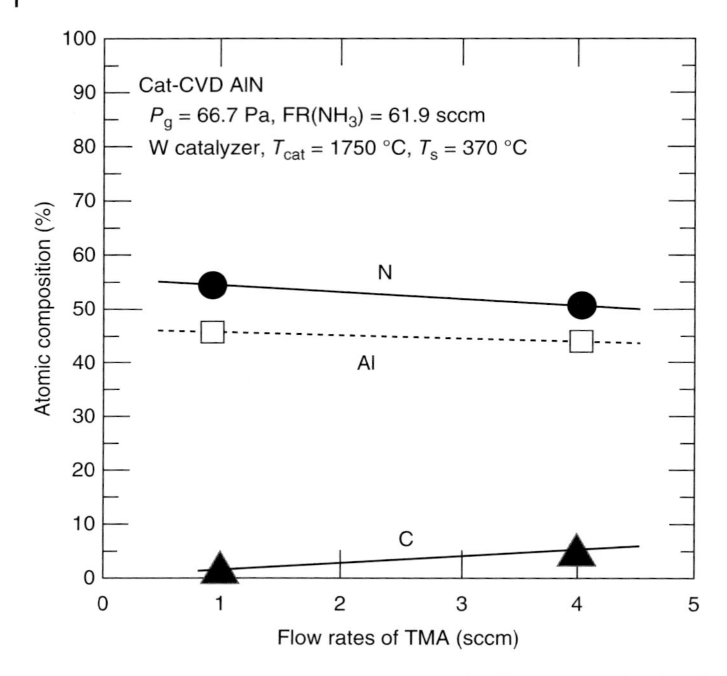

Figure 5.62 Atomic composition in Cat-CVD AlN films prepared with different flow rates of TMA. Source: Data from Dupuie and Gulari 1992 [72].

sample, and thus, AlN films are more densely packed. Actually, this is also known from the comparison of atomic H contents.

From IR spectra, the H contents are evaluated from the concentration of N—H bonds, which have an absorption peak at a wavenumber of about 3300 cm^{-1}. According to their report [72], the N–H concentration of 370 °C sample is about 6×10^{21} cm^{-3}, and the N–H concentration of 460 °C sample is 1.8×10^{21} cm^{-3}. That is, AlN films of 460 °C is highly dense, highly packed, and lose the freedom of vibration. Then, the IR absorption peak becomes sharper than that of the 370 °C sample.

The atomic composition of Cat-CVD AlN is shown in Figure 5.62 as a function of the flow rate of TMA, FR(TMA), for a fixed flow rate of NH_3 at 61.9 sccm. The AlN films appear slightly N-rich with respect to the stoichiometric value. When FR(TMA) increases, the N composition ratio is likely to decrease slightly, whereas the C composition ratio increases, although Al composition ratio appears to stay constant. All these data demonstrate that AlN films can be deposited by Cat-CVD using TMA and NH_3. The work appears to further expand the feasibility of Cat-CVD technology.

5.8 Summary of Cat-CVD Inorganic Films

As explained above, many kinds of films can be made by Cat-CVD by using various source gases. Here, we summarize the kinds of inorganic films prepared by

Table 5.14 Various inorganic films prepared by Cat-CVD.

Films	Catalyzer	Temperature of catalyzer	Source gases	References
a-Si	W, Ta	1700–2000 °C	SiH_4, H_2	[19, 21, 73, 74]
μc-Si, poly-Si	W, Ta	1700–2000 °C	SiH_4, H_2	[44, 75–77]
Si epi-growth	W, Ta	1800 °C	SiH_4, H_2	[57, 78–80]
a-SiC, poly-SiC	W, Ta, TaC, C, Re	1700–2000 °C	SiH_4, C_nH_m, H_2	[78, 81–83]
a-SiGe	W, Ta	(1100[a]–) 1700–2000 °C	SiH_4, GeH_4, H_2	[84–87]
Si_3N_4 (SiN_x)	W, Ru	(1180[a]–) 1700–2000 °C	SiH_4, NH_3, H_2 HMDS, NH_3, H_2	[62, 63, 65, 66, 88, 89]
SiO_2	Ir, W	1700–2000 °C	SiH_4, N_2O, H_2 TEOS, N_2O, H_2	[69, 90]
Al_2O_3	Ir, W, Chromel, SUS-304	200–900 °C	TMA, O_2, H_2	[70, 91, 92]
AlN	W	1700–2000 °C	TMA, NH_3	[71, 72]

Much more additional references are introduced in Chapter 8.
a) Only Matsumura's initial reports on a-SiGe [84] and SiN_x [88] are using lower catalyzer temperatures.

Cat-CVD in Table 5.14. The table offers a broad outline of the possible thin film materials. For further reading, we refer to the references. Deposition of organic films is summarized in the next Chapter 6.

References

1 Chittick, R.C., Alexander, J.H., and Sterling, H.F. (1969). The preparation and properties of amorphous-silicon. *J. Electrochem. Soc.* 116: 77–81.

2 Spear, W.E. and Le Comber, P.G. (1975). Substitutional doping of amorphous-silicon. *Solid State Commun.* 17: 1193–1196.

3 Taguchi, M., Yano, A., Tohoda, S. et al. (2014). 24.7% record efficiency HIT solar cell on thin silicon wafer. *IEEE J. Photovoltaics* 4: 96–99. https://doi.org/10.1109/JPHOTOV.2013.2282737.

4 Yoshikawa, K., Kawasaki, H., Yoshida, W. et al. (2017). Silicon heterojunction solar cell with interdigitated back contacts for a photo-conversion efficiency over 26%. *Nat. Energy* 2: 17032-1–17032-8. https://doi.org/10.1038/nenergy.2017.32.

5 Le Comber, P.G., Spear, W.E., and Ghaith, A. (1979). Amorphous-silicon field-effect devices and possible application. *Electron. Lett* 15: 179–181. https://doi.org/10.1049/el:19790126.

6 Mott, N.F. and Davis, E.A. (1979). Non-crystalline semiconductors, Chapter 6. In: *Electronic Processes in Non-crystalline Materials*, 2e. Clarendon Press-Oxford. ISBN: 0-19-851288-0.

7 Lucovsky, G. and Hayes, T.M. (1979). Chapter 8: Short-range order in amorphous semiconductors. In: *Amorphous Semiconductors*, Topics in Applied Physics, vol. 36 (ed. M.H. Brodsky), 215–250. Berlin: Springer-Verlag. ISBN: 3-540-09496-2 and ISBN 0-587-09496-2.

8 Kittel, C. (1986). Free Electron Fermi Gas, Chapter 6. In: *Introduction to Solid State Physics*, 6e. New York: Wiley. ISBN: 0-471-87474-4.

9 Ovshinsky, S.R. and Madan, A. (1978). A new amorphous silicon-based alloy for electronic applications. *Nature* 276: 482–484.

10 Madan, A., Ovshinsky, S.R., and Benn, E. (1979). Electrical and optical properties of amorphous Si:F:H alloys. *Philos. Mag. B* 40: 259–277.

11 Matsumura, H., Nakagome, Y., and Furukawa, S. (1980). A heat-resting new amorphous silicon. *Appl. Phys. Lett.* 36: 439–440.

12 Kittel, C. (1986). Semiconductor crystal, Chapter 8. In: *Introduction to Solid State Physics*, 6e. New York: Wiley. ISBN: 0-471-87474-4.

13 Matsumura, H., Ihara, H., and Tachibana, H. (1985). Hydro-fluorinated amorphous silicon made by thermal CVD (chemical vapor deposition) method. *Proceedings of 18th IEEE Photovoltaic Specialist Conference*, Las Vegas, Nevada, USA (21–25 October 1985). pp. 1277–1282.

14 Tsai, C.C. and Fritzsche, H. (1979). Effect of annealing of the optical properties of plasma deposited amorphous hydrogenated silicon. *Sol. Energy Mater.* 1: 29–42.

15 Yamaguchi, M., Matsumura, H., and Morigaki, K. (1990). Photoluminescence of a-Si:H films prepared by catalytic chemical vapor deposition. *Jpn. J. Appl. Phys.* 29 (8): L1366–L1368.

16 Boccara, A.C., Foumier, D., Jackson, W., and Amer, N.M. (1980). Sensitive photothermal deflection technique for measuring absorption in optically thin media. *Opt. Lett.* 5: 377–379.

17 Knief, S. and von Niessen, W. (1999). Disorder, defects, and optical absorption in a-Si and a-Si:H. *Phys. Rev. B: Condens. Matter* 59: 12940–12946.

18 https://www.iap.tuwien.ac.at/www/surface/vapor_pressure, Institut fur Angeweandte Physik der Technichen Universitat Wien, "Vapor Pressure Calculator", or https://www.powerstream.com/vapor-pressure.htm.

19 Mahan, A.H., Carapella, J., Nelson, B.P. et al. (1991). Deposition of device quality, low H content amorphous silicon. *J. Appl. Phys.* 69: 6728–6730.

20 Schropp, R.E.I. (2004). Present status of micro- and polycrystalline silicon solar cells made by hot-wire chemical vapor deposition. *Thin Solid Films* 451–452: 455–465.

21 Heintze, M., Zedlitz, R., Wanka, H.N., and Schubert, M.B. (1996). Amorphous and microcrystalline silicon by hot wire chemical vapor deposition. *J. Appl. Phys.* 79: 2699–2706.

22 Sakai, M., Tsutsumi, T., Yoshioka, T. et al. (2001). High performance amorphous-silicon thin film transistors prepared by catalytic chemical vapor deposition with high deposition rate. *Thin Solid Films* 395: 330–334.

23 Nishizaki, S., Ohdaira, K., and Matsumura, H. (2008). Study on stability of amorphous-silicon thin-film transistors prepared by catalytic chemical vapor deposition. *Jpn. J. Appl. Phys.* 47: 8700–8706.

24 Lucovsky, G., Nemanich, R.J., and Knights, J.C. (1979). Structural interpretation of vibrational spectra of a-Si:H alloys. *Phys. Rev. B: Condens. Matter* 19: 2064–2073.

25 Brodsky, M.H., Cardona, M., and Cuomo, J.J. (1977). Infrared and Raman spectra of the silicon–hydrogen bonds in amorphous silicon prepared by glow discharge and sputtering. *Phys. Rev. B: Condens. Matter* 16: 3556–3571.

26 Langford, A.A., Fleet, M.L., Nelson, B.P. et al. (1992). Infrared absorption strength and hydrogen content of hydrogenated amorphous silicon. *Phys. Rev. B: Condens. Matter* 45: 13367–13377.

27 Crandall, R.S., Mahan, A.H., Nelson, B.P. et al. (1992). Properties of hydrogenated amorphous silicon produced at high temperature. *AIP Conf. Proc.* 268: 81–87.

28 Matsumura, H., Masuda, A., and Izumi, A. (1999). Cat-CVD process and its application to preparation of Si-based thin films. *Mater. Res. Soc. Symp. Proc.* 557: 67–78.

29 Veprek, S., Sarrott, F.A., Rambert, S., and Taglauer, E. (1989). Surface hydrogen content and passivation of silicon deposited by plasma induced chemical vapor deposition from silane and the implications for the reaction mechanism. *J. Vac. Sci. Technol., A* 7: 2614–2624.

30 Matsuda, A. (2004). Thin-film silicon—growth process and solar cell application. *Jpn. J. Appl. Phys.* 43: 7909–7920.

31 Matsumura, H. (1992). A new amorphous-silicon with low hydrogen content. *Oyo Buturi* 61: 1013–1019. [In Japanese].

32 Shirai, H., Hanna, J.-i., and Shimizu, I. (1991). Role of atomic hydrogen during growth of hydrogenated amorphous silicon in the chemical annealing. *Jpn. J. Appl. Phys.* 30: L679–L682.

33 Matsumura, H. (1998). Formation of silicon-based thin films prepared by catalytic chemical vapor deposition (Cat-CVD) method. *Jpn. J. Appl. Phys.* 37: 3175–3187.

34 Nelson, B.P., Xu, Y., Mahan, A.H. et al. (2000). Hydrogenated amorphous-silicon grown by hot-wire CVD at deposition rates up to 1 μm/minute. *Mater. Res. Soc. Symp. Proc.* 609: A22.8-1–A22.8-6.

35 Nishikawa, S., Kakinuma, H., Watanabe, T., and Nihei, K. (1985). Influence of deposition conditions on properties of hydrogenated amorphous silicon prepared by RF glow discharge. *Jpn. J. Appl. Phys.* 24: 639–645.

36 Inoue, H., Tanaka, K., Sano, Y. et al. (2011). High-rate deposition of amorphous silicon films by microwave-excited high-density plasma. *Jpn. J. Appl. Phys.* 50: 036502-1–036502-6.

37 Staebler, D.L. and Wronski, C.R. (1977). Reversible conductivity changes in discharge produced amorphous Si. *Appl. Phys. Lett.* 31: 292–294.

38 Matsumura, H. (2001). Summary of research in NEDO Cat-CVD project in Japan. *Thin Solid Films* 395: 1–11.

39 Mahan, A.H. and Vanecek, M. (1991). A reduction in the Stableer–Wronski effect observed in low H content a-Si:H films deposited by the hot-wire technique. In: *Tech. Digest of 3rd Sunshine Workshop on Solar Cells*, vol. 1991, 3–10. Pacifico Yokohama, Yokohama, Japan: NEDO (New Energy Development Organization).

40 Shimuzu, S., Kondo, M., and Matsuda, A. (2011). Fabrication of the hydrogenated amorphous silicon films exhibiting high stability against light soaking. In: *Solar Cells – Thin Film Technologies* (ed. L. Kosyachenko), 303–318. Europe: In Tech. ISBN: 978-953-307-570-9.

41 Butler, J.E., Mankelevich, Y.A., Cheeseman, A. et al. (2009). Understanding the chemical vapor deposition of diamond: recent progress. *J. Phys. Condens. Matter.* 21: 364202-1–364202-19. https://doi.org/10.1088/0953-8984/21/36/364201.

42 Oikawa, S., Ohtsuka, S., and Tsuda, M. (1992). Elementary processes of surface reaction in amorphous silicon film growth. *Appl. Surf. Sci.* 60–61: 29–38.

43 Usui, S. and Kikuchi, M. (1979). Properties of heavily doped GD-Si with low resistivity. *J. Non-Cryst. Solids* 34: 1–11.

44 Matsumura, H. (1991). Formation of polysilicon films by catalytic chemical vapor deposition (Cat-CVD) method. *Jpn. J. Appl. Phys.* 30: L1522–L1524.

45 Wikipedia Scherrer equation. https://en.wikipedia.org/wiki/Scherrer_equation (accessed 23 October 2018).

46 Kittel, C. (1986). Reciprocal lattice, Chapter 2. In: *Introduction to Solid State Physics*, 6e. New York: Wiley. ISBN: 0-471-87474-4.

47 Viera, C., Huet, S., and Boufendi, L. (2001). Crystal size and temperature measurements in nanostructured silicon using Raman spectroscopy. *J. Appl. Phys.* 90: 4175–4183.

48 Richter, H., Wang, Z.P., and Ley, L. (1981). The one phonon Raman spectrum in microcrystalline silicon. *Solid State Commun.* 39: 625–629.

49 Heya, A., Masuda, A., and Matsumura, H. (1999). Low temperature crystallization of amorphous silicon using atomic hydrogen generated by catalytic reaction on heated tungsten. *Appl. Phys. Lett.* 74: 2143–2146.

50 Eaglesham, D.J., Unterwald, F.C., Lufman, H. et al. (1989). Effect of H on Si molecular-beam epitaxy. *J. Appl. Phys.* 74: 6515–6517.

51 Matsumura, H., Umemoto, H., Izumi, A., and Masuda, A. (2003). Recent progress of Cat-CVD research in Japan—bridging between the first and second Cat-CVD conferences. *Thin Solid Films* 430: 7–14.

52 Kasai, H., Kusumoto, N., Yamanaka, H. et al. (2001). Fabrication of high mobility poly-Si TFT by Cat-CVD. In: *Tech. Reports of IEICE, ED2001-4*, 19–25. Japan: The Institute of Electronics, Information and Communication Engineering (Written in Japanese, the contents are introduced in Ref. [51]).

53 Van der Pauw, L.J. (1958). A method of measuring specific resistivity and Hall effect of discs of arbitrary shape. *Philips Res. Rep.* 13: 1–9.

54 Morin, F.J. and Maita, J.P. (1958). Electrical properties of silicon containing arsenic and boron. *Phys. Rev.* 96: 28–35.

55 Matsumura, H., Tashiro, Y., Sasaki, K., and Furukawa, S. (1994). Hall mobility of low-temperature-deposited polysilicon films by catalytic chemical vapor deposition method. *Jpn. J. Appl. Phys.* 33: L1209–L1211.

56 Kamins, T. (1994). Electrical properties, Chapter 5. In: *Polycrystalline Silicon for Integrated Circuit Application*. Kluwer Academic Publishers. ISBN: 0-89838-259-9.

57 Yamoto, H., Yamanaka, H., Yagi, H. et al. (1999). Low temperature Si epitaxial growth by Cat-CVD method. In: *Ext. Abstract of International Pre-Workshop on Cat-CVD (Hot-Wire CVD) Process*, 61–63. Ishikawa, Japan: Organizing Committee of International Pre-Workshop on Cat-CVD Process.
58 Schropp, R.E.I., Nishizaki, S., Housweling, Z.S. et al. (2008). All hot wire CVD TFTs with high deposition rate silicon nitride (3 nm/s). *Solid-State Electron.* 52: 427–431.
59 Parsons, G.N., Souk, J.H., and Batey, J. (1991). Low hydrogen content stoichiometric silicon nitride films deposited by plasma-enhanced chemical vapor deposition. *J. Appl. Phys.* 70: 1553–1560.
60 Yin, Z. and Smith, F.W. (1990). Optical dielectric function and infrared absorption of hydrogenated amorphous silicon nitride films: experimental results and effective-medium-approximation analysis. *Phys. Rev. B: Condens. Matter* 42: 3666–3674.
61 Lanford, W.A. and Rand, M.J. (1978). The hydrogen content of plasma-deposited silicon nitride. *J. Appl. Phys.* 49: 2473–2477.
62 Osono, T., Heya, A., Niki, T. et al. (2006). High-rate deposition of SiN_x films over 100 nm/min by Cat-CVD method at low temperatures below 80 °C. *Thin Solid Films* 501: 55–57.
63 Wang, Q., Ward, S., Gedvilas, L., and Keyes, B. (2004). Conformal thin-film silicon nitride deposited by hot-wire chemical vapor deposition. *Appl. Phys. Lett.* 84: 338–340.
64 Fujinaga, T., Kitazoe, M., Ymamoto, Y. et al. (2007). Development of Cat-CVD apparatus. *ULVAC Tech. J.* (67): 30–34. [In Japanese].
65 Verlaan, V., van der Werf, C.H.M., Houweling, Z.S. et al. (2007). Multi-crystalline silicon solar cells with very fast deposited (180 nm/min) passivating hot-wire CVD silicon nitride antireflection coating. *Prog. Photovoltaics Res. Appl.* 15: 563.
66 Izumi, A. and Oda, K. (2006). Deposition of SiCN films using organic liquid materials by HWCVD method. *Thin Solid Films* 501: 195–197.
67 Ogawa, Y., Ohdaira, K., Oyaidu, T., and Matsumura, H. (2008). Protection of organic light-emitting diodes over 50,000 hours by Cat-CVD SiN_x/SiO_xN_y stacked thin films. *Thin Solid Films* 516: 611–614.
68 Oyaidu, T., Ogawa, Y., Tsurumaki, K. et al. (2008). Formation of gas barrier films by Cat-CVD method using organic silicon compound. *Thin Solid Films* 516: 604–606.
69 Saito, K., Uchiyama, Y., and Abe, K. (2003). Preparation of SiO_2 thin films using the Cat-CVD method. *Thin Solid Films* 430: 287–291.
70 Ogita, Y.-I. and Tomita, T. (2006). The mechanism of alumina formation from TMA and molecular oxygen using catalytic-CVD with an iridium catalyzer. *Thin Solid Films* 501: 35–38.
71 Dupuie, J.L. and Gulari, E. (1991). Hot filament enhanced chemical vapor deposition of AlN thin films. *Appl. Phys. Lett.* 59: 549–551.
72 Dupuie, J.L. and Gulari, E. (1992). The low temperature catalyzed chemical vapor deposition and characterization of aluminum nitride thin films. *J. Vac. Sci. Technol., A* 10: 18–28.

73 Matsumura, H. (1986). Catalytic chemical vapor deposition (CTL-CVD) method producing high quality hydrogenated amorphous silicon. *Jpn. J. Appl. Phys.* 25: L949–L953.

74 Doyle, J., Robertson, R., Lin, G.H. et al. (1988). Production of high-quality amorphous silicon films by evaporative silane surface decomposition. *J. Appl. Phys.* 64: 3215–3223.

75 Schropp, R.E.I. and Rath, J.K. (1999). Novel profiled thin film polycrystalline silicon solar cells on stainless steel substrates, Special issue IEEE Transaction on Electron Devices, on Progress and Opportunities in Photovoltaic Solar Cells Science & Engineering, . *IEEE Trans. Electron Devices* 46: 2069–2071.

76 Bouree, J.E. (2001). Correlated structural and electronic properties of microcrystalline silicon films deposited at low temperature by catalytic-CVD. *Thin Solid Films* 395: 157–162.

77 Finger, F., Mai, Y., Klein, S., and Carius, R. (2008). High efficiency microcrystalline silicon solar cells with hot-wire CVD buffer layer. *Thin Solid Films* 516: 728–732.

78 Watahiki, T., Abe, K., Tamura, H. et al. (2001). Low temperature epitaxial growth of Si and $Si_{1-y}C_y$ films by hot wire cell method. *Thin Solid Films* 395: 221–234.

79 Mason, M.S., Chen, C.M., and Atwater, H.A. (2003). Hot-wire chemical vapor deposition for epitaxial silicon growth on large grained polysilicon templates. *Thin Solid Films* 430: 54–57.

80 Teplin, C.W., Wang, Q., Iwaniczko, E. et al. (2006). Low-temperature silicon homoepitaxy by hot-wire chemical vapor deposition with a Ta filament. *J. Cryst. Growth* 287: 414–418.

81 Klein, S., Carius, R., Finger, F., and Houben, L. (2006). Low substrate temperature deposition of crystalline SiC using HWCVD. *Thin Solid Films* 501: 169–172.

82 Mori, M., Tabata, A., and Mizutani, T. (2006). Properties of hydrogenated amorphous silicon carbide films prepared at various hydrogen gas flow rates by hot-wire chemical vapor deposition. *Thin Solid Films* 501: 177–180.

83 Itoh, T., Kawasaki, T., Takai, Y. et al. (2008). Properties of hetero-structured SiC_x films deposited by hot-wire CVD using SiH_3CH_3 as carbon source. *Thin Solid Films* 516: 641–643.

84 Matsumura, H. (1987). High-quality silicon–germanium produced by catalytic chemical vapor deposition. *Appl. Phys. Lett.* 51: 804–805.

85 Nelson, B.P., Xu, Y., Williamson, D.L. et al. (1998). Hydrogenated amorphous silicon–germanium alloys grown by the hot-wire chemical vapor deposition technique. *Mater. Res. Soc. Symp. Proc.* 507: 447–452.

86 Datta, S., Xu, Y., Mahan, A.H. et al. (2006). Superior structural and electronic properties for amorphous silicon–germanium alloys deposited by a low temperature hot wire chemical vapor deposition process. *J. Non-Cryst. Solids* 352: 1250–1254.

87 Schropp, R.E.I., Li, H., Franken, R.H. et al. (2008). Nanostructured thin films for multibandgap silicon triple junction solar cells. *Thin Solid Films* 516: 6818–6823.

88 Matsumura, H. (1989). Silicon nitride produced by catalytic chemical vapor deposition method. *J. Appl. Phys.* 66: 3512–3617.

89 Okada, S. and Matsumura, H. (1997). Improved properties of silicon nitride films prepared by the catalytic chemical vapor deposition method. *Jpn. J. Appl. Phys.* 36: 7035–7040.

90 Matsumoto, Y. (2006). Hot wire-CVD deposited a-SiO_2 and its characterization. *Thin Solid Films* 501: 95–97.

91 Ogita, Y.-I. and Tomita, T. (2006). The mechanism of alumina formation from TMA and molecular oxygen using catalytic-CVD with a tungsten catalyzer. *Thin Solid Films* 501: 39–42.

92 Ogita, Y.-I., Kudoh, T., and Iwai, R. (2009). Low temperature decomposition of large molecules of TMA using catalyzers with resistance to oxidation in catalytic-CVD. *Thin Solid Films* 517: 3439–3442.

6

Organic Polymer Synthesis by Cat-CVD-Related Technology – Initiated CVD (iCVD)

Catalytic chemical vapor deposition (Cat-CVD) is also applicable for the preparation of organic films. As the previous explanation has been mainly for inorganic film deposition, the topic of deposition of organic films appears quite new. Thus, here, as one development form of Cat-CVD technology, we introduce a new CVD technology named as initiated chemical vapor deposition (iCVD) here, which is the most suitable technique for high-quality organic thin films.

6.1 Introduction

Extending the precision of chemical vapor deposition (CVD) methods developed for inorganic layers to organic polymer thin films requires the use of appropriate precursors and reactor conditions [1]. The CVD surface modification seeks to alter the overall interaction of the coated part with the environment while not damaging the properties of the bulk object. The later goal motivates the use of solvent-free and low substrate-temperature processes. The organic functional groups of CVD polymers also offer the opportunity to provide chemical and biological specificity in the form of conformal thin films for ready integration onto surfaces of complex geometry and into a variety of devices.

CVD polymerization methods augment the possibilities for surface modification by polymers based on solution and melt processing. The CVD method is particularly well suited to polymers that have limited or no solubility in common solvents, such as fluoropolymers and highly cross-linked organic networks. The latter offers a highly desirable combination of physical robustness and mechanical flexibility. High-purity CVD polymers, without residual solvents or other additives, are ideal for optoelectronic devices and biomedical applications.

The CVD polymer surfaces containing one or more organic functionalities have been designed to address a wide range of fundamental research questions and technical applications, including but not limited to

- systematically controlling wettability, adhesion, fouling resistance with respect to water, oil, ice, inorganic scale, hydrates, proteins, or microbes;
- functionalization of surfaces for chemical and biological specificity and reactivity, including the postdeposition attachment of growth factors, antibodies, dyes, and nanoparticles;

Catalytic Chemical Vapor Deposition: Technology and Applications of Cat-CVD,
First Edition. Hideki Matsumura, Hironobu Umemoto, Karen K. Gleason, and Ruud E.I. Schropp.

- protection of the underlying substrate from environmental stresses such as moisture, oxygen, solvents, biological environments, or UV light;
- encapsulation of machined, molded, and 3D printed parts having complex geometries;
- ultrathin (<10 nm) and ultrasmooth (rms roughness <1 nm) pinhole-free coatings;
- modification of the surface chemistry and pore geometry of membranes;
- tunable film properties for integration into lightweight and flexible optoelectronic and microfluidic devices;
- fabrication of photovoltaics, logic circuits, memory, and medical diagnostics on textile and paper substrates;
- engineering of the interfaces between the substrate and film to create robust adhesion; and
- the enabling of switching behavior that is responsive to external stimuli such as light, temperature, pH, or chemical species.

In situ monitoring provides precise control over the CVD polymer film thickness, even for ultrathin films (<10 nm). Dewetting effects present in traditional solution-based polymer film formation methods make it challenging to produce such ultrathin films free of pinhole defects. Micro- and nanoscale features in substrates become obscured or can be bridged-over using solution-based methods because of dewetting and surface tension effects. In contrast, the CVD methods described in this chapter can provide "conformal" coverage in which a film of uniform thickness exactly follows the surfaces of complex structures and porous media [2]. Conformal coverage is desirable for the integration of organic coatings into a wide range of real-world applications. The high deposition rates for conformal coverage differentiate CVD polymerization from either atomic layer deposition (ALD)/molecular layer deposition (MLD) and layer-by-layer (LBL) deposition. The absence of dewetting effects also enables the iCVD formation of ultrathin (<10 nm) films that are free of pinhole defects.

As CVD films grow from the substrate up, the strategic modification of the substrate before film growth can enable covalent chemical bond formation across the interface, producing robustly adhered films, another key attribute required for most commercial applications. Additionally, by changing the ratio of gases in the inlet feed, the composition of the iCVD film can be graded through the thickness of the layer.

This chapter will describe CVD methods for the single-step synthesis of organic films from vapor-phase monomers, achieved using resistively heated wires/filaments/metal pins suspended inside reactor chambers held at modest vacuum. Polymer thin films deposit on a cooled surface. These methods contain a mechanistic step for initiating the growth of the macromolecular chains. In some cases, the initiating species is a product of decomposition of the monomeric reactant (i.e. self-initiation), denoted as hot-wire chemical vapor deposition (HWCVD) in this chapter. For some, but not all, HWCVD polymer processes, the chemical composition of the resistively heated filament wire can provide a catalytic effect (e.g. Cat-CVD). Alternatively, the initiating species can form as a result of the thermal decomposition of an intentionally introduced initiator at or near the heated filament. Using a separate initiator provides a higher degree of control over the CVD process and can lead to greatly enhanced film deposition rates. In this chapter, metering in a separate initiator into the vacuum chamber is termed iCVD. When referring to HWCVD, Cat-CVD, and iCVD collectively, the term CVD polymerization is used.

This chapter will describe the first HWCVD/iCVD polymer to be discovered, polytetrafluoroethylene (PTFE), along with the scientific principles underlying its growth process. The central sections of this chapter will overview the fundamental knowledge that enabled the rapid expansion of iCVD film compositions and applications. Finally, this chapter will conclude with a discussion of reactors, including large-scale systems used in commercial production and directions for future developments.

6.2 PTFE Synthesis by Cat-CVD-Related Technology

The fluoropolymer, PTFE, is a linear chain macromolecule composed of —CF_2-monomeric units. Widely known by the Dupont trade name, Teflon™, PTFE possesses excellent chemical resistance and high thermal stability for a polymeric material. Additional technological applications of PTFE stem from its low dielectric constant, low refractive index, low coefficient of friction (COF), and low surface energy (Table 6.1).

Sintering of conventionally synthesized PTFE power at elevated temperature (~400 °C) can produce thick films (~25 μm thick), such as those found on non-stick cookware. However, because of its low solubility in common solvents, PTFE powders are difficult to process into thin film form. In contrast, CVD eliminates the need to dissolve macromolecules and provides for simultaneous polymerization and thin film formation. The earliest demonstration of HWCVD polymerization was the growth of PTFE thin films. Several comprehensive reviews of CVD PTFE have appeared [3, 4].

Ultrathin (<10 nm) CVD PTFE films have been demonstrated. The substrate remains near room temperature during the CVD PTFE process, a critical characteristic that permits film growth on thermally sensitive substrates, including paper, plastic foils, and membranes, which would not survive the sintering step

Table 6.1 Properties of bulk and iCVD polytetrafluoroethylene (PTFE).

Property	Bulk PTFE	iCVD PTFE
Coefficient of friction (COF)	0.07–0.30	0.03–0.20
Static water contact angle (WCA) (°)	110	120–150
Refractive index (RI)	1.35–1.38	1.35–1.38
Dielectric constant (k)	2.1	1.7–2.1
Dielectric strength (V/m)	4500	4000–7000
Density (g/cm^3)	2.2	1.5–2.2

Source: Adapted from Ref. [1], Chapter 19.

of the conventional PTFE "spray-and-bake" application process. Additionally, the solvent-free CVD process avoids swelling of delicate substrates. Another distinction is that CVD PTFE is free of the surfactants, such as perfluorooctanoic acid (PFOA), commonly found in conventionally synthesized PTFE. Because it bioaccumulates in the environment, avoiding PFOA is desirable.

For HWCVD PTFE, thermal energy, typically delivered using resistively heated filament wire, is used to decompose the precursor gas, hexafluoropropylene oxide (HFPO), into trifluoracetyl fluoride and difluorocarbene, which is predominately in the more stable singlet state:

$$\underset{\text{(epoxide ring: O bridging } CF_2 \text{ and } CF)}{CF_2\!:\!-CF-CF_3} \xrightarrow{\Delta} CF_2 + \overset{O}{\overset{\|}{C}}F-CF_3 \tag{6.1}$$

The gas-phase CF_2: produced from the thermal decomposition of HFPO has been directly observed using laser-induced fluorescence and UV absorption spectroscopy. The ease of breaking the strained epoxide ring in the HFPO allows Reaction (6.1) to proceed at relatively modest filament temperatures (~150 to ~800 °C).

Once produced by Reaction (6.1), the CF_2: either polymerizes (Reaction (6.2)) or recombines (Reaction (6.3)) to form tetrafluoroethylene (TFE). The overall polymerization reaction to form thin solid films is

$$nCF_2 : \rightarrow (-CF_2-)_{n\,(\text{film})} \tag{6.2}$$

The measured trends in CF_2: gas-phase concentrations with reactor conditions, observed deposition rates of PTFE films and powders, and the calculation of effective sticking coefficients all support the hypothesis that CF_2: forms gas-phase oligomers, $(CF_2)_x$, as an intermediate to the PTFE film deposition [5].

The recombination reaction for difluorocarbene produces TFE:

$$2CF_2 : \rightarrow CF_2{=}CF_2 \tag{6.3}$$

Interestingly, TFE is the monomer used for the conventional bulk synthesis of PTFE. However, TFE polymerization is not favored during HWCVD, as the activation energy of TFE polymerization is significantly higher than the activation energy for the thermal decomposition of HFPO (Reaction (6.1)). The hypothesis that for HWCVD PTFE CF_2: is the monomer species undergoing polymerization, not TFE, is also supported by density functional theory (DFT) analysis. TFE (a product of Reaction (6.3)) and trifluoroacetyl fluoride (a product of Reaction (6.1)) are primary effluent species from HWCVD detected by gas-phase Fourier transform infrared (FTIR). Decreasing the filament to substrate standoff favors the pathway for film formation (Reaction (6.2)) over the recombination to gaseous TFE (Reaction (6.3)). Thus, filaments are often held only 1–2 cm away from the deposition surface.

6.2.1 Select Characteristics and Applications of CVD PTFE Films

The chemical signatures of the bonding environments in conventional bulk PTFE are also observed in HWCVD PTFE films (Figure 6.1). These features

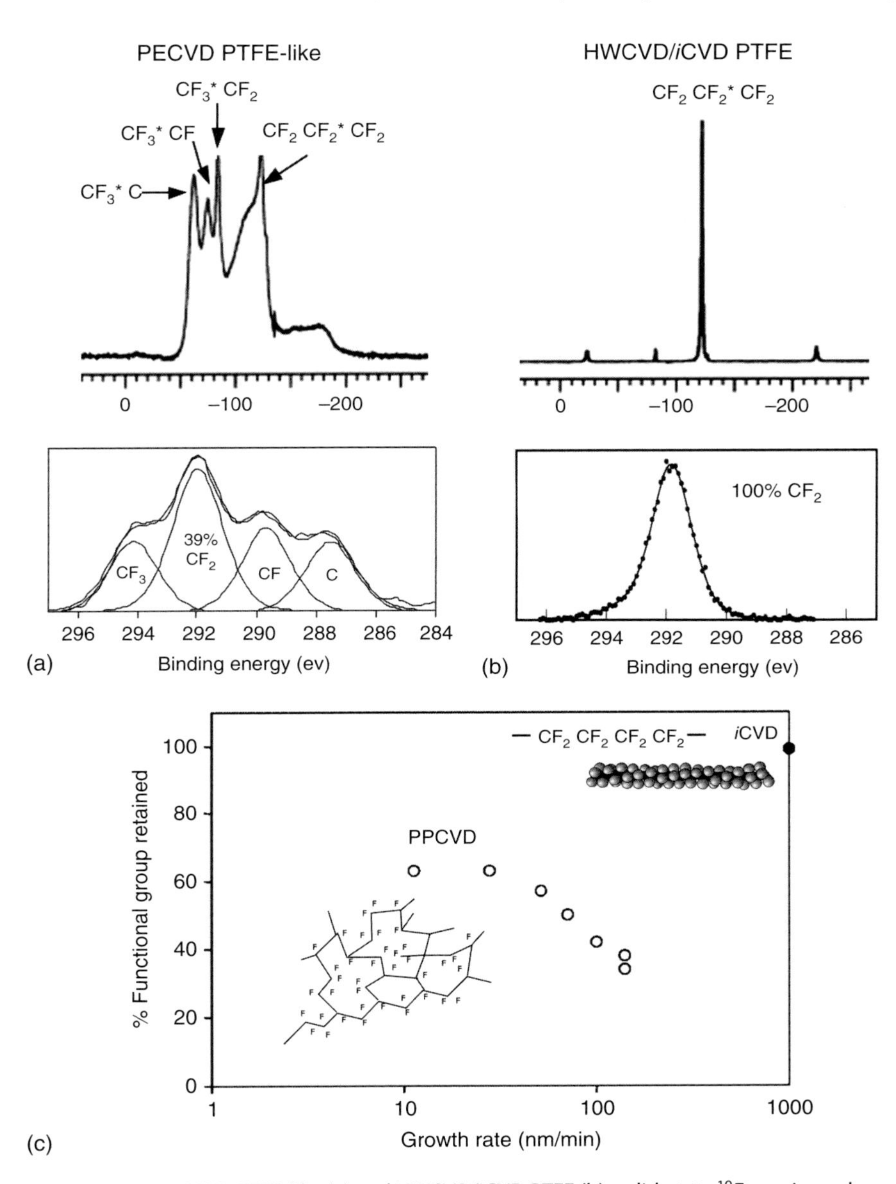

Figure 6.1 For PECVD PTFE-like (a) and HWCVD/iCVD PTFE (b), solid-state ^{19}F magic angle spinning NMR (top) and C1s XPS (bottom). For pulsed PECVD (PPCVD), there is an incomplete retention of $—CF_2—$ moieties, which decreases with increasing deposition rate (c). For iCVD PTFE, 100% functional retention is achieved at all deposition rates, even at rates exceeding 1 μm/min. Source: Tenhaeff and Gleason 2008 [6]. Reprinted with permission of John Wiley & Sons.

included the symmetric (1155 cm^{-1}) and asymmetric (1215 cm^{-1}) stretching modes —CF_2— in the FTIR spectra. Additionally, one dominant —CF_2— peak appears in the solid-state magic angle spinning ^{19}F nuclear magnetic resonance (NMR) and in the carbon 1s X-ray photoelectron spectroscopy (XPS) spectra. In contrast, plasma-enhanced chemical vapor deposition (PECVD) films provide less of the —CF_2— functionality than in PTFE while also containing >C<, CF, and CF_3 functional groups not found in bulk PTFE. The dangling bond defect density, as determined by electron spin resonance (ESR), is much lower in HWCVD PTFE films than in PECVD fluoropolymers. When plasma excitation is utilized to create reactive gas-phase species for a CVD process, there is typically a trade-off between the percentage of —CF_2— functional groups in the film and the overall film growth rate [6]. In contrast, HWCVD can rapidly deposit films, in some cases at growth rates >1 μm/min, composed of essentially 100% —CF_2— groups.

Crystallites have been observed by X-ray diffraction (XRD) in as-deposited and annealed CVD PTFE films. Slowing the CVD PTFE growth rates increases the degree of crystallinity. Crystallinity also results in changes to the —CF_2— wagging (641 and 629 cm^{-1}), deformation (555 cm^{-1}), and rocking (523 and 530 cm^{-1}) modes in the FTIR spectra. Spherulites, up to ~1 mm in diameter, have been observed by optical microscopy. The PTFE crystallites are anisotropic, and inefficient packing can lead to porosity up to 30 vol% as characterized by variable angle spectroscopic ellipsometry (VASE) and to significant surface roughness as imaged by atomic force microscopy (AFM). Higher growth rates typically reduce both the degree of crystallinity and porosity.

Bulk PTFE displays high lubricity, but the COF of CVD PTFE films can be even lower (Table 6.1). The low COF results from the ease by which the linear PTFE chains can slide with respect to one another. In contrast, the COF of PECVD fluoropolymers is often poor because cross-links prevent molecular segments from slipping past one another. The cross-links, >C< and CF, also substantially reduce the thermal stability to as low as ~250 °C for the PECVD fluoropolymers, as compared to ~400 °C for the linear PTFE chains [7].

The surface energy of PTFE is quite low, ca. 20 mN/m. Ultrathin CVD PTFE films displaying roughness or grown conformally over nanostructured surfaces can display superhydrophobic behavior (Figure 6.2a,b) [8]. The CVD PTFE films, which can be >1 μm thick, can be deposited conformally around macroscale structures, such as lead wire for implantation into the cortex (Figure 6.2c) [9]. The observed wetting behavior stems from the combination of the low surface energy and the crystallinity-driven texture of CVD PTFE films (Figure 6.2d) [10]. A "nano shish kabab" hierarchical structure for iCVD PTFE on carbon nanotubes was able to be converted from lotus-type to rose-petal-type wetting upon the application of an electric field [11]. The low surface temperature, often 25 °C, during the process even allows the tissue paper to survive the CVD surface modification. When applied to an ordinary tissue paper, CVD PTFE prevents the absorption of water. It is interesting to note that as compared to bulk PTFE, PECVD fluoropolymer films can have even lower surface energy because they contain some —CF_3 groups at the surface, not just —CF_2— units.

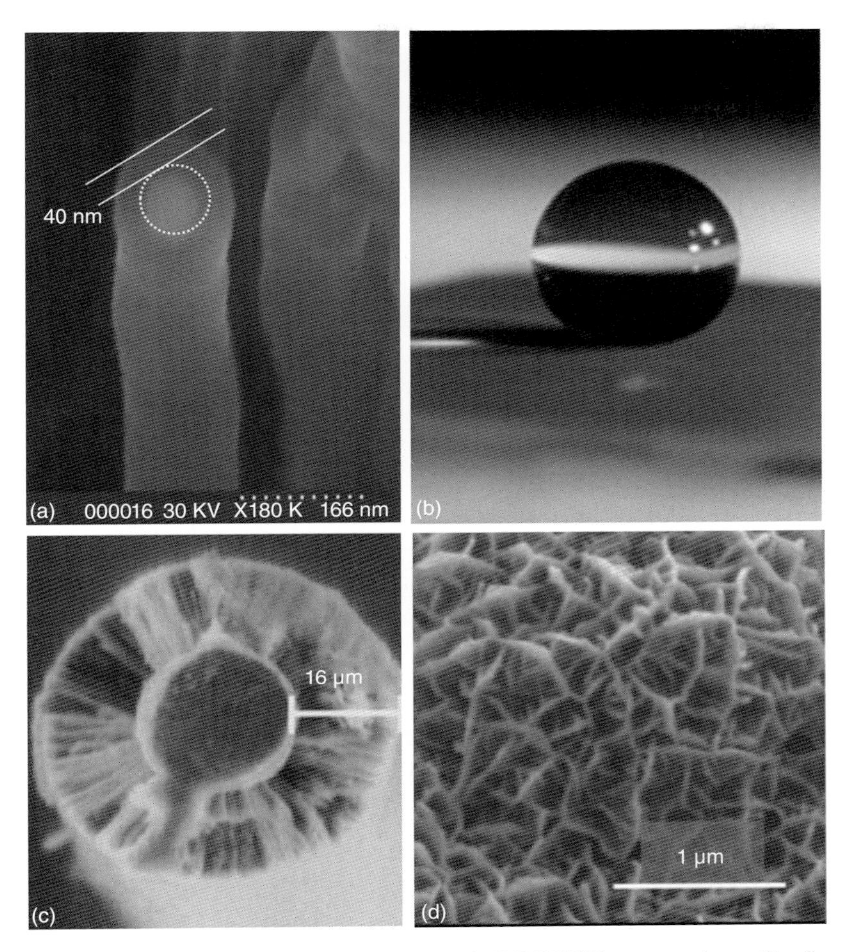

Figure 6.2 (a) Ultrathin and conformal (~40 nm) iCVD PTFE films, around a multiwall carbon nanotube (~50 nm diameter, ~2 μm height) in a vertically aligned "forest," yielding (b) a superhydrophobic surface displaying advancing and receding water contact angles of 170°and 160°, respectively. Source: Lau et al. 2003 [8]. Reprinted with permission of American Chemical Society. (c) A 16-μm-thick HWCVD PTFE conformal coating of fine metal wire used for neural implants. Source: Limb et al. 1996 [9]. Reprinted with permission of AIP Publishing. (d) The semicrystalline nature of HWCVD PTFE can lead to rough surface morphology. Source: Thieme et al. 2013 [10]. Reprinted with permission of Elsevier.

The combination of low COF, good chemical and thermal stability, durability, adhesion, ability to scale to large deposition areas, and conformal coverage of complex features has enabled the commercialization of HWCVD PTFE as a release coating for multiple applications [12]. For example, heating and pressurization of a rubber blank inside of the molds used to induce vulcanization and form the final geometry of a vehicle tire, including its tread pattern. If the rubber sticks to the surface of the mold, the manufacturing process slows down, requiring more time for demolding the tire. The increased forces required for removing the rubber from the mold can also result in misshapen tires. Spray on release agents, such as silicone oils, has traditionally been used to avoid sticking of the rubber to the mold. However, the spray must be reapplied every few molding cycles, causing residue from the spray to build up on the mold's surface. This residue can exacerbate sticking problems to the point where the manufacturing line must be shut down to remove the mold for cleaning. In contrast, CVD PTFE releases coating and can survive thousands of tire molding cycles while keeping the tire production rate high, as demolding remains quick and manufacturing shutdowns caused by residue build up on the molds are avoided. Keeping the demolding force low also keeps the quality of the tires produced high. The CVD PTFE release coatings deposit conformally over all the geometric features of a tire mold, including the complex tread patterns utilized in advanced tire designs. The CVD PTFE coatings survive several thousand molding cycles, demonstrating the excellent adhesion and cohesion achievable.

6.2.2 Influence of the Catalyzing Materials for PTFE Deposition

In early studies, with all other conditions being identical, nichrome (NiCr) filaments were observed to provide much higher HWCVD PTFE deposition rates (10–100 nm/min) as compared to the rates observed using a heated alumina pyrolysis surface (<10 nm/min). The higher growth rates associated with the NiCr suggest the hypothesis of a catalytic effect associated with this filament material.

Figure 6.3 compares the FTIR absorption spectra of several PTFE films deposited on crystalline silicon (c-Si) substrates by Cat-CVD. The catalyzing materials employed were NiCr, Inconel-600, stainless steel (SUS)-304, iron (Fe), molybdenum (Mo), nickel (Ni), titanium(Ti), tantalum (Ta), and tungsten (W) [13]. The temperatures of catalyzing wires, T_{cat}, were also varied over the range of 800–1200 °C. Each deposition utilized the identical HFPO gas flow rate (16 sccm), total chamber pressure, P_g, at 1.5×10^{-3} Pa, and substrate temperature, T_s, near room temperature. The IR absorption peaks in Figure 6.3 are characteristic of CF_2 bonding vibrations such as the symmetric or asymmetric stretching modes and the rocking and wagging vibrations. This FTIR analysis confirms that the films are PTFE.

For each film, the deposition time was 30 minutes. Thus, the films with higher FTIR peaks correspond to thicker films prepared with high deposition rates. Thus, Figure 6.3 clearly demonstrates that the choice of filament material strongly impacts the deposition rate of PTFE films. The NiCr catalyzer yields the highest deposition rate. Inconel-600 and SUS-304, both of which contain Ni atoms in alloys, also show relatively high deposition rates. However, the deposition rates prepared with pure metal catalyzers, even pure Ni, are not as fast as those obtained with Ni alloys. For instance, the deposition rate with Ni catalyzer is about 1/5 of that with NiCr catalyzer. Using W catalyzer, as is often employed for a-Si or SiN_x depositions, the PTFE deposition rate becomes almost 1/10 of the rate obtained by a NiCr catalyzer.

These observations led to speculation that the best catalyzer surfaces for the dissociative adsorption of an HFPO molecule possess two different types of sites. For example, one site may be suitable to accept CF_n (n = 2 or 3), whereas the

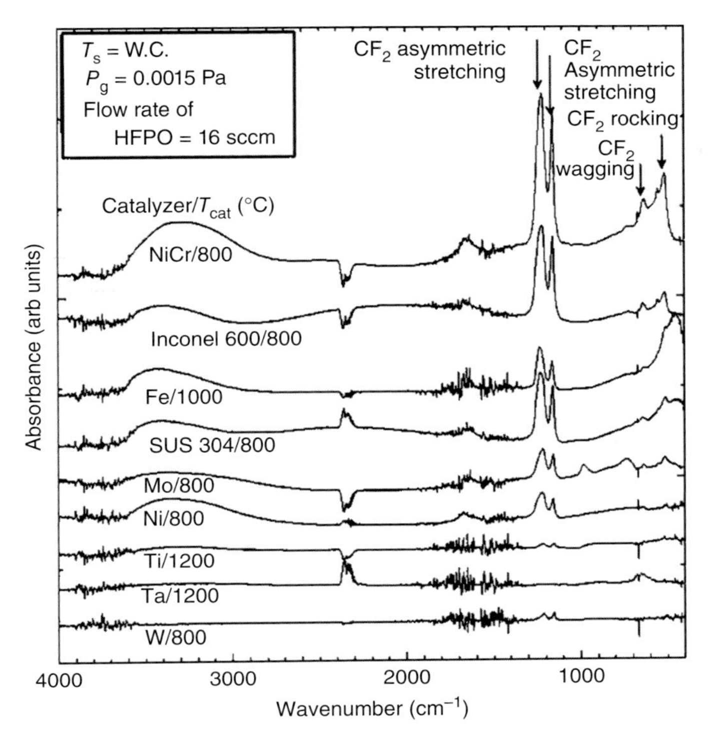

Figure 6.3 FTIR spectra of HWCVD PTFE films prepared with various catalyzing materials and temperatures. Source: Matsumura et al. 2017 [13]. Reprinted with permission of Journal of Vacuum Science & Technology A.

second site accepts other decomposition products of HFPO. Thus, having two types of sites may enhance the velocity of catalytic cracking compared to filament surfaces possessing only a single type of site. The results strongly suggest that the decomposition is not carried out by the simple thermal process but by the catalytic cracking process even for PTFE deposition from HFPO with T_{cat} over 800 °C.

6.3 Mechanistic Principles of iCVD

The large and growing number of homopolymers, random copolymers, and alternating copolymers synthesized by iCVD stems in part from a fundamental understanding of the principles that govern the process. For homopolymers, quantitative models have been developed to predict the film growth rate and the corresponding number-average molecular weight of its polymeric chains. For copolymers, quantitative modeling allows prediction of the iCVD film composition from the monomer ratio in the gas phase. Models for the degree of conformal coverage also provide quantitative predictions.

6.3.1 Initiators and Inhibitors

In conventional chain growth polymerization, an initiating species (I^*) reacts with the monomers, M, to form macromolecular chains. In this case, the first steps of film formation process are the surface reactions

$$I^* + M \rightarrow IM^* \text{ (initiation)} \tag{6.4}$$

and

$$IM^* + (n-1)M \rightarrow IM_n^* \text{ (propagation)} \tag{6.5}$$

The first intentionally introduced initiator used for CVD PTFE was perfluorooctanesulfonyl fluoride (PFOSF) [14]. Use of a small concentration of PFOSF along with HFPO resulted in substantial enhancement in growth rates, achieving values >1 μm/min. The thermal decomposition of PFOSF over the heated filaments in the vacuum chamber was proposed to produce $CF_3(CF_2)_7\cdot$ radicals as the initiating species. This higher molecular weight, less volatile radical, having eight carbons (C8), was hypothesized to adsorb to the growth surface much more readily than CF_2, giving rise to the observed acceleration in film growth rate. If there is a desire to avoid using a bioaccumulating C8 species, perfluorobutanesulfonylfluoride can be utilized as the initiator to yield the more volatile and less bioaccumulating C4 species, $CF_3(CF_2)_3$, as the initiating radical [8].

The concept of feeding an initiator along with one or more monomers into CVD polymer reactors to enhance the growth rate rapidly expanded to vinyl polymerization. At a fixed filament temperature, using an initiator typically greatly enhances the polymer film deposition rates. Alternatively, using an initiator allows the use of lower filament temperature while still maintaining desirable deposition rates. Lowering the filament temperature provides the benefit of reducing the heat load on the substrate, thus reducing the active cooling requirement for the growth susceptor.

The free radical initiators, I_2, for vinyl polymerization by iCVD include tert-butyl peroxide (TBPO), tert-amyl peroxide, and tert-butyl peroxybenzoate (TBPOB). These initiators contain highly labile peroxy (—O—O—) bonds that readily undergo thermal decomposition at or near the heated filaments to yield initiator radicals, I^*. At the modest filament temperatures, 150–350 °C, many monomers, M, are thermally stable. Figure 6.4 shows a schematic of the iCVD process along with the observed growth rates as a function of filament temperature for two different initiators with the vinyl monomer cyclohexyl methacrylate (CHMA) [15]. The TBPOB initiator is observed to decompose at lower filament temperatures than TBPO. Lowering the filament temperature reduces the heat load on the growth surface. Reports to date have primarily utilized nichrome (80/20) and SUS filaments to decompose the peroxide initiators. No observation to date is consistent with a catalytic role for the metal filament in the iCVD polymerization of vinyl monomers.

The TBPO initiator also serves as an initiator for ring-opening iCVD polymerization [16]. Triethylamine (TEA) is another free radical initiator for iCVD but requires the use of higher filament temperatures (>450 °C), which can induce the undesirable fragmentation of some monomers [17].

Although free radical initiators can be used to produce a wide variety of polymers, some macromolecular compositions require other synthesis strategies.

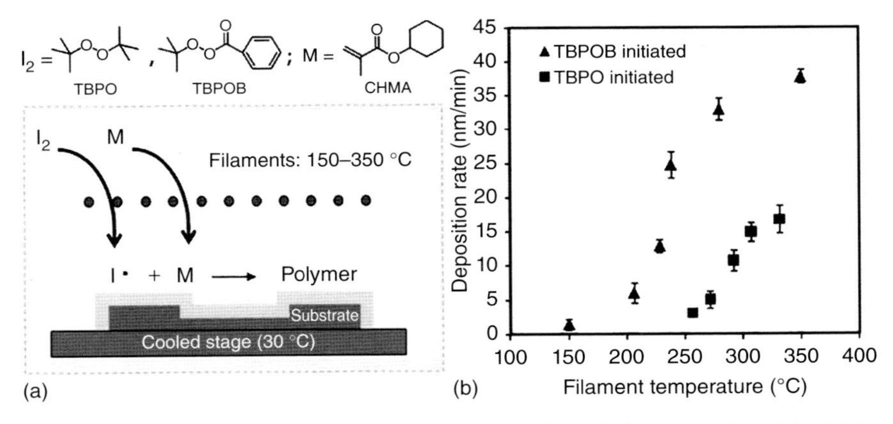

Figure 6.4 Schematic of the iCVD process (a), showing thermal decomposition of the initiator, I_2, at or near the heated filaments, to form initiating radicals, I*. The monomers, M, react with the initiator radicals at the cooled surface to produce a polymer thin film that conformally covers nonplanar substrates; (b) and the observed deposition rates of polymer films as a function of filament temperatures using two different initiators, tert-butyl peroxide (TBPO) and tert-butyl peroxybenzoate with the same monomer, cyclohexyl methacrylate (CHMA). Source: Xu and Gleason 2010 [15]. Reprinted with permission of American Chemical Society.

Recently, the iCVD method was demonstrated to be compatible with cationic initiators. In this case, boron trifluoride complexed with diethyl etherate was employed as the cationic initiator for the ring-opening polymerization of ethylene oxide, yielding thin films of polyethylene oxide (PEO) [18]. Cationic initiation for the iCVD of polystyrene (PS) and polydivinylbenzene (PDVB) has been achieved using $TiCl_4$, a strong Lewis acid, in combination with a hydrogen donor, H_2O [19].

Metal salts, such as $CuCl_2$ and $Cu(NO_3)_2$, applied to the substrate, inhibit iCVD film growth, analogous to their effect on solution free radical polymerization. Selective application of the metal salts on the substrate can be used to achieve patterned iCVD films [20]. Inhibitors, based on (2,2,6,6-tetramethylpiperidin-1-yl)oxyl (TEMPO) are often added to liquid monomers to prevent their polymerization in the source jar.

6.3.2 Monomer Adsorption

Monomers for iCVD should be stable enough to pass through the heated filament array without decomposing. Additionally, the monomers should be volatile enough to exist as a vapor in the vacuum chamber. However, their volatility cannot be too high, or the monomers will not absorb on the cooled growth stage. The iCVD deposition rate will depend on the concentration of the adsorbed monomer, which has been described using the Brunauer–Emmett–Teller (BET) adsorption isotherm:

$$V_{ad} = \frac{V_{ml}c(P_M/P_{sat})}{(1 - P_M/P_{sat})[1 - (1 - c)(P_M/P_{sat})]}, \tag{6.1}$$

where V_{ad} is the total adsorbed volume, V_{ml} is the monolayer adsorbed volume, and c is the BET constant, which depends on the magnitude of the

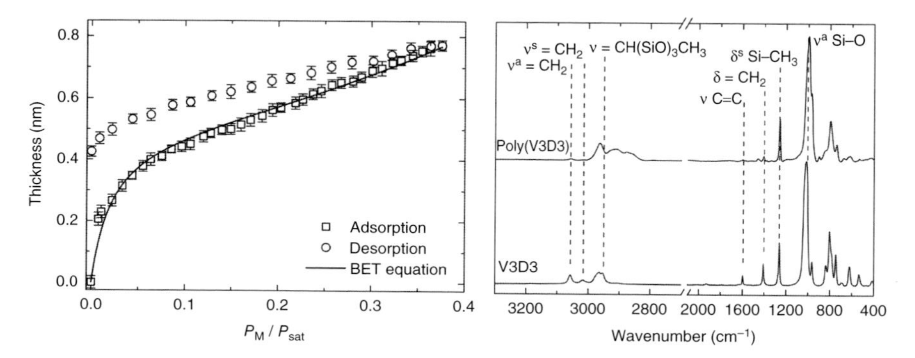

Figure 6.5 Adsorption data (squares, right) of the V3D3 monomer at a fixed temperature of 40 °C, as a function of the saturation ratio, P_M/P_{sat}. These data are well described by the BET equation (solid line, Eq. (6.1)). The desorption data (circles) display hysteresis. (left) The FTIR spectra for the V3D3 monomer (bottom) and for the corresponding iCVD homopolymer (top), showing the consumption of vinyl moieties (—CH=CH_2) upon polymerization, while maintaining the SiO and Si—CH_3 functionalities. Source: Aresta et al. 2012 [21]. Reprinted with permission of American Chemical Society.

surface/adsorbate interaction. The dimensionless ratio of the monomer's partial pressure, P_M, to its saturation pressure, P_{sat}, can be seen to be a key parameter in the BET isotherm.

For iCVD, a commonly used range is $0.3 < P_M/P_{sat} < 0.5$, which typically brackets monolayer coverage. It is possible to grow iCVD films at both lower and higher values of P_M/P_{sat}. Knowledge of the vapor pressure data (P_{sat}) for a new monomer enables prediction of the desired partial pressure range in advance of any experimentation. Tabulation of vapor pressure values and correlations to predict P_{sat} as a function of temperature are widely available for many monomers. Even for less common molecules, vendors typically provide either the P_{sat} values at room temperature or the heat of vaporization, ΔH_{vap}, and also report the boiling temperature, for which $P_{sat} = 1$ atm. From this data, P_{sat} at any other temperature can be estimated using the Clausius–Clapeyron equation:

$$P_{sat}(T_2) = P_{sat}(T_1)\ \exp[-(\Delta H_{vap}/R)(1/T_2 - 1/T_1)]. \tag{6.2}$$

Figure 6.5 shows the experimentally measured data points for the adsorption of the monomer, 1,3,5-trivinyl-1,3,5-trimethylcyclotrisiloxane (V3D3), as a function of the ratio P_M/P_{sat} [21]. The solid line is the best fit to the BET isotherm (Eq. (6.1)), which provides an excellent match to the data. The BET isotherm also provided an excellent $R^2 > 0.99$ for the absorption the monomers, ethyl acrylate [22] and butyl acrylate [23]. The best-fit parameters from the BET equation were used to estimate monolayer coverages of 8.01×10^{14} cm^{-2} for ethyl acrylate and 5.59×10^{14} cm$-^{2}$ for butyl acrylate.

At low values of P_M/P_{sat}, typically valid up to ~0.25, Eq. (6.1) simplifies to

$$[M] \sim V_{ad}|_{P_M/P_{sat}\to 0} = V_{ml}c(P_M/P_{sat}), \tag{6.3}$$

representing a linear relationship between the concentration of monomer adsorbed on a surface at temperature T, to the partial pressure of the monomer in the gas phase. This limit is often called Henry's law.

6.3.3 Deposition Rate and Molecular Weight

The most common iCVD mechanism is vinyl polymerization and is consistent with the observation of the loss of vinyl peaks in the FTIR spectra between the monomer V3D3 and its iCVD polymer, PV3D3 (Figure 6.5) [19]. In these spectra, the most intense peaks are similar between the monomer and the polymer confirming the retention of other organic functionalities, such as the six-member siloxane ring of the V3D3 monomer.

For acrylates and methacrylates, the rate of chain propagation is fast, often making monomer adsorption the rate-limiting step for film deposition. In this regime, iCVD film deposition rates are observed to increase as the substrate temperature is lowered, by increasing the concentration of monomer on the growth surface. No change in iCVD growth rate occurs when overall reaction pressure is varied while holding P_M fixed [24], consistent with the hypothesis that homogeneous gas-phase processes are not rate controlling. Both iCVD film deposition rate and number-average molecular weight typically increase with increasing P_M/P_{sat} and match well with the quantitative results of kinetic modeling (Figure 6.6). The measured polydispersity index (PDI) for iCVD films, which reflects the distribution of molecular weight for polymer chains, is also consistent with a free radical polymerization mechanism.

In some cases, iCVD growth rate increases after the deposition of some initial film thickness. Potentially, the second and faster regime of growth results from a reservoir of a monomer created by monomers diffusing into the just-deposited material [25]. By growing on top of a thin iCVD cross-linked underlayer, high growth rate deposition of very smooth films at conditions where $P_M/P_{sat} > 1$ (supersaturation) [26]. At the same processing conditions without the underlayer, the monomer condenses into liquid droplets on the substrate, which then solidify into an undulating surface upon polymerization. The difference in morphology may be related to the adsorption of monomer into the underlayer.

Although there is significant absorption of the monomers, M, on the surface, the initiator radicals, I*, are much more volatile and have a much shorter lifetime on the surface. It has been proposed that the initiator radicals impinging on the surface from the gas phase do not adsorb on the surface, but instead react directly with the vinyl bond of an absorbed monomer, known as an Eley–Rideal (E–R) mechanism. Such reactions have near-zero activation barriers. Indeed, little variation in deposition rate occurs when varying substrate temperature while holding P_M/P_{sat} fixed [27].

6.3.4 Copolymerization

By using two or more monomers, iCVD process copolymerization is possible. Spectroscopic methods validate the formation of the copolymer as opposed to the simultaneous growth of two different homopolymers. As the fraction of the monomer methacrylic acid increases relative to the monomer ethyl acrylate, the position of the very strong carbonyl stretch in the FTIR shifts monotonically from 1733 to 1738 cm^{-1}, reflecting a change in electronic density along the backbone polymer chain resulting from copolymerization. Although NMR provides specific information on the distribution of triad groups (e.g. AAA, ABA, and ABB)

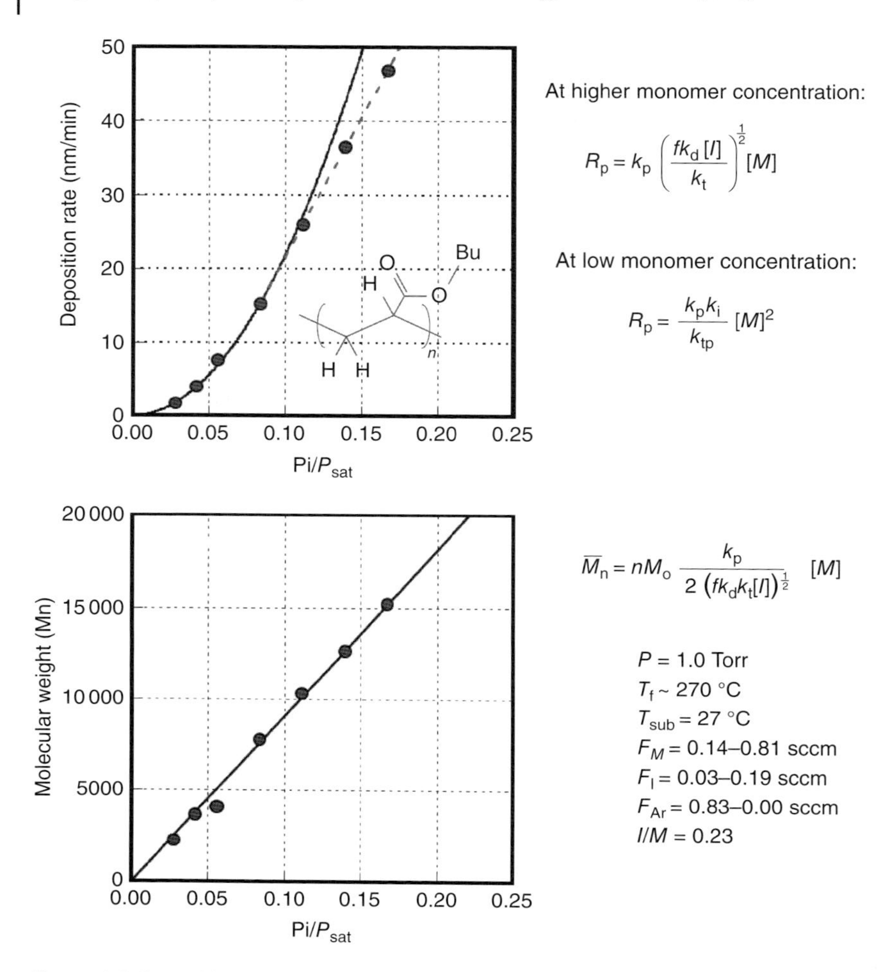

Figure 6.6 Deposition rate, R_p, (top) and number-average molecular weight, M_n, (bottom) for iCVD polybutyl acrylate (PBuA) as a function of the saturation ratio, P_M/P_{sat}. All other process conditions were held at the fix values listed in the bottom right corner. In both graphs, solid circles are experimental measurements, and the solid lines are the predictions of the full kinetic model (first and third equations on the right side). The dotted line in the top graph is the prediction in the limit of low monomer concentration (second equation on the right). Source: Lau and Gleason 2006 [23]. Reprinted with permission of American Chemical Society.

in iCVD copolymers [28], this analysis requires a relatively large amount of film and a film composition capable of being solubilized. FTIR or XPS analysis frequently allows the determination of the iCVD copolymer composition utilizing peak area ratios.

For traditional solution synthesis methods, the kinetics of copolymerization relies on process conditions including the reaction temperature, the concentration of the monomers, and the reactivity ratios of the monomers. These same factors determine the kinetics of iCVD copolymerization. As the iCVD polymerization takes place on the surface, surface temperature and absorbed monomer concentrations determine the kinetics. The surface temperature can be directly controlled by cooling or heating the growth stage. The surface concentrations are indirectly controlled by adjusting the partial pressures of monomers in the reactant chamber and can be estimated at low P_M/P_{sat} values by Henry's law (Eq. (6.3)) or for all P_M/P_{sat} by the BET isotherm (Eq. (6.1)). A growing copolymer chain adds either monomer A or monomer B. The reactivity ratio, r_A, is the probability that a dyad of AA forms divided by the probability of forming an AB dyad. Similarly, r_B is the ratio of the BB and BA dyad probabilities. When both reactivity ratios are unity, each monomer reacts equally well with itself or with the other monomer, leading to a random distribution of both monomers along the chain (e.g. —ABBABAA—). When both reactivity ratios are zero, only alternation of the two monomer types occurs (e.g. ABABABAB—). For $r_A > 1$ with $r_B < 1$, growing copolymers become difficult, as chain compositions close to homopolymer A becomes favored. The Q-e scheme is an empirical method for estimating the reactivity ratios before attempting copolymerization [29].

Considering the impact of Henry's law and the reactivity ratios means that feeding a 1 : 1 ratio of two monomer vapors into the vacuum chamber will typically not result in 1 : 1 incorporation of monomers into the iCVD film. The feed composition is most often related to the iCVD film composition by the Fineman–Ross copolymerization equation [30]. The Fineman–Ross equations allow extraction of reactivity ratios from experimental data. Kelen–Tudos plots are an alternative means of determining reactivity ratios from measured results [28].

6.3.5 Conformality

The probability that an initiator radical, I*, will adsorb or react when it collides with the surface, rather than reflecting away, is known as the sticking probability, Γ. As a result of detailed experimentation and modeling [24, 31], Γ has been hypothesized as the key factor controlling the degree of conformality resulting from the iCVD growth of acrylate and methacrylate monomers. Step coverage, S, the ratio of the film thickness at the top and bottom of a trench feature, is a simple quantitative measure of the degree of conformality. A two-dimensional model applies to trenches that are much longer than either their width, W, or depth, L. If, additionally, the gas-phase diffusion rate of the monomer is rapid relative to the film deposition rate, S depends only on Γ and the aspect ratio, L/W:

$$\mathrm{Ln}(S) = -0.48\,\Gamma\,(L/W)^2. \tag{6.4}$$

Regression analysis using Eq. (6.4) with experimental measurements of S for trenches having different L/W ratios allows values for Γ to be estimated [27]. Typical values of Γ for iCVD process fall in the range of 0.01–0.1. Values of Γ increase as P_M/P_{sat} increases, which is consistent with the proposed E–R model, in which the probability of an initiator radical reacting on the surface increases as the surface coverage of vinyl bonds increases. Also consistent with the hypothesized E–R mechanism is that at the same P_M/P_{sat}, Γ is ~2× higher using a divinyl monomer than a monovinyl one and shows little variation with substrate temperature.

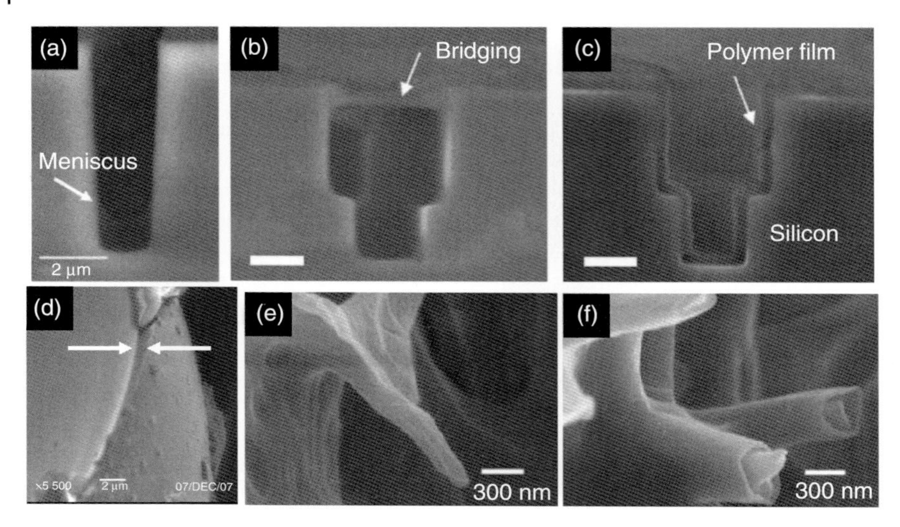

Figure 6.7 Cross-sectional scanning electron micrographs (SEMs) of microtrench features in silicon wafers (a–c). Spin-coating of a polymer coating from solution results in nonconformal coverage with (a) showing surface tension driven meniscus formation and essentially no coverage of the slide walls and (b) displaying liquid bridging across the feature. In contrast, (c) shows the conformal coverage achieved by iCVD, where the same coating thickness is achieved along the top surface of the wafer and over all the interior surfaces of the feature. Source: (a) Tenhaeff and Gleason 2008 [6]. Reprinted with permission of John Wiley & Sons. (b, c) Reeja-Jayan et al. 2014 [32]. Reprinted with permission of John Wiley & Sons. SEMs showing interior surface of a polyvinylidene fluoride (PVDF) membrane obtained by fracturing (d) before and (e) after conformal coating by iCVD. SEM of a fractured microparticle, revealing a conformal iCVD coating over the external surface (f). Source: (d) Baxamusa et al. 2008 [33]. Reprinted with permission of American Chemical Society. (e, f) Sun et al. 2016 [34]. Reprinted with permission of Royal Society of Chemistry.

For other geometries, such as square or cylindrical pores, a geometric prefactor, typically between 0.1 and 1, is used to multiply the right-hand side of Eq. (6.4) [1].

Lowering P_M/P_{sat} increases conformality with the trade-off of lower deposition rate. Under conditions of low Γ, conformal coverage results over all the interior and exterior surfaces of even high aspect ratio features (Figure 6.7) [6, 32–34]. Conformality is also important for fully encapsulating and protecting the external surfaces of complex geometries at both the nano- and microscales. The conformality achievable by iCVD is another differentiating feature from PECVD, in which the directionality of ion bombardment resulting from the electric field produces variation in thickness coating over the geometry of a feature.

6.4 Functional, Surface-Reactive, and Responsive Organic Films Prepared by iCVD

Tables 6.2 and 6.3 show the breadth of monomers employed for iCVD to date. More details of the iCVD process for the monomers and the properties and applications of the corresponding homopolymers and copolymers appear in previous

Table 6.2 Example monomers for iCVD listed alphabetically by their acronyms.

Acronym	Monomer name (nonvinyl polymerization type, if applicable: RO = ring opening; A = alkyne)	Functionality	Acronym	Monomer name (nonvinyl polymerization type, if applicable: RO = ring opening; A = alkyne)	Functionality
1V2P	1-Vinyl-2-pyrrolidone	Pyrrolidone	MAA	Methacrylic acid	Carboxylic acid
4AS	4-Aminostyrene	Primary amine	MAH	Methacrylic anhydride	Cross-linking
4VBC	4-Vinyl benzyl chloride	Chloro	MA	Maleic anhydride	Anhydride
4VP	4-Vinyl pyridine	Pyridiene	MCA	Methyl α-chloroacrylate	Chloro
AA	Acrylic acid	Carboxylic acid	MMA	Methyl methacrylate	Methyl
AAm	Allyl amine	Primary amine	*m*DEB	*m*-Diethynyl benzene(A)	Aromatic hydrocarbon
BMA	Benzyl methacrylate	Benzyl	MDO	2-Methylidene-1,3-dioxepane (RO)	Ring
C6PFA	1*H*,1*H*,2*H*,2*H* perfluorooctyl acrylate	Fluoroalkane	*n*BA	*n*-Butyl acrylate	Alkyl
CEA	2-Chloroethyl acrylate	Chloro	*n*BMA	*n*-Butyl methacrylate	Alkyl
CHMA	Cyclohexyl methacrylate	Cycloalkyl	neoMA	Neopentyl methacrylate	Alkyl
CNEA	2-Cyanoethyl acrylate	Cyanato	*n*HA	*n*-Hexyl acrylate	Alkyl
D3	Hexamethylcyclotrisiloxane (RO)	Siloxane	NIPAAm	*N*-Isopropyl acrylamine	Secondary amine
D4	Octamethylcyclotetrasiloxane (RO)	Siloxane	*n*PA	*n*-Pentyl acrylate	Alkyl
DEAAm	Diethyl acrylamide	Tertiary amine	*n*PrA	*n*-Propyl acrylate	Alkyl
DEAEMA	Diethylaminoethylmethacrylate		*n*PrgA	Propargyl acrylate	Proparyl
DEAEA	Diethylaminoethyl acrylate	Tertiary amine	NVCL	*N*-Vinylcaprolactam	Tertiary amine
DEGDVE	Di(ethylene glycol) divinyl ether	Cross-linking	*o*NBMA	*o*-Nitrobenzyl methacrylate	*o*-Nitrobenzyl
DFHA	1*H*,1*H*,2*H*,2*H*,7*H* perfluorododecaheptyl acrylate	Fluoroalkyl	PEGMA	Polyethylene glycol methacrylate	Fluoroalkyl
DMA	Dodecyl methacrylate	Alkyl	PFDA	1*H*,1*H*,2*H*,2*H*-Perfluorodecyl acrylate	Fluoroalkyl

(Continued)

Table 6.2 (Continued)

Acronym	Monomer name (nonvinyl polymerization type, if applicable: RO = ring opening; A = alkyne)	Functionality	Acronym	Monomer name (nonvinyl polymerization type, if applicable: RO = ring opening; A = alkyne)	Functionality
DMAAm	*N*,*N*-Dimethylacrylamide	Tertiary amine	PFDMA	1*H*,1*H*,2*H*,2*H*-Perfluorodecyl methacrylate	Fluoroalkyl
DMAMS	Dimethylaminostyrene	Tertiary amine	PFM	Pentafluorophenyl methacrylate	Pentafluoro-phenyl
DMAEMA	*N*,*N*-Dimethylaminoethyl methacrylate	Tertiary amine	PhAc	Phenyl acetylene (A)	Phenyl
DPAEMA	2-Diisopropylaminoethyl methacrylate	Tertiary amine	PrgMA	Propargyl methacrylate	Proparyl
DVB	Divinylbenzene	Cross-linking	S	Styrene	Phenyl
EA	Ethyl acrylate	Alkyl	*t*BA	t-Butyl acrylate	Alkyl
EGDA	Ethylene glycol diacrylate	Cross-linking	tHEN	1,2,3,4-Tetrahydro-1, 4-epoxynaphthalene (RO)	Furan
EGDMA	Ethylene glycol dimethacrylate	Cross-linking	TrOx	Trioxane (RO)	Ring
EO	Ethylene oxide (RO)	Ether	TVTSO	1,1,3,5,5 Pentamethyl-1,3,5 trivinyltrisiloxane	Siloxane
FMA	Furfuryl methyacrylate	Furfuryl	V3D3	Trivinyltrimethyl-cyclotrisiloxane	Siloxane
GMA	Glycidyl methacrylate	Epoxy	V3N3	Trivinyltrimethyl-cyclotrisilazane	Silazane
HEMA	Hydroxyethyl methacrylate	Hydroxyl	V4D4	Tetravinyltetramethyl-cyclotetrasiloxane	Siloxane
HDFDMA	1*H*,1*H*,2*H*,2*H* heptadecadecyl methacrylate	Fluoroalkane	V4N4	Tetravinyltetramethyl-cyclotetrasilazane	Silazane
HVDS	Hexavinyldisiloxane	Siloxane	VCin	Vinyl cinnamate	Cinnamate
iBA	Isobornyl acrylate	Isobornyl	VI	1-Vinylimidazole	Imidazole
IBF	Isobenzofuran	Furan			
IEM	Isocyanatoethyl methacrylate	Isocyanate			

Table 6.3 Properties for selected iCVD monomers.

	Molecular weight (g/mol)	Normal boiling point (°C)	ΔH_{vap} (kJ/mol)	P_{sat} at 25 °C (Torr)	Chemical functionality	Characteristics of the corresponding iCVD homopolymer
Monomers with one vinyl bond forming linear polymer chains						
MA	86.1	160.5	43.8	3.03	Carboxylic acid	Hydrophilic, pH responsive
4VP	105.1	173.6	39.3	3.88	Tertiary amine	Converts to quatenary amine and zwitterionic functionalities
GMA	142.1	189.0	42.5	1.72	Epoxy	Reactive toward amines, thiols
MAA	90.1	202.0	43.8	1.05	Anhydride	Reacts with amines, thiol, and hydroxyls, alternating polymerization with styrenes
NIPAAm	113.1	225.1	46.2	0.42	Secondary amine	Thermally responsive, switching from hydrophilic to hydrophobic
HEMA	129.1	261.0	49.8	0.02	Hydroxy	Hydrophilic, swellable hydrogel
PFDA	518.2	268.0	47.0	0.15	Perfluoroalkyl (C8)	Hydrophobic
*o*NBMA	221.2	318.8	56.0	0.01	*o*-Nitrobenzyl	UV light responsive
Monomers with multiple vinyl bonds forming smooth, mechanical robust organic covalent networks						
DVB	130.0	195.0	57.9	1.50	Aromatic hydrocarbon	Hydrophobic, dielectric, topcoat for directed self-assembly
V3D3	258.0	201.9	42.5	1.38	Cyclic siloxane	Biopassivation, flexible ultrathin dielectric
V4D4	344.1	247.8	46.5	0.25	Cyclic siloxane	Electrolyte for lithium ion conduction
EGDA	170.2	223.9	42.0	0.45	Ether	Sacrificial layer
EGDMA	198.2	260.6	43.8	0.11	Ether, methyl	Flexible ultrathin dielectric

reviews [1, 3, 4, 35] and the sections that follow in this chapter. The majority of monomers used are liquids, having atmospheric boiling points in the range of ~50–270 °C, allowing the delivery of these compounds to the reactor through an external feed system. Solid monomers can be delivered in the same way providing that they can be heated in the external jar to above their melting point. Less volatile monomers, such as solid-phase metalloporphyrins, can be evaporated using crucibles inside the chamber [36].

Pendant functional groups from these monomers incorporate with nearly 100% retention into the iCVD polymeric films. In contrast, PECVD produces "polymer-like" films that have only partial retention of functional groups. For PECVD, there is often a trade-off between the growth rate and the degree of functional retention (Figure 6.1c). In contrast, iCVD provides 100% functional retention at all growth rates, from slow rates used to controllably produce ultrathin films (<10 nm) to fast rates for encapsulation applications (>1 μm).

The complexity of the PECVD mechanism also creates new types of functional groups and dangling bond defects that can have a detrimental impact on the thermal, electrical, and mechanical properties of the films. PECVD is limited to relatively simple functional groups, such as –OH, –COOH, and –NH [37]. In contrast, iCVD polymeric functional group retention has enabled complex functionalization of surfaces by epoxy and anhydride groups, the formation of surface zwitterionic moieties, and surfaces for click chemistry and Diels–Alder reactions. The full retention of functional groups by iCVD is also essential to achieve responsive properties needed for light-sensitive patterning, swelling transitions because of changes in pH or temperature, and shape memory behavior.

Several categories of monomers appear in Figures 6.8 and 6.9. Within each category, the monomer's chemical structure and acronym are shown (Table 6.2). Each monomer illustrated has been successfully reported in the synthesis of an iCVD or HWCVD homopolymer or copolymer. In many cases, a monomer of interest also belongs to one or more additional categories. For example, 4-aminostyrene (4AS) appears in the N-containing category (Figure 6.9c) but could have appeared with the other styrenes (Figure 6.8d). For two of the monomers, divinyl benzene (DVB) (Figure 6.8c) and dimethylaminostyrene (DMAMS) (Figure 6.9c), the structures are drawn to indicate a mixture of isomers, which reflect their typically available commercial form. In these cases, it is difficult to carry out the separation for producing pure components because of the chemical similarity of the isomers. Later, in Section 6.4.4, the chemical structures of Si-containing monomers possessing multiple vinyl bonds will be shown.

Acrylates have the chemical structure $CH_2{=}CH{—}(C{=}O){—}O{—}R$, where R is the functional group. Acrylates often display the fastest kinetics of polymer chain propagation, enabling high iCVD film growth rates. Figure 6.8a shows three acrylates with hydrocarbon pendant groups, *n*-butyl, *t*-butyl, and isobornyl, and two acrylates with *1H,1H,2H,2H* perfluorocarbon pendant chains, $—CH_2CH_2C_6F_{13}$ and $—CH_2CH_2C_8F_{17}$, having six and eight perfluorinated carbon chains, respectively. The $—CH_2CH_2—$ portion of these monomers is a result of their synthesis procedure and enhances thermal stability. Monomers with perfluorinated functionalities often provide for hydrophobic iCVD surface modification.

Methacrylates are structurally similar to the acrylates but have an additional methyl ($—CH_3$) group, $CH_2{=}C(CH_3){—}(C{=}O){—}O{—}R$. The additional methyl group tends to reduce the rate of chain propagation as compared to acrylates. Additionally, the methacrylate polymers are slightly more hydrophobic than their acrylate counterpart. Figure 6.8b shows methacrylates having *n*-butyl, cyclohexyl, benzyl, and neopentyl hydrocarbon functional groups.

Successful iCVD acrylate and methacrylate monomers include those with hydrocarbon pendant groups ranging from $—CH_3$ to $—C_{10}H_{21}$. Note that the

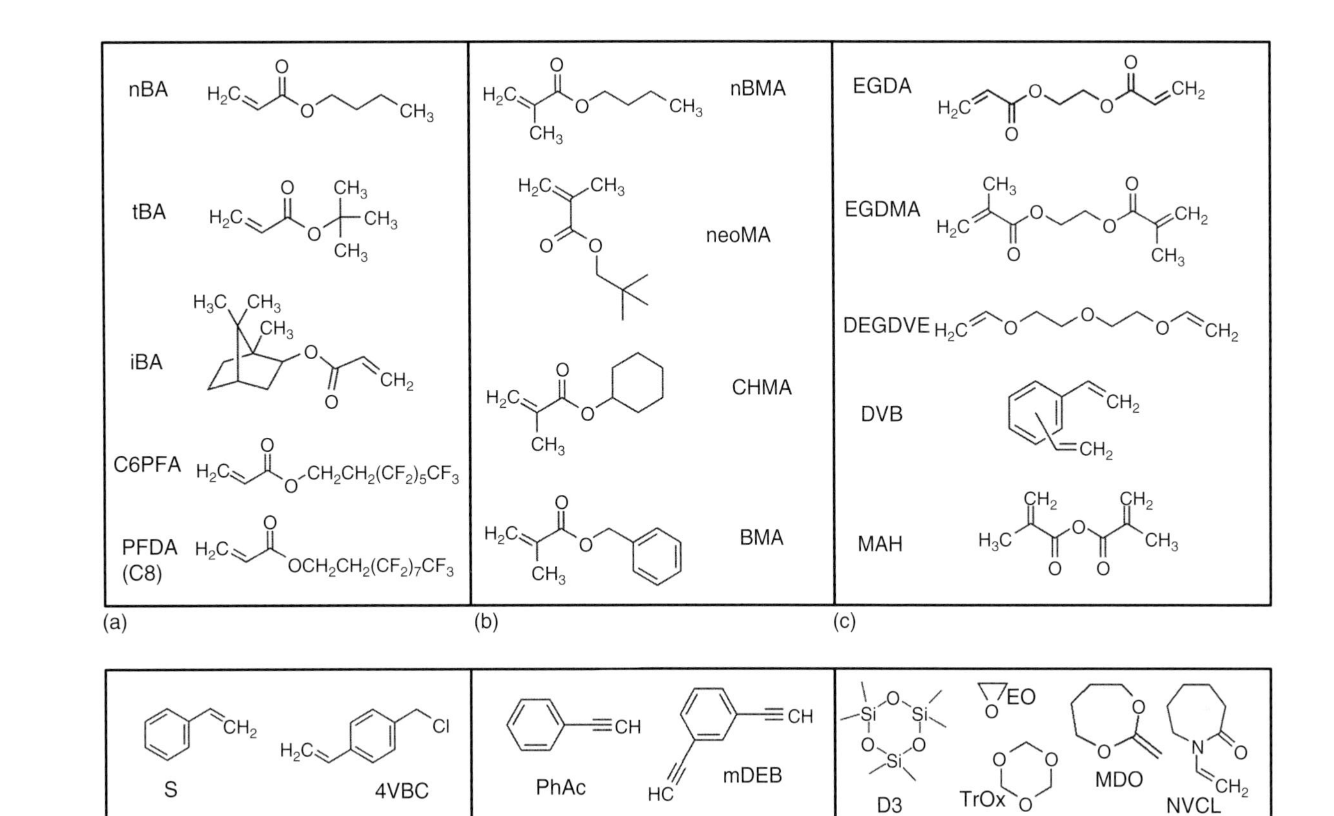

Figure 6.8 Examples of monomers, along with their acronyms (Table 6.2), which have been successfully used for the synthesis of iCVD/HWCVD homopolymers and copolymers. The acrylates, methacrylates, cross-linkers, and styrenes all polymerize through their vinyl (C=C) bonds. Monomers that polymerize through their alkyne (C≡C) or by a ring opening mechanism are also shown. (a) Acrylates, (b) methacrylates, (c) cross-linkers, (d) styrenes, (e) alkynes, and (f) ring opening.

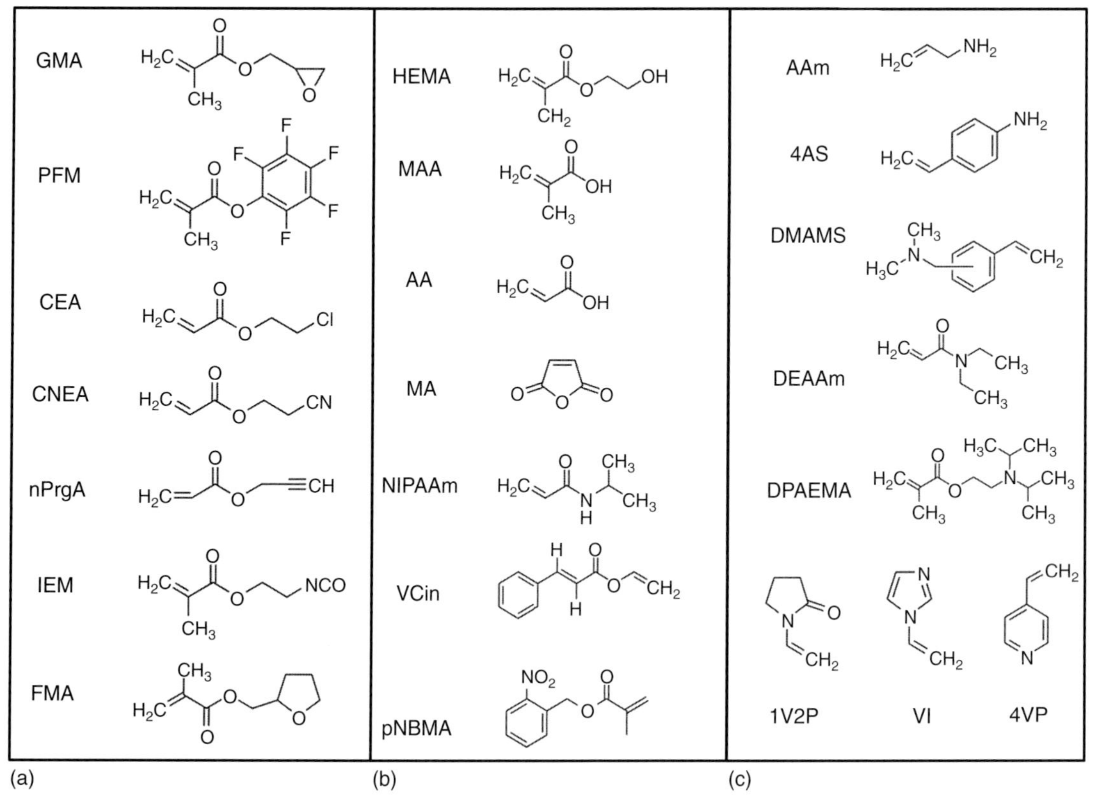

Figure 6.9 Examples of monomers, along with their acronyms (Table 6.2), which have been successfully used for the iCVD vinyl polymerization of homopolymers and copolymers. Their specific organic functional groups determine the resulting properties of the iCVD polymer films, including reactivity and responsive behavior. (a) Reactive, (b) responsive, and (c) N-containing.

number of possible isomers grows as the number of carbons in the functional group increases. Even variation of only the hydrocarbon functional group achieves significant design differences in the corresponding iCVD surface modification.

Although *n*-butyl acrylate and *t*-butyl acrylate, *n*BA and *t*BA, respectively, are isomers (Figure 6.8a), their corresponding homopolymer properties are quite different. For example, the glass transition temperature, T_g, of poly(*n*-butyl acrylate) (P*n*BA) is −54 °C, making it a soft, rubbery material at room temperature. In contrast, the range of T_g values measured of poly(*t*-butyl acrylate) (P*t*BA) ranges from 43 to 107 °C [38], corresponding to a hard, glassy material at room temperature. For comparison, polymerizing *n*-butyl methacrylate (*n*BMA, Figure 6.8b) results in a polymer having a T_g of 20 °C.

Additional polymer properties, including refractive index (RI), strongly depend on the chemical structure of the functional group. For example, the RI of the methacrylate homopolymers synthesized from *n*BMA, CHMA, and BMA (benzyl methacrylate) (Figure 6.8b), are 1.453, 1.507, 1.589, respectively. The highest RI, 1.589, results from the high electron density in the aromatic ring in the BMA monomer unit.

Neopentyl methacrylate (neoMA, Figure 6.8b) iCVD films have been studied intensively as sacrificial layers for forming air gap structures. The iCVD films based on neoMA decompose in a desirable temperature range for use in integrated circuit processing and leave behind little char after thermal annealing [39].

Cross-linkers, each having two polymerizable vinyl bonds (>C=C<), are shown in Figure 6.8c. When both vinyl bonds reacted forming a cross-linking site, typically, both the mechanical properties and smoothness of the resulting films improve.

Styrenes (Figure 6.8d) have substantially lower rates of free radical chain propagation than acrylates or methacrylates. The slow deposition rates can be beneficial for achieving careful control over thickness in the sub-10-nm range. Higher deposition rates with styrenes can occur using iCVD copolymerization or cationic initiation schemes.

Alkynes (Figure 6.8e) polymerize through acetylenic (—C≡C—) bonds in contrast to the examples of vinyl polymerization discussed so far. Although vinyl polymerization converts sp^2 carbon bonding in the monomer to sp^3 carbon in the polymer, polymerization of the alkynes converts sp carbon bonding in the monomer to sp^2 carbon in the polymer. Hence, conjugated polymers can result from iCVD alkyne polymerization [40], having the potential for semiconducting or conducting electrical properties.

Ring-opening polymerization by HWCVD utilizes no intentionally introduced initiator. The first demonstration of this method used the monomers hexamethylcyclotrisiloxane (D3) and octamethylcyclotetrasiloxane (D4) (Figure 6.8f). HWCVD growth from D3 and D4 required filament temperatures between 800 and 1200 °C [41]. Subsequently, HWCVD from trioxane occurred at filament temperatures of 700 °C [42]. HWCVD using filament temperatures between 680 and 800 °C resulted in polyisobenzofuran films from the monomer 1,2,3,4-tetrahydro-1,4-epoxynaphthalene (tHEN) [43].

By using a free radical initiator, the iCVD ring-opening polymerization of 2-methylidene-1,3-dioxepane (MDO) [16] and *N*-vinylcaprolactam (NVCL) [44] proceeded at substantially lower filament temperatures, 225 and 200 °C, respectively. Using cationic initiation allowed the polymerization of ethylene oxide (EO) [18].

Reactive monomers (Figure 6.9a) have chemical functional groups that provide specific reactivity toward complementary moieties. This reactivity can be used to create adhesion to the underlying substrate, to cause reactions within the bulk of the iCVD film for producing cross-linking or for creating new functional groups, or for allowing postdeposition functionalization of the surface. One specific example is the rapid and specific reaction of iCVD films containing propargyl groups (—C≡CH) with azide (—N=N=N) bearing dye molecules, a so-called click chemistry reaction [45]. A second example is the Diels–Alder reaction of the furan group in an iCVD polyfurfuryl methacrylate film with a conjugated diene to form cyclohexene [46]. This highly selective reaction employed with carbohydrate–diene or peptide–diene conjugates for the immobilization of these biomolecules on a solid surface. As will be described in subsequent sections, many additional functionalization possibilities are available for the covalent attachment of fluorescent dyes; the binding of biomolecules, such as antibodies, DNA, or cell growth factors; and the tethering of nanoparticles.

Some strategies require two postdeposition functionalization steps. For example, in a first step, surface anhydride groups of an iCVD film react with the amine functionality of the small-molecule cysteamine ($H_2N—CH_2—CH_2—SH$) achieving a thiol (—SH)-functionalized surface. The surface thiol groups provide for the covalent attachment of CdSe/ZnS nanoparticles [47].

Responsive iCVD films exhibit a change in response to an external stimulus. Incorporating the monomers shown in Figure 6.9b into iCVD homopolymers or copolymers can result in responsive behavior. The hydroxyl (–OH) group of hydroxyethyl methacrylate (HEMA), methacrylic acid (MAA), and acrylic acid (AA) is the source of films that respond upon contact with liquid water, exposure to humidity, and changes in pH. The anhydride group of maleic anhydride (MA) also provides a basis for a response to pH. Thermally responsive films have been iCVD synthesized from *N*-isopropyl acrylamine (NIPAAm) and diethyl acrylamide (DEAAm) (Figure 6.9b). The monomers vinyl cinnamate (VCin) and *n*-nitrobenzyl methacrylate (*n*NBMA) are useful for iCVD films sensitive to UV irradiation and hence useful for creating patterned surfaces.

Nitrogen-containing iCVD monomers in Figure 6.9c represent a diverse array of chemical structures and functionality. The intermolecular hydrogen bonding in monomers with primary amine groups ($–NH_2$) and secondary amine groups (>NH) often make them less volatile than hydrogen-bond-free monomers of similar molecular weight. Some of the monomer units in iCVD films containing tertiary amines (>N–) react further to quaternary amine groups (>N+<) as desired for the capture of nucleic acids [48] and also for some cell culture [49], antimicrobial [50], and zwitterionic antibiofouling [51] applications. Annealing was shown to cause reaction of tertiary amines with alkyl chloride across the interface to induce the lamination of two opposing surfaces coated with the iCVD copolymer P(DMAEMA-*co*-CEA) [52]. Additionally, amine-containing iCVD layers create a dipole layer at the interface with conductive materials including metals and graphene [53]. The amine-containing iCVD layers serve as electron injection layers to shift the work function of the electrode.

6.4.1 Polyglycidol Methacrylate (PGMA): Properties and Applications

Demonstrating the first example of vinyl polymerization by iCVD utilized the monomer glycidyl methacrylate (GMA). As will be described in more detail at the end of this section, the epoxy functional group of GMA is useful for postdeposition functionalization by dyes and biomolecules, for creating adhesive bonds, and as a resist material for use with electron beam lithography. The film growth rate was only ~5 nm/min with self-initiation induced by the fragments of the GMA monomer, which partially decomposed over the heated filament. Using otherwise identical process conditions, substantially higher growth rates of >200 nm/min resulted from using TBPO as an intentionally introduced initiator [54]. Ultrathin and conformal films formed on multiwall carbon nanotubes. Both the FTIR and NMR spectra confirmed full retention of the pendant epoxy functional groups from the GMA monomer. Utilizing supersaturated vapors of GMA monomer gave rise to deposition rate exceeding 600 nm/min [26].

Opening of the strained epoxide rings at the surface of the as-deposited iCVD polyglycidol methacrylate (PGMA) films, particularly through nucleophilic addition reactions by amine groups (Reaction (6.6)), is the basis for multiple schemes for achieving postdeposition surface functionalization, including

$$\text{(epoxide)} + H_2N\text{–}R \longrightarrow \text{(OH, HN–R)} \tag{6.6}$$

Figure 6.10 spectroscopically demonstrates the successful surface functionalization of iCVD PGMA with hexamethylene diamine [55]. Functionalizing with a –NH containing dye fluoresceine-5-thiosemicarbazide (FTSC) resulting in fluorescent green layers that enabled the imaging of conformal iCVD PGMA around particle substrates [56]. Conformal iCVD PGMA has been demonstrated on rough anodized titanium implants to improve protein attachment, most likely by attachment through –NH functional groups on the protein [57]. Long-term neural cell culture was demonstrated on a bilayer structure grown using a two-step iCVD synthesis process. First, iCVD PGMA was grown. Next, without breaking vacuum, the DMAEMA monomer was added to the iCVD chamber. The reaction is expected between the epoxy functional groups of the GMA and the tertiary amine functional groups of DMAEMA to produce quaternary ammonium functionality with the acetylcholine-like surface [49]. The epoxy functionality of the iCVD PGMA is also reactive toward thiol (–SH) and hydroxyl groups (–OH). Indeed, both thiophenol and 2-napthalenethiol have been successfully conjugated to iCVD PGMA [58].

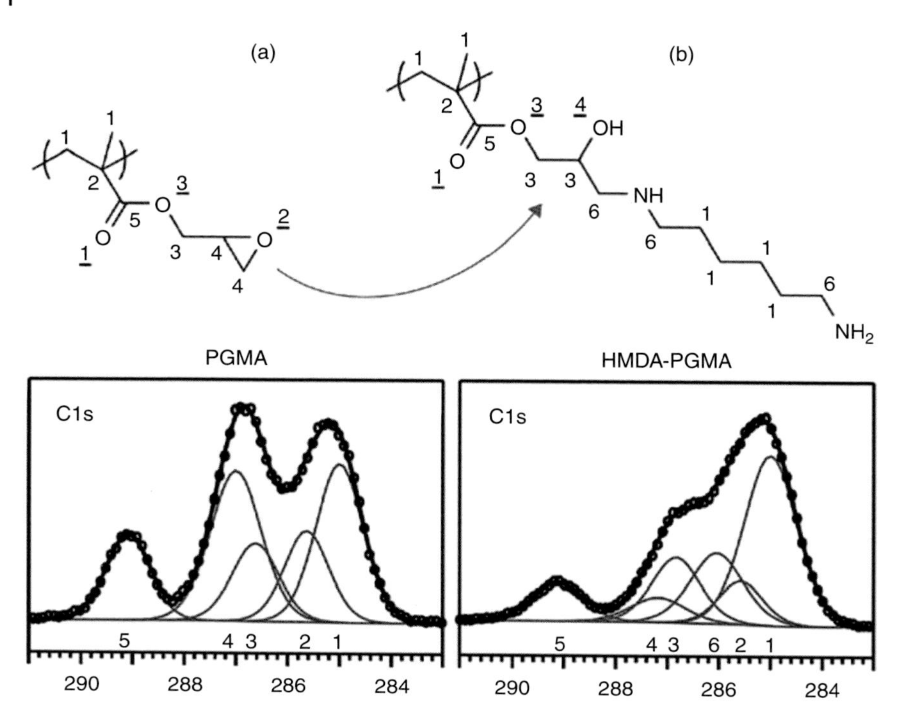

Figure 6.10 As-deposited iCVD PGMA (a) surface functionalized with hexamethylene diamine (HMDA) (b), confirmed via C1s X-ray photoelectron spectroscopy (XPS). Source: Lau and Gleason 2008 [55]. Reprinted with permission of Elsevier.

Multiple strategies for nanoadhesion rely on iCVD PGMA. For nanoadhesion, the goal is to form strong covalent chemical bonds across the interface between two parts. For example, one part can have the epoxy functionalization from the iCVD PGMA, whereas the other part has the amine functionalization [59]. Placing these two surfaces in contact and heating induces bonding by Reaction (6.6). Alternatively, both parts can be surface modified with a copolymer of GMA and an amine-containing monomer [60]. In a third approach, bonding of two PGMA-coated parts can be achieved using a diamine to link between both surfaces [61]. The addition reaction between the epoxy and amine functional groups (Reaction (6.6)) produces no outgassing of by-products. Hence, bubble-free interfacial bonding can be achieved, even over large areas.

The iCVD PGMA-based nanoadhesives have achieved the successful bonding of microelectrical and mechanical systems (MEMS) devices based on polydimethylsiloxane (PDMS) and other bulk polymers. Specific examples include an oxygen-free flow lithography device containing complex 3D internal structures [62], nanoglue for nonideal surfaces, including rough metal foils [63], and solvent-free wafer-scale bonding [64]. The temperature utilized for wafer bonding determined the type of the bond achieved. Below the glass transition temperature (T_g) of PGMA, a temporary thermoplastic bond forms from chain entanglement across the interface. Above T_g, a permanent thermoset bond results from ring-opening reactions of PGMAs epoxy functional groups, leading to covalent chemical bonding across the interface. Conformal coating of multiwall carbon nanotube arrays with ~50 nm of iCVD PGMA provided alignment for transfer to a different substrate using a low-temperature flip-over method [65].

The smoothness and conformality of the iCVD PGMA films allow them to function as barrier coatings on microspheres [66]. Low roughness combined with high adhesion were key attributes of integrating iCVD PGMA with HWCVD silicon nitride to form multilayer barrier coatings [67]. The epoxy functional groups in the PGMA films are anticipated to provide excellent adhesion to the silicon nitride layers, which are expected to have some surface –NH functionalities. Copolymerizing the GMA monomer with other vinyl-containing monomers followed by annealing produced cross-links between the epoxy functional groups of the GMA and improved their mechanical properties [68]. The copolymerization of GMA with the amine-containing monomer diethylaminoethylmethacrylate (DEAEMA) promotes cross-linking in the resultant iCVD surface modification layer on a membrane [69].

For the fabrication of integrated circuits, iCVD PGMA performed successfully as a negative-tone resist for electron beam lithography, compatible with both conventional and an all-dry supercritical CO_2 pattern development step [70]. The application of the iCVD PGMA avoided the solution-based spin-on method used to apply conventional resists. Utilizing the conformality of iCVD PGMA resists allowed for patterning the surfaces of three-dimensional substrates [71].

6.4.2 iCVD Films with Perfluoroalkyl Functional Groups: Properties and Applications

Because fluorinated monomers can be difficult or impossible to solubilize in common solvents, the iCVD method represents an attractive alternative synthesis route for forming thin fluoropolymer films. Polymers having pendant perfluoroalkane functional groups can have exceptionally low surface energies, 5.6–7.8 mN/m, as a result of the terminal —CF_3 of the perfluorinated moieties.

The iCVD homopolymers from 1*H*,1*H*,2*H*,2*H*-perfluorodecyl acrylate (PFDA) display full retention of the pendant —$(CF_2)_7$—CF_3 functional groups from the monomer and no measurable cross-linking. As has been observed for other iCVD monomers, increasing the partial pressure of PFDA increases the deposition rate and number-average molecular weight of the resulting iCVD homopolymer films, with highest reported values being 375 nm/min and 177 300, respectively [72]. The refractive index (1.36–1.37) and the static contact angle with water ($120 \pm 1.2°$) for the polyperfluorodecylacrylate (PPFDA) films are relatively insensitive to changes in the iCVD process conditions.

Hydrophobic [73], oleophobic [73], icephobic [74], and hydrate-phobic [75, 76] results from the low surface energy of the iCVD PPFDA homopolymer or a copolymer of PFDA with a cross-linker: ethylene glycol diacrylate (EGDA), ethylene glycol dimethacrylate (EGDMA), or DVB. The observed properties are often enhanced by surface morphology arising from the inherent roughness of the iCVD film and/or the texture of the underlying substrate.

Conformal coverage of features in the substrate by iCVD PPFDA has been demonstrated inside high-aspect pores in polymeric membranes [31, 77], and over porous, fiber-based substrates including paper [78] and electrospun mats [79]. Micromolds [80], 3D printed parts, and microfluidic devices, based on PDMS [81] and paper [82], have been conformally modified with low surface energy iCVD fluoropolymers. Freestanding films have been obtained with PPFDA and P(PFDA-*co*-EGDA) using droplets of ionic liquids as the iCVD substrates [83, 84]. Additionally, porous films grown by iCVD on liquids can be subsequently coated with PPFDA [85]. Conformal iCVD PPFDA polymer coatings on aligned carbon nanotube stamps prevent densification upon drying, enabling high-speed flexographic printing with nanoparticle inks [86, 87]. The diverse applications of iCVD films synthesized from the methacrylate analog of PFDA, 1*H*,1*H*,2*H*,2*H*-perfluorodecyl methacrylate include Janus membranes for enhancing skin regeneration [88], separating lipids from algae [89], and creating oleophobic sponges [90].

The $-(CF_2)_7-CF_3$ functional group possesses eight perfluorinated carbon atoms and is sometimes referred to by the shorthand name "C8." The linear helical structure of the C8 chains often crystallizes, which limits reorientation of the side chains away from the interface when exposed to water, decreasing contact angle hysteresis. For some iCVD PPFDA films, the C8 pendant groups were observed to crystallize into smectic B bilayers having a periodicity of 3.24 nm with hexagonal packing within the layers having a 0.64 nm lattice parameter. Holding all process conditions constant except for changing the filament temperature from 240 to 300 °C, the orientation of the perfluoro side chains with respect to the surface switched from parallel to perpendicular [91]. The parallel orientation produced relative smooth surfaces with moderate hydrophobicity, having an advancing water contact angle (WCA) ~130 °C. The perpendicular orientation yielded significant rougher and more hydrophobic surfaces (advancing WCA ~150 °C). The hypothesis for the observed morphology is the known increase in the decomposition of the tert-butoxy initiator radicals to methyl radicals as the filament temperature increases from 240 to 300 °C. Use of a trichlorovinylsilane pretreatment resulted in the grafting of the iCVD PPFDA chains and produced a high degree of crystallinity, the perpendicular orientation of the perfluoro side chains, and surface roughness [92]. The grafted iCVD PPFDA displayed an advancing WCA of 160° with extremely low hysteresis (5°) and showed oleophobicity (120° advancing contact angle with mineral oil).

The hydrophobicity and surface roughness of PPFDA homopolymer iCVD films decreased upon annealing [93]. Copolymerization with EGDMA decreases as-deposited film roughness, as the crystallinity of the C8 side chains is interrupted. The incorporation of EGDMA also makes the hydrophobicity and surface roughness stable against thermal cycling. Figure 6.11a displays the systematic change in the coefficient of thermal expansion (CTE) with EGDMA % in both high- and low-temperature regions. The CTE of P(PFDA-*co*-EGDMA) films can be seen to be significantly lower than that of the homopolymer. Thus, the copolymers developed significantly less thermal stress upon temperature cycling, making them more resistant to crack formation.

The condensation of water vapor is an essential step in electrical power generation and other industrial processes. Water condensing on bare metal

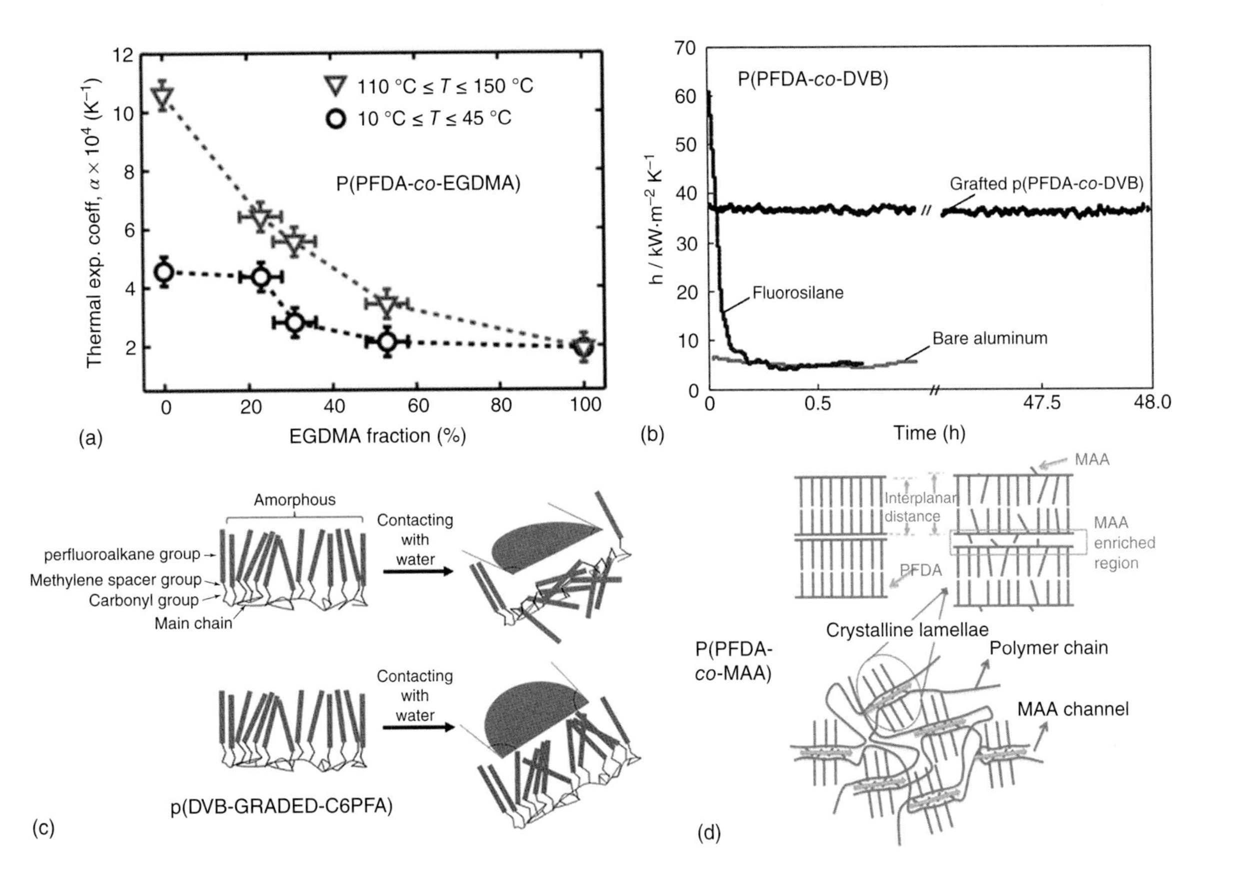

Figure 6.11 Characteristics of iCVD films synthesized using the monomer 1*H*,1*H*,2*H*,2*H* perfluorodecylacrylate (PFDA): (a) coefficient of thermal expansion (CTE) for P(PFDA-*co*-EGDMA) reduces as the cross-linker content increases, reducing the driving force for crack formation upon thermal cycling; (b) heat transfer coefficient for aluminum shows great initial enhancement from both an ultrathin coating of P(PFDA-*co*-DVB) and a fluorosilane treatment, but at 48 hours, only the iCVD film is robust enough to maintain the enhancement; (c) schematic of C6 perfluoroalkane pendent group orientation in response to water for PC6PFA (top) and for cross-linked P(DVB-graded-C6PFA) (bottom); and (d) schematic of proposed water channel formation in P(PFDA-*co*-MAA) fuel cell membranes. Source: Reprinted with permission from Refs. [93] (a), [94] (b), [95] (c), and [96] (d).

fully wets the surface, and this blanket of liquid water represents a barrier to heat transfer. If the water instead condenses as drops, leaving the bare metal in contact with the steam, the heat transfer coefficient and overall process efficiency are significantly higher. Hydrophobic iCVD P(PFDA-*co*-DVB) applied to a metal surface was able to achieve dropwise condensation [94]. The iCVD films must be ultrathin (~30 nm) to avoid causing a significant resistance to heat transfer. For practical applications, the hydrophobic treatment must be durable. Figure 6.11b shows that the high heat transfer coefficient of iCVD-coated metal is unchanged over 48 hours, whereas a liquid-applied fluorosilane hydrophobic surface treatment completely losses its efficacy in only a few minutes. The ultrathin iCVD P(PFDA-*co*-DVB) layer applied to the outer surface of a metal tube provided a prototype demonstration of an industrial steam condenser tube.

Coating the inner surfaces of stainless microtubes (ID < 1 mm, length 4 mm) with 50–160 nm thick iCVD PFDA reliably enhanced their boiling heat transfer performance by up to 60%, in part by providing more surface sites for the nucleation and release of bubbles [97]. Further enhancement of the boiling heat transfer resulted from grading the wettability inside the microtubes by feeding the PFDA monomer from one end while feeding the hydrophilic monomer HEMA from the opposite end during the iCVD process [98].

Environmental concerns resulting from the bioaccumulation of C8 compounds drive interest utilizing lower molecular weight pendant groups, such as C6, which display lower levels of bioaccumulation. However, the C6 side chains are less likely to crystallize, and thus, higher values of contact angle hysteresis are often observed. The monomer 1*H*,1*H*,2*H*,2*H*-perfluorooctyl acrylate (C6PFA) successfully demonstrated thin film formation by iCVD synthesis, with full retention of the perfluorinated C6 pendant groups [95]. The iCVD polymerization of the C6PFA with DVB and grading of the deposited layer from pure DVB to predominately C6PFA were two successful strategies for reducing the contact angle hysteresis for the iCVD films with the perfluoro C6 pendant groups (Figure 6.11c).

Films of iCVD P(PFDA-*co*-MAA) perform as proton exchange membranes in fuel cells [96, 99]. Conventional liquid-phase polymerization of the hydrophobic PFDA monomer with the hydrophilic MAA would be very difficult because a common solvent for the two monomers would be required. When wetted, the MAA domains swelled, resulting in ionic conductivity value of up to 70 mS/cm, which is on par with commercial Nafion membranes. XRD studies revealed that the bilayer structure of the C8 pendant groups found in PPFDA also formed in the copolymer, but with a slightly larger periodic (3.56 nm) and without the ordered hexagonal packing of the chains within the layer (e.g. a smectic A phase, Figure 6.11d). The expanded bilayer structures were hypothesized to be the channels for ion transport.

6.4.3 Polyhydroxyethylacrylate (PHEMA) and Its Copolymers: Properties and Applications

Utilizing iCVD synthesis, the desirable properties of bulk polyhydroxyethylacrylate (PHEMA), such as biocompatibility and resistance to biofouling, become available in an ultrathin and conformal thin film form. As a result of the ~100%

retention of the hydroxyl (–OH) functional groups from the HEMA monomer, iCVD PHEMA displays hydrophilic behavior: the contact angle measurements with water give an advancing angle of 37° and an ultimate receding angle of 17° [100]. The iCVD PHEMA displays the characteristic responsive behavior of a hydrogel, as is also observed in its conventionally synthesized bulk counterpart. When exposed to water, the iCVD PHEMA films increase in volume because of the absorption water, or the films can even dissolve. Wrinkling behavior results from the swelling of iCVD PHEMA films that are strongly adhered to a non-swelling substrate (Figure 6.12a) [101].

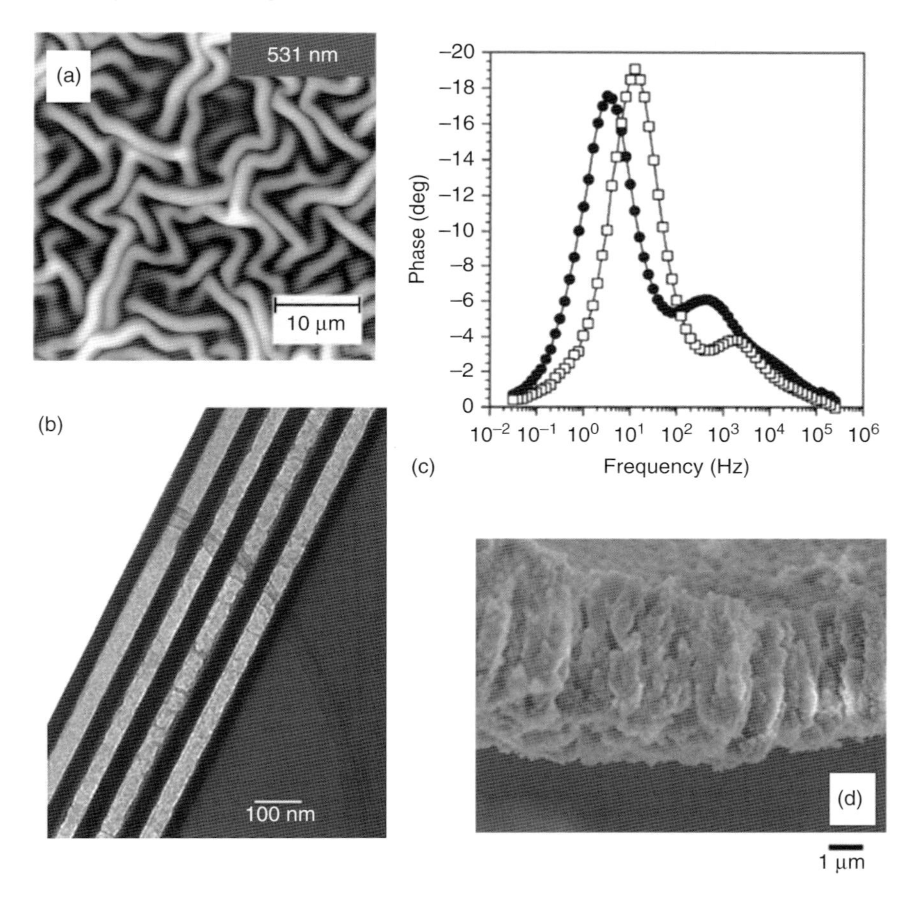

Figure 6.12 Results obtained by iCVD using the monomer hydroxyethyl methacrylate (HEMA). (a) Wrinkles form when PHEMA swells while being well adhered to a nonswelling substrate. Source: Christian et al. 2016 [101]. Reprinted with permission of American Chemical Society. (b) Bode plot reveals the decrease in charge recombination at the electrode–electrolyte interface for dye-sensitized solar cells (DSSCs) fabricated using PHEMA as a solid electrolyte to infill gaps between titania particles (filled circles) as compared to a liquid electrolyte (open circles). Source: Karaman et al. 2008 [102]. Reprinted with permission of American Chemical Society. (c) alternation of ultrathin swellable organic PHEMA (light color) with inorganic titania layers, forming a mechanical flexible Bragg mirror that is responsive to changes in humidity. Source: Nejati and Lau 2011 [103]. Reprinted with permission of American Chemical Society. (d) A freestanding porous P(HEMA-*co*-EGDA) film grown directly on low-volatility silicone oil. Source: Haller et al. 2013 [84]. Reprinted with permission of American Chemical Society.

The deposition rate for iCVD PHEMA films increases with the partial pressure of HEMA in the reactor. As the partial pressure of HEMA nears its saturation value, deposition rates increase to over 1.5 μm/min, allowing ready fabrication freestanding films of 0.1–0.35 mm [104]. The use of high HEMA partial pressures also results in ultrahigh number-average molecular weights, with reported values up to ~820 000. At such high molecular weights, the chains have sufficient entanglement that the PHEMA homopolymer films do not dissolve in water.

In multiple applications, the integration of iCVD PHEMA into optoelectronic devices has advanced their performance. Figure 6.12b shows the alternation of the ultrathin organic PHEMA films with inorganic titania layers to form the stacked structure of a Bragg mirror [102]. Exposure to humidity causes the hydrogel layers to swell, linearly shifting the frequency of the high absorptivity peak of the Bragg mirror to higher wavelengths. A reversible color change response is achieved between green in the dry state and red when the device is exposed to at 1 mol% water vapor in nitrogen. Conformal iCVD PHEMA has been infiltrated into mesoporous titania electrodes. As compared to a liquid electrolyte, this iCVD solid/gel electrolyte reduces interfacial charge recombination (Figure 6.12c) in the resulting dye-sensitized solar cells (DSSCs) [103].

Cross-linking of PHEMA is often employed to control the degree of swelling and to improve film stability in aqueous environments. Cross-linked HEMA-based iCVD hydrogels result from copolymerization with either EGDA or EGDMA. The choice of cross-linker can be influenced by multiple factors, including the ease of delivering its vapor to the CVD chamber, its cost, and the targeted surface energy of the resultant film. Note that as a result of its two additional methyl groups, the EGDMA is more hydrophobic than EGDA. Freestanding microstructured films resulted when a nonvolatile liquid was employed as a substrate (Figure 6.12d) [85].

Figure 6.13a displays the fractional change in the thickness of iCVD films with the duration of exposure to relative humidity levels [106]. The swelling of the P(HEMA-*co*-EGDMA) films can be seen to be systematically tunable by incorporating varying amounts of the EGDMA cross-linker. Several minutes were required to achieve the full swelling response of the ~100-nm-thick copolymer films. Because water vapor enters the iCVD layers through diffusion, shorter response times would occur with thinner films. The swelling response of the copolymer films falls in between that of two homopolymers, where polyethylene glycol dimethacrylate (PEGDMA) displays no swelling, whereas PHEMA shows more than a 25% thickness increase. The density of $1.26 \pm 0.09\ g/cm^3$ of dry iCVD PHEMA homopolymer determined from X-ray reflectivity measurements falls at the high end of the $1.15–1.27\ g/cm^3$ range for conventionally synthesized PHEMA. At their equilibrium swelling condition, calculations based on Flory–Rehner theory determined the range of mesh sizes for the P(HEMA-*co*-EGDMA) hydrogels to be from 0.6 nm down to 0.4 nm, as cross-linking density increased. This result is comparable to the 0.75–0.34 nm range found in a previous study of iCVD P(HEMA-*co*-EGDMA) [107]. The architecture and composition of the hydrogel mesh determine the ease of permeability of different molecular species of varied size, shape, and polarity. The iCVD P(HEMA-*co*-EGDA) copolymers provide larger mesh sizes. As the

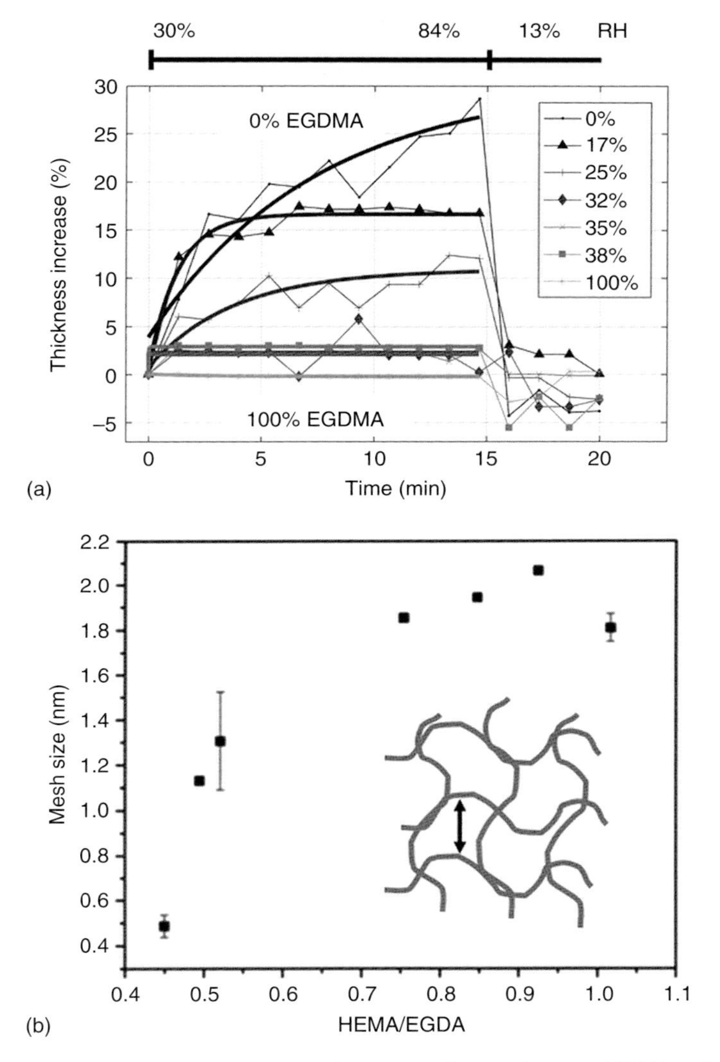

Figure 6.13 Systematic tuning of properties for copolymers of HEMA with varying amounts of a cross-linking comonomer: (a) swelling dynamics of P(HEMA-*co*-EGDMA) upon exposure to humidity and (b) mesh size, calculated from experimental data for P(HEMA-*co*-EGDA). Source: (a) Courtesy of Prof. Anna Marie Coclite, TU Graz; (b) Reprinted with permission from Ref. [105].

content of the EGDA cross-linker decreases, the mesh size extends from ~0.5 up to ~2.0 nm (Figure 6.13b) [105].

Thin films of iCVD PHEMA and its copolymers both permit the selective permeation of small molecules across the thickness of the layers and impart resistance to biofouling. This combination of properties is desirable for the encapsulation of sensors [108], including sensors for implantation into animals [109]. Responsive P(HEMA-*co*-EGDA) nanotubes of high aspect ratios were grown using anodized aluminum oxide (AAO) membranes as templates [110].

The conformal and biocompatible nature of iCVD P(HEMA-*co*-EGDMA) enables molecular imprinting of specific recognition sites for immunoglobulin [111]. When grown on the inner wall of microtubes at thicknesses ranging from 50 to 150 nm, enhancement occurred in the flow boiling heat transfer coefficient of the microtube [112].

Vapor-phase deposition makes it possible to copolymerize monomers that do not have a common solvent. Thus, it is possible to copolymerize HEMA with fluorine-containing monomers. Vapor deposition from mixtures of HEMA and PFDA results in random copolymers with molecular-scale compositional heterogeneity that enhance the resistance against protein absorption as compared to either of the homopolymers [113]. Copolymerization of HEMA, EGDA, and PFM (pentafluorophenyl methacrylate) provides a hydrogel layer that can be readily functionalized with signaling peptides, providing an advantageous chemical and mechanical microenvironment for culturing cells [114]. Incorporation of PFM reduces the degree of swelling of the hydrogel. As the substitution reaction of PFM with amine groups of the functionalization moiety

$$+ \; H_2N{-}R \longrightarrow \; + \qquad (6.7)$$

occurs only in the near-surface region, the swelling behavior of the majority of the hydrogel can be preserved if the reactive pentafluorophenyl leaving groups are concentrated near the surface. This type of graded composition can be achieved by metering in the PFM monomer only toward the conclusion of the iCVD process [115].

6.4.4 Organosilicon and Organosilazanes: Properties and Applications

Organosilicon polymers grown by iCVD have been evaluated for a range of applications including the biopassivation of implantable devices [116], encapsulation/barrier layers for electronic protection [117], flexible and ultrathin dielectric films, low dielectric constant materials [118], pore sealing of low dielectric constant layers [119, 120], and ultrathin solid-state electrolytes for lithium ions [121]. The iCVD of organosilicon films has most commonly employed the cyclic, multivinyl monomers V3D3 and 1,3,5,7-tetravinyl-1,3,5,7-tetramethylcyclotetrasiloxane (V4D4) (Figure 6.14). The V3D3 rings are planar and can have the three vinyl moieties in either the cis or trans configuration, whereas the V4D4 rings have a puckered structure. The growth of organosilazane iCVD films has been achieved from the structural similar monomers, trivinyltrimethyl-cyclotrisilazane (V3N3) and tetravinyltetramethyl-cyclotetrasilazane (V4N4), where –NH-groups occur in the place of oxygen. Successful iCVD organosilicon film deposition with a tunable degree of cross-linking has also been reported from the monomers

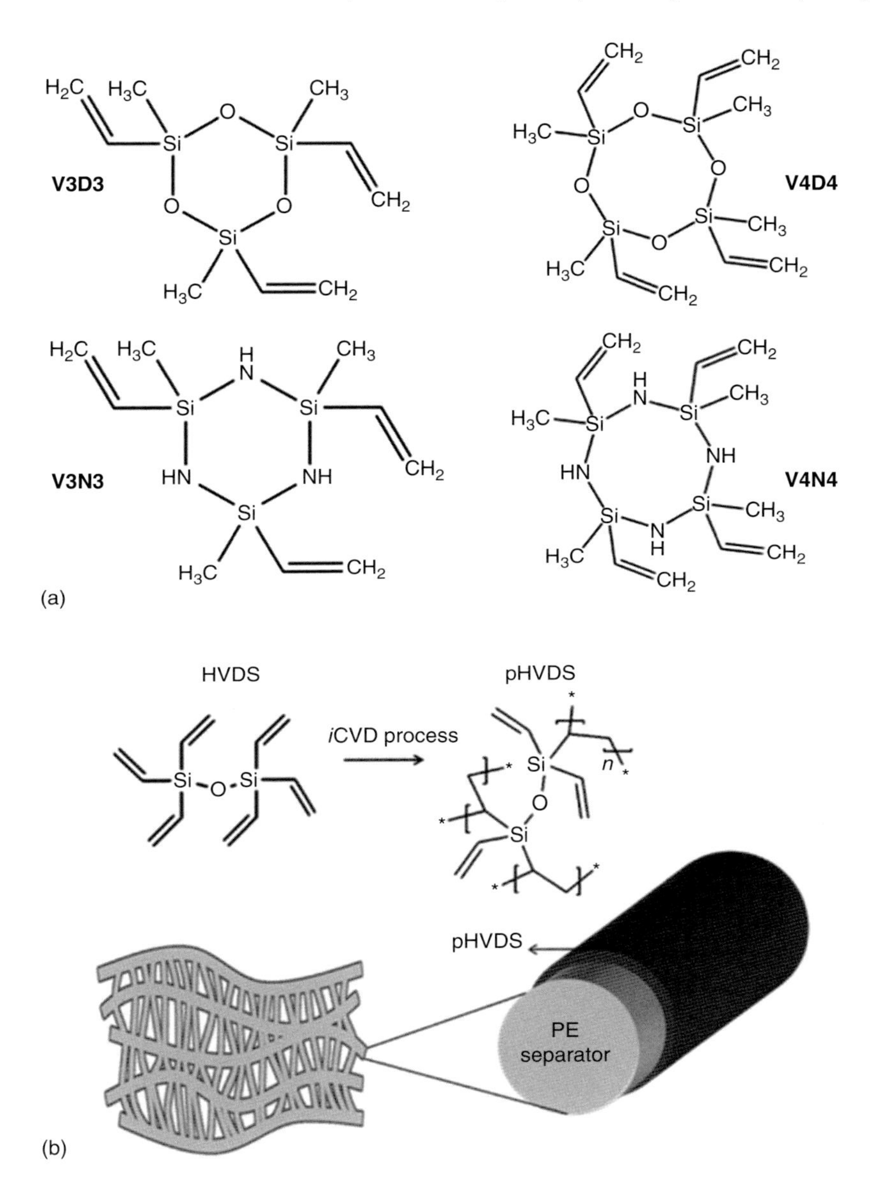

Figure 6.14 Example multivinyl, Si-containing iCVD monomers for the deposition of cross-linked organic covalent networks (refer to Table 6.2 for full chemical names): (a) cyclic siloxanes and silazanes and (b) HVDS, along with schematics of iCVD synthesis of PHVDS and the formation of conformal PHVDS on polyethylene (PE) fibrils for improved battery separator performance. Source: (a) Reeja-Jayan et al. 2015 [121]. Reprinted with permission of American Chemical Society. (b) Yoo et al. 2015 [122]. Reprinted with permission of American Chemical Society.

hexavinyldisiloxane (HVDS) [122, 123], V3D3 copolymerized with HVDS [124], and 1,1,3,5,5-pentamethyl-1,3,5-trivinyltrisiloxane (TVTSO) [125].

Initiation and propagation of the V3D3 and V4D4 monomers result in hydrocarbon backbone chains with siloxane rings as pendant groups [118, 126]. The growing chains are terminated by initiator radicals or other propagating chains, resulting in the cross-linked network structures, facilitated by the multivinyl nature of the monomeric units. Postdeposition thermal treatment of PV4D4 films drives the formation of silsesquioxane cages having intrinsic porosity and resulting in annealed films with dielectric constants as low as 2.15. At filament temperatures >600 °C, HWCVD organosilicon films grow without the use of an initiator, most likely by the ring-opening polymerization of the cyclic monomer. Most commonly, the initiator TBPO is employed at filament temperatures 300–500 °C. The resulting initiator radicals react with the vinyl groups of the monomer. The loss of vinyl bonds through initiation and propagation to form the iCVD layer is clearly evidenced by FTIR spectroscopy (Figure 6.5). At the modest filament temperature used, the FTIR also reveals that the V3D3 organosilicon rings present in the monomers are fully retained in the films. At some process conditions, unreacted vinyl moieties remain because of steric hindrance to reaction. Indeed, cysteine-linked proteins were covalently linked to the surface of PV4D4 utilizing the unreacted vinyl groups using a UV-assisted thiol-ene click reaction [127].

$$\mathrm{H(X)C{=}CH_2} + \mathrm{HS{-}Y} \xrightarrow{\mathrm{UV}} \mathrm{H_2C{-}CH_2(S{-}Y)}\ \text{(with X on } \mathrm{H_2C)} \tag{6.8}$$

The same strategy was used to attach DNA primer end functionalized with thiol (–SH) as a fabrication step for medical diagnostic device to rapidly detect the Middle East respiratory syndrome (MERS) corona virus [128].

The cross-linked covalent network formed by iCVD from the multivinyl monomer V3D3 creates an exceptionally smooth film. For example, at a thickness of ~250 nm, an rms roughness of 0.4 nm with a maximum peak to valley excursion of 0.9 nm was observed [116]. For comparison, the silicon wafer substrate had an rms roughness of 0.15 nm. This low roughness is essential for avoiding the pinhole defects required for excellent dielectric performance. The measured optoelectrical properties of the iCVD PV3D3 films were relatively insensitive to deposition conditions. For iCVD PV3D3, the resistivity, dielectric constant, and refractive index have reported values of $4 \pm 2 \times 10^{15}\ \Omega/\mathrm{cm}$, 2.5 ± 0.2, and 1.465 ± 0.01, respectively [129]. The band gap of iCVD PV3D3 is 8.25 eV with a highest occupied molecular orbital (HOMO) level of 9.45 eV [130].

The electrical resistance and film adhesion were demonstrated to be stable for over 2.5 years while the film was submerged in physiological saline and subjected to electrical bias cycled between +5 and −5 V. Film stability was also demonstrated for a range of solvents and against immersion in boiling water in shorter duration tests. Biocompatibility testing confirmed that the iCVD PV3D3 films are noncytotoxic and do not influence the proliferation of PC12 neurons [116]. The perfluoro C8 initiator radical produced by thermal decomposition of the PFOSF initiator reacts with pendant vinyl groups on the V3D3 monomer units resulting in biopassive films having a combination of organosilicon and fluoropolymer characteristics [131].

Remarkable breakthroughs in energy-efficient flexible electronics facilitated the successful integration of ultrasmooth, ultrathin (<10 nm), pinhole-free, and mechanical flexible dielectric iCVD PV3D3 into multiple device architectures. Even ultrathin (6 nm) PV3D3 layers (Figure 6.15a) exhibit exceptionally low leakage currents ($<10^{-9}$ A/cm^2 at 3 MV/cm). Oxygen plasma treatment to form a SiOx skin layer can further enhance the electrical characteristics of iCVD P3D3 film [134]. Flexible and foldable organic flash memory has been integrated onto multiple substrates, even on disposable ones such as paper [135]. This breakthrough exploited the exceptional insulating properties of iCVD PV3D3 to fabricate thin film transistors (TFTs) with long retention time. This work builds on series of earlier publications which optimized the chemistry of the iCVD PV3D3 dielectric and the associated device steps for fabricating high-performance individual TFTs and arrays of TFTs on soft substrates [130]. By varying the ratio of monomers incorporated into iCVD P(V3D3-*co*-1-vinylimidazole(VI)), the threshold voltage of organic TFTs was systematically tuned (Figure 6.15c) [136]. Additionally, high-quality iCVD V3D3 dielectric layers have enabled high-speed organic photomemory [137] and been integrated together with carbon nanotubes into TFTs for logic circuits [138]. The iCVD PV3D3 was also utilized as a tunneling dielectric and combined with MoS_2 to create low-power nonvolatile charge storage memory [132] and combined with amorphous In–Zn–Sn–O to create flexible memristive nonvolatile logic-in-memory circuits [139].

ALD is a desirable method for growing high-quality conformal inorganic films. There are multiple applications for hybrid multilayer structures composed of inorganic ALD films combined with iCVD organic polymer layers. For example, hybrid films with alternating organic and inorganic layers have been extensively explored as flexible multilayer barriers against atmospheric oxygen for the protection of flexible organic devices such as LEDs. A chamber has been designed (Section 6.6) to carry out both ALD and iCVD without breaking vacuum [140]. The main hardware change is using two feed systems, one for each process. Typically, ALD utilizes high substrate temperatures, whereas iCVD substrate temperatures are typically near room temperature. Modification of both processes to work at the identical substrate temperature of 90 °C speeds the growth of hybrid films, as there is no delay time required for heating or cooling the substrate for the repeated cycling of ALD and iCVD process steps [141].

Surface passivation of the hydroxyl groups on the surface of the ALD grown inorganic high dielectric constant (high k) material, Al_2O_3, was demonstrated by a nonpolar iCVD layer grown from V3D3. The passivation reduced hysteresis, mobility decreases, and shifts of the threshold voltage in organic TFTs. Additionally, the resulting ultrathin (<15 nm) hybrid film survived tensile strains of up to 3.3%, as desired for the fabrication of flexible electronics [142].

For the fabrication of high energy density 3D batteries, nanoscale and conformal thin solid film electrolytes are required to cover nonplanar electrode arrays. Ultrathin (25 nm), ultrasmooth, and pinhole-free iCVD PV4D4 layers

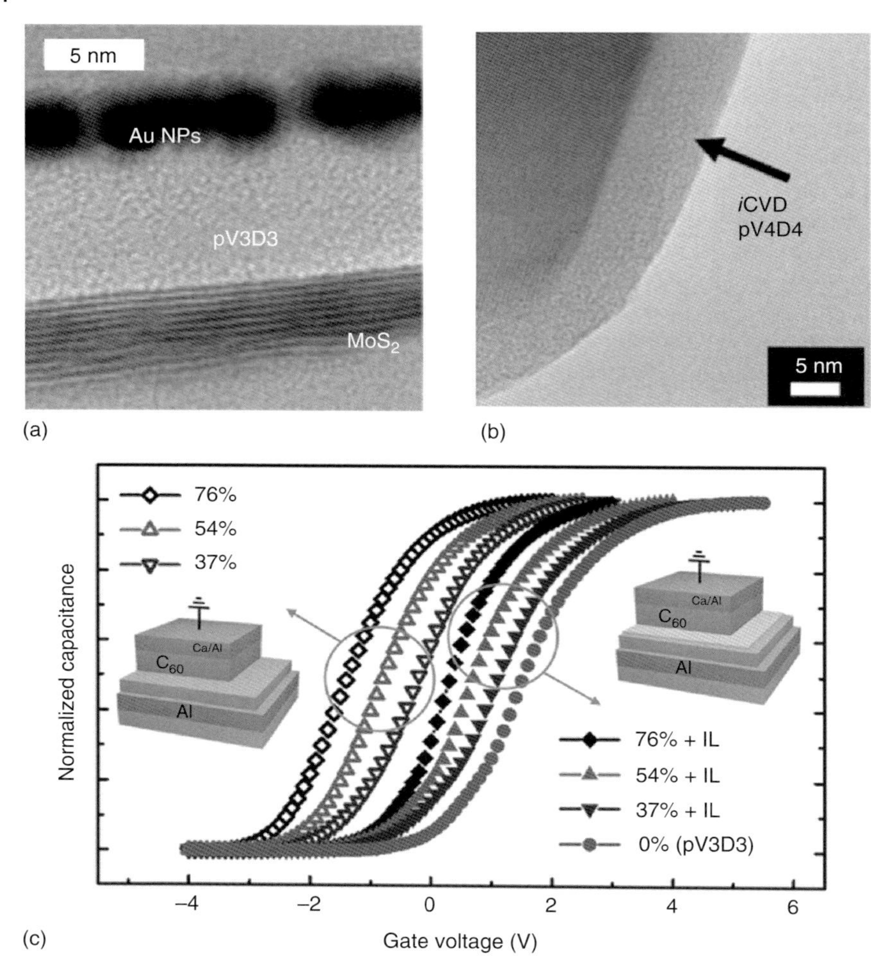

Figure 6.15 Structures and device performance obtained for iCVD films based on cyclic siloxane monomers: (a) ultrathin and smooth PV3D3 gate dielectric between MoS_2 and Au nanoparticle layers; (b) ultrathin and smooth PV4D4 solid electrolyte for lithium ions grown conformally around a lithium spinel oxide particle; and (c) systematic tuning of the gate voltage of a multilayered electronic device achieved by varying the composition of the P(V3D3-*co*-VI) copolymer gate dielectric. Source: Reprinted with permission from Refs. [132] (a) and [133] (b). (c) Courtesy of Prof. Sung Gap Im, KAIST.

can undergo postdeposition lithiation with $LiClO_4$ to yield a room temperature lithium-ion conductivity of $7.5 \pm 4.5 \times 10^{-8}$ S/cm [121]. These modest conductivities are more than offset by the short diffusion length across the ultrathin layer, resulting in the orders-of-magnitude lower diffusion times for Li^+ than in bulk electrolytes. DFT calculations support the hypothesis that the electronegative oxygen atoms in the siloxane ring are favorable for binding the Li+ ions while minimization of the electronic repulsion for the ClO_4^- counterions occurs within the siloxane rings. Related iCVD films deposited from the related monomers, V3D3, V3N3, and V4N4, also exhibit lithium-ion conduction, but at reduced conductivity values. Nanoscale conformal coverage of a spinal powder by PV4D4 is shown in Figure 6.15b [133].

6.4.5 iCVD of Styrene, 4-Aminostyrene, and Divinylbenzene: Properties and Applications

The iCVD deposition rate of the homopolymer PS using the TBPO-free radical initiator is slow as compared to acrylates and methacrylates [28]. This experimental observation is consistent with the lower rate constant for chain propagation, k_p, for styrenic vinyl bonds. The slow rate polymerization provides the control desired for fabricating ultrathin layers. Indeed, ultrathin and conformal iCVD PS layers proved to enhance the performance of a room temperature chemiresistor, where the high surface area was achieved through the use of vertically aligned arrays of carbon nanotubes (Figure 6.16a) [143].

Copolymerization of the styrene monomer with MA resulted in greatly enhanced film deposition rates. The copolymerization did not give a random copolymer, as was the case for all previous iCVD polymerizations. In a random copolymer, the composition of the films changes as the ratio of the two monomers in the feed is varied. However, in an alternating copolymer, a one-to-one ratio of the S and MA monomeric units in the films results for all gas feed ratios used. Spectroscopic analysis confirmed that the electron donating styrene reacted much more frequently with the electropositive MA than either monomer reacted with itself, resulting in alternating copolymer P(S-alt-Ma). Conformal P(S-alt-Ma) coatings on paper substrates provided a high density of anhydride functional groups as a base for biofunctionalization by poly-L-lysine (PLL). The functionalized paper enabled tracheal reconstruction utilizing an origami-based tissue engineering paradigm [146].

Synthesis of alternating copolymers also resulted from iCVD synthesis from the monomers 4-AS and MA [147]. Annealing the P(4AS-alt-MA) caused the amine groups from one polymer chain reacting with the MA functional groups on another chain (Figure 6.16b), creating a massively cross-linked network that remains flexible, with the Young's modulus exceeding 20 GPa, an order of magnitude higher than is typical for polymeric materials.

Its volatility and low cost makes DVB an excellent monomer for iCVD synthesis. The iCVD PDVB homopolymers are smooth, transparent, thermally stable, mechanically robust, hydrophobic, low stress, and cross-linked layers. As compared to PECVD materials, iCVD PDVB displays an excellent oxidative and photochemical stability[148]. Additionally, DVB serves as a cross-linker when copolymerized with other iCVD monomers, as has been described in previous sections.

The vinyl bonds of DVB have lower reactivity as compared to acrylates and methacrylates. Thus, some unreacted vinyl bonds remain in the PDVB films. FTIR analysis has been used to quantify the percentages of vinyl bonds that remain unreacted in the iCVD films [149]. This analysis takes into account that the DVB monomer is actually a mixture of meta- and para-isomers, as separating m-DVB from p-DVB is difficult. The unreacted vinyl bonds in the as-deposited films can react when exposed to oxygen or light, producing aging of PDVB properties over time. *In situ* [150] and *ex situ* [151] annealing successfully stabilize the properties of iCVD PDVB. Heating to a temperature greater than the boiling point of the DVB monomer also ensures that no residual monomer remains in the film [152].

Applications of the iCVD PDVB include the encapsulation of optoelectronic devices [153], serving as an ablator material for laser targets for nuclear fusion [154], and forming a planarizing layer for paper-based devices [135]. Additionally, iCVD PDVB performed as a top coat for the directed self-assembly of sub-10 μm line and space structures (Figure 6.16c) [144]. The iCVD topcoat, as

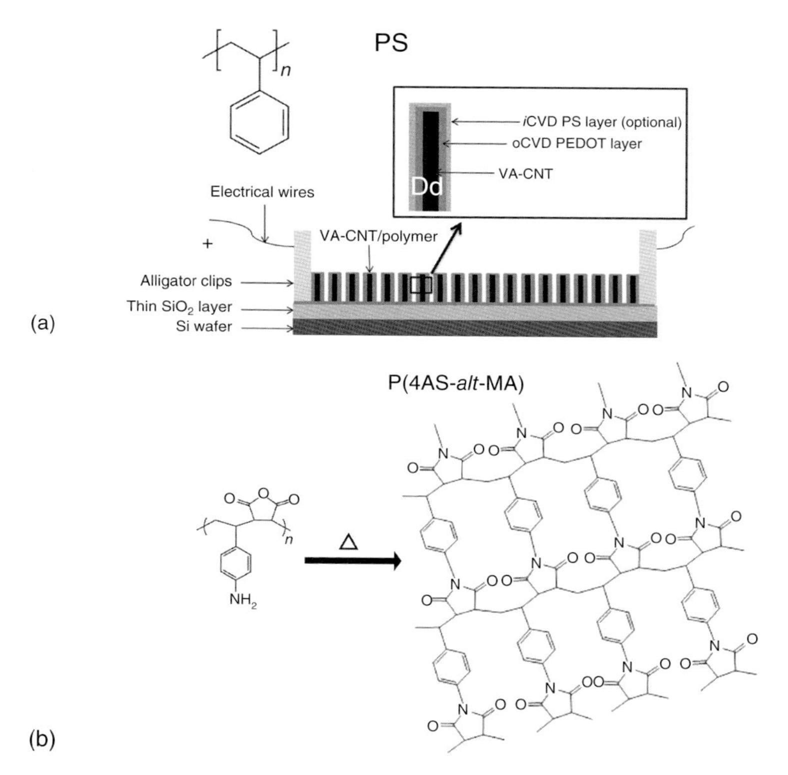

Figure 6.16 Example chemical structures and applications for iCVD films based on styrenic monomers; (a) ultrathin and conformal coverage of a vertically aligned carbon nanotube (VACNT) array by polyethylene dioxythiophene (PEDOT) grown by oxidative CVD and then by iCVD PS to form a gas sensor. Source: Wang et al. 2016 [143]. Reprinted with permission of American Chemical Society. (b) The alternating copolymer from 4AS and MA is annealed, causing reaction between amine and anhydride functionalities on different polymer chains, resulting in a cross-linked covalent network with outstanding mechanical properties. (c) An ultrathin and smooth layer of PDVB is used to aid the vertical alignment of the block copolymer underneath, enabling the dual-scale patterning of 8-nm lines and spaces along with ~100 nm scale features. Source: Suh et al. 2017 [144]. Reprinted with permission of Springer Nature. (d) Conformal modification of the interior surfaces of a hydrophilic porous membrane by iCVD PDVB, rendering the membrane hydrophobic as desired for the membrane desalination of water. Source: Servi et al. 2016 [145]. Reprinted with permission of Elsevier.

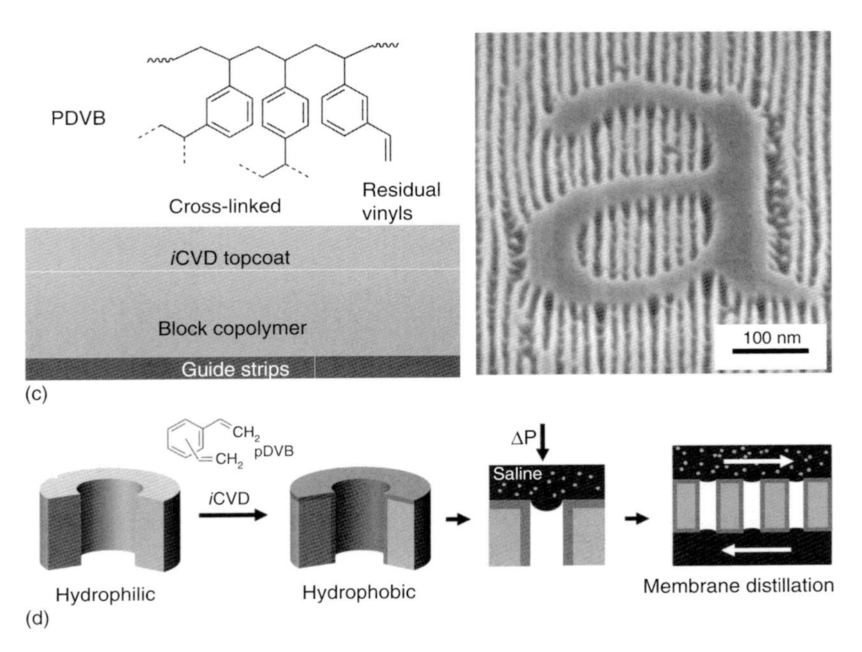

Figure 6.16 (*Continued*)

thin as 7 nm, controls the interfacial properties of the block copolymer films below, inducing the formation of domains having the perpendicular orientation needed for pattern transfer. Applying conformal PDVB to hydrophilic nylon membranes of uniform pore size renders it sufficiently hydrophobic for membrane desalination applications (Figure 6.16d) [145]. This approach avoids the use of fluorine in the hydrophobic surface modification layer and in the bulk membrane. Conformal iCVD DVB has also been used to narrow the diameter of high aspect ratio pores [31].

Using a cationic initiation scheme instead of a free radical initiator greatly accelerated the CVD growth rates for PS and PDVB films [19]. For the cationic initiation, $TiCl_4$, a strong Lewis acid, is combined with a hydrogen donor, H_2O. The reaction of $TiCl_4$ with water does not require a filament.

6.4.6 iCVD of EGDA and EGDMA: Properties and Applications

In addition to acting as cross-linkers with other iCVD monomers, the multivinyl monomers, EGDA and EGDMA, produce smooth, pinhole-free iCVD homopolymer layers, similar to the results for the multivinyl organosiloxane monomers and DVB. Because its thermal decomposition leaves behind little char, poly(ethylene glycol) diacrylate (PEGDA) has been used as a sacrificial layer to create air gaps in microelectronic structures (Figure 6.17a) [155]. Depositing iCVD PEGDA on the surface of an ionic liquid creates a freestanding film with a graded composition. The top surface is the cross-linked organic homopolymer, whereas in the droplets, a copolymer forms between EGDA and the ionic liquid,

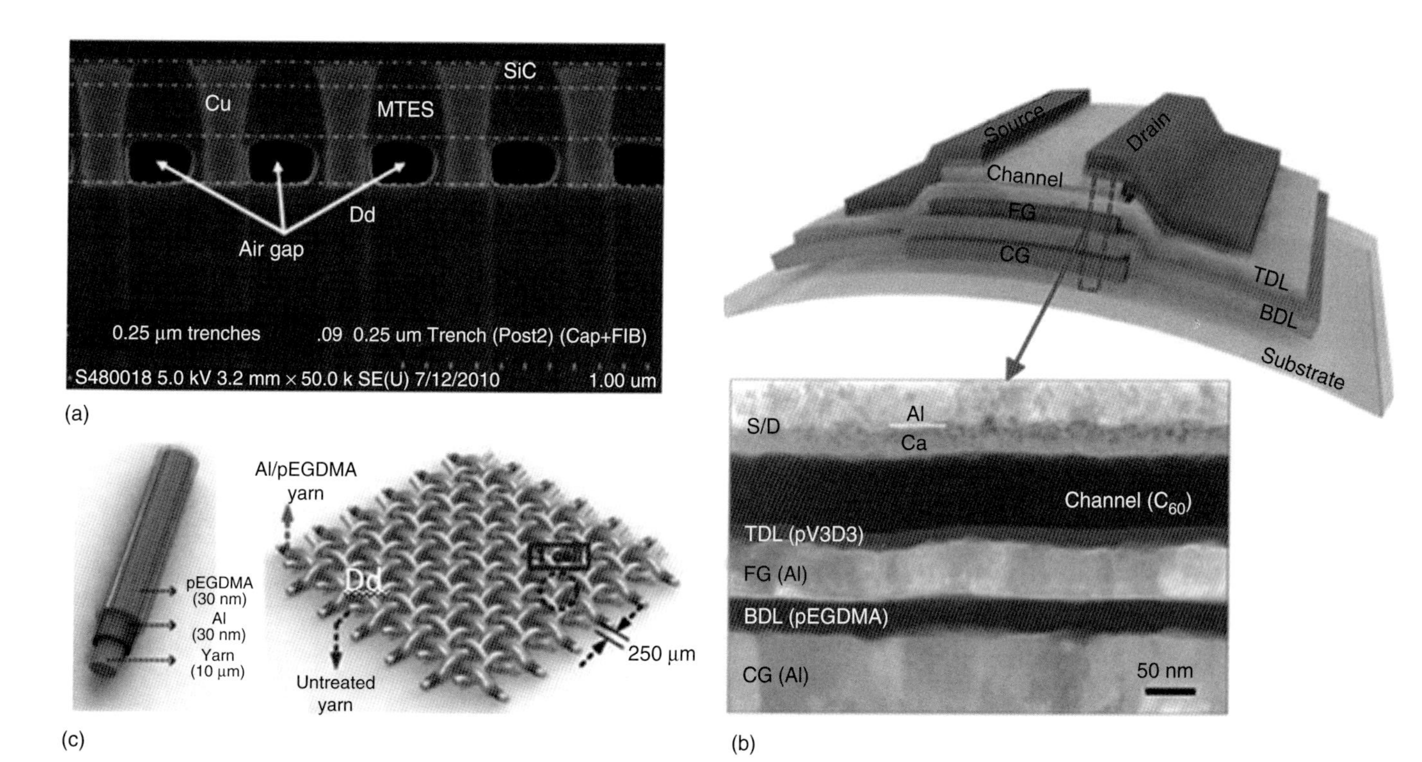

Figure 6.17 Homopolymerization of the divinyl iCVD monomers EGDA and EGDMA utilized for (a) air gap feature created by annealing a PEGDA sacrificial layer. Source: Lee et al. 2011 [155]. Reprinted with permission of American Chemical Society. (b) Ultrathin PEGDMA blocking dielectric layer (BDL) in a multilayer, flexible TFT organic flash memory device. (c) Conformal PEGDMA on yarn, woven into memristors for electronic textiles. Source: From Refs. [135] (b), and [156] (c).

1-ethyl-3-vinyl imidazolium bis(trifluoromethyl sulfonyl)imide [157]. Such ion liquid gels have potential as ion conduction membranes and catalytic supports.

The low substrate temperature and conformal nature of iCVD method make it compatible with the fabrication of electronics on paper and textiles. Ultrathin PEGDMA layers were integrated into TFT-based organic flash memories as the blocking dielectric layer. A cross-sectional transmission electron micrograph (TEM) of this multilayer device (Figure 6.17b) includes layers of aluminum for the control gate and floating gate, iCVD PV3D3 as the tunneling layer, and a channel layer comprising C_{60}. The resulting flash memory operated even after bending down to a 300 μm radius and was compatible with fabrication on inexpensive and disposable paper substrates [135]. The conformal coatings of yarns with iCVD PEGDMA and overcoated with alumina were woven into a textile to create logic circuits including NOT, NOR, OR, AND, and NAND logic circuits (Figure 6.17c) [156]. Organic memory circuits on paper were fabricated using iCVD PEGDMA as the resistive switching layer [138]. The memory performance was retained even after repeated physical bending and folding of the circuit.

6.4.7 Zwitterionic and Polyionic iCVD Films: Properties and Applications

Avoiding the unintended and undesirable accumulation of biomolecules and microbes is crucial for numerous real-world biomedical, industrial, and marine applications [158]. Surface modification to resist biofouling has been explored using a variety of film-forming techniques (Table 6.4), with CVD processing offering a number of benefits as compared to solution-based thin film forming techniques. The iCVD surface modification imparted antifouling performance of reverse osmosis membranes used for water desalination. The iCVD layer must be ultrathin (<30 nm) to maintain a high flux of water across the membrane.

Zwitterionic-modified surfaces are one of the most promising candidates for achieving ultralow fouling. The paired positive and negative charges that

Table 6.4 Comparison of techniques for forming polymer coatings.

Method	SAMs	Grafting-to	Grafting-from	Spin-coating	PECVD	iCVD
Solvent free					•	•
Long-term stability		•	•	•	•	•
Conformality	•	•	•			•
Nanometer thickness control	•	•	•		•	•
Substrate independent					•	•
Functional groups fully retained						•
Rapid synthesis speed				•	•	•
Scalable				•	•	•
Single-step process					•	•

Source: Adapted from Ref. [155].

comprise the neutral zwitterion display high levels of water uptake. Displacing this water by absorption of a biofoulant is thermodynamically unfavorable and is the postulated mechanism for the observed anti-biofouling behavior. Underwater oleophobicity has also been demonstrated [159].

To begin the iCVD zwitterion synthesis, a copolymer, P(DMAEMA-*co*-EGDMA) [160], P(4VP-*co*-DVB) [51, 161], or P(4VP-*co*-EGDA) [162], is formed. The first monomer of each of these copolymers contains a tertiary amine functional group, whereas the second monomer is a cross-linker for enhancing film stability. The P(4VP-*co*-DVB) film contains no labile oxygen bonds, and the corresponding zwitterionic film proves to be resistant against attack by chlorine. Immediately following film deposition, vapors of 1,3 propane sultone are introduced into the iCVD chamber and can react with the tertiary amines at the surface to form a sulfobetaine zwitterion:

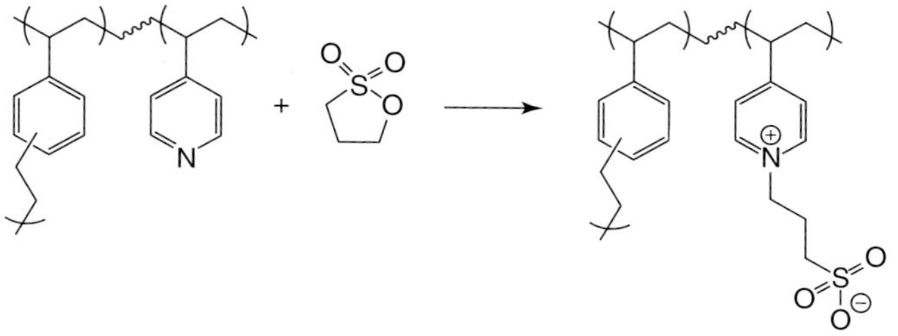

(6.9)

As the reaction with 1,3 propane sultone is diffusion limited, the functionalization to produce quaternary amine occurs only in the near-surface region (~10 nm) as was confirmed by angle-resolved XPS spectra (Figure 6.18). The iCVD synthesis method produces the zwitterions precisely at the surface as desired for antifouling applications. For other synthesis methods, the surfaces must be soaked in water for approximately one day to allow the zwitterions to reorient to the top surface. Reacting the surface amine groups with 3-bromopropionic acid instead of 1,3-propane sultone yields carboxybetaine zwitterions [162].

Anti-biofouling behavior is also exhibited by the iCVD polyionic films P(MAA-*co*-EGDA), P(DMAEMA-*co*-MAA-*co*-EGDA), and P(DMAEMA-*co*-EGDA). All of the iCVD polyionic films reduced the attachment of microglia cells and protein bovine serum albumin (BSA). These polyionic thin films were successful in conformally coating neural probes [163].

6.4.8 iCVD "Smart Surfaces": Properties and Applications

Multiple iCVD chemistries can enable the design of films that display responsive behavior to external stimuli, such as exposure to UV light or changes in temperature or pH. For chemical and biological sensing applications, as has been previously reviewed [3], the response of polymers to exposure to molecules and microbes is desired. Freestanding shape memory polymeric films were demonstrated using iCVD P(tBA-*co*-DEGDVE) [164].

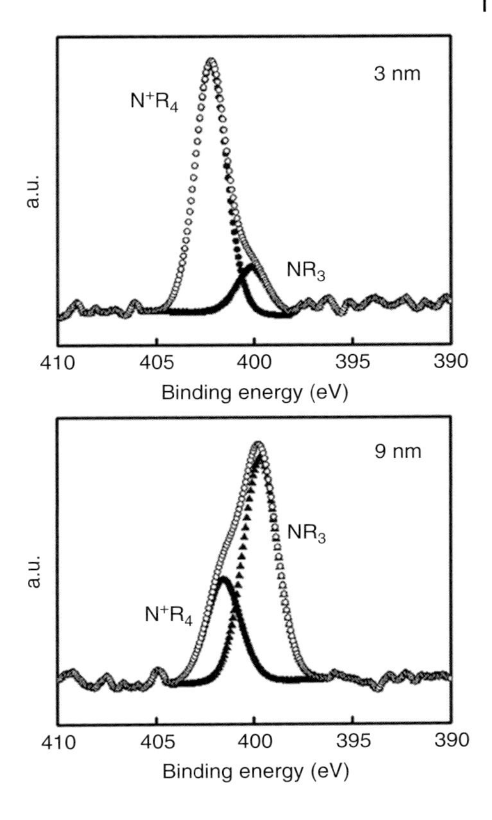

Figure 6.18 Angle-resolved high-resolution N1s X-ray photoelectron spectroscopy (ARXPS) of zwitterionic surface based on iCVD P(DMAEMA-*co*-EGDMA) functionalization *in situ* by propane sultone at take-off angles of (left) 19.5° and (right) 90° corresponding to sample depths of 3 and 9 nm, respectively. The quaternary amine functionality (N + R4) of the zwitterions are observed to be concentrated in the top few nanometers near the surface, as desired for antibiofouling applications. Source: Yang and Gleason 2012 [160]. Reprinted with permission of American Chemical Society.

Light-responsive iCVD layers enable patterning of multiple types of substrates that are not readily compatible with existing photolithography methods. For instance, patterning capability becomes extensible to nonplanar surfaces ranging from hemispherical lens to fiber-based papers and textiles. These challenges are addressable by designing photosensitive iCVD, which display conformal coverage over complex geometries, porous materials, and rough surfaces.

Complexing the photoactive molecule 10,12-tricosanoic acid (TDA) to iCVD P4VP films forms UV-sensitive layers. Exposure to 254 nm UV light creates polydiacetylene (PTDA) over the P4VP layer, which is resistant to dissolution. Negative tone patterns result when regions of the P4VP/TDA without UV exposure are selectively solvated by ethanol. The sensitivity of this iCVD photosensitive layer to UV irradiation was comparable to that of commercially available spin-on resists. Well-defined patterns and pattern transfer were demonstrated over large areas and on substrates having a wide range of curvature[165].

The ability to produce contrasting regions of material properties was demonstrated using iCVD PVCin [166], which undergoes a cross-linking reaction in response to 254 nm UV light exposure. Reversible responsive texturing with water immersion and drying cycles was demonstrated for iCVD P(VCin-*co*-NIPAAm) layers, utilizing postdeposition exposure to UV irradiation through a mask.

Lithographic patterning on paper substrates offers the possibility of fabricating mechanically flexible, inexpensive, and disposable devices. The potentially low cost, ease of portability, and simplicity of use, paper-based microfluidics are attractive for medical diagnostics applications. Patterns on chromatography paper were achieved using poly(*o*-nitrobenzyl methacrylate) (PoNBMA) as a hydrophobic conformal photoresist (Figure 6.19a–f) [167]. This iCVD film penetrated through the thickness of the paper without occluding its pores. Exposure to 254 nm UV light causes cleavage of the *o*-nitrobenzene moieties to form a carboxylic acid (–COOH) group. Thus, the UV treatment converts PoNBMA to polymethyl methacrylate (PMMA) (Figure 6.19d). UV exposure of the PoNBMA through a mask was demonstrated for defining hydrophilic channels for wetting and the flow of fluids (Figure 6.19f) [82] and for creating responsive switches [168] to control the flow of analyte solutions. These methods are readily extensible to other rough and flexible substrates, such as membranes, filters, and textiles. The iCVD synthesis of PoNBMA also provided the outer shell of core–shell nanoparticles [169]. Exposure to UV light allowed subsequent dissolution of the shell layer.

Photoactive layers can result from the postdeposition functionalization of iCVD films, for example, an iCVD PPFM-reactive layer on top of P(HEMA-*co*-EGDMA) functionalized (Reaction (6.7)) by azobenzene after deposition [170]. The result was a UV-responsive hydrogel (Figure 6.19g,h), where the degree of swelling could be reversibly controlled by the UV exposure.

Thermally responsive iCVD hydrogels [171, 172] display different degrees of swelling in water when above or below their lower critical solution temperature (LCST). Below the LCST, the films fully swell with water and are hydrophilic. Above the LCST, intramolecular hydrogen bonding of the polymer becomes thermodynamically favorable over the hydrogen bonding of the polymer to water. Thus, above the LCST water is expelled and polymers collapse into a globular state, and the surface becomes hydrophobic.

Monomers including NIPAAm, DEAAm, DMAAm (*N*,*N*-dimethylacrylamide), and DMAEMA (*N*,*N*-dimethylaminoethyl methacrylate) have been successfully used to deposit thermally responsive iCVD layers. The ability to switch surface properties from hydrophilic to hydrophobic with temperature has been used to "print" arrays of gold nanoparticles [173] using iCVD NIPAAm (Figure 6.20a). Copolymerization with a cross-linker, such as EGDA, helps the films resist dissolution when immersed in water. Temperature-switchable, controlled drug release has been demonstrated using iCVD P(NIPAAm-*co*-DEGDVE) [176].

The composition of the iCVD film can be used to tune the LCST. A library of 12 different iCVD thermosensitive thin films was synthesized and demonstrated to systematically tune the LCST from 32.3 °C for P(NIPAAm(90%)-*co*-EGDA) to 40.1 °C for P(DEAAm(70%)-*co*-DMAAm(20%)-*co*-EGDA) [177]. This temperature range is of interest for biological applications, such as the release

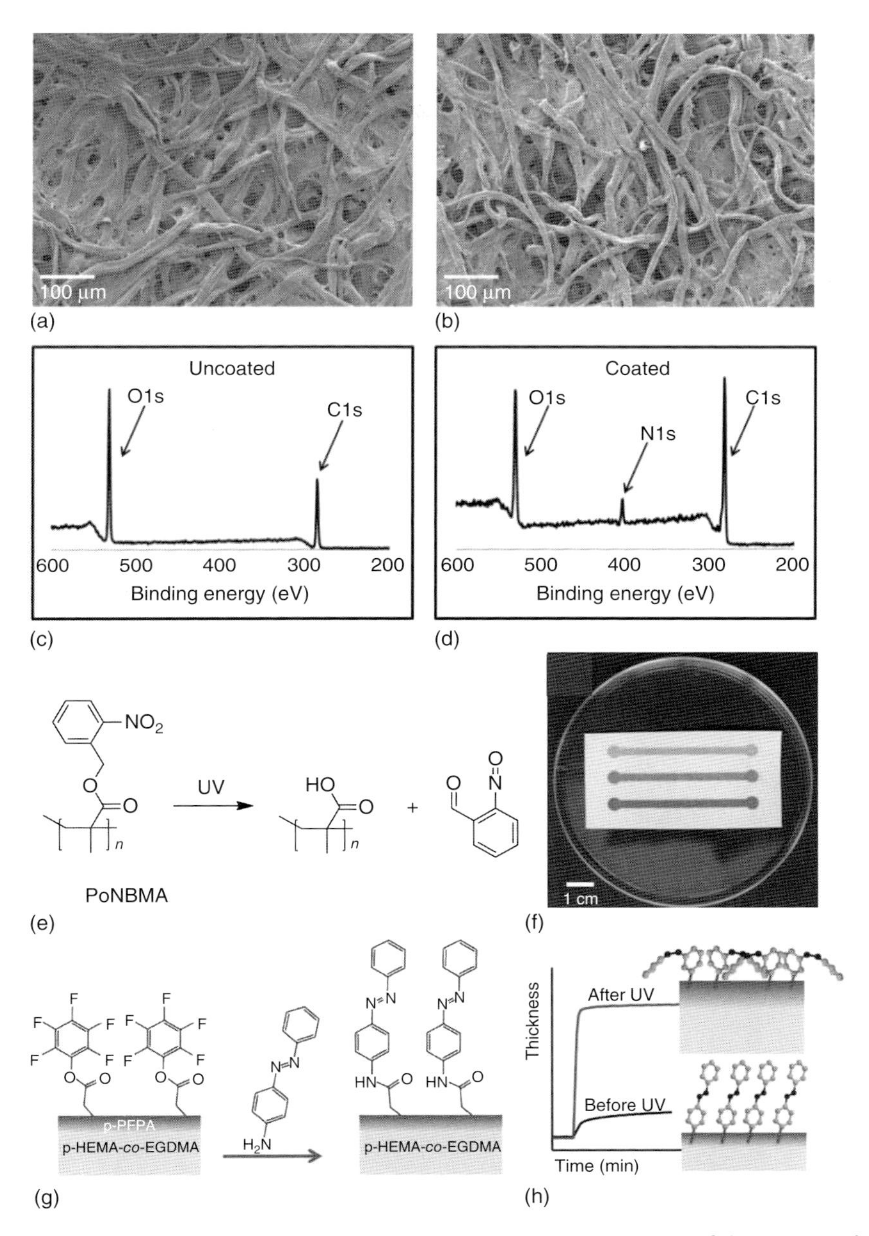

Figure 6.19 UV-responsive iCVD surface modification: top-down SEMs of chromatography paper (a) uncoated and (b) conformally covered by iCVD PoNBMA with their survey X-ray photoelectron spectroscopy (XPS) spectra shown in (c) and (d), respectively. The N1s peak is present only in the coated sample (d). Exposure to 254 nm UV (e) converts PoNBMA to PMMA. Colored dyes (f) are contained in channels in the chromatography paper defined by patterning of the PoNBMA. Source: Haller et al. 2011 [167]. Reprinted with permission of Royal Society of Chemistry. (g) An ultrathin layer of iCVD PPFM grown on a iCVD P(HEMA-*co*-EGDA) hydrogel layer functionalized with azobenzene and (h) displays UV switchable and reversible swelling. Source: Kwong and Gupta 2012 [168] Reprinted with permission of American Chemical Society.

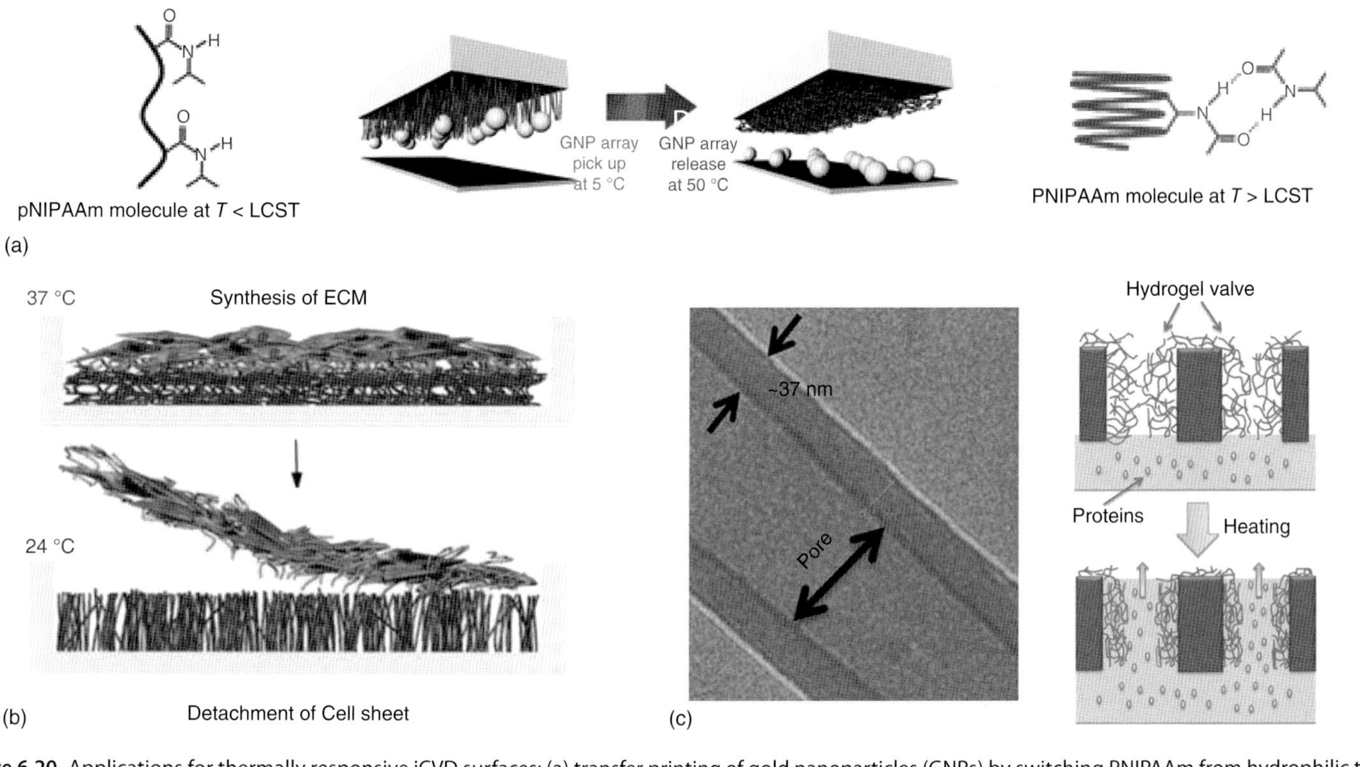

Figure 6.20 Applications for thermally responsive iCVD surfaces: (a) transfer printing of gold nanoparticles (GNPs) by switching PNIPAAm from hydrophilic to hydrophobic raising the temperature above the LCST. Source: Abkenar et al. 2017 [173]. Reprinted with permission of Royal Society of Chemistry. (b) Release of extracellular matrix (ECM) grown at 37 °C on P(NIPAAm-*co*-EDGMA) by dropping the temperature below the LCST. Source: Bakirci et al. 2017 [174]. Reprinted with permission of IOP Publishing. (c) Transmission electron micrograph (TEM) of ultrathin P(DMAEMA-*co*-EGDA) on the inner surface of the pore, visualized by etching away the polyethylene track etch membrane and a schematic of the thermoresponsive switching of pore size enabled by the conformal iCVD coating. Source: Tufani and Ince 2017 [175]. Reprinted with permission of Elsevier.

of cell constructs and sheets (Figure 6.20b) [174, 178]. Conformal application of P(DMAEMA-*co*-EGDA) onto track-etched polycarbonate membranes resulted in a thermally responsive value for protein separation. The ultrathin thermosensitive hydrogel layer that conformally coated the interior walls of the pores was revealed by etching away the track-etched polycarbonate membrane (Figure 6.20c). Also shown is a schematic of the opening of the nanovalve upon heating to above the LCST. The ionization of the tertiary amine resulted in an impressive swelling ratio of 15.4 for P(DMAEMA-*co*-EGDA) with an EGDA to DMAEMA ratio of 0.24. The separation performance for the protein BSA was switched by alternating the temperatures above and below the LCST. The swellability of iCVD hydrogels based on DMAEMA was exploited to create freestanding Janus films [179]. Using a single-step iCVD process, sequential introduction of the monomers build up layers of P(DMAEMA-*co*-EGDA), PEGDA, P(PFDA-*co*-EGDA), and PFDA. Thus, the film's interface with its substrate is hydrophilic while its top surface is hydrophobic. Immersion in water causes the hydrogel layer to swell, and the resulting stress produces the exfoliation of the asymmetric iCVD layer from its substrate.

pH-responsive iCVD films are often copolymers synthesized using MAA as one of the monomers. The pH response of the MAA units is due to their carboxylic acid (–COOH) functional group. The iCVD P(MAA-*co*-EA) and P(MAA-*co*-EGDA) resulted in enteric encapsulation of pharmaceuticals for pH-responsive controlled drug release [180]. Conformal coatings onto vertically aligned carbon nanotubes by P(MAA-*co*-EGDA) displayed swelling transitions upon changing the pH [181]. Smart, pH-responsive membranes were fabricated from P(MAA-*co*-EGDMA) conformally templated into the pores of AAO [175].

Nanotubes capable of tuning the rate of controlled release for large molecules have been fabricated using multiple types of stimuli-responsive iCVD polymers [182]. The nanotubes are templated by the conformal growth of the iCVD films into the straight pores of AAO membranes and then selectively etching away the AAO. Sequential deposition was used to fabricate thermal-responsive PNIPAAm outer layers having pH-responsive inner layers of either PMAA and PHEMA.

Another pH-switchable surface is iCVD PDPAEMA. On a flat silicon surface, reversible switching of the contact angle with water between 87° at high pH to 28° at low contact is achieved through multiple cycles [183]. The conformal and nondamaging nature of iCVD process allowed surface modification of fragile PMMA electrospun fiber mats by PDPAEMA. When applied to the rough surfaces of the mats, the iCVD PDPMA demonstrated switching behavior from superhydrophobic ($155 \pm 3°$) to superhydrophilic ($22 \pm 5°$), at high and low pH, respectively.

6.5 Interfacial Engineering with iCVD: Adhesion and Grafting

As the growth of iCVD films proceeds upward from the substrate, it offers the opportunity for interfacial engineering before beginning iCVD synthesis.

Most strategies to achieve robust interfaces rely on reactive sites present on the surface of the substrate. These sites can be created by pretreatments external to the iCVD chamber (*ex situ*) or inside the iCVD reactor immediately before film growth (*in situ*). In some cases, a "linker molecule" is reacted with the surface to create a reactive functional group, such as vinyl moiety. In other cases, atoms are abstracted directly from the substrate to create surface free radicals, allowing for linker-free grafting. In all cases, the possible adhesion promotion strategies and their optimization will depend on the specific identities of the substrate and the iCVD layers, as well as the environmental conditions to which the coating will be exposed.

Some substrates inherently possess reactive sites. For instance, copper surfaces undergo metal chelating reactions with the iCVD monomer 4VP, and thus, iCVD P4VP promotes the adhesion of copper layers to various types of substrates from which it would otherwise delaminate [184]. Another example is wool, possessing primary amine ($-NH_2$) groups at its surface, which have been demonstrated to react directly with the isocyanato functional group (—N=C=O) of the iCVD monomeric unit isocyanatoethyl methacrylate (IEM) [185]. The isocyanato moiety is also reactive toward hydroxyl groups (–OH).

For other substrates, a surface modification process is required to generate the reactive sites desired for adhesion. The reactive surface sites on the substrate can be organic functional groups that undergo addition or substitution reactions with complementary functional groups of the iCVD monomers or polymer. Surface vinyl groups are also used to promote adhesion as these bonds can participate directly in the growth of iCVD polymer chains tethered to the substrate.

Ex situ modification of the substrate's surface occurs before placing the substrate in the iCVD reactor. For example, the substrate can be immersed in a reactive liquid solution, exposure to a vapor of a silane coupling agent, or treated in a stand-alone oxygen plasma cleaning unit. For such strategies, the time between the pretreatment and the iCVD deposition and the environmental conditions of the laboratory, such as temperature and humidity, are variables that can impact the final results achieved.

The *ex situ* modification of silicon by the silane coupling 3-aminopropyldimethylmonoethoxysilane (3-AMS) yields a surface terminated in primary amine groups. The 3-AMS pretreatment affords the opportunity to improve interfacial adhesion to iCVD polymers containing functional moieties that react with $-NH_2$. For example, anhydride groups, such as those found in MA monomer units, undergo a nucleophilic substitution reaction with amines. The interfacial grafting reaction between amine and anhydride functional groups successfully enhanced adhesion and prevented delamination of the iCVD ionic hydrogel P(MA-*co*-DMAA-*co*-DEGDVE) upon water soaking [28]. In this case, achieving grafting is particularly significant, as these ionic terpolymer hydrogels can swell by >10× in water, resulting in a large degree of interfacial stress. Without pretreatment with the 3-AMS surface linking agent, the iCVD ionic hydrogel simply delaminated when immersed in water.

Another silane coupling agent, trichorovinyl silane (TVS), is often utilized *ex situ* to create substrate surfaces that are functionalized by vinyl groups [186]. During the iCVD process, the surface-anchored vinyl groups react with monomers to form surface-tethered polymer chains.[3] These grafted sites improve the adhesion of the chains directly bound to the interface of the subsequently grown macromolecules that become physically entangled with the grafted polymer chains. The grafted iCVD interfaces resulting from using the TVS linking agent have shown great stability under conditions that mimic industrial operation [94] and physiological conditions [34]. A TVS pretreatment was also used to create covalent linkages of iCVD PEGDA to the surface of PDMS elastomeric sheet. By depositing onto a stretched substrate and then controllably releasing the strain produces herringbone wrinkle patterns, evidence of mechanically pinning (i.e. grafting) at the interface [187].

In situ modification of the substrate is carried out in the deposition chamber immediately before growth and without breaking vacuum before beginning the iCVD process. This approach eliminates the variables associated with the delay and atmospheric exposure between pretreatment and deposition. For example, the intrinsically amine-functionalized surface of reverse osmosis membranes can be converted to a vinyl-functionalized surface through exposure to vapors of pure MA, allowing the MA to act as a linker molecule between the surface and the iCVD polymer [188].

Linker-free grafting can be achieved *in situ* immediately before the iCVD growth to surfaces by creating free radical sites directly on the surface. Indeed, the *in situ* generation of reactive free radicals on the surface of the substrate can be carried out nonselectively in iCVD chambers equipped with plasma capabilities. Alternatively, the photoinitiator, benzophenone, has been used to create free radical sites selectively on parylene surfaces [189]. A third method is selectively creating active sites on the surfaces containing X—H bond exposure to thermally decomposing TPBO initiator in the absence of monomer inside the iCVD chamber. At filament temperatures, ~270–300 °C, the TBPO fragmentation to tert-butoxy radicals proceeds further to produce methyl radicals ($\cdot CH_3$). The methyl radicals can abstract hydrogen from the surface X—H groups, leaving a dangling bond-free radical at the interface. The surface free radical, X·, is capable of reacting with vinyl groups upon the introduction of monomer (M) vapors into the chamber, thus forming a covalent bond across the interface and a surface vinyl radical that propagates by subsequent reactions with additional monomer units:

$$\text{Hydrogen abstraction—XH}_{(\text{surface})} + \text{CH}_{3(\text{gas})} \rightarrow \text{—X}\cdot_{(\text{surface})} + \text{CH}_{4(\text{gas})} \tag{6.10}$$

$$\text{Initiation of grafting—X}_{(\text{surface})} + \text{M}_{(\text{adsorbed})} \rightarrow \text{—X—M}_{(\text{surface})} \tag{6.11}$$

and

$$\text{Propagation of the grafted chain—X—M}_{(\text{surface})} + (n-1)\text{M}_{(\text{absorbed})} \rightarrow \text{—X—M}_n \tag{6.12}$$

The *in situ* grafting scenario mentioned above produced hydrogen-passivated silicon [190] (e.g. X—H = —Si—H) and provided adhesion to steel and surfaces treated with an oxygen plasma [74] (e.g. X—H = —O—H). It is likely also applicable for other types of —X—H bonds. Linker-free grafted cross-linked PDVB layers display outstanding robustness and have served as a base layer for a covalent-attached top layer of iCVD PFDA. The grafted PDVB/PPFDA bilayers provide resistance against the attachment of ice and natural gas hydrates [75, 76, 191].

6.6 Reactors for Synthesizing Organic Films by iCVD

Reactors for iCVD polymerization have many of the same primary components as reactors for the CVD of inorganic films: a delivery system for the reactants, a reaction chamber, process control components, and an effluent handing system. Process monitoring of film growth rate and gas-phase composition are common add-ons. Two example iCVD reactor configurations are shown in Figure 6.21.

The reactants for iCVD are most often liquid held in jars outside of the reaction chamber. Other delivery methods and considerations for solid monomers have been previously reviewed. The initiators can be quite volatile, and its flow into the chamber can be controlled using a needle valve or a mass flow controller. Heating of the jar holding the monomer may need to develop significant vapor pressure in the headspace for delivery to the reaction chamber. An inhibitor, such as those based on TEMPO (e.g. 4-hydroxy TEMPO or polystyrene-bound TEMPO), can be added to the monomer jar to avoid free radical polymerization of the liquid in the jar. When the monomer vapor expands when the pressure drops, Joule–Thompson cooling can occur. A large degree of expansion may produce undesirable condensation of the monomer.

Employing a batch deposition technique eliminates the need to establish a steady-state flow of monomeric reactants or to optimize their delivery parameters to obtain uniform film deposition while avoiding condensation of the monomer. The batch deposition approach reduces the cost of the iCVD reaction equipment as this approach has a simpler feed delivery system and can use a smaller vacuum pump. The operating costs are also lower, as there is a higher degree of conversion of the monomer to film during the batch iCVD process. This latter aspect is particularly important for expensive monomers [16].

The deposition surface in the iCVD chamber should ideally be the coldest surface to avoid unwanted condensation of the monomer vapors elsewhere in the reactor chamber. Backside cooling of the growth surface is generally required to remove the heat load associated with the heated filaments. As filament temperatures for iCVD polymerization are relatively low, the cooling requirement for the growth susceptor is correspondingly modest.

The reactor chamber typically operates at modest vacuum (0.1–1 Torr), which can be achieved by a mechanical pump, although more sophisticated pumping packages improve the base pressure. Traps installed between the chamber and the pumping system are often essential for prolonging the life of the pump. Trapping monomer using activated carbon or a cold trap before it reaches the pump avoids the potential for polymerization inside the pump.

The heated wire array inside the reaction chamber produces a temperature gradient, which can lead to recirculation of gases by natural convection. The

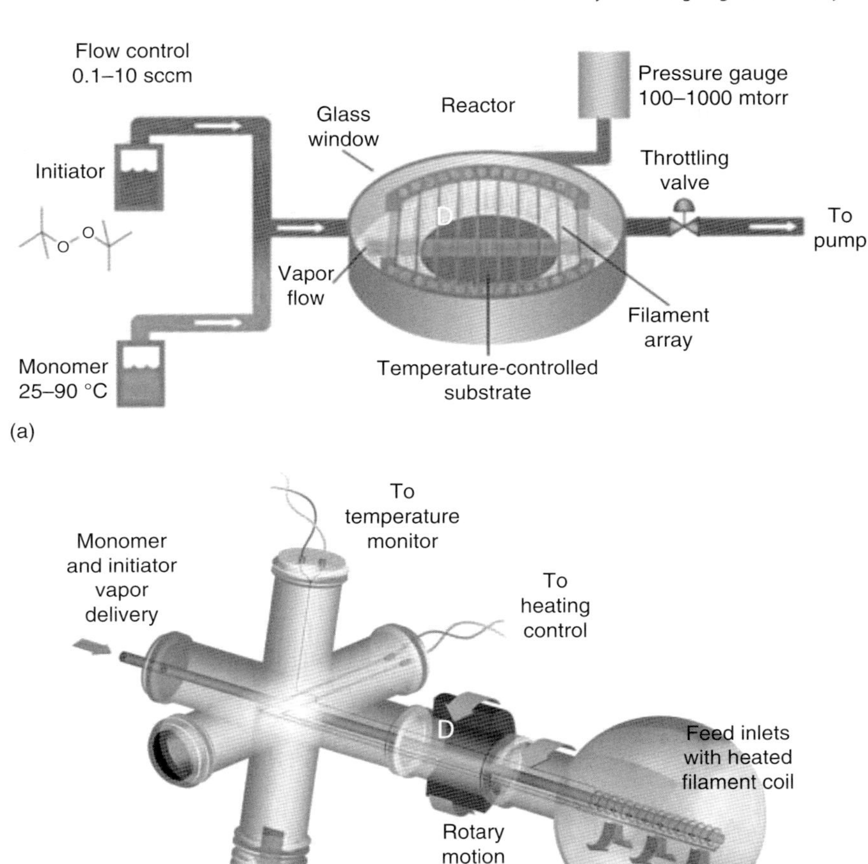

Figure 6.21 Examples of HWCVD/iCVD reactors: (a) pancake-style, 200-mm diameter chamber with resistively heated filament wire array having the reactant flow parallel to the growth surface. Source: Reeja-Jayan et al. 2014 [32]. Reprinted with permission of John Wiley & Sons. (b) Off-the-shelf rotary evaporator modified for the iCVD coating of microparticles. Source: Lau and Gleason 2006 [56]. Reprinted with permission of John Wiley & Sons.

recirculation cells occur in localized areas and keep reaction by-products in the reactor for longer than average, which have negative consequences for the film uniformity and film quality, respectively. Use of a pancake geometry avoids the effects of natural convection (Figure 6.21a). However, the first HWCVD PTFE was performed in a vertical barrel reactor originally designed for capacitively coupled PECVD [9], and a similar style reactor has more recently been used for iCVD research [21].

The iCVD process has been adapted to a bell jar configuration [63], a potentially lower cost solution for laboratory research. Vials of the monomer, GMA, and the initiator, TBPO, are contained inside the glass dome, eliminating the need for mass flow controllers. Instead of filaments, heated pins can achieve the 250–300 °C temperatures required for decomposing the initiator.

Recently, a combined reactor for iCVD and ALD has been reported [140]. A small chamber height is used to avoid the effects of natural convection. The delivery system for the ALD precursors is the primary hardware addition to a typical design of a rectangular iCVD chamber.

An iCVD reactor for coating of particles is shown in Figure 6.21b [56]. In this design, the vacuum system is a standard and relatively inexpensive laboratory rotary evaporator. Most commonly, a rotary evaporator removes solvent from samples without the need to heat the sample. A "rotovap" adapted for iCVD resulted from inserting a ceramic tube down its axis. The tube is used to deliver the iCVD reactants that exit from the tube into the reactor through holes drilled into the tube. Additionally, the tube serves as the mechanical support for a spiral wound filament. The clear round-bottom flask contains the particles to be coated. Immersing the flask in a cooling bath provide control over the substrate temperature. The rotation of the flask, a standard feature of rotovaps, provides agitation and ensures that each particle in the powder is coated.

Scale-up of reactors for commercialization is essential as increasing the coating area can dramatically reduce coating cost. Moving from 200 mm diameter lab-scale reactor to a reactor with 10× the coating area reduces the cost of deposition per area for CVD PTFE by approximately a factor of 10 [1]. At 60× the coating area of the lab-scale system, the cost is ~100 times lower.

Figure 6.22 shows scale-up to commercial iCVD PTFE reactors, produced by GVD Corporation. Figure 6.22a is a commercial lab-scale chamber. Although obscured by the white rack around it, the geometry of the chamber shown in Figure 6.22b is a rectangular pancake reactor, 1.2 m across. Figure 6.22c shows a reactor over 1 m in diameter. Roll-to-roll reactors have also been demonstrated for iCVD and rules for adjusting the conditions used in the small batch-scale system (Figure 6.22d) for use in the roll-to-roll systems [192]. Using semicontinuous roll-to-roll processing improves the economics of film production, as much more of the roll-to-roll system time is used for depositions because many fewer start-up/shutdown cycles are required than for a batch system.

6.7 Summary and Future Prospects for iCVD

The CVD of polymeric films couples the versatility of organic chemistry with the purity and process control of all-dry vacuum processing. This chapter has reviewed the developments in this expanding field since the first report of HWCVD in 1996 and the first report of vinyl polymerization by iCVD in 2004. Applications for iCVD polymer films exploit one or more signature characteristics of the vacuum deposition process, such as ultrathin, pinhole-free deposition, low substrate temperature deposition, high-purity reactants and films, unique composition, or conformal coverage of the external and internal surfaces of substrates having nonplanar geometries.

Figure 6.22 Commercial iCVD/HWCVD reactors: (a) pancake-style, 200-mm diameter chamber; (b) batch production system with front loading of a 1.2-m-wide rectangular pancake-style chamber; (c) batch production system for coating molds used for manufacturing tires; and (d) chamber for semicontinuous roll-to-roll operation. Source: Courtesy of GVD Corporation.

It is hoped that the multitude of organic film compositions, applications, and custom reactor designs described in this chapter will provide a strong foundation for inspiring further innovation. New iCVD compositions can result from the identification of additional monomers with the appropriate volatility and reactivity, or by choosing new combinations of known monomers to form novel copolymers. Both fundamental understanding and the practical optimization of copolymers benefits from the simplicity of tuning properties such as wettability or surface reactivity through systematic variation of the ratio of monomer compositions in the vapor feed. Changing the ratios of the vapor feeds during the course of a single deposition affords the opportunity to grade the film composition throughout its thickness. Introducing reactant vapor feeds having distinct compositions at different locations in the reactor can produce compositional gradients along the surface of the film.

Postdeposition functionalization of the reactant groups incorporated from the monomers into the iCVD organic films provides an additional means of customizing the surface. The chemical and biological specificity of the as-deposited and functionalized iCVD surfaces is especially attractive for biomedical and sensing applications.

Achieving durability through interfacial engineering to increase adhesion, or by achieving covalent grafting across the interface, is often the key to the successful demonstration of prototypes and moving forward on the path to commercial applications. New avenues are opened up as a result of combining the conformal and low substrate temperature iCVD process with unusual substrates, such as nonvolatile liquids, textiles, particles, and nanostructures.

The ability of the iCVD polymers to readily integrate with other processes utilized in the microelectronics industry makes the fabrication of unique optoelectronic devices possible. In some cases, the same chamber can be used for both the growth of the iCVD organic layer and for the deposition of the inorganic layer, facilitating the alternation of these layers to form a hybrid stack.

The iCVD films often bring a degree of mechanical flexibility to the overall multilayer device stack, as is highly desirable for flexible electronics. The low substrate temperature of iCVD finds utility for paper-based microfluidics and optoelectronic circuits. The iCVD process is also compatible with other substrates having limited thermal stability and resistance to solvents, such as membranes and textiles.

A reoccurring theme in this chapter has been the utility of ultrathin and conformal iCVD films in a diverse array of applications including flexible electronics, energy storage devices, biomedical diagnostics, sensing, and membrane modification. The ability to achieve defect-free layers <100 nm or even <10 nm thick is an important differentiator for the iCVD technology. To achieve such thin and pinhole-free films requires minimization of surface roughness. For this reason, multivinyl monomers are preferred because of their ability to yield extremely smooth iCVD homopolymers or copolymers. The cross-linking afforded by using the multivinyl monomers also tends to improve the robustness of these layers.

The iCVD process is a platform technology. Thus, an advance in understanding achieved for one specific film composition can typically be extended to a multitude of other iCVD compositions. The same concept applies to reactor design. Large-scale batch and roll-to-roll reactors designed with one iCVD process in mind can often be utilized with many additional types of monomers. Scaling up of the reactor size dramatically reduces the cost of film per square area.

The availability of commercial lab-scale iCVD systems reduces the barrier for the technology to move into new laboratories. At the lab scale, retrofitting of reactors developed for inorganic CVD or PECVD of polymer-like films for iCVD research can be as simple as adding electrical feedthroughs for powering a filament array. Work is currently underway to simplify lab-scale reactors for iCVD to reduce the cost of purchasing and maintaining them, while at the same time making them easier to use. Moving in this direction will enable characteristics of iCVD films to be exploited by laboratories that do not have a core competency in vacuum technology.

Identifying new applications for existing iCVD polymers is also an important avenue for innovation, as the desirable film properties find utility for a diverse array of applications. For example, as was detailed in Section 6.4.4, the monomer, V3D3, was first synthesized by iCVD to achieve ultrastable biopassivation coatings for biomedical implants. More recently, iCVD PV3D3 has found utility as ultrasmooth and ultrathin (<10 nm thick) dielectric layers in flexible organic devices, including for circuits fabricated on paper and textiles. The same iCVD V3D3 covalent organic networks also serve as a conformal electrolyte for conducting lithium ions, as desired for integration with high-surface-area electrodes for three-dimensional batteries. For new applications, first, the requirements need to be detailed. This includes such properties as thermal stability, solvent resistance, thickness, roughness, surface energy, chemical reactivity, responsiveness, refractive index, and electronic band gap. Based on the property compilations for bulk polymers and the iCVD literature, the large library of potential monomers can be down-selected to a few leading candidates for experimental research.

References

1 Gleason, K.K. (ed.) (2015). *CVD Polymers: Fabrication of Organic Surfaces and Devices*. Wiley-VCH.

2 Moni, P., Al-Obeidi, A., and Gleason, K.K. (2017). Vapor deposition routes to conformal polymer thin films. *Beilstein J. Nanotechnol.* 8: 723–735.

3 Coclite, A.M., Howden, R.M., Borrelli, D.C. et al. (2013). 25th Anniversary article: CVD polymers: a new paradigm for surface modification and device fabrication. *Adv. Mater.* 25 (38): 5392–5422.

4 Alf, M.E., Asatekin, A., Barr, M.C. et al. (2010). Chemical vapor deposition of conformal, functional, and responsive polymer films. *Adv. Mater.* 22 (18): 1993–2027.

5 Cruden, B.A., Gleason, K.K., and Sawin, H.H. (2002). Ultraviolet absorption measurements of CF_2 in the parallel plate pyrolytic chemical vapour deposition process. *J. Phys. D: Appl. Phys.* 35 (5): 480–486.

6 Tenhaeff, W.E. and Gleason, K.K. (2008). Initiated and oxidative chemical vapor deposition of polymeric thin films: iCVD and oCVD. *Adv. Funct. Mater.* 18 (7): 979–992.

7 Cruden, B., Chu, K., Gleason, K., and Sawin, H. (1999). Thermal decomposition of low dielectric constant pulsed plasma fluorocarbon films – II. Effect of postdeposition annealing and ambients. *J. Electrochem. Soc.* 146 (12): 4597–4604.

8 Lau, K.K.S., Bico, J., Teo, K.B.K. et al. (2003). Superhydrophobic carbon nanotube forests. *Nano Lett.* 3 (12): 1701–1705.

9 Limb, S.J., Labelle, C.B., Gleason, K.K. et al. (1996). Growth of fluorocarbon polymer thin films with high CF_2 fractions and low dangling bond concentrations by thermal chemical vapor deposition. *Appl. Phys. Lett.* 68 (20): 2810–2812.

10 Thieme, M., Streller, F., Simon, F. et al. (2013). Superhydrophobic aluminium-based surfaces: wetting and wear properties of different CVD-generated coating types. *Appl. Surf. Sci.* 283: 1041–1050.

11 Laird, E.D., Bose, R.K., Qi, H. et al. (2013). Electric field-induced, reversible lotus-to-rose transition in nanohybrid shish kebab paper with hierarchical roughness. *ACS Appl. Mater. Interfaces* 5 (22): 12089–12098.

12 Lewis, H.G.P., Bansal, N.P., White, A.J., and Handy, E.S. (2009). HWCVD of polymers: commercialization and scale-up. *Thin Solid Films* 517 (12): 3551–3554.

13 Matsumura, H., Mishiro, M., Takachi, M., and Ohdaira, K. (2017). Super water-repellent treatment of various cloths by deposition of catalytic-CVD polytetrafluoroethylene films. *J. Vac. Sci. Tech., A* 35 (6): 061514.

14 Lewis, H.G.P., Caulfield, J.A., and Gleason, K.K. (2001). Perfluorooctane sulfonyl fluoride as an initiator in hot-filament chemical vapor deposition of fluorocarbon thin films. *Langmuir* 17 (24): 7652–7655.

15 Xu, J.J. and Gleason, K.K. (2010). Conformal, amine-functionalized thin films by initiated chemical vapor deposition (iCVD) for hydrolytically stable microfluidic devices. *Chem. Mater.* 22 (5): 1732–1738.

16 Petruczok, C.D., Chen, N., and Gleason, K.K. (2014). Closed batch initiated chemical vapor deposition of ultrathin, functional, and conformal polymer films. *Langmuir* 30 (16): 4830–4837.

17 Chan, K. and Gleason, K.K. (2005). Initiated CVD of poly(methyl methacrylate) thin films. *Chem. Vap. Deposition* 11 (10): 437–443.

18 Bose, R.K., Nejati, S., Stufflet, D.R., and Lau, K.K.S. (2012). Graft polymerization of anti-fouling PEO surfaces by liquid-free initiated chemical vapor deposition. *Macromolecules* 45 (17): 6915–6922.

19 Gao, Y.F., Cole, B., and Tenhaeff, W.E. (2018). Chemical vapor deposition of polymer thin films using cationic initiation. *Macromol. Mater. Eng.* 303 (2): 1700425.

20 Kwong, P., Flowers, C.A., and Gupta, M. (2011). Directed deposition of functional polymers onto porous substrates using metal salt inhibitors. *Langmuir* 27 (17): 10634–10641.

21 Aresta, G., Palmans, J., van de Sanden, M.C.M., and Creatore, M. (2012). Initiated-chemical vapor deposition of organosilicon layers: monomer adsorption, bulk growth, and process window definition. *J. Vac. Sci. Technol., A* 30 (4): 041503.

22 Lau, K.K.S. and Gleason, K.K. (2006). Initiated chemical vapor deposition (iCVD) of poly(alkyl acrylates): an experimental study. *Macromolecules* 39 (10): 3688–3694.

23 Lau, K.K.S. and Gleason, K.K. (2006). Initiated chemical vapor deposition (iCVD) of poly(alkyl acrylates): a kinetic model. *Macromolecules* 39 (10): 3695–3703.

24 Baxamusa, S.H. and Gleason, K.K. (2008). Thin polymer films with high step coverage in microtrenches by initiated CVD. *Chem. Vap. Deposition* 14 (9–10): 313–318.

25 Bonnet, L., Altemus, B., Scarazzini, R. et al. (2017). Initiated-chemical vapor deposition of polymer thin films: unexpected two-regime growth. *Macromol. Mater. Eng.* 302 (12): 1700315.

26 Tao, R. and Anthamatten, M. (2012). Condensation and polymerization of supersaturated monomer vapor. *Langmuir* 28 (48): 16580–16587.

27 Ozaydin-Ince, G. and Gleason, K.K. (2010). Tunable conformality of polymer coatings on high aspect ratio features. *Chem. Vap. Deposition* 16 (1-3): 100–105.

28 Tenhaeff, W.E. and Gleason, K.K. (2007). Initiated chemical vapor deposition of alternating copolymers of styrene and maleic anhydride. *Langmuir* 23 (12): 6624–6630.

29 Greenley, R.Z. (1975, 505). Determination of Q-values and e-values by a least-squares technique. *J. Macromol. Sci., Chem.* 9 (4): –516.

30 Fineman, M. and Ross, S.D. (1950). Linear method for determining monomer reactivity ratios in copolymerization. *J. Polym. Sci.* 5 (2): 259–262.

31 Asatekin, A. and Gleason, K.K. (2011). Polymeric nanopore membranes for hydrophobicity-based separations by conformal initiated chemical vapor deposition. *Nano Lett.* 11 (2): 677–686.

32 Reeja-Jayan, B., Kovacik, P., Yang, R. et al. (2014). A route towards sustainability through engineered polymeric interfaces. *Adv. Mater. Interfaces* 1 (4): n/a.

33 Baxamusa, S.H., Montero, L., Dubach, J.M. et al. (2008). Protection of sensors for biological applications by photoinitiated chemical vapor deposition of hydrogel thin films. *Biomacromolecules* 9 (10): 2857–2862.

34 Sun, M., Wu, Q.Y., Xu, J. et al. (2016). Vapor-based grafting of crosslinked poly(*N*-vinyl pyrrolidone) coatings with tuned hydrophilicity and anti-biofouling properties. *J. Mater. Chem. B* 4 (15): 2669–2678.

35 Yu, S.J., Pak, K., Kwak, M.J. et al. (2017). Initiated chemical vapor deposition: a versatile tool for various device applications. *Adv. Eng. Mater.* 20 (3): 1700622.

36 Boscher, N.D., Wang, M.H., Perrotta, A. et al. (2016). Metal-organic covalent network chemical vapor deposition for gas separation. *Adv. Mater.* 28 (34): 7479–7485.

37 Hetemi, D. and Pinson, J. (2017). Surface functionalisation of polymers. *Chem. Soc. Rev.* 46 (19): 5701–5713.

38 Aldrich Polymer Products Application and Reference Information. Thermal transitions of homopolymers: glass transition and melting point. https://www3.nd.edu/~hgao/thermal_transitions_of_homopolymers.pdf (accessed 2018).

39 Lee, L.H. and Gleason, K.K. (2008). Cross-linked organic sacrificial material for air gap formation by initiated chemical vapor deposition. *J. Electrochem. Soc.* 155 (4): G78–G86.

40 Reeja-Jayan, B., Moni, P., and Gleason, K.K. (2015). Synthesis of insulating and semiconducting polymer films via initiated chemical vapor deposition. *Nanosci. Nanotechnol. Lett.* 7 (1): 33–38.

41 Lewis, H.G.P., Casserly, T.B., and Gleason, K.K. (2001). Hot-filament chemical vapor deposition of organosilicon thin films from hexamethylcyclotrisiloxane and octamethylcyclotetrasiloxane. *J. Electrochem. Soc.* 148 (12): F212–F220.
42 Loo, L.S. and Gleason, K.K. (2001). Hot filament chemical vapor deposition of polyoxymethylene as a sacrificial layer for fabricating air gaps. *Electrochem. Solid-State Lett.* 4 (11): G81–G84.
43 Choi, H.G., Amara, J.P., Martin, T.P. et al. (2006). Structure and morphology of poly(isobenzofuran) films grown by hot-filament chemical vapor deposition. *Chem. Mater.* 18 (26): 6339–6344.
44 Lee, B., Jiao, A., Yu, S. et al. (2013). Initiated chemical vapor deposition of thermoresponsive poly(*N*-vinylcaprolactam) thin films for cell sheet engineering. *Acta Biomater.* 9 (8): 7691–7698.
45 Im, S.G., Bong, K.W., Kim, B.S. et al. (2008). Patterning nanodomains with orthogonal functionalities: solventless synthesis of self-sorting surfaces. *J. Am. Chem. Soc.* 130 (44): 14424–14425.
46 Chen, G.H., Gupta, M., Chan, K., and Gleason, K.K. (2007). Initiated chemical vapor deposition of poly(furfuryl methacrylate). *Macromol. Rapid Commun.* 28 (23): 2205–2209.
47 Tenhaeff, W.E. and Gleason, K.K. (2009). Surface-tethered pH-responsive hydrogel thin films as size-selective layers on nanoporous asymmetric membranes. *Chem. Mater.* 21 (18): 4323–4331.
48 You, J.B., Kim, Y.T., Lee, K.G. et al. (2017). Surface-modified mesh filter for direct nucleic acid extraction and its application to gene expression analysis. *Adv. Healthcare Mater.* 6 (20): 1700642.
49 Yu, S.B., Baek, J., Choi, M. et al. (2016). Polymer thin films with tunable acetylcholine-like functionality enable long-term culture of primary hippocampal neurons. *ACS Nano* 10 (11): 9909–9918.
50 Martin, T.P., Kooi, S.E., Chang, S.H. et al. (2007). Initiated chemical vapor deposition of antimicrobial polymer coatings. *Biomaterials* 28 (6): 909–915.
51 Yang, R., Goktekin, E., Wang, M., and Gleason, K.K. (2014). Molecular fouling resistance of zwitterionic and amphiphilic initiated chemically vapor-deposited (iCVD) thin films. *J. Biomater. Sci., Polym. Ed.* 25 (14-15): 1687–1702.
52 Joo, M., Kwak, M.J., Moon, H. et al. (2017). Thermally fast-curable, "sticky" nanoadhesive for strong adhesion on arbitrary substrates. *ACS Appl. Mater. Interfaces* 9 (46): 40868–40877.
53 Baek, J., Lee, J., Joo, M. et al. (2016). Tuning the electrode work function via a vapor-phase deposited ultrathin polymer film. *J. Mater. Chem. C* 4 (4): 831–839.
54 Mao, Y. and Gleason, K.K. (2004). Hot filament chemical vapor deposition of poly(glycidyl methacrylate) thin films using tert-butyl peroxide as an initiator. *Langmuir* 20 (6): 2484–2488.
55 Lau, K.K.S. and Gleason, K.K. (2008). Initiated chemical vapor deposition (iCVD) of copolymer thin films. *Thin Solid Films* 678–680.
56 Lau, K.K.S. and Gleason, K.K. (2006). Particle surface design using an all-dry encapsulation method. *Adv. Mater.* 18 (15): 1972–1977.

57 Park, S.W., Lee, D., Lee, H.R. et al. (2015). Generation of functionalized polymer nanolayer on implant surface via initiated chemical vapor deposition (iCVD). *J. Colloid Interface Sci.* 439: 34–41.

58 Lau, K.K.S. and Gleason, K.K. (2007). Particle functionalization and encapsulation by initiated chemical vapor deposition (iCVD). *Surf. Coat. Technol.* 201: 9189–9194.

59 Im, S.G., Bong, K.W., Lee, C.H. et al. (2009). A conformal nano-adhesive via initiated chemical vapor deposition for microfluidic devices. *Lab Chip* 9 (3): 411–416.

60 Kwak, M.J., Kim, D.H., You, J.B. et al. (2018). A sub-minute curable nanoadhesive with high transparency, strong adhesion, and excellent flexibility. *Macromolecules* 51 (3): 992–1001.

61 You, J.B., Min, K.I., Lee, B. et al. (2013). A doubly cross-linked nano-adhesive for the reliable sealing of flexible microfluidic devices. *Lab Chip* 13 (7): 1266–1272.

62 Bong, K.W., Xu, J.J., Kim, J.H. et al. (2012). Non-polydimethylsiloxane devices for oxygen-free flow lithography. *Nat. Commun.* 3.

63 Randall, G.C., Gonzalez, L., Petzoldt, R., and Elsner, F. (2017). An evaporative initiated chemical vapor deposition coater for nanoglue bonding. *Adv. Eng. Mater.* 20: 1700839.

64 Jeevendrakumar, V.J.B., Pascual, D.N., and Bergkvist, M. (2015). Wafer scale solventless adhesive bonding with iCVD polyglycidylmethacrylate: effects of bonding parameters on adhesion energies. *Adv. Mater. Interfaces* 2 (9): 1500076.

65 Ye, Y.M., Mao, Y., Wang, F. et al. (2011). Solvent-free functionalization and transfer of aligned carbon nanotubes with vapor-deposited polymer nanocoatings. *J. Mater. Chem.* 21 (3): 837–842.

66 Parker, T.C., Baechle, D., and Demaree, J.D. (2011). Polymeric barrier coatings via initiated chemical vapor deposition. *Surf. Coat. Technol.* 206 (7): 1680–1683.

67 Spee, D.A., Rath, J.K., and Schropp, R.E.I. (2015). Using hot wire and initiated chemical vapor deposition for gas barrier thin film encapsulation. *Thin Solid Films* 575: 67–71.

68 Mao, Y. and Gleason, K.K. (2006). Vapor-deposited fluorinated glycidyl copolymer thin films with low surface energy and improved mechanical properties. *Macromolecules* 39 (11): 3895–3900.

69 Sariipek, F. and Karaman, M. (2014). Initiated CVD of tertiary amine-containing glycidyl methacrylate copolymer thin films for low temperature aqueous chemical functionalization. *Chem. Vap. Deposition* 20 (10–12): 373–379.

70 Mao, Y., Felix, N.M., Nguyen, P.T. et al. (2004). Towards all-dry lithography: electron-beam patternable poly(glycidyl methacrylate) thin films from hot filament chemical vapor deposition. *J. Vac. Sci. Technol., B* 22 (5): 2473–2478.

71 Yoshida, S., Kobayashi, T., Kumano, M., and Esashi, M. (2012). Conformal coating of poly-glycidyl methacrylate as lithographic polymer via initiated chemical vapor deposition. *J. Micro/Nanolithogr. MEMS MOEMS* 11 (2): 023001.

72 Gupta, M. and Gleason, K.K. (2006). Initiated chemical vapor deposition of poly(1*H*,1*H*,2*H*,2*H*-perfluorodecyl acrylate) thin films. *Langmuir* 22 (24): 10047–10052.

73 Ma, M.L., Gupta, M., Li, Z. et al. (2007). Decorated electrospun fibers exhibiting superhydrophobicity. *Adv. Mater.* 19 (2): 255–259.

74 Sojoudi, H., McKinley, G.H., and Gleason, K.K. (2015). Linker-free grafting of fluorinated polymeric cross-linked network bilayers for durable reduction of ice adhesion. *Mater. Horiz.* 2 (1): 91–99.

75 Sojoudi, H., Walsh, M.R., Gleason, K.K., and McKinley, G.H. (2015). Designing durable vapor-deposited surfaces for reduced hydrate adhesion. *Adv. Mater. Interfaces* 2 (6): 1500003.

76 Sojoudi, H., Walsh, M.R., Gleason, K.K., and McKinley, G.H. (2015). Investigation into the formation and adhesion of cyclopentane hydrates on mechanically robust vapor-deposited polymeric coatings. *Langmuir* 31 (22): 6186–6196.

77 Gupta, M., Kapur, V., Pinkerton, N.M., and Gleason, K.K. (2008). Initiated chemical vapor deposition (iCVD) of conformal polymeric nanocoatings for the surface modification of high-aspect-ratio pores. *Chem. Mater.* 20 (4): 1646–1651.

78 Barr, M.C., Rowehl, J.A., Lunt, R.R. et al. (2011). Direct monolithic integration of organic photovoltaic circuits on unmodified paper. *Adv. Mater.* 23 (31): 3499.

79 Cai, J.C., Liu, X.H., Zhao, Y.M., and Guo, F. (2018). Membrane desalination using surface fluorination treated electrospun polyacrylonitrile membranes with nonwoven structure and quasi-parallel fibrous structure. *Desalination* 429: 70–75.

80 Karaman, M., Cabuk, N., Ozyurt, D., and Koysuren, O. (2012). Self-supporting superhydrophobic thin polymer sheets that mimic the nature's petal effect. *Appl. Surf. Sci.* 259: 542–546.

81 Riche, C.T., Roberts, E.J., Gupta, M. et al. (2016). Flow invariant droplet formation for stable parallel microreactors. *Nat. Commun.* 7: 10780.

82 Chen, B., Kwong, P., and Gupta, M. (2013). Patterned fluoropolymer barriers for containment of organic solvents within paper-based microfluidic devices. *ACS Appl. Mater. Interfaces* 5 (23): 12701–12707.

83 Bradley, L.C. and Gupta, M. (2012). Encapsulation of ionic liquids within polymer shells via vapor phase deposition. *Langmuir* 28 (27): 10276–10280.

84 Haller, P.D., Bradley, L.C., and Gupta, M. (2013). Effect of surface tension, viscosity, and process conditions on polymer morphology deposited at the liquid–vapor interface. *Langmuir* 29 (37): 11640–11645.

85 Bradley, L.C. and Gupta, M. (2015). Microstructured films formed on liquid substrates via initiated chemical vapor deposition of cross-linked polymers. *Langmuir* 31 (29): 7999–8005.

86 Kim, S., Sojoudi, H., Zhao, H. et al. (2016). Ultrathin high-resolution flexographic printing using nanoporous stamps. *Sci. Adv.* 2 (12): e1601660.

87 Sojoudi, H., Kim, S., Zhao, H.B. et al. (2017). Stable wettability control of nanoporous microstructures by iCVD coating of carbon nanotubes. *ACS Appl. Mater. Interfaces* 9 (49): 43287–43299.

88 An, Y.H., Yu, S.J., Kim, I.S. et al. (2017). Hydrogel functionalized Janus membrane for skin regeneration. *Adv. Healthcare Mater.* 6 (5): 1600795.
89 Kwak, M.J., Yoo, Y., Lee, H.S. et al. (2016). A simple, cost-efficient method to separate microalgal lipids from wet biomass using surface energy-modified membranes. *ACS Appl. Mater. Interfaces* 8 (1): 600–608.
90 Kim, D., Im, H., Kwak, M.J. et al. (2016). A superamphiphobic sponge with mechanical durability and a self-cleaning effect. *Sci. Rep.* 6: 29993.
91 Coclite, A.M., Shi, Y.J., and Gleason, K.K. (2012). Controlling the degree of crystallinity and preferred crystallographic orientation in poly-perfluorodecylacrylate thin films by initiated chemical vapor deposition. *Adv. Funct. Mater.* 22 (10): 2167–2176.
92 Coclite, A.M., Shi, Y.J., and Gleason, K.K. (2012). Grafted crystalline poly-perfluoroacrylate structures for superhydrophobic and oleophobic functional coatings. *Adv. Mater.* 24 (33): 4534–4539.
93 Christian, P. and Coclite, A.M. (2017). Vapor-phase-synthesized fluoroacrylate polymer thin films: thermal stability and structural properties. *Beilstein J. Nanotechnol.* 8: 933–942.
94 Paxson, A.T., Yague, J.L., Gleason, K.K., and Varanasi, K.K. (2014). Stable dropwise condensation for enhancing heat transfer via the initiated chemical vapor deposition (iCVD) of grafted polymer films. *Adv. Mater.* 26 (3): 418–423.
95 Liu, A., Goktekin, E., and Gleason, K.K. (2014). Cross-linking and ultrathin grafted gradation of fluorinated polymers synthesized via initiated chemical vapor deposition to prevent surface reconstruction. *Langmuir* 30 (47): 14189–14194.
96 Coclite, A.M., Lund, P., Di Mundo, R., and Palumbo, F. (2013). Novel hybrid fluoro-carboxylated copolymers deposited by initiated chemical vapor deposition as protonic membranes. *Polymer* 54 (1): 24–30.
97 Nedaei, M., Motezakker, A.R., Zeybek, M.C. et al. (2017). Subcooled flow boiling heat transfer enhancement using polyperfluorodecylacrylate (pPFDA) coated microtubes with different coating thicknesses. *Exp. Therm. Fluid Sci.* 86: 130–140.
98 Nedaei, M., Armagan, E., Sezen, M. et al. (2016). Enhancement of flow boiling heat transfer in pHEMA/pPFDA coated microtubes with longitudinal variations in wettability. *AIP Adv.* 6 (3): 035212.
99 Ranacher, C., Resel, R., Moni, P. et al. (2015). Layered nanostructures in proton conductive polymers obtained by initiated chemical vapor deposition. *Macromolecules* 48 (17): 6177–6185.
100 Chan, K. and Gleason, K.K. (2005). Initiated chemical vapor deposition of linear and cross-linked poly(2-hydroxyethyl methacrylate) for use as thin-film hydrogels. *Langmuir* 21 (19): 8930–8939.
101 Christian, P., Ehmann, H.M.A., Coclite, A.M., and Werzer, O. (2016). Polymer encapsulation of an amorphous pharmaceutical by initiated chemical vapor deposition for enhanced stability. *ACS Appl. Mater. Interfaces* 8 (33): 21177–21184.
102 Karaman, M., Kooi, S.E., and Gleason, K.K. (2008). Vapor deposition of hybrid organic-inorganic dielectric Bragg mirrors having rapid and reversibly tunable optical reflectance. *Chem. Mater.* 20 (6): 2262–2267.

103 Nejati, S. and Lau, K.K.S. (2011). Pore filling of nanostructured electrodes in dye sensitized solar cells by initiated chemical vapor deposition. *Nano Lett.* 11 (2): 419–423.

104 Bose, R.K. and Lau, K.K.S. (2010). Mechanical properties of ultrahigh molecular weight PHEMA hydrogels synthesized using initiated chemical vapor deposition. *Biomacromolecules* 11 (8): 2116–2122.

105 Yague, J.L. and Gleason, K.K. (2012). Systematic control of mesh size in hydrogels by initiated chemical vapor deposition. *Soft Matter* 8 (10): 2890–2894.

106 Unger, K., Resel, R., and Coclite, A.M. (2016). Dynamic studies on the response to humidity of poly(2-hydroxyethyl methacrylate) hydrogels produced by initiated chemical vapor deposition. *Macromol. Chem. Phys.* 217 (21): 2372–2379.

107 Tufani, A. and Ince, G.O. (2015). Permeability of small molecules through vapor deposited polymer membranes. *J. Appl. Polym. Sci.* 132 (34): n/a.

108 Montero, L., Gabriel, G., Guimera, A. et al. (2012). Increasing biosensor response through hydrogel thin film deposition: Influence of hydrogel thickness. *Vacuum* 86 (12): 2102–2104.

109 Ozaydin-Ince, G., Dubach, J.M., Gleason, K.K., and Clark, H.A. (2011). Microworm optode sensors limit particle diffusion to enable in vivo measurements. *Proc. Natl. Acad. Sci. U.S.A.* 108 (7): 2656–2661.

110 Ince, G.O., Demirel, G., Gleason, K.K., and Demirel, M.C. (2010). Highly swellable free-standing hydrogel nanotube forests. *Soft Matter* 6 (8): 1635–1639.

111 Ince, G.O., Armagan, E., Erdogan, H. et al. (2013). One-dimensional surface-imprinted polymeric nanotubes for specific biorecognition by initiated chemical vapor deposition (iCVD). *ACS Appl. Mater. Interfaces* 5 (14): 6447–6452.

112 Cikim, T., Armagan, E., Ince, G.O., and Kosar, A. (2014). Flow boiling enhancement in microtubes with crosslinked pHEMA coatings and the effect of coating thickness. *J. Heat Transfer Trans. ASME* 136 (8).

113 Baxamusa, S.H. and Gleason, K.K. (2009). Random copolymer films with molecular-scale compositional heterogeneities that interfere with protein adsorption. *Adv. Funct. Mater.* 19 (21): 3489–3496.

114 Mari-Buye, N., O'Shaughnessy, S., Colominas, C. et al. (2009). Functionalized, swellable hydrogel layers as a platform for cell studies. *Adv. Funct. Mater.* 19 (8): 1276–1286.

115 Montero, L., Baxamusa, S.H., Borros, S., and Gleason, K.K. (2009). Thin hydrogel films with nanoconfined surface reactivity by photoinitiated chemical vapor deposition. *Chem. Mater.* 21 (2): 399–403.

116 O'Shaughnessy, W.S., Murthy, S.K., Edell, D.J., and Gleason, K.K. (2007). Stable biopassive insulation synthesized by initiated chemical vapor deposition of poly(1,3,5-trivinyltrimethylcyclotrisiloxane). *Biomacromolecules* 8 (8): 2564–2570.

117 Coclite, A.M., Ozaydin-Ince, G., Palumbo, F. et al. (2010). Single-chamber deposition of multi layer barriers by plasma enhanced and initiated chemical vapor deposition of organosilicones. *Plasma Processes Polym.* 7 (7): 561–570.

118 Trujillo, N.J., Wu, Q.G., and Gleason, K.K. (2010). Ultralow dielectric constant tetravinyltetramethylcyclotetrasiloxane films deposited by initiated chemical vapor deposition (iCVD). *Adv. Funct. Mater.* 20 (4): 607–616.

119 Yoon, S.J., Pak, K., Nam, T. et al. (2017). Surface-localized sealing of porous ultralow-k dielectric films with ultrathin (<2 nm) polymer coating. *ACS Nano* 11 (8): 7841–7847.

120 Aresta, G., Palmans, J., van de Sanden, M.C.M., and Creatore, M. (2012). Evidence of the filling of nano-porosity in SiO_2-like layers by an initiated-CVD monomer. *Microporous Mesoporous Mater.* 151: 434–439.

121 Reeja-Jayan, B., Chen, N., Lau, J. et al. (2015). A group of cyclic siloxane and silazane polymer films as nanoscale electrolytes for microbattery architectures. *Macromolecules* 48 (15): 5222–5229.

122 Yoo, Y., Kim, B.G., Pak, K. et al. (2015). Initiated chemical vapor deposition (iCVD) of highly cross-linked polymer films for advanced lithium-ion battery separators. *ACS Appl. Mater. Interfaces* 7 (33): 18849–18855.

123 Coclite, A.M., Ozaydin-Ince, G., d'Agostino, R., and Gleason, K.K. (2009). Flexible cross-linked organosilicon thin films by initiated chemical vapor deposition. *Macromolecules* 42 (21): 8138–8145.

124 Achyuta, A.K.H., White, A.J., Lewis, H.G.P., and Murthy, S.K. (2009). Incorporation of linear spacer molecules in vapor-deposited silicone polymer thin films. *Macromolecules* 42 (6): 1970–1978.

125 Perrotta, A., Aresta, G., van Beekum, E.R.J. et al. (2015). The impact of the nano-pore filling on the performance of organosilicon-based moisture barriers. *Thin Solid Films* 595: 251–257.

126 O'Shaughnessy, W.S., Gao, M.L., and Gleason, K.K. (2006). Initiated chemical vapor deposition of trivinyltrimethylcyclotrisiloxane for biomaterial coatings. *Langmuir* 22 (16): 7021–7026.

127 Jeong, G.M., Seong, H., Kim, Y.S. et al. (2014). Site-specific immobilization of proteins on non-conventional substrates via solvent-free initiated chemical vapour deposition (iCVD) process. *Polym. Chem.* 5 (15): 4459–4465.

128 Jung, I.Y., You, J.B., Choi, B.R. et al. (2016). A highly sensitive molecular detection platform for robust and facile diagnosis of Middle East respiratory syndrome (MERS) corona virus. *Adv. Healthcare Mater.* 5 (17): 2168–2173.

129 O'Shaughnessy, W.S., Edell, D.J., and Gleason, K.K. (2009). Initiated chemical vapor deposition of a siloxane coating for insulation of neural probes. *Thin Solid Films* 517 (12): 3612–3614.

130 Moon, H., Seong, H., Shin, W.C. et al. (2015). Synthesis of ultrathin polymer insulating layers by initiated chemical vapour deposition for low-power soft electronics. *Nat. Mater.* 14 (6): 628–635.

131 Murthy, S.K., Olsen, B.D., and Gleason, K.K. (2004). Effect of filament temperature on the chemical vapor deposition of fluorocarbon-organosilicon copolymers. *J. Appl. Polym. Sci.* 91 (4): 2176–2185.

132 Woo, M.H., Jang, B.C., Choi, J. et al. (2017). Low-power nonvolatile charge storage memory based on MoS_2 and an ultrathin polymer tunneling dielectric. *Adv. Funct. Mater.* 27 (43): 1703545.

133 Wang, M.H., Wang, X.X., Moni, P. et al. (2017). CVD polymers for devices and device fabrication. *Adv. Mater.* 29 (11): 1604606.

134 Seong, H., Baek, J., Pak, K., and Im, S.G. (2015). A surface tailoring method of ultrathin polymer gate dielectrics for organic transistors: improved device performance and the thermal stability thereof. *Adv. Funct. Mater.* 25 (28): 4462–4469.

135 Lee, S., Seong, H., Im, S.G. et al. (2017). Organic flash memory on various flexible substrates for foldable and disposable electronics. *Nat. Commun.* 8: 725.

136 Pak, K., Seong, H., Choi, J. et al. (2016). Synthesis of ultrathin, homogeneous copolymer dielectrics to control the threshold voltage of organic thin-film transistors. *Adv. Funct. Mater.* 26 (36): 6574–6582.

137 Kim, M., Seong, H., Lee, S. et al. (2016). Efficient organic photomemory with photography-ready programming speed. *Sci. Rep.* 6: 30536.

138 Lee, D., Yoon, J., Lee, J. et al. (2016). Logic circuits composed of flexible carbon nanotube thin-film transistor and ultra-thin polymer gate dielectric. *Sci. Rep.* 6: 26121.

139 Jang, B.C., Nam, Y., Koo, B.J. et al. (2018). Memristive logic-in-memory integrated circuits for energy-efficient flexible electronics. *Adv. Funct. Mater.* 28 (2): 1704725.

140 Kim, B.J., Park, H., Seong, H. et al. (2017). A single-chamber system of initiated chemical vapor deposition and atomic layer deposition for fabrication of organic/inorganic multilayer films. *Adv. Eng. Mater.* 19 (6): 1600819.

141 Kim, B.J., Seong, H., Shim, H. et al. (2017). Initiated chemical vapor deposition of polymer films at high process temperature for the fabrication of organic/inorganic multilayer thin film encapsulation. *Adv. Eng. Mater.* 19 (7): 1600870.

142 Seong, H., Choi, J., Kim, B.J. et al. (2017). Vapor-phase synthesis of sub-15 nm hybrid gate dielectrics for organic thin film transistors. *J. Mater. Chem. C* 5 (18): 4463–4470.

143 Wang, X., Ugar, A., Goktas, H. et al. (2016). Room temperature resistive volatile organic compound sensing materials based on hybrid structure of vertically aligned carbon nanotubes and conformal oCVD/iCVD polymer coatings. *ACS Sens.* 1: 374–383.

144 Suh, H.S., Kim, D.H., Moni, P. et al. (2017). Sub-10-nm patterning via directed self-assembly of block copolymer films with a vapour-phase deposited topcoat. *Nat. Nanotechnol.* 12 (6): 575–581.

145 Servi, A.T., Kharraz, J., Klee, D. et al. (2016). A systematic study of the impact of hydrophobicity on the wetting of MD membranes. *J. Membr. Sci.* 520: 850–859.

146 Kim, S.H., Lee, H.R., Yu, S.J. et al. (2015). Hydrogel-laden paper scaffold system for origami-based tissue engineering. *Proc. Natl. Acad. Sci. U.S.A.* 112 (50): 15426–15431.

147 Xu, J.J., Asatekin, A., and Gleason, K.K. (2012). The design and synthesis of hard and impermeable, yet flexible, conformal organic coatings. *Adv. Mater.* 24 (27): 3692–3696.

148 Baxamusa, S.H., Suresh, A., Ehrmann, P. et al. (2015). Photo-oxidation of polymers synthesized by plasma and initiated CVD. *Chem. Vap. Deposition* 21 (10-12): 267–274.
149 Petruczok, C.D., Yang, R., and Gleason, K.K. (2013). Controllable cross-linking of vapor-deposited polymer thin films and impact on material properties. *Macromolecules* 46 (5): 1832–1840.
150 Zhao, J., Wang, M., and Gleason, K. (2017). Stabilizing the wettability of initiated chemical vapor deposited (iCVD) polydivinylbenzene thin films by thermal annealing. *Adv. Mater. Interfaces* 4 (18): 1700270.
151 Lepro, X., Ehrmann, P., Rodriguez, J., and Baxamusa, S. (2018). Enhancing the oxidation stability of polydivinylbenzene films via residual pendant vinyl passivation. *ChemistrySelect* 3 (2): 500–506.
152 Lepro, X., Ehrmann, P., Menapace, J. et al. (2017). Ultralow stress, thermally stable cross-linked polymer films of polydivinylbenzene (PDVB). *Langmuir* 33 (21): 5204–5212.
153 Chen, N., Kovacik, P., Howden, R.M. et al. (2015). Low substrate temperature encapsulation for flexible electrodes and organic photovoltaics. *Adv. Energy Mater.* 5 (6): 1401442.
154 Baxamusa, S.H., Lepro, X., Lee, T. et al. (2017). Initiated chemical vapor deposition polymers for high peak-power laser targets. *Thin Solid Films* 635: 37–41.
155 Lee, E., Faguet, J., Brcka, J. et al. (2011). Single-chamber filament-assisted chemical vapor deposition of polymer and organosilicate films for air gap interconnect formation. *Thin Solid Films* 519 (14): 4571–4573.
156 Bae, H., Jang, B.C., Park, H. et al. (2017). Functional circuitry on commercial fabric via textile-compatible nanoscale film coating process for fibertronics. *Nano Lett.* 17 (10): 6443–6452.
157 Bradley, L.C. and Gupta, M. (2014). Copolymerization of 1-ethyl-3-vinylimidazolium bis(trifluoromethylsulfonyl)imide via initiated chemical vapor deposition. *Macromolecules* 47 (19): 6657–6663.
158 Yang, R., Asatekin, A., and Gleason, K.K. (2012). Design of conformal, substrate-independent surface modification for controlled protein adsorption by chemical vapor deposition (CVD). *Soft Matter* 8 (1): 31–43.
159 Yang, R., Moni, P., and Gleason, K.K. (2015). Ultrathin zwitterionic coatings for roughness-independent underwater superoleophobicity and gravity-driven oil-water separation. *Adv. Mater. Interfaces* 2 (2): 1400489.
160 Yang, R. and Gleason, K.K. (2012). Ultrathin antifouling coatings with stable surface zwitterionic functionality by initiated chemical vapor deposition (iCVD). *Langmuir* 28 (33): 12266–12274.
161 Yang, R., Jang, H., Stocker, R., and Gleason, K.K. (2014). Synergistic prevention of biofouling in seawater desalination by zwitterionic surfaces and low-level chlorination. *Adv. Mater.* 26 (11): 1711–1718.
162 Shafi, H.Z., Khan, Z., Yang, R., and Gleason, K.K. (2015). Surface modification of reverse osmosis membranes with zwitterionic coating for improved resistance to fouling. *Desalination* 362: 93–103.

163 Zhi, B., Song, Q., and Mao, Y. (2018). Vapor deposition of polyionic nanocoatings for reduction of microglia adhesion. *RSC Adv.* 8 (9): 4779–4785.

164 Kramer, N.J., Sachteleben, E., Ozaydin-Ince, G. et al. (2010). Shape memory polymer thin films deposited by initiated chemical vapor deposition. *Macromolecules* 43 (20): 8344–8347.

165 Petruczok, C.D. and Gleason, K.K. (2012). Initiated chemical vapor deposition-based method for patterning polymer and metal microstructures on curved substrates. *Adv. Mater.* 24 (48): 6445–6450.

166 Petruczok, C.D., Armagan, E., Ince, G.O., and Gleason, K.K. (2014). Initiated chemical vapor deposition and light-responsive cross-linking of poly(vinyl cinnamate) thin films. *Macromol. Rapid Commun.* 35 (15): 1345–1350.

167 Haller, P.D., Flowers, C.A., and Gupta, M. (2011). Three-dimensional patterning of porous materials using vapor phase polymerization. *Soft Matter* 7 (6): 2428–2432.

168 Kwong, P. and Gupta, M. (2012). Vapor phase deposition of functional polymers onto paper-based microfluidic devices for advanced unit operations. *Anal. Chem.* 84 (22): 10129–10135.

169 Frank-Finney, R.J. and Gupta, M. (2016). Two-stage growth of polymer nanoparticles at the liquid-vapor interface by vapor-phase polymerization. *Langmuir* 32 (42): 11014–11020.

170 Unger, K., Salzmann, P., Masciullo, C. et al. (2017). Novel light-responsive biocompatible hydrogels produced by initiated chemical vapor deposition. *ACS Appl. Mater. Interfaces* 9 (20): 17409–17417.

171 Alf, M.E., Hatton, T.A., and Gleason, K.K. (2011). Insights into thin, thermally responsive polymer layers through quartz crystal microbalance with dissipation. *Langmuir* 27 (17): 10691–10698.

172 Alf, M.E., Godfrin, P.D., Hatton, T.A., and Gleason, K.K. (2010). Sharp hydrophilicity switching and conformality on nanostructured surfaces prepared via initiated chemical vapor deposition (iCVD) of a novel thermally responsive copolymer. *Macromol. Rapid Commun.* 31 (24): 2166–2172.

173 Abkenar, S.K., Tufani, A., Ince, G.O. et al. (2017). Transfer printing gold nanoparticle arrays by tuning the surface hydrophilicity of thermo-responsive poly *N*-isopropylacrylamide (pNIPAAm). *Nanoscale* 9 (9): 2969–2973.

174 Bakirci, E., Toprakhisar, B., Zeybek, M.C. et al. (2017). Cell sheet based bioink for 3D bioprinting applications. *Biofabrication* 9: 024105.

175 Tufani, A. and Ince, G.O. (2017). Smart membranes with pH-responsive control of macromolecule permeability. *J. Membr. Sci.* 537: 255–262.

176 McInnes, S.J.P., Szili, E.J., Al-Bataineh, S.A. et al. (2016). Fabrication and characterization of a porous silicon drug delivery system with an initiated chemical vapor deposition temperature-responsive coating. *Langmuir* 32 (1): 301–308.

177 Pena-Francesch, A., Montero, L., and Borros, S. (2014). Tailoring the LCST of thermosensitive hydrogel thin films deposited by iCVD. *Langmuir* 30 (24): 7162–7167.

178 Tekin, H., Ozaydin-Ince, G., Tsinman, T. et al. (2011). Responsive microgrooves for the formation of harvestable tissue constructs. *Langmuir* 27 (9): 5671–5679.
179 Ye, Y.M. and Mao, Y. (2017). Vapor-based synthesis and micropatterning of Janus thin films with distinct surface wettability and mechanical robustness. *RSC Adv.* 7 (40): 24569–24575.
180 Lau, K.K.S. and Gleason, K.K. (2007). All-dry synthesis and coating of methacrylic acid copolymers for controlled release. *Macromol. Biosci.* 7 (4): 429–434.
181 Ye, Y.M., Mao, Y., Wang, H.Z., and Ren, Z.F. (2012). Hybrid structure of pH-responsive hydrogel and carbon nanotube array with superwettability. *J. Mater. Chem.* 22 (6): 2449–2455.
182 Armagan, E. and Ince, G.O. (2015). Coaxial nanotubes of stimuli responsive polymers with tunable release kinetics. *Soft Matter* 11 (41): 8069–8075.
183 Karaman, M. and Cabuk, N. (2012). Initiated chemical vapor deposition of pH responsive poly(2-diisopropylamino)ethyl methacrylate thin films. *Thin Solid Films* 520 (21): 6484–6488.
184 You, J.B., Kim, S.Y., Park, Y.J. et al. (2014). A vapor-phase deposited polymer film to improve the adhesion of electroless-deposited copper layer onto various kinds of substrates. *Langmuir* 30 (3): 916–921.
185 Feng, J.G., Sun, M., and Ye, Y.M. (2017). Ultradurable underwater superoleophobic surfaces obtained by vapor-synthesized layered polymer nanocoatings for highly efficient oil-water separation. *J. Mater. Chem. A* 5 (29): 14990–14995.
186 Trujillo, N.J., Baxamusa, S., and Gleason, K.K. (2009). Grafted polymeric nanostructures patterned bottom-up by colloidal lithography and initiated chemical vapor deposition (iCVD). *Thin Solid Films* 517 (12): 3615–3618.
187 Yin, J., Yague, J.L., Eggenspieler, D. et al. (2012). Deterministic order in surface micro-topologies through sequential wrinkling. *Adv. Mater.* 24 (40): 5441–5446.
188 Yang, R., Xu, J.J., Ozaydin-Ince, G. et al. (2011). Surface-tethered zwitterionic ultrathin antifouling coatings on reverse osmosis membranes by initiated chemical vapor deposition. *Chem. Mater.* 23 (5): 1263–1272.
189 De Luna, M.M., Chen, B., Bradley, L.C. et al. (2016). Solventless grafting of functional polymer coatings onto Parylene C. *J. Vac. Sci. Tech., A* 34 (4): 041403.
190 Yang, R., Buonassisi, T., and Gleason, K.K. (2013). Organic vapor passivation of silicon at room temperature. *Adv. Mater.* 25 (14): 2078–2083.
191 Sojoudi, H., Arabnejad, H., Raiyan, A. et al. (2018). Scalable and durable polymeric icephobic and hydrate-phobic coatings. *Soft Matter* 14 (18): 3443–3454.
192 Gupta, M. and Gleason, K.K. (2006). Large-scale initiated chemical vapor deposition of poly(glycidyl methacrylate) thin films. *Thin Solid Films* 515 (4): 1579–1584.

7

Physics and Technologies for Operating Cat-CVD Apparatus

In this chapter, realistic topics that are important for industrial implementation of catalytic chemical vapor deposition (Cat-CVD) technology are summarized. The influence of thermal radiation from heated catalyzers and the contamination from heated metals are discussed along with methods to overcome the problems associated with these aspects. The points relevant to building mass production machines are introduced and discussed. This chapter may be useful for people who venture to construct a Cat-CVD apparatus and for people who like to know the practical issues of Cat-CVD technology for actual application.

7.1 Influence of Gas Flow in Cat-CVD Apparatus

7.1.1 Experiment Using a Long Cylindrical Chamber for Establishing Quasi-laminar Flow

In atmospheric pressure CVD, the flow of gas is a key factor deciding film uniformity. Low-pressure CVD is a method to improve uniformity, by introducing surface reaction control mechanisms during film formation. Cat-CVD is also one of the low-pressure CVD methods. However, contrary to simple thermal CVD, active species are already produced at the surface of the catalyzer, and such species have to be transported to the substrate. In Cat-CVD, the influence of the flow of decomposed species is likely one of the factors to decide the film uniformity. Thus, here, at first, the influence of the gas flow is studied, using a special deposition chamber [1].

Figure 7.1 shows a schematic diagram of the experimental chamber. In this case, a long cylindrical chamber with an inner diameter of 5 cm (diameter of X cm is often denoted as X cmØ in this book) and a length of 25 cm is used to make quasi-laminar gas flow. At the middle of the cylindrical chamber, a W catalyzing wire with a diameter of 0.4 mm and a length of 20 cm is mounted normal to the laminar flow. The W catalyzing wire is heated to 1850 °C by supplying electric current directly to it.

At the bottom of the chamber, a glass plate with a width of 1 cm, length of 10 cm, and thickness of 0.7 mm is placed to determine the thickness distribution of the deposited films. If the film deposition is not influenced by the gas flow, the film should distribute symmetrically around the center position of the catalyzer. However, if the gas flow has an influence, the film distribution should be distorted

Catalytic Chemical Vapor Deposition: Technology and Applications of Cat-CVD,
First Edition. Hideki Matsumura, Hironobu Umemoto, Karen K. Gleason, and Ruud E.I. Schropp.

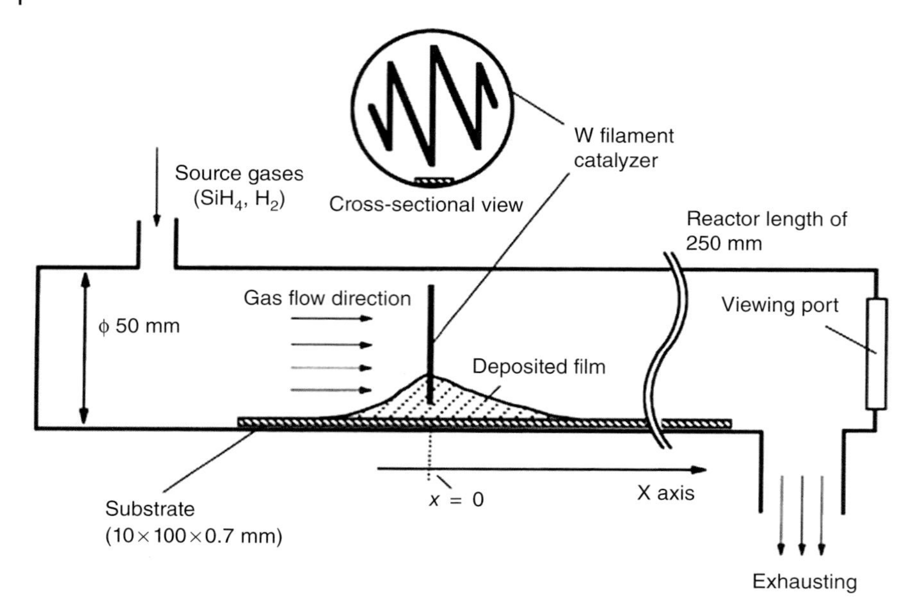

Figure 7.1 A schematic diagram of experimental Cat-CVD chamber of a shape of a long cylinder with an inner diameter of 5 cm and a length of 5 cm. Source: Reprinted with permission from Ref. [1].

from symmetry. In this experiment, pure SiH_4 or a mixture of SiH_4 and H_2 is used as source gases and amorphous-silicon (a-Si) film deposition is used to determine the film distribution.

The results are shown in Figure 7.2a,b. The catalyzer is set at the center, $x = 0$. In this experiment, the flow rate of SiH_4 is only 0.05 of the H_2 flow rate, resulting in a large total flow rate. In Figure 7.2a, the distribution of deposition rates for a fixed total flow rate of 42 sccm is shown, taking gas pressure as a parameter.

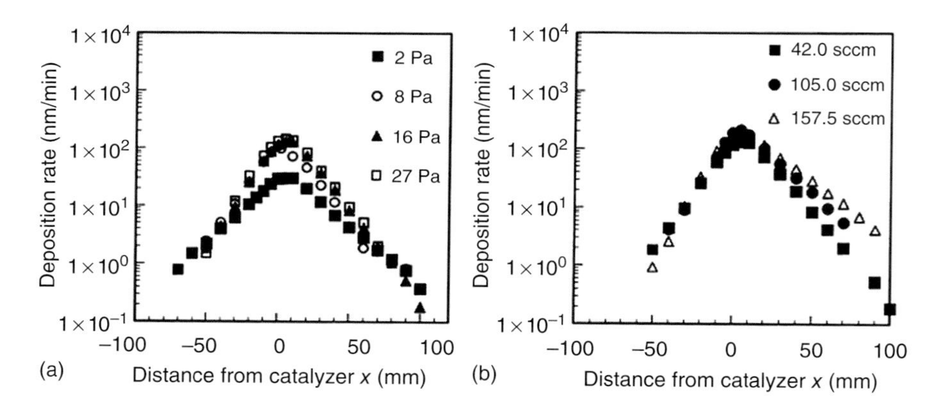

Figure 7.2 Distribution of a-Si deposition rates. At the position of $x = 0$, a catalyzer is set keeping normal to the direction of parallel gas flow. (a) The distribution when total gas flow rate of SiH_4 and H_2 mixture is fixed at 42 sccm and (b) the distribution when gas pressure is fixed at 16 Pa. Source: Reprinted with permission from Ref. [1].

The deposition rate takes a maximum at the center, and it distributes symmetrically around the center, under these experimental conditions. Figure 7.2b shows the distribution for the case of a fixed gas pressure of 16 Pa, as at P_g = 16 Pa, the distribution is symmetric, as shown in Figure 7.2a. In this case, the total flow rates are taken as a parameter. Although for the flow rate of 42 sccm the distribution is symmetric, it starts to be distorted from symmetry when the total flow rates exceed over 105 sccm. The distribution is apparently distorted from symmetry, expanding along the direction of gas flow.

The flow rate of 42 sccm in a cylinder with 5 cmØ is approximately equivalent to 1890 sccm for a 30 cmØ cylindrical chamber and 5250 sccm for a 50 cmØ chamber, although a quasi-laminar flow cannot be expected in the latter two, more common, chambers. Such large gas flow rates are never used. In addition, as already shown in Chapter 5, most of the Cat-CVD films are deposited at a gas pressure less than 10 Pa. The experiment implies that in most of the Cat-CVD conditions, the influence of gas flow is negligible. However, it should be noted that film deposition is apparently influenced in a chamber in which a gas inlet is placed near to the gas outlet, or pumping port. The conclusion above may be applicable when the gas flow introduced in the chamber does not have nonuniformity initially.

7.1.2 Dissociation Probability of SiH_4 Derived from a Cylindrical Chamber

From the experiment using a long cylindrical chamber, we can derive another important physical factor in a-Si deposition. In the experiment, all Si atoms used for film formation can be counted from the integrated volume of the distributed a-Si films. If we count the total Si atoms included in the SiH_4 gas and compare them with the total of Si atoms included in the a-Si films, we can estimate the efficiency of gas use, or gas utilization γ.

Figure 7.3 shows the profiles of the deposition rates of a-Si films vs. the position from the catalyzing wire. In this case, a-Si film is deposited on the inner wall of 5 cmØ cylindrical chamber with T_{cat} = 1850 °C, FR(SiH_4) = 2 sccm, and P_g = 0.2 Pa. The thickness of a-Si film, deposited on a long thin plate placed on

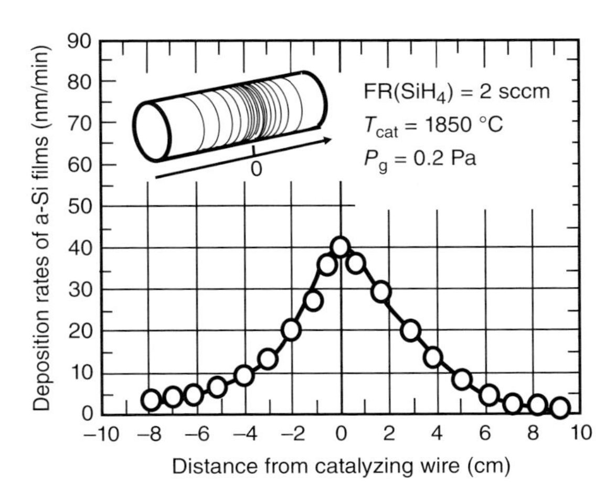

Figure 7.3 The profile of deposition rate of a-Si on cylindrical chamber wall.

the wall of chamber, is measured using a step profiler, after etching the a-Si film at particular positions. The data points in the figure correspond to the measured positions. The efficiency of gas use can be estimated from the total number of Si atoms in the a-Si film, which can be evaluated by integrating the profile.

In the case of crystalline-silicon (c-Si), the atomic density is known to be $5 \times 10^{22}\,cm^{-3}$; however, the atomic density of a-Si is usually slightly smaller because of the incorporation of H atoms. Here, we roughly assume $4.5 \times 10^{22}\,cm^{-3}$ as the Si atomic density in a-Si, taking into account the H content in a-Si. Figure 7.3 implies that the deposition rate expressing the total volume of a-Si deposited on the chamber wall per minute is $3.82 \times 10^{-4}\,cm^3/min$, and thus, the number of Si atoms deposited is 1.72×10^{19} atoms/min. As FR(SiH_4) is fixed at 2 sccm, the total number of Si atoms supplied into the Cat-CVD chamber per minute is $2\,cm^3/min \times 2.69 \times 10^{19}\,cm^{-3}$ (molecular density at 0 °C shown in Table 2.1) = 5.38×10^{19}/min. Comparing this number with the Si atoms in a-Si films, the efficiency of gas use is evaluated to be 0.32, although higher values of the efficiency of gas use have been reported for other conditions [2, 3].

The inner volume of the chamber, V_{ch}, is about 491 cm^3 and the surface area of the catalyzer, S_{cat}, is about 2.5 cm^2 in this experiment. The estimation of gas temperatures in this case has a lot of ambiguities; however, we roughly assume that the gas temperature is 77 °C (350 K), as the temperature of the cylinder wall is measured as about 70–80 °C when the catalyzing wire is heated. In this case, the density of SiH_4 molecules of $P_g = 0.2$ Pa is evaluated to be $4.14 \times 10^{13}\,cm^{-3}$ from Eq. (2.1), and their thermal velocity is 4.81×10^4 cm/s, as deduced from Eq. (2.2). As $S_{cat} = 2.5\,cm^2$, thus, the number of collisions of SiH_4 molecules with the catalyzing wire is evaluated as 1.24×10^{18} times/s.

On the other hand, as the inner volume of the chamber, V_{ch}, is 491 cm^3 ($=2.5^2 \times \pi \times 25$), the residence time of a SiH_4 molecule is estimated to be 0.029 seconds from Eq. (2.8), for FR(SiH_4) = 2 sccm and $P_g = 0.2$ Pa. As the total

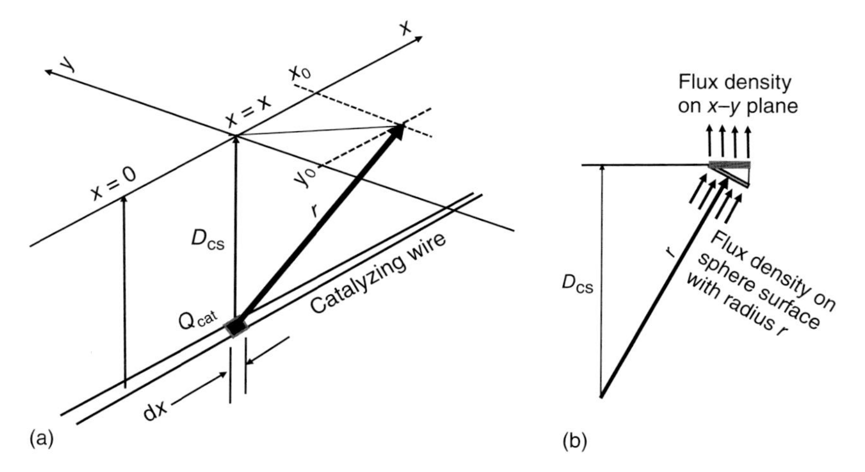

Figure 7.4 Geometrical relation between an emission point on a straight catalyzing wire and the plate of substrate. (a) Shows an overall view of geometrical relation and (b) shows the relation between the flux density on the *x*–*y* plane and that on the sphere surface.

number of molecules existing inside the chamber, 4.14×10^{13} cm^{-3} $\times$ 491 cm^3 = 2.03×10^{16}, is renewed in every 0.029 seconds, $2.03 \times 10^{16}/0.029 = 7.01 \times 10^{17}$ molecules are renewed every second. During a period of 1 second, SiH_4 molecules collide with the catalyzer 1.24×10^{18} times/s, as mentioned above. That is, a SiH_4 molecule collides with the catalyzer 1.77 times before evacuation.

If the probability of SiH_4 decomposition at the collision with a W catalyzer and also the number of collisions that a SiH_4 molecule experiences before evacuation are described by α and A, respectively, and the total efficiency of gas use, γ, has a relationship of $(1-\gamma) = (1-\alpha)^A$. A was evaluated to be 1.77 above. α is evaluated to be 0.20 from $(1-0.32) = (1-\alpha)^{1.77}$.

After the report cited above [1], there have been some systematic reports on this dissociation probability of SiH_4 for various deposition parameters [4–6]. Such reported data appear to support the result described here. The α appears to vary from 0.15 to 0.40, depending on the variation of T_{cat} from 1600 to 2000 °C. These data are summarized in Figure 7.27 later in the discussion on the use of tantalum carbide (TaC) as a new catalyzer [6].

7.2 Factors Deciding Film Uniformity

7.2.1 Equation Expressing the Geometrical Relation Between Catalyzer and Substrates

As mentioned above, when the source gases introduced into the deposition chamber have a quasi-laminar flow, the film uniformity is not decided by the gas flow. This is the case under ordinary Cat-CVD conditions. From many experiments and reports, it is known that the film uniformity is simply decided by the geometrical distances between heated catalyzing wires and substrates [7–9].

Species decomposed at the catalyzer surface are emitted into the space in all directions, and during their travel toward the substrate, they may react in the gas phase to form the final precursors used for deposition at the substrate. As the gas phase reactions are isotropic, the species generated in the gas phase reaction may not be affected to the geometrical relation between the catalyzing wire and the substrate. However, in most of the experiments, the uniformity estimated from the geometrical relation is a good measure for prediction of film uniformity in advance.

Here, let us consider the geometrical relation of a catalyzing wire with a substrate on the x-y plane, as shown in Figure 7.4. From a point of a small-length dx on a straight and long catalyzing wire, a quantity of species of $Q_{cat}\,dx$ is emitted evenly distributed over all directions. When the x–y plane is considered at a distance D_{cs} from such a point, the flux of radiated species used for film deposition is estimated to be according to Eq. (7.1), providing that the length of a wire is $2L$ and a middle of the wire is set at $x = 0$.

$$\text{Flux density at } (x_0, y_0) = \int_{-L}^{L} Q_{cat}\,dx \times \frac{1}{4\pi r^2} \times \frac{D_{cs}}{r} dx$$
$$= \frac{Q_{cat} D_{cs}}{4\pi} \int_{-L}^{L} \frac{dx}{\left\{\{x - x_0\}^2 + y_0^2 + D_{cs}^2\right\}^{3/2}}. \quad (7.1)$$

The flux density of species forming the film is described equivalently to the density of radiated energy, resulting in a similar equation. If we assume an infinite length of the wire, $L = \infty$, the flux density becomes proportional to the inverse of the square of r, as shown in Eq. (7.2).

$$\text{Flux density (at } y = y_0) = \frac{1}{y_0^2 + D_{cs}^2} \tag{7.2}$$

This equation shows the profile of flux density or the profile of film thickness when a long enough catalyzing wire along the x-axis is considered. Of course, near the end of the catalyzing wire, we have to perform more precise calculations. However, to a first-order approximation, this simple expression can be used to estimate the film uniformity. To know the profile of film thickness when many wires are placed in parallel, the flux density is summed along the y-axis.

7.2.2 Example of Estimation of Uniformity of Film Thickness

From the above discussions, it is clear that we can simply deduce the thickness profile under the assumption that the film thickness is simply proportional to the flux density of precursors. Eq. (7.2) tells us that if two long straight catalyzing wires are mounted parallel to the x-axis keeping the distance of the catalyzing wires at D_{cc}, we can estimate the flux density of film precursors at any point on the substrate as shown in Eq. (7.3).

$$\text{Flux density } (y = y_0) = \frac{1}{\left(y_0 - \frac{D_{cc}}{2}\right)^2 + D_{cs}^2} + \frac{1}{\left(y_0 + \frac{D_{cc}}{2}\right)^2 + D_{cs}^2}. \tag{7.3}$$

Figure 7.5 shows the theoretically estimated distribution of film thickness under the assumption of straight and infinite length of catalyzing wires along with experimental results of the distribution of deposited micro-crystalline silicon (μc-Si) films. The theoretically estimated distribution of film thickness is equivalent to the theoretically estimated distribution of flux of deposition species. Theoretical results are expressed with broken curves, and experimental results are denoted with closed circles and open quadrangles and solid curves as a guide to the eye. In both theoretical and experimental cases, D_{cs} is fixed at 4 cm, and the case of two different D_{cc}'s of 4 and 5 cm is also included. In the experimental case, the length of two parallel wires is 12 cm each. In the experiment, μc-Si films are deposited with $T_{cat} = 1800\,°C$ and $P_g = 30$ Pa from a mixture of SiH_4 and H_2. The figure is drawn by using the data shown in Ref. [8]. It is demonstrated that uniform deposition is possible by adjusting the geometrical position of the catalyzing wires with respect to the substrates. Very roughly speaking, when D_{cc} is equivalent to or about 10–20% larger than D_{cs}, the thickness fluctuation can be less than 5% along the direction normal to the catalyzing wire.

In this calculation, we simply estimate the uniformity of the thickness by estimating the flux density of species generated on the catalyzer. This is adequate in

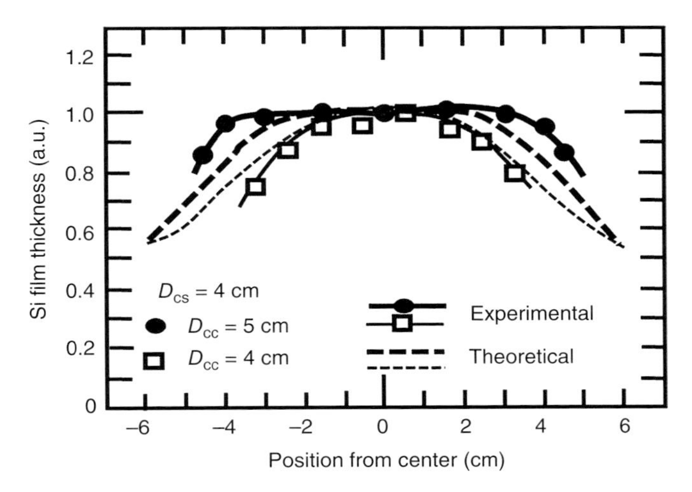

Figure 7.5 Theoretical estimation of the distribution of film thickness (or the distribution of flux density), denoted by broken curves, and experimental results for the distribution of deposited Si film thickness, denoted by solid curves as an eye guide and closed circles and open quadrangles for $D_{cc} = 4$ or 5 cm and for $D_{cs} = 4$ cm. Source: Figure is newly drawn by using the data shown in Ref. [8].

the case of deposition of a single-element film such as a-Si or μc-Si. However, in the case of compound materials such as SiN_x, the situation becomes a little bit more complicated. As explained in Chapter 4, NH_3 is decomposed to $NH_2 + H$ on the W catalyzer and other species such as NH and N are produced through the reactions in the gas phase. Of course, SiH_3, which is a key precursor for a-Si deposition, is also produced in the gas phase through the reaction between SiH_4 and H; however, the mixing ratio of N-related species to Si-related species changes depending on the distance between the catalyzer and the particular position on the substrate. Thus, exactly speaking, the film quality should depend on the distance between the catalyzer and the substrate. That is, the quality of SiN_x films is a mix of qualities, and thus, an averaged quality results. However, taking this into consideration, the above method for estimating the uniformity can still be used as a first order approximation.

7.3 Limit of Packing Density of Catalyzing Wires

When the total surface area of a catalyzer is increased by increasing the packing density of catalyzing wires, the deposition rates usually increase, although the influence of thermal radiation increases too. In this case, it is interesting to address the question what is the limit of packing density of catalyzing wires, ignoring the effects of an increase of thermal radiation? Simple speculation leads to the conclusion that if the distance between parallel wires is shorter than the mean free path of source gases, the increase of packing density of wires does not directly work to increase the deposition rate proportionally.

Here, we like to consider the flow of molecules or species near to the catalyzer surface. In Chapter 2, we have already explained that the number of collisions with catalyzing wires can be expressed by Eq. (2.6), using the density of gas molecules, n_g, and their thermal velocity. In the derivation of Eq. (2.6), we did not take into account the collisions in the gas phase. That is, the equation tells us the phenomena occurring at the areas just near to the surface of the catalyzer, within the mean free path. If all molecules recoil from the surface of the catalyzer without receiving any additional energy, the number of molecules does not change and there is no particular fluctuation in their distribution. However, actually, after colliding with the catalyzer, molecules are decomposed with the probability of α, and the residual nondecomposed molecules get the energy from the heated catalyzer and thus usually accelerate to a higher thermal velocity.

For instance, when we consider SiH_4 molecules and T_{cat} =1850 °C, about 20% of the SiH_4 molecules hitting the surface of the catalyzer are decomposed, but the other 80% of the SiH_4 molecules recoil while receiving an energy equivalent to the temperature of 1850 °C, although the gas temperature of SiH_4 molecules before collision with the catalyzer is in thermal equilibrium with the temperature of the chamber wall. In addition, decomposed SiH_4 molecules generate Si and H atoms with an energy equivalent to about 1000 °C, as explained in Chapter 4. The mean free path of original SiH_4 molecules is about 0.677–1.18 cm, as seen in Table 2.4. This is, say, about 1 cm, while that of recoiled SiH_4 molecules is about 4–5 cm, that of H atoms with 1000 °C is about 35 cm, and that of Si atoms is about 7 cm at P_g = 1 Pa. Considering the influence of recoiled species, the space within a distance of several centimeters from the catalyzer, the mean free path of recoiled species, is not an equilibrium space. Even if a multitude of catalyzing wires is confined within this space, the surface area should not be simply added.

The mathematical approach to exactly determine the nonequilibrium area has not been developed; however, the mean free path is a good measure for this discussion.

7.4 Thermal Radiation from a Heated Catalyzer

7.4.1 Fundamentals of Thermal Radiation

One of the most critical aspects in Cat-CVD may be the influence of thermal radiation from a heated catalyzer. In the initial stage of Cat-CVD research, when we reported that the W catalyzing wire was heated up to 1800 °C for a-Si deposition, people expressed concerns about keeping the real substrate temperatures at 200 or 300 °C under the strong thermal radiation. The total energy $Q_{rad\ total}$ emitted by thermal radiation from a heated catalyzer with a temperature T_{cat} and a surface area S_{cat} is expressed by Eq. (7.4):

$$Q_{rad\ total} = \varepsilon_{em} S_{cat} \sigma T_{cat}^{\ 4} \tag{7.4}$$

where ε_{em} and σ are an emissivity of the catalyzing material and the Stefan Boltzmann constant, respectively, and T_{cat} is the absolute temperature. The Stefan Boltzmann constant σ is a proportionality constant for radiation emitted from

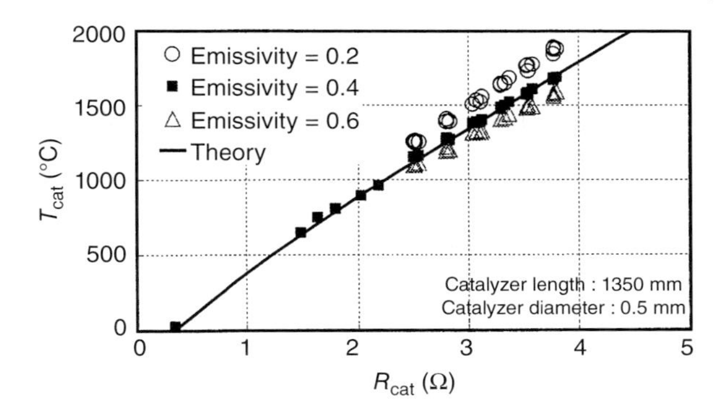

Figure 7.6 The relation between catalyzer temperature T_{cat} and the resistance R_{cat}, taking emissivity of W wire as a fitting parameter.

a black body. It is about 5.67×10^{-8} W/m^2 K^4. However, as actual materials are not perfect black bodies, the radiation is adjusted by a number between 0 and 1. This number is called the emissivity of the material. The emissivity depends on the nature of the surface, the temperature of the material, and the wavelength of emitted radiation. In the case of a tungsten wire, for $T_{cat} = 1800\,°C$, the emissivity roughly varies from about 0.46 for the wavelength of 500 nm to 0.38 for 1000 nm [10].

Figure 7.6 shows the relation between the estimated temperature T_{cat} of a W wire of a length of 1350 mm and a diameter of 0.5 mmØ and its resistance R_{cat}. The resistivity of a metal wire depends on its temperature, and thus, the temperature of such a metal wire can be estimated from the value of the resistance of the wire. This property is often used for monitoring and controlling the temperature of the catalyzer in Cat-CVD. The surface temperature of the catalyzer can also be known from the measurement by a radiation thermometer using an appropriate value for the emissivity of the metal wire. In the figure, the measured values are also plotted for various emissivity values. By fitting a solid curve to the data, the emissivity of a W wire is derived to be about 0.4 under the most common Cat-CVD conditions.

7.4.2 Control of Substrate Temperatures in Thermal Radiation

The density of radiation follows the same relation as for the flux density of species that was used to derive the film thickness uniformity. In practice, the radiation emitted toward the substrate is an important figure. Here, the radiation is described as Q_{rad}.

In Cat-CVD, sample substrates are usually placed on a substrate holder and the temperature of the substrate holder is usually controlled by heaters installed inside the holder or by coolant flow inside it. When the samples are mounted to the substrate holder, the difference between the temperature of the sample and that of the substrate holder depends on the heat transfer coefficient Λ. When radiation Q_{rad} is incident on the sample, the temperature of the sample T_s and the temperature of the substrate holder T_{holder} have a relation expressed by Eq. (7.5).

$$Q_{rad} = \Lambda\,(T_s - T_{holder}). \tag{7.5}$$

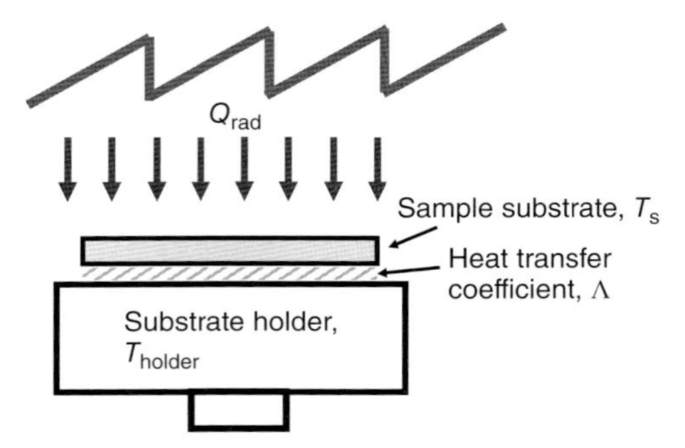

Figure 7.7 Schematic illustration of thermal radiation from catalyzing heaters.

The situation is schematically illustrated in Figure 7.7. When the radiation is large, the difference between the temperature of the sample and that of the holder becomes large, if Λ is constant. However, even in this case, the temperature difference can be lowered, if Λ is large enough.

Figure 7.8 shows the experimental results for measurement of both T_s of a Si wafer and T_{holder} of a stainless steel substrate holder. T_s is measured by inserting a thermocouple in the Si wafer. Initially, the Si wafer is heated to 300 °C by the heater inside the holder [11]. At that time, T_s is almost equal to T_{holder}. However, when the catalyzer is switched on, the Si wafer starts to absorb the radiation from the catalyzer and T_s quickly jumps up and saturates at a value of about 380 °C. The increase in temperature of about 80 °C is caused by the radiation. Then, here, when T_{holder} starts to decrease, T_s also starts to decrease, while the temperature difference between T_{holder} and T_s stays constant at 80 °C. This experiment clearly confirms the validity of Eq. (7.5).

In principle, if T_{holder} can be cooled down to 0 °C, deposition at T_s less than 100 °C is possible, even under strong radiation. Actually, as mentioned later in Chapter 8, the film coating on organic devices that cannot survive at temperatures over 100–120 °C can be successfully performed by cooling the substrate holder to −20 °C by using circulating coolant.

When Λ can be kept large, the control of T_s becomes even easier under strong thermal radiation. As a method to increase Λ, electrostatic chucking (ESC) is a widely used method. In ESC, the electrodes of metal plates are embedded in insulators and wafers are placed on such an insulating plate. When a relatively

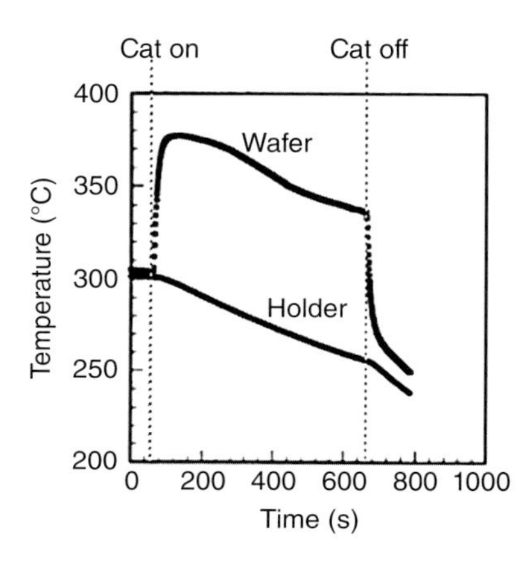

Figure 7.8 Variation of the temperatures of Si wafers and the substrate holder when the catalyzing wires are turned on or off. Source: Karasawa et al. 2001 [11]. Reprinted with permission of Elsevier.

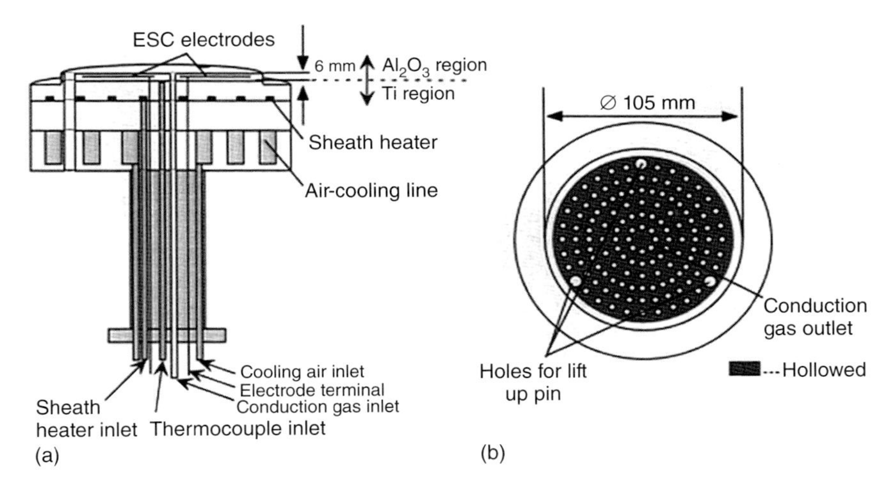

Figure 7.9 Schematic illustration of electrostatic chucking (ESC) systems. (a) Side cross-sectional view and (b) top plan view. Source: Karasawa et al. 2001 [11]. Reprinted with permission of Elsevier.

high voltage of around several 100 V is applied to the metal electrodes, an electrostatic force is induced between the wafer and the insulating substrate holder. The structure of ESC is schematically illustrated in Figure 7.9. In addition, in the small gap that still remains between the wafer and the substrate holder, a gas is introduced and the gap is filled by the gas. In this way, the thermal conductance between the wafer and the substrate holder is increased. As such a gas helium (He) is best for its high thermal conductivity. However, because of cost issues, H_2 or N_2 are also used for this purpose.

Figure 7.10 clearly demonstrates the evidence for the effect of this ESC system. In the figure, a Si wafer and a GaAs wafer are set on the ESC substrate holder. The results for the Si wafer are shown in Figure 7.10a and those for the GaAs wafer in Figure 7.10b. In the experiment, to clearly demonstrate the effect of ESC, the

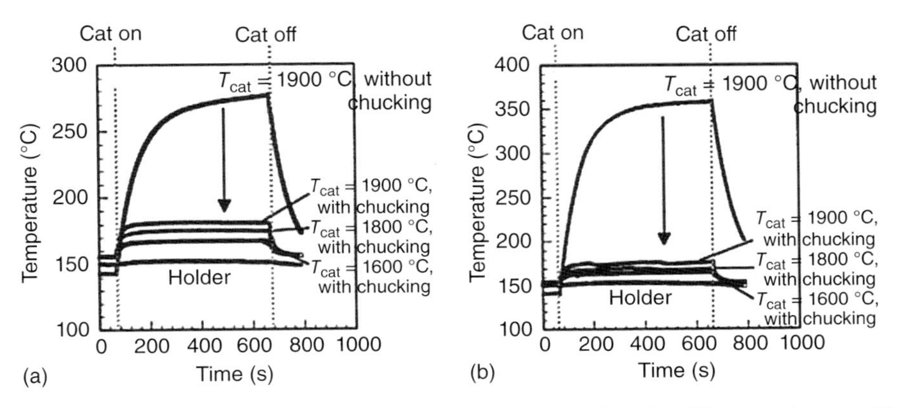

Figure 7.10 Temperatures of Si wafer (a) and GaAs wafer (b) when the ESC system is on or off. The case when the catalyzer temperature, T_{cat}, is 1600, 1800, or 2000 °C is demonstrated. Source: Karasawa et al. 2001 [11]. Reprinted with permission of Elsevier.

distance between the catalyzer and the substrate is kept at only 3 cm. The effect of the ESC system is apparent. Without ESC, when the catalyzer is heated, T_s jumps to 270 °C from initial 150 °C for the Si wafer and to 350 °C from 150 °C for the GaAs wafer by radiation from the catalyzer. However, when the ESC system turns on with a H_2 gas flow of several tens standard cubic centimeters per minute (sccm), the temperature difference between T_s and T_{holder} drops to only 30 °C for the Si wafer and to about 20 °C for the GaAs wafer.

The ESC system is a powerful tool, and it is used in mass production machines for dry-etching, as the radio frequency (RF) power is increased to increase the etching speed, and thus, the elevation of the temperature of samples makes it a serious challenge to keep the photoresist behaving normally. However, the ESC system sometimes has problems. The elimination of electrostatic force from wafers sometimes is a problem, as the elimination takes time, and as a final operation, the mechanical removal of wafers from the ESC substrate holder using pins sometimes becomes necessary.

Without this ESC system, if we cool down the substrate holder sufficiently by a coolant, low T_s deposition can still be carried out.

7.4.3 Thermal Radiation in CVD Systems

In the previous section, we concentrated our discussion on the influence of thermal radiation from the heated catalyzer. However, the surface area of the catalyzer is limited to, may be, around 50 cm^2 in the case of the chamber shown in Figure 4.2. In the chamber, the substrate holder with a diameter of 30 cm is set at the center of a cylindrical Cat-CVD chamber with an inner diameter of 50 cm and a height of 50 cm. When the catalyzer turns on and the film deposition is carried out continuously, the temperature of the chamber wall gradually increases because of the thermal radiation from the heated catalyzer. According to our system shown in Figure 4.2, after continuous deposition for more than 30 minutes keeping $T_{hloder} = 250$ °C and $T_{cat} = 1800$ °C, the temperature of the chamber wall approaches about 80 °C and saturates at that temperature.

If the top and side walls of the chamber are heated to 80 °C, and if the bottom of the chamber is replaced by a 30 cmØ substrate holder of 250 °C, the total thermal radiation emitted from the 80 °C chamber wall and 250 °C substrate holder is $\{(25\ \text{cm} \times 25\ \text{cm} \times \pi) + (50\ \text{cm diameter} \times \pi \times 50\ \text{cm height})\} \times \varepsilon_{em\text{-}SUS} \times \sigma \times (353\ \text{K})^4 + (15\ \text{cm} \times 15\ \text{cm} \times \pi) \times \varepsilon_{em\text{-}SUS} \times \sigma \times (523\ \text{K})^4 = 2.05 \times 10^{14}\ \text{cm}^2\ \text{K}^4 \times \varepsilon_{em\text{-}SUS} \times \sigma$. On the other hand, when the catalyzer with $S_{cat} = 50\ \text{cm}^2$ is heated to 1800 °C, similar radiation becomes $50\ \text{cm}^2 \times \varepsilon_{em\text{-}W} \times \sigma \times (2073\ \text{K})^4 = 9.23 \times 10^{14}\ \text{cm}^2\ \text{K}^4 \times \varepsilon_{em\text{-}W} \times \sigma$. The emissivity of stainless steel (SUS), $\varepsilon_{em\text{-}SUS}$, which is the chamber material, is known as approximately 0.8 when it is oxidized as usual [12], but the emissivity of W, $\varepsilon_{em\text{-}W}$, is about 0.4 as mentioned in Section 7.4.1. The total emission from the chamber is $1.6 \times 10^{14}\ \text{cm}^2\ \text{K}^4 \times \sigma$, and the total emission from the heated catalyzer is $3.7 \times 10^{14}\ \text{cm}^2\ \text{K}^4 \times \sigma$. In addition, only half of the catalyzer is facing the substrate, and thus the total radiation that the samples on the substrate holder receive from the chamber is about $1.6 \times 10^{14}\ \text{cm}^2\ \text{K}^4 \times \sigma$, and the total radiation that the sample receives from the catalyzer is about $1.9 \times 10^{14}\ \text{cm}^2\ \text{K}^4 \times \sigma$.

Surprisingly, the thermal radiation from the heated catalyzer is actually almost equal to that emitted from the chamber. The samples set on a similar substrate holder in plasma enhanced chemical vapor deposition (PECVD) also receive a similar amount of thermal radiation from the chamber and the substrate holder. That is, the thermal radiation in Cat-CVD is not particularly serious compared with PECVD. Actually, as introduced above, the dry-etching system using a plasma requires a cooling mechanism in their machine. The temperature elevation in PECVD is sometimes a serious issue to overcome.

7.5 Contamination from a Heated Catalyzer

7.5.1 Contamination of Catalyzing Materials

In the initial stage of Cat-CVD research, possible contamination originating from the heated catalyzing metal wires was one of the issues of concern. When we deposited SiN_x films as passivation films on GaAs devices by Cat-CVD using a W catalyzer, the expected contamination from the heated catalyzer became the first bottleneck toward acceptance of Cat-CVD for SiN_x deposition. At that time, engineers and scientists working on semiconductor devices were extremely careful to avoid pollution of their test line from unknown levels of contamination.

Therefore, the W contamination in Cat-CVD a-Si from the heated W catalyzer itself was checked by secondary ion mass spectrometry (SIMS), Rutherford backscattering (RBS), and total reflection X-ray fluorescence (TXRF). Figure 7.11

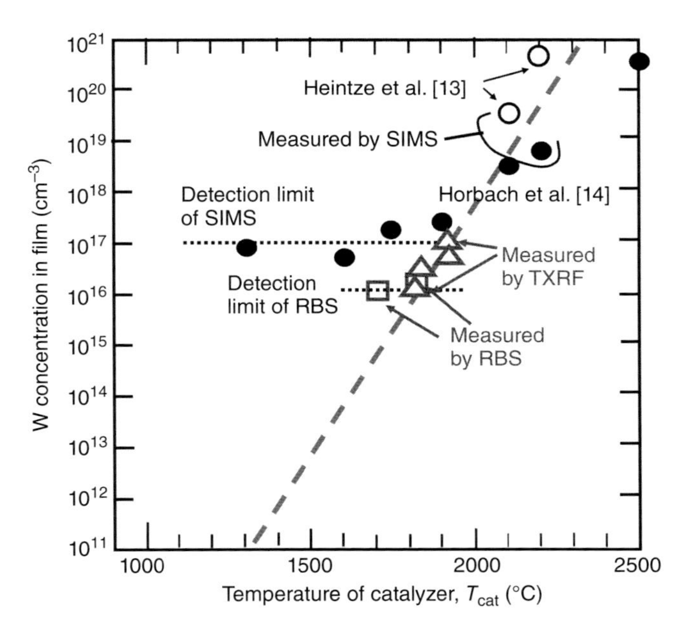

Figure 7.11 W concentration unintentionally incorporated in a-Si films, as a function of temperature of catalyzer T_{cat}. Source: Matsumura et al. 2014 [13]. Reprinted with permission of The Japan Society of Applied Physics.

shows the W density incorporated in a-Si films as a function of temperature of the catalyzer T_{cat}, based on the reports of SIMS analysis by Heintze et al. [14] and Horbach et al. [15] and also based on our RBS and TXRF data [13].

Here, SIMS is a method to detect impurities contained in the film by observing the mass number of emitted secondary ions of impurities from the film when probe ions are incident onto the film. The detection limit of this method depends on the environment surrounding the detected impurities and usually is 10^{17} cm^{-3} for W atoms in a-Si. TXRF is a method to detect impurities by observing the characteristic X-rays of impurities when high-energy X-rays used for excitation are directed at the surface of the film under a glancing angle.

The measured values in TXRF are all expressed as intensity/unit area, say, intensity/cm^2. Usually, the penetration depth of such glancing X-rays is about 10 nm or less in Si films. The concentration of detected impurities is evaluated by dividing the observed intensity by 10 nm. The detection limit is usually around 10^{16} cm^{-3} or less.

The RBS is a method detecting heavy impurities embedded in films composed of relatively light elements. The mass number of impurities is known by observing the energy of backscattered probe ions such as mega electron volt high-energy He ions, as incident probe ions are backscattered by elastic collisions with heavy impurity atoms and the energy of the backscattered ions depends on the mass number of the impurity. W atoms are much heavier than Si atoms, and the detection limit of W atoms in a-Si is therefore as low as 10^{16} cm^{-3}.

In Figure 7.11, a-Si films were prepared under various conditions. For instance, the deposition rates varied from 0.5 to 4.0 nm/s, T_s was 150–400 °C, P_g 1–5 Pa, D_{cs} 1–5 cm, and the area of catalyzer S_{cat} was also different in each experiment. Nevertheless, when the W concentration is plotted as a function of $1/T_{cat}$, all data are approximately on the same line. This means that the ratio between the flux density of deposition species of a-Si and that of W contaminants for a certain fixed T_{cat} is almost constant at the substrate surface in all investigated cases. For instance, when T_{cat} is fixed at 2000 °C, from Figure 7.11, it is seen that the W concentration is about 10^{17} cm^{-3}. Thus, as the Si atomic density in a-Si is very roughly approximated as the same as that of c-Si, 5×10^{22} cm^{-3}, the deposition species of a-Si always includes 2 ppm W contaminants for $T_{cat} = 2000$ °C, and they include 0.2 ppm W contaminants for $T_{cat} = 1800$ °C.

The figure demonstrates that W contamination is likely to increase as T_{cat} increases; however, it also shows that, when T_{cat} is kept at 1800 °C or less, the flux density of W contaminants in the flux of deposition species can be kept low and the W concentration in the films can be suppressed to less than the detection limit. That is, if T_{cat} is 1800 °C or less, W contamination is negligible in practice.

7.5.2 Contamination from Other Impurities

When we deposited SiN_x films as passivation films on GaAs devices by Cat-CVD using the W catalyzer, the contamination from the heated catalyzer into GaAs was also a concern. Although W contamination is negligible as mentioned above, aren't there contaminations by other impurities? We immediately measured

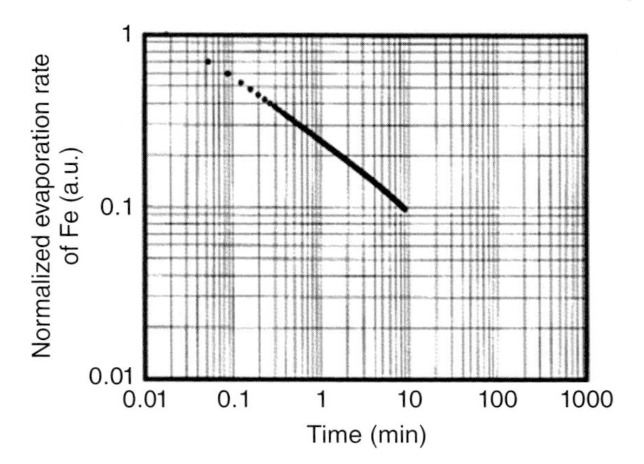

Figure 7.12 Simulation of Fe evaporation from W filament of 0.5 mmØ as a function of annealing time. Source: Ishibashi 2001 [16]. Reprinted with permission of Elsevier.

impurity contamination for various impurities in Cat-CVD SiN_x films. As results, other impurities such as iron (Fe), nickel (Ni), and chromium (Cr) were detected at the level of 10^{17} cm^{-3}. These metals are typical components of stainless steel (SUS), which is used in the Cat-CVD chamber. At first, we suspected the SUS chamber to be an origin of the contamination.

However, soon we understood that all contaminations were impurities present in the W wires. At that time, the W wires that were provided were filaments of light bulbs. A purity level equivalent to semiconductor devices is not required for such filaments, and therefore, minimal care to reduce the impurity contamination was given. Fortunately, all contaminating materials could diffuse out of the wires by evaporation at high temperatures, but lower than the melting point of the W wires.

Figure 7.12 shows the simulated result of Fe concentration incorporated in W wires during heating at 2500 °C [16]. Fe is out-diffused quickly from W wires with a diameter of 0.5 mm. To check this, the impurity concentration in Cat-CVD SiN_x films, which were prepared with W catalyzers on GaAs wafers, was measured by SIMS. The results for Fe, Ni, and Cr are shown in Figure 7.13a–c, respectively [16]. After seven minutes of baking the W catalyzing wire with a diameter of 0.5 mm, the impurity concentration could be reduced by 1–2 orders of magnitude. In this measurement, the detection limit of SIMS appears better than in the case of measurement in a-Si. It is confirmed that the initial concentration of Fe, Ni, and Cr at about 10^{17}–10^{18} cm^{-3} can be decreased to 10^{15}–10^{16} cm^{-3} by a short time baking period of the W catalyzer.

Table 7.1 shows the data of contaminants in the conventional W wires and in the carefully made high-purity W wires, which were provided by Allied Materials Corp. Japan. Until the research of Cat-CVD started, the main market of W wires had been the filaments used in electric light bulbs, and the inclusion of contaminants had been positively accepted as they help to keep the shape of filaments unchanged.

Although pure W wires can be obtained, even conventional W wires can also be used, provided that we bake them in the Cat-CVD chamber with H_2 gas at about 10–100 Pa and a temperature of 2000–2100 °C, which is 200–300 °C higher than the catalyzer temperatures used during the deposition process. Because of the

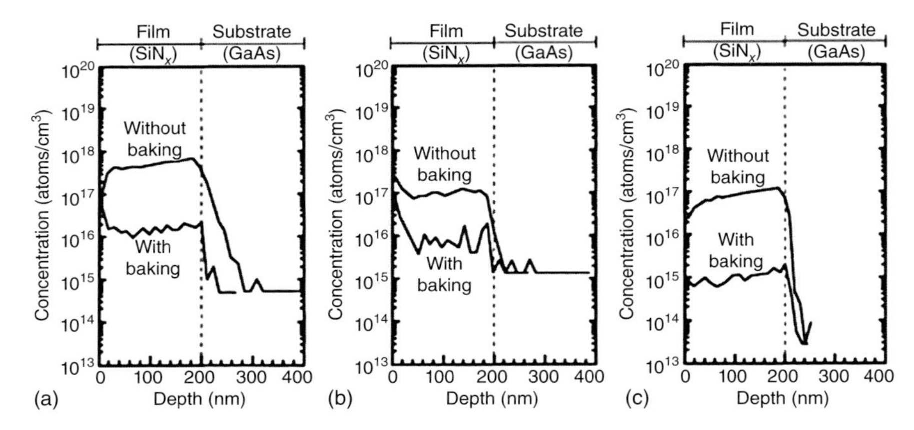

Figure 7.13 SIMS profiles of (a) Fe, (b) Ni, and (c) Cr in Cat-CVD SiN_x films deposited on GaAs before and after baking of catalyzing W wires. Source: Ishibashi 2001 [16]. Reprinted with permission of Elsevier.

Table 7.1 Impurities included in the conventional W wires and in the high-purity W wires.

	Contents (ppm)	
Elements	**Conventional W**	**Purified W**
Al	9	<2
C		<50
Ca	1	<2
Cr	4	<2
Cu	<1	<2
Fe	17	<3
K	52	<2
Mg	<1	<2
Mn	1	<2
Mo	8	<2
N		<30
Na		<2
Ni	1	<2
O		<50
Si	<5	<2
Sn	<2	<2
Th	Not detected	<1
U		<1

pre-bake procedure, the contamination in the deposited films by the impurities in the W catalyzer can be suppressed. Only one thing to take care of is the fact that W metal grains are likely to grow at these high annealing temperatures, making the wire fragile, because of the reduction of the number of complicated grain boundaries and the concomitant simplification of grain boundary lines.

7.5.3 Flux Density of Impurities Emitted from Heated Catalyzers

Using purified W wires, SiN_x films were deposited on 40 Si wafers, and the contamination for all 40 samples was checked by the TXRF method. T_{cat} was 1800 °C, D_{cs} = 4 cm, and S_{cat} = 88 cm^2 to reach easily detectable contamination levels. The actual parameters for making device quality SiN_x are much more tailored to reduce the contamination, as explained in Chapter 5. The results are demonstrated in Figure 7.14 [13]. Only a small amount of contamination by metals such as W, Fe, vanadium (V), Ni, manganese (Mn), zinc (Zn), Cr, copper (Cu), and titanium (Ti) could be detected, whereas others were not detected. There is some fluctuation in the data; however, the amount is always less than 6×10^{10} atoms/cm^2. In TXRF, the measured depth is believed to be about 10 nm because of the penetration depth of X-rays with glancing incident angle. Thus, the value is equivalent to a concentration of about 6×10^{16} atoms/cm^3, which appears almost equivalent to the concentration shown in Figure 7.11 for the case of a-Si. This level of contamination is sometimes observed in other processes such as PECVD, and thus, the contamination cannot be considered a drawback of Cat-CVD.

The atomic density of Cat-CVD SiN_x is relatively high, at about 2.8 g/cm^3 as shown in Figure 5.41. Assuming stoichiometric Si_3N_4, the total atomic density of the sum of Si and N atoms is estimated to be 8.4×10^{22} cm^{-3}. If we also assume

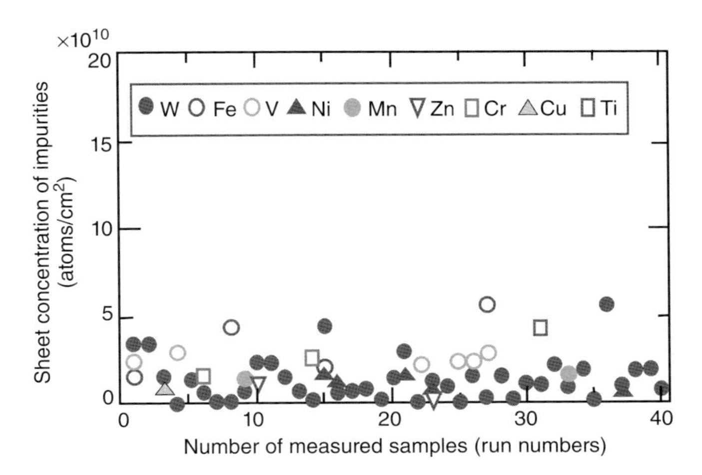

Figure 7.14 Sheet concentration of all impurities in Cat-CVD SiN_x, detected by TXRF measurement for 40 samples. Source: Matsumura et al. 2014 [13]. Reprinted with permission of The Japan Society of Applied Physics.

that all contaminants are included in the flux of deposition species of SiN_x films at the same ratio of the concentration of contaminants in the films, the flux includes 0.7 ppm of contaminants at its maximum, at $T_{cat} = 1800\,°C$. The value appears slightly larger than that for a-Si deposition. However, if we take an average concentration of impurities for the 40 samples of Figure 7.14, the concentration of contaminants is $2 \times 10^{16}\,cm^{-3}$, and the flux of contaminants becomes 0.24 ppm, reaching the value of a-Si.

The real flux density of contaminants is also discussed in Chapter 9, when we study the application of the radical flux generated in Cat-CVD.

7.6 Lifetime of Catalyzing Wires and Techniques to Expand Their Lifetimes

7.6.1 Introduction

One of the most serious problems remaining in Cat-CVD is the lifetime of the catalyzing wires. According to textbooks, the catalyzer should not be involved in the chemical product but only help to promote chemical reactions without undergoing any change itself. This is the definition of an ideal case of a catalytic process. In practice, there have not been any catalyzers that do not change in the process at all. All catalyzers are ultimately facing inactiveness, by degradation of the surface of the catalyzer because of contamination. If a catalyzer is used in organic synthesis (e.g. in initiated chemical vapor deposition, *i*CVD), care must be taken to avoid carburization, and in mass production systems of organic materials, the catalyzer has to be refreshed every several thousand hours. The extension of the life of the catalyzer is still a serious research target.

In Cat-CVD, the situation is more serious, as in semiconductor industry, we are using gases that are more reactive, such as SiH_4. As already explained in Chapter 4, SiH_4 is decomposed to Si + 4H on heated W surface, but it requires higher temperatures to desorb Si atoms from Si—W bonds. If the temperature is not high enough, the surface of W gradually becomes covered with W-silicide. This silicide formation decides the lifetime of the catalyzer. There have been some illustrating reports on the formation of this W-silicide and on the study of the composition of these silicides [17]. However, here, we first start with an explanation of the lifetime of the W catalyzer when silicide formation occurs, based on the report by K. Honda et al., as they have studied the silicidation phenomena systematically [18].

7.6.2 Silicide Formation of W Catalyzer

Figure 7.15 shows the specially designed experimental apparatus for studying the lifetime of W catalyzers. A tungsten wire with a length of 66 cm and a diameter of 0.5 mmØ is fixed to the terminals, keeping it straight. The terminal that is also drawn in the figure has a unique structure. As the heat of a wire escapes to the electrode metal, the temperature drops in the vicinity of the terminal, and thus, in the region of the wire near to the terminal, silicide layers are easily grown. The

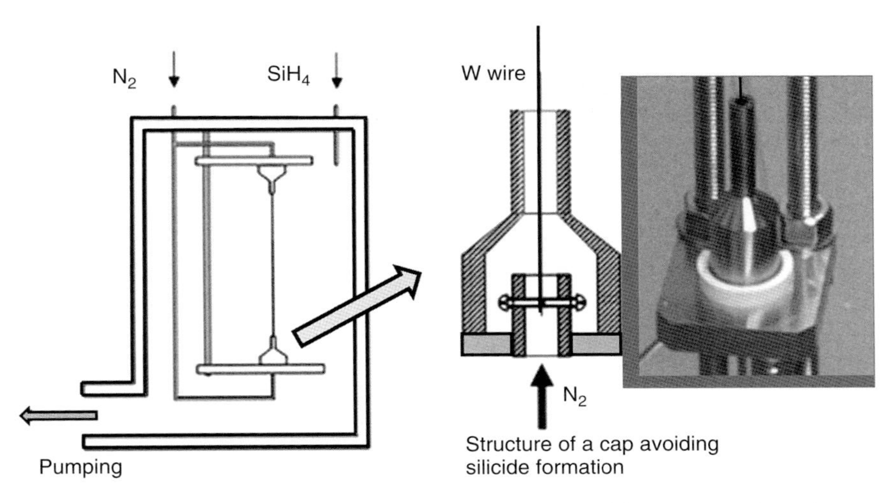

Figure 7.15 Schematic diagram of experimental apparatus and the structure of capping system to avoid silicide formation at the terminal of W wire and its photograph. Source: Reprinted with permission from Honda et al. [18]. Copyright 2008, The Japan Society of Applied Physics.

terminal has a double-capping structure. A tungsten wire is fixed to the inner terminal, and through the hole in the inner terminal, N_2 gas is introduced. This part is covered by an outer cap, and thus, the N_2 gas fills the space inside the outer cap. By this way, the SiH_4 concentration penetrating into the cap is lowered to very small values, and the actual silicidation in the vicinity of the terminal can be avoided by this system.

The effect of this terminal cap system is demonstrated in Figure 7.16. In this case, the temperature of the W catalyzing wire is kept relatively low, to induce the formation of a silicide; however, even so, the W wire inside the cap is not silicided. The system is utilized in many practical cases of Cat-CVD reactors, to extend the life of the catalyzer, as silicidation that has started at the terminal may spread to the whole wire quickly.

The formation of silicide is mainly studied by taking samples from the middle of the wire, but the results are not significantly different from those of other parts of the wire. SiH_4 gas with FR(SiH_4) = 50 sccm is introduced into the experimental chamber shown in Figure 7.15, keeping P_g at 10 Pa. After 30 minutes, the W wire is taken out and cut at the middle of the wire. The cross section of the W wire is observed by scanning electron microscopy (SEM). The results are demonstrated in Figure 7.17. Figure 7.17a shows the cross-sectional view of the original W wire and Figure 7.17b shows the cross section after exposure to SiH_4 at T_{cat} = 1850 °C. Figure 7.17c–f show the cross-sectional views after SiH_4 exposure at T_{cat} of 1750, 1650, 1550, and 1450 °C, respectively. When the W wire is heated at 1850 °C, no change is observed in the W wire after 30 minutes of exposure to the SiH_4 gas. However, for T_{cat} lower than 1750 °C, the surface of W wire changes apparently and some unknown layers are formed on the W surface. The thickness of this changed layer on the W wire is plotted as a function of the root of the exposure

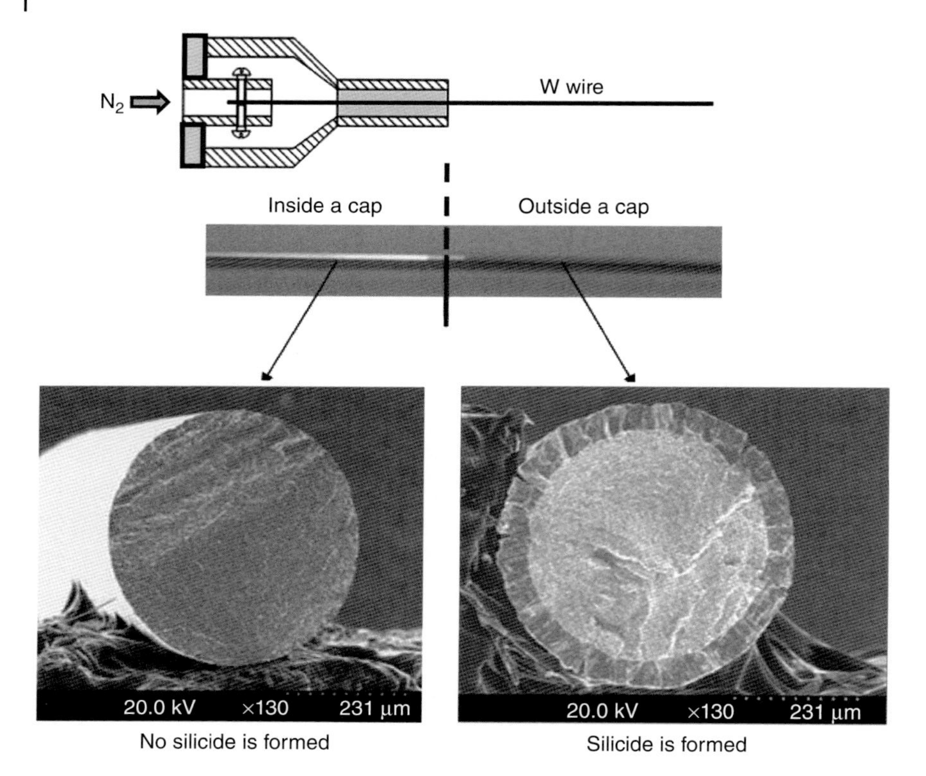

Figure 7.16 The effect of the cap system at the terminal of W wire. At the area inside the cap silicide formation is avoided. Source: Honda 2008 [19]. Reprinted with permission of IOP Publishing.

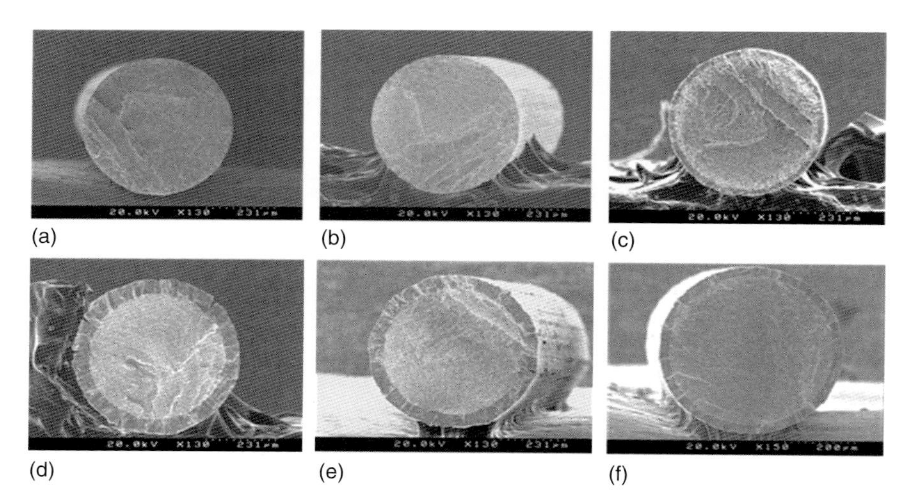

Figure 7.17 Cross-sectional SEM images of W catalyzing wires after using for 30 minutes in SiH_4 ambient of 10 Pa. The temperatures of W wires are varied from (b) at 1850 °C, (c) 1750 °C, (d) 1650 °C, (e) 1550 °C, and (f) 1450 °C. Panel (a) shows the image of W catalyzing wire before exposure. Source: Reprinted with permission from Honda et al. [18]. Copyright 2008, The Japan Society of Applied Physics.

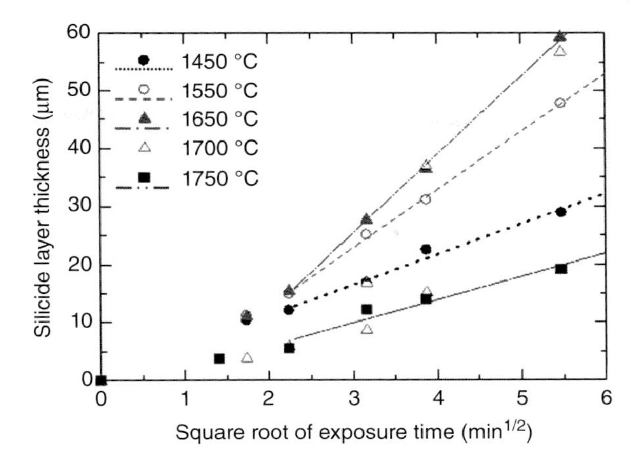

Figure 7.18 Thickness of surface layer (silicide layer) vs. square root of process times. Source: Reprinted with permission from Honda et al. [18]. Copyright 2008, The Japan Society of Applied Physics.

time. The results are shown in Figure 7.18, taking T_{cat} as a parameter. The relation is quite simple. This demonstrates that the thickness of the new formed layer is proportional to the square root of the exposure time.

What is formed on the W surface? Next, we have checked the material formed on the W wire by electron probe microanalysis (EPMA). EPMA is the method to characterize the materials detected in the SEM image, using X-rays emitted from the materials when they are hit by probe electrons. Figure 7.19 shows the cross-sectional images of the W wire exposed to SiH_4 for 30 minutes at (a) $T_{cat} = 1750\,°C$ and (b) 1450 °C. In this figure, in addition to the SEM images, the results of material analysis are demonstrated to show the components within the surface layers. It is confirmed from the figure that the surface layer consists of W-silicide and that the composition of the silicide formed at $T_{cat} = 1750\,°C$ is apparently different from that formed at $T_{cat} = 1450\,°C$. When $T_{cat} = 1450\,°C$, WSi_2 silicide is formed, but when $T_{cat} = 1750\,°C$, W_5Si_3 silicide appears to be formed, considering the data shown in the figure in combination with the phase diagram of W-silicide.

The composition of the silicide layer after 30 minutes of exposure to SiH_4 at 10 Pa is plotted as a function of T_{cat} in Figure 7.20. When T_{cat} is lower than 1650 °C, WSi_2 is formed, but when T_{cat} is higher than 1750 °C, it becomes W_5Si_3. Probably, the initial stage of silicide formation also involves W_5Si_3, which includes fewer Si than WSi_2.

Here, let us consider a growth mechanism of W-silicide on W wires based on Grove's model, which has been successfully used for the explanation of thermally grown oxide layers on Si substrates [7]. Assuming that the reaction to form a silicide takes place at the surface of the silicide or at the boundary between the silicide and the W-wire, the diffusion of species that facilitate silicide growth has to diffuse through the silicide layer already formed. In this situation, this implies for the thickness of the silicide layer that the square of $t_{silicide}$ has to be proportional to the process time $T_{process}$ as shown in Eq. (7.6).

$$(t_{silicide})^2 = BT_{process} \tag{7.6}$$

where the proportional constant, B, is often called the parabolic rate constant. If B is large, the growth speed of silicide is large, and if B is small, silicide formation

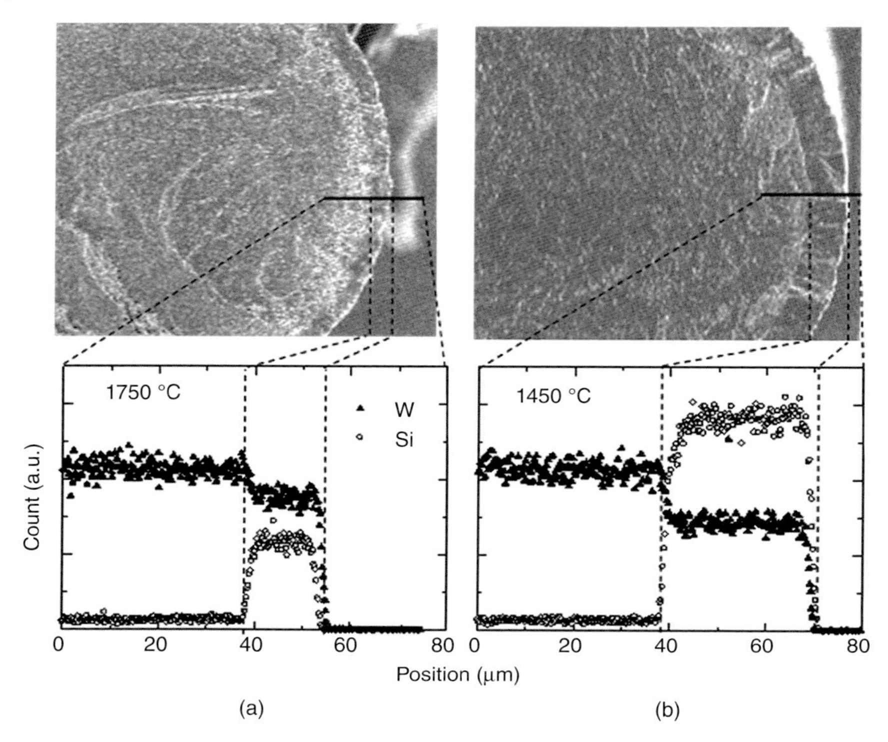

Figure 7.19 Results of EPMA (electron probe microanalysis) observation of surface of catalyzing W wires after heating at 1750 and 1450 °C. Source: Reprinted with permission from Honda et al. [18]. Copyright 2008, The Japan Society of Applied Physics.

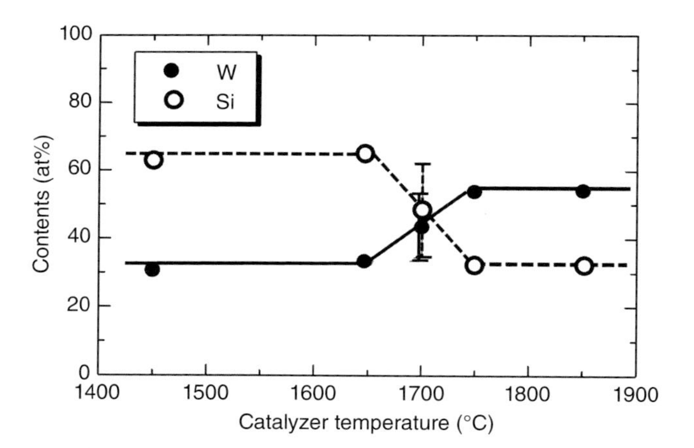

Figure 7.20 Contents in W-silicide formed on the surface of W wire as a function of catalyzer temperatures. Source: Reprinted with permission from Honda et al. [18]. Copyright 2008, The Japan Society of Applied Physics.

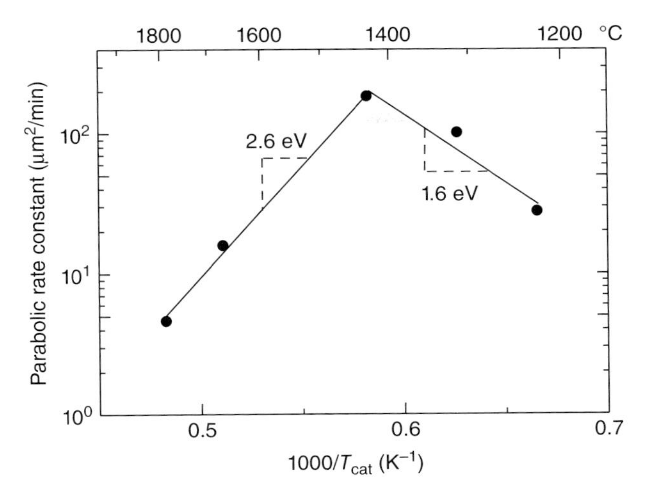

Figure 7.21 The parabolic rate constant of silicide formation as a function of reciprocal of catalyzer temperature T_{cat}. Source: Reprinted with permission from Honda et al. [18]. Copyright 2008, The Japan Society of Applied Physics.

becomes very slow. The parabolic rate constant of W-silicide is demonstrated in Figure 7.21 for formation of W_5Si_3 by keeping T_{cat} over 1850 °C.

From the figure, it is seen that when T_{cat} is 1850 °C, B is about 4 μm²/min. Therefore, it takes about 25 minutes to grow the 10-μm-thick W_5Si_3 layer. In the experiment, the conditions are the same as in the experiments mentioned above, that is, FR(SiH_4) is 50 sccm and P_g is 10 Pa. As shown in Table 4.1, P_g is usually kept at around 1 Pa to obtain device quality a-Si films at a deposition rate of 1–3 nm/s, say, 2 nm/s. When P_g is 0.1 Pa, as the number of collisions with catalyzer becomes 1/10 smaller, the growth rate of the silicide layer can be simply 1/10 of the result shown here. That is, for such conditions, B becomes 0.04 μm²/min. In this case, it takes 2500 min = 41.7 hours to grow a 10-μm-thick W_5Si_2 silicide.

Supposing the 2 nm/s deposition rate for a-Si, during this period, 0.3-mm-thick a-Si films can be deposited. If a-Si is used for fabrication of a-Si/c-Si heterojunction solar cells, as the total thickness of a-Si films on both sides of c-Si wafers required for a solar cell is at most about 50 nm, the value is equivalent to making 6000 coatings. If 100 solar cells are made at the same time on a single tray, this is equivalent to fabrication of 600 000 cells.

To use a W catalyzer for very long times, regeneration for the catalyzer is a useful approach. In the following, we discuss the removal of Si atoms from the silicide to recover a pure W catalyzer. There are various methods to remove Si atoms from a solid. When H atoms attack Si wafers, under certain conditions, Si is etched to produce volatile SiH_4 gas, as described later in Chapter 9. H atoms have the ability to remove Si atoms. Similarly, when silicide is exposed to H_2 gas at high temperatures, H_2 is decomposed to H atoms, and these H atoms are expected to remove Si atoms from the silicide.

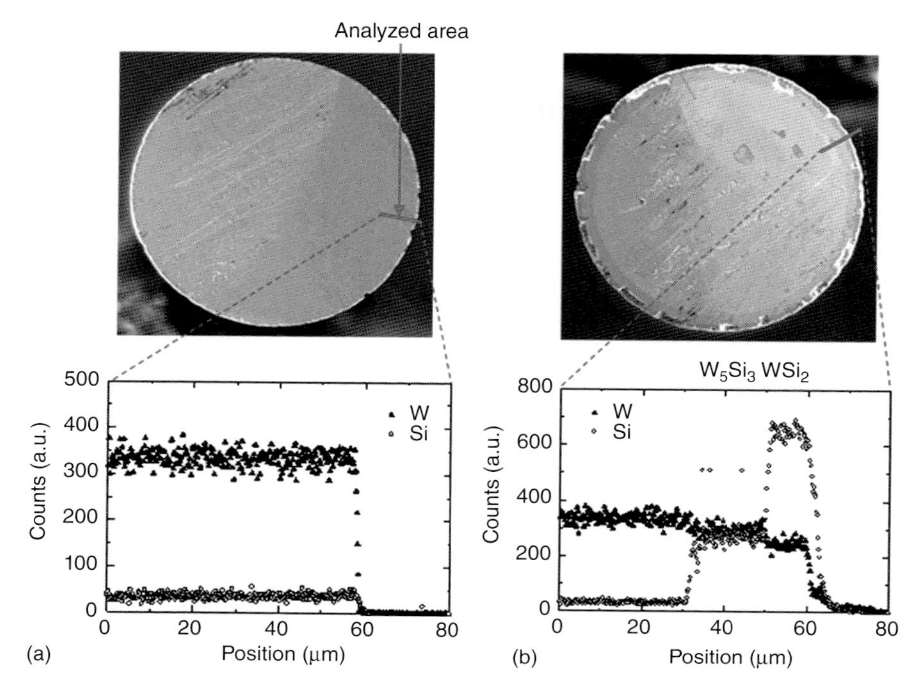

Figure 7.22 Variation of silicide layers after heating in vacuum for one hour at 1850 °C. (a) For W_5Si_3 and (b) for WSi_2. Source: Honda 2008 [19]. Reprinted with permission of IOP Publishing.

Another simple approach is to just evaporate Si atoms from the silicide by heating the catalyzing wire. The Si removal rates may be lower than in the case of using H atoms because there is no help by chemical reactions. However, just for showing this possibility for recovering catalyzing W wires, we show the results of Si removal by a simple heating treatment after stopping the flow of the SiH_4 gas.

In this experiment, 10-μm-thick W_5Si_3 and 30-μm-thick-WSi_2 layers are formed as experimental samples, and these samples are annealed at 1850 and 2100 °C for one hour. Figure 7.22 shows the results of the EPMA observation of samples after heating at 1850 °C, and Figure 7.23 shows the results for 2100 °C heating. It is clearly demonstrated that 10-μm-thick W_5Si_3 layer returns to pure W after the treatment at 1850 °C for one hour but that the WSi_2 layer returns to W_5Si_3 layer at the surface and WSi_2 layer remains in the deeper region. After heating at 2100 °C for one hour, even the WSi_2 layer returns to pure W. However, it should be noticed that once WSi_2 is formed, the volume has expanded, and then, even if Si atoms are removed, the surface of W cannot return to its original smooth state. It is another point to regenerate the W catalyzer early, to avoid this volume expansion.

Although the experimental data are not yet complete, we expect to be able to use the same W catalyzing wire for a long time, provided that we refresh the catalyzer by heating at 1850 °C for 1 hour after every 40 hours of deposition of a-Si at T_{cat} at 1850 °C. In addition, if we remove Si atoms by introducing H_2, as

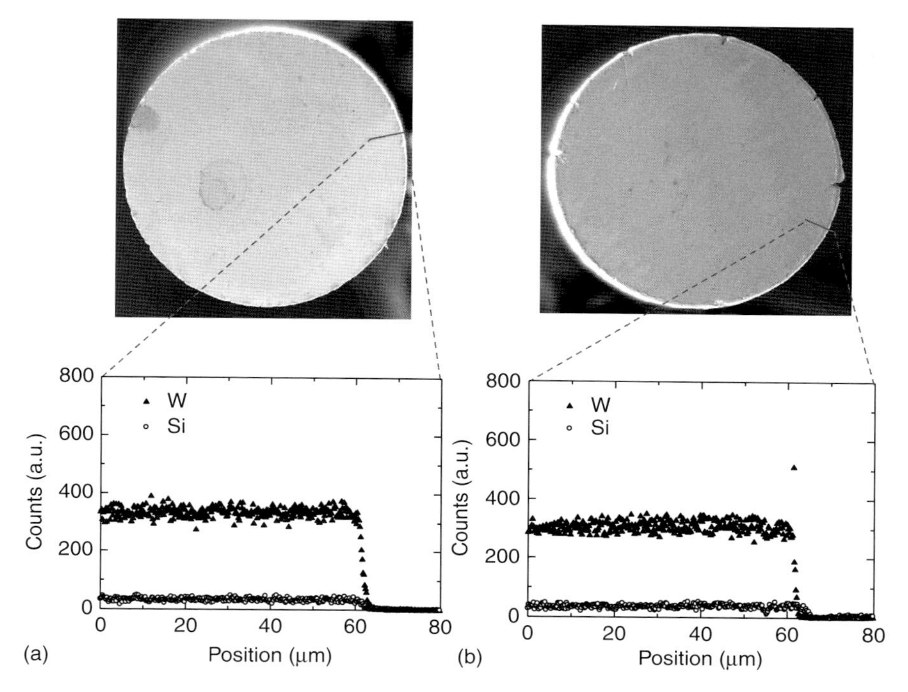

Figure 7.23 Variation of silicide layers after heating in vacuum for one hour at 2100 °C. (a) For W_5Si_3 and (b) for WSi_2. Source: Honda 2008 [19]. Reprinted with permission of IOP Publishing.

H atoms have an ability of chamber cleaning as mentioned later, we can clean the chamber walls simultaneously with the wire regeneration.

7.6.3 Silicide Formation of Ta Catalyzer

Tantalum (Ta) is another candidate for the catalyzing wires for a-Si deposition, as T_{cat} can be lowered slightly compared to the W catalyzer. Silicide formation and the recovering method after silicidation of Ta wires have been extensively studied by the group of R.E.I. Schropp at Utrecht University, the Netherlands [20], and by the group of B. Schroeder, at University of Kaiserslautern, Germany [21]. Here, we introduce the work of the Utrecht group [20].

They observed the structural change of the Ta catalyzing wires with a diameter of 0.5 mm and a length of 15 cm by optical microscopy (OMS) and SEM, after a total six-hour exposure to SiH_4 gas at $T_{cat} = 1750\,°C$. Figure 7.24a shows an OMS cross-sectional image of the center part of the 15 cm Ta wire and Figure 7.24b shows a magnification of the same wire. A shell with a thickness of 20 μm had formed. According to the X-ray diffraction (XRD) measurement of the shell, it was found to be Ta_3Si_5, similar to the compound W_3Si_5, which forms on a W surface when W is heated in SiH_4 to temperatures over 1750 °C. Figure 7.24c,d show similar images after heating the Ta_3Si_5-formed wire at 2100–2200 °C in vacuum for 10 minutes. It is seen that a new shell appears at the surface. The XRD measurement reveals that the new outer shell consists of pure Ta. Figure 7.24e,f show

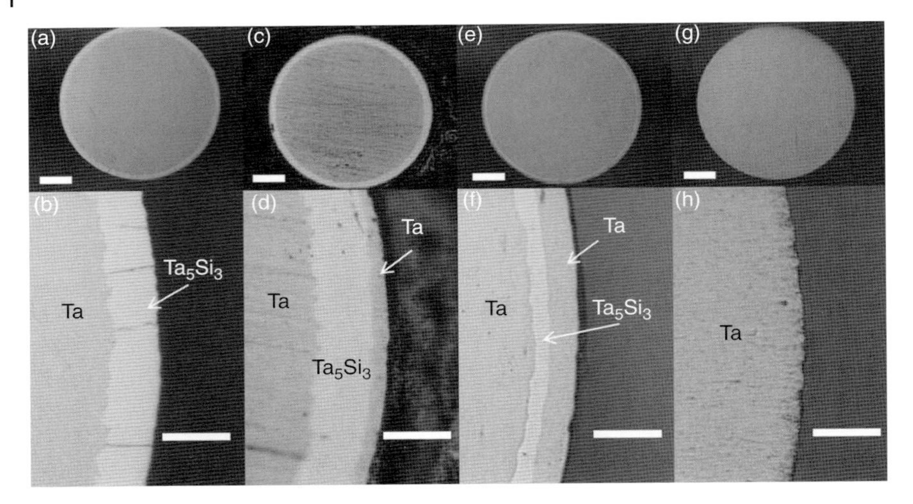

Figure 7.24 The cross-sectional images of Ta catalyzing wires, (a) and (b) after 6-hour exposure in SiH_4 at $T_{cat} = 1750\,°C$, (c) and (d) after 10 minutes, (e) and (f) after 1 hour, and (g) and (h) after 4 hours of heating at $T_{cat} = 2100–2200\,°C$ in vacuum. Source: van der Werf et al. 2009 [20]. Reprinted with permission of Elsevier.

similar images after one hour, and Figure 7.24g,h after four hours of annealing at 2100–2200 °C. Similar to W, Ta_3Si_5 can be completely recovered to pure Ta. Reuse of Ta catalyzing wires appears possible just like W catalyzing wires.

7.6.4 Suppression of Silicide Formation by Carburization of W Surface

(As mentioned already in Chapter 5, in this book, the term "carburization" or "carburized" is used instead of using "carbidation" or "carbided.")

By controlling the W catalyzer temperature to keep it always over 1850 °C and regenerating it periodically, we can use the same W catalyzer for a-Si deposition for a long time. However, if some parts of the W catalyzer cannot be kept at these high temperatures, the lifetime of the W catalyzer is shortened easily. For more secure use of the W catalyzing wire, it would be best to reduce the growth rate of W-silicide itself by altering the surface of the catalyzer. One possible way is to form other layers on the W wire that reduce the growth rate of silicide. There have not been many reports on this type of solution. The data we can present here appear limited.

Among the few data, we show the results of carburization of the W surface to reduce the growth rate of silicide on W. In this case, the W surface is covered with a carburized layer. Such a carburized layer is formed by exposing W wires to methane (CH_4) ambient in the Cat-CVD chamber. The carburization is carried out at $T_{cat} = 1900\,°C$ and $P_g = 10\,Pa$ in 100% CH_4 gas for five minutes. In this way, about 10-μm-thick W_2C layer can be formed. After that, SiH_4 gas is introduced into the same chamber at $T_{cat} = 1850\,°C$ and $P_g = 10\,Pa$.

Figure 7.25 shows the silicide thickness as a function of time of exposure of the catalyzer to SiH_4 gas while T_{cat} is kept at 1850 °C and P_g at 10 Pa. When the

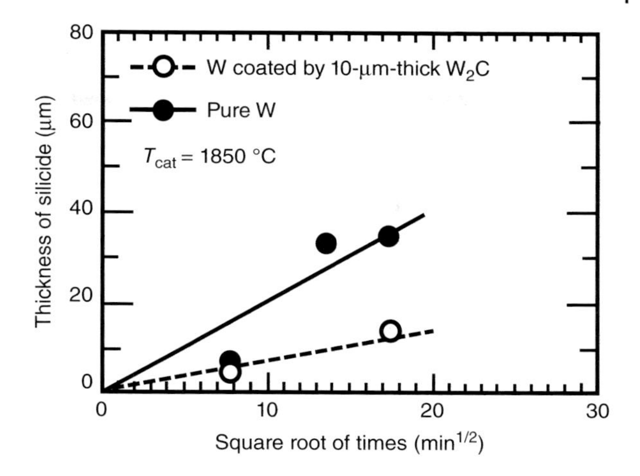

Figure 7.25 Thickness of W-silicide formed on the W catalyzer for the cases when the surface of catalyzer is carbonized or not. Source: Honda 2008 [19]. Reprinted with permission of IOP Publishing.

pure W catalyzer is exposed to SiH_4 gas, the surface of the W catalyzer becomes a W-silicide with a parabolic rate constant B of about 4 μm²/min, as presented in Figure 7.21. However, when the W catalyzer whose surface is covered with a 10-μm-thick W_2C carburized layer is used, the thickness of W-silicide is apparently reduced to about a factor 0.4 [19]. The parabolic rate constant B in this case is evaluated to be about 0.64 μm²/min carburized. The time until regeneration can be extended from 40 hours as mentioned above to 260 hours when a-Si is deposited at P_g = 1 Pa with FR(SiH_4) = 50 sccm. Thus, the lifetime of this carburized W catalyzer is greatly extended.

We have checked the quality of the a-Si films deposited by using the carburized W catalyzer, showing no difference with films made by using a pure W catalyzer. Although the data are not extensive, this shows the feasibility to extend the lifetime of a W catalyzer.

7.6.5 Ta Catalyzer and Method for Extension of Its Lifetime

As already mentioned in Chapter 4, instead of W, Ta is often used as a catalyzing material as the temperature to avoid silicide formation appears by 100 °C lower than W [13]. In addition, W becomes brittle after use, although Ta is likely to keep softness even after use. Therefore, there have been many reports to mention a-Si deposition by using a Ta catalyzer instead of a W catalyzer. However, Ta is very soft metal, and keeping its shape in high temperatures is difficult compared with W.

There have not been so many reports about the lifetimes of Ta catalyzer, except for a structural study on Ta-silicide [22]. There has been no systematic study on the growth rate of silicide like the study discussed above for W-silicide. D. Knoesen et al. [22] claim that a Ta catalyzer cannot be used for more than three to five hours of a-Si deposition in SiH_4 ambient at T_{cat} = 1600 °C; however, when H_2 treatment at T_{cat} = 1600 °C for five minutes was applied before and after a-Si deposition, it could be used for more than 320 hours. Anyway, it should be noted that T_{cat} can be lowered when using Ta instead of W. If we were to use W at

T_{cat} = 1600 °C, it would immediately modify to silicide. Ta is soft and bendable. Because of these advantages, many people like to use Ta instead of W.

7.6.6 Lifetime Extension by Using TaC

I.T. Martin et al. reported that when TaC is used as a catalyzer instead of W or Ta, the lifetime of the catalyzer can be enormously extended [6]. In their case, at first, graphite rods with a diameter of about 1.6 mmØ were prepared, mounted in a hot wall conventional thermal chemical vapor deposition (thermal CVD) system, and heated to over 2200 °C. This method is quite common to produce graphite coatings from metal–organic source gases. Finally, the diameter of the TaC/C rods becomes 1.63 mmØ because of the TaC coating. This TaC/C rod is shown in a photograph in their report [6], and here, it is reproduced as Figure 7.26. The surface is shiny and golden colored. This does not change even when it is exposed to SiH_4 ambient for a wide range of T_{cat}. The temperature of the TaC/C rod is monitored by thermal radiation, assuming an emissivity of 0.6. The a-Si films prepared by the W catalyzer are compared with those prepared by TaC/C rods under the same conditions. They did not show the exact lifetimes of a TaC/C catalyzer; however, they claim that the extension of lifetimes in SiH_4 ambient with respect to the case using a W catalyzer is clearly observed. The lifetime of TaC/C rods is limited by the silicide formation at the end terminals of the rod, where T_{cat} is low and thus silicide is easily grown. It can be expected that if they had used a capping system as shown in Figures 7.15 and 7.16, the lifetimes would have been extended more.

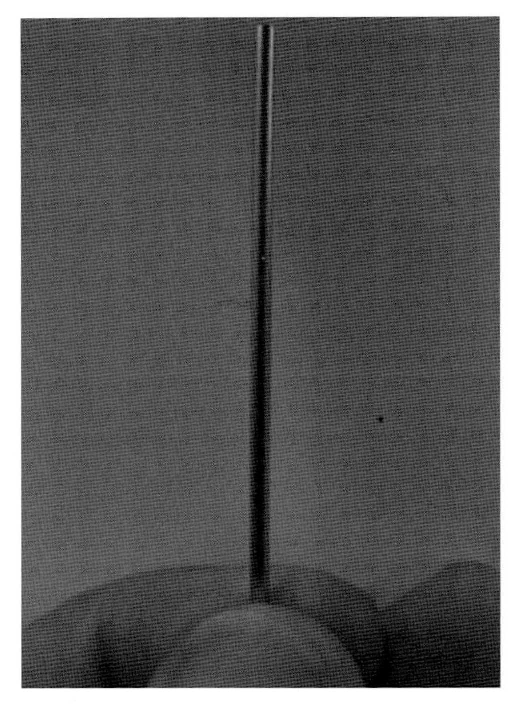

Figure 7.26 A photograph of TaC/C rod as a new type of the catalyzer in Cat-CVD. Source: Martin et al. 2011 [6]. Reprinted with permission of Elsevier.

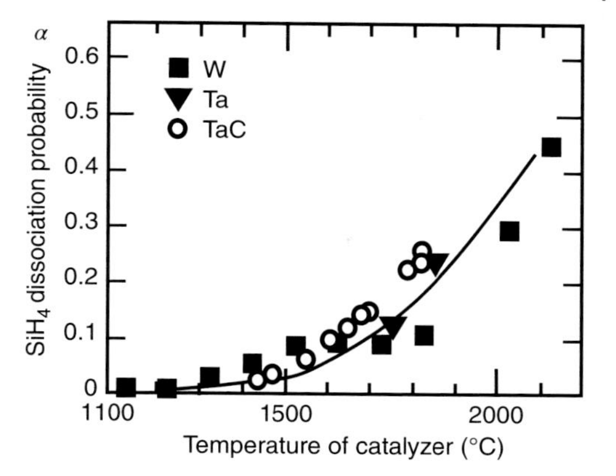

Figure 7.27 SiH_4 dissociation probability, a, as a function of the temperature of catalyzer, T_{cat}, for W, Ta, and TaC/C catalyzers. Source: Martin et al. 2011 [6]. Reprinted with permission of Elsevier.

They also studied the SiH_4 dissociation probability α for TaC/C rods and compared them with the reported values for W or Ta catalyzers. The dissociation probability is not so different from those of W or Ta catalyzers, as shown in Figure 7.27. In the figure, it is shown that α for W at $T_{cat} = 1850\,°C$ is about 0.2. This value is exactly the same as the value derived in Section 7.1.2.

The use of a TaC/C catalyzer does not meet many challenges. As the TaC/C rod system is stable for SiH_4 decomposition and it is also stable mechanically but the length of the rods is limited, it may be used for deposition of films over a limited area such as on the inside of poly-ethylene terephthalate (PET) bottles.

7.6.7 Lifetime Extension by Using Other Ta Alloys

There are some reports or patents for the extension of lifetimes of a catalyzer by using Ta alloys apart from TaC. As Ta is a soft metal, the formation of alloys is useful if mechanical stability is required. Several methods have been published for preparing Ta alloys in order to make deposition of Cat-CVD stable. Figure 7.28

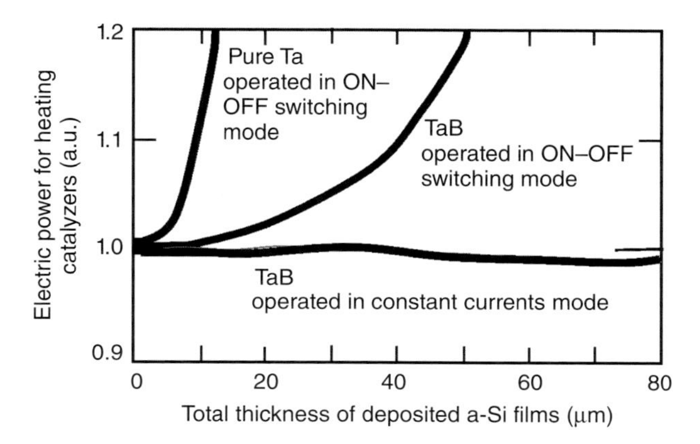

Figure 7.28 Stability of electric power supply for heating up catalyzing wires at more than 1750 °C. When the electric power is kept constant, it means that the catalyzer is stable. Source: Reprinted with permission from Ref. [23].

shows one of those methods. The stability of electric power supplied to heat the catalyzers to a certain temperature is demonstrated as a function of cumulative thickness of deposited a-Si films using the catalyzer [23]. The results of two modes for electric power supply are shown. One is the case in which the power is supplied by applying a constant current. The other is the case in which the power is not continuously supplied but periodically switched on and off states every few minutes. Probably because of mechanical stress, when the power is supplied periodically, the catalyzing wire appears to be unstable. However, when the power is supplied constantly, the catalyzer can be used until the total thickness of a-Si films exceeds 80 μm at least.

7.6.8 Lifetimes of W Catalyzer in Carbon-Containing Gases

As described in Section 5.3.6, SiN_x films can be obtained by using safe hexa-methyl-disilazane (HMDS) gas as a source gas providing Si and N atoms. However, as shown in Figure 5.47, the electric currents to heat the W catalyzer show very high instability. This is caused by carburization of the W catalyzer. We have earlier listed the typical conditions to avoid carburization in Table 5.7, and here, we return to this topic for SiN_xC_y deposition by organic sources as an approach to extend the lifetimes of the catalyzer. The deposition conditions are almost the same as in Table 5.7; however, here, we show it again to understand which parameters are varied with respect to Table 5.7. Conclusions obtained from the study are also briefly summarized in Table 7.2.

Figure 7.29 shows the XRD patters of the W catalyzer after deposition of SiN_x films, when they are deposited without addition of NH_3 and H_2. Only in this case, the flow rate of FR(HMDS) is fixed at 1 sccm, but in all other experiments, it is kept at 0.5 sccm. In the figure, T_{cat} is taken as a parameter and varied from 1200–2200 °C. P_g is fixed at 1 Pa. The figure shows that W surface is carburized

Table 7.2 Deposition conditions to prepare SiN_x films by using HMDS and searching conditions to find the conditions by which carburization of a W catalyzer is avoided.

	Study on finding noncarburized conditions	**Conclusions obtained from study for avoiding carburization**	**For avoiding W carburization**
Catalyzer	W		W
FR(HMDS)	0.5 sccm (only for T_{cat} variation: 1 sccm)		0.5 sccm
FR(NH_3)	0–200 sccm	Higher FR(NH_3) is effective	50 sccm or more
FR(H_2)	0–200 sccm	Higher FR(H_2) helps NH_3 effect	40 sccm or more
T_{cat}	1200–2200 °C	No effect in T_{cat} variation	1900 °C
P_g	1–130 Pa	Higher P_g is effective	50 Pa or more
D_{cs}	8 cm		8 cm
T_s	<100 °C		<100 °C

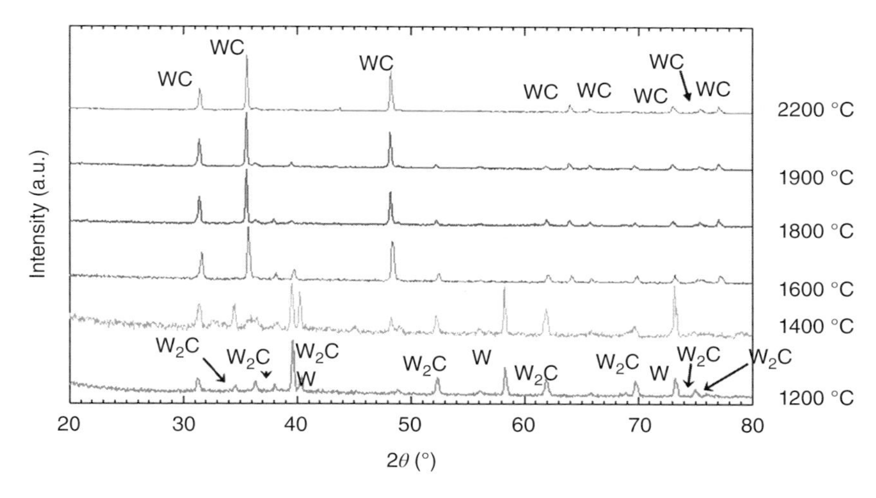

Figure 7.29 XRD spectra of W catalyzer after depositing SiN_x films by HMDS, taking T_{cat} as a parameter.

for any T_{cat}, although higher T_{cat} appears to make WC instead of W_2C. This figure shows that the variation of T_{cat} is not effective in suppressing carburization.

As there is no particular value of T_{cat} that suppresses carburization, it is fixed at 1900 °C, and next, the flow rate of NH_3, $FR(NH_3)$, varied at a fixed P_g of 21 Pa, without introducing H_2. Figure 7.30 shows the XRD patterns of the W catalyzer after deposition of SiN_x films. As $FR(NH_3)$ increases, the type of carburized W changes from WC to W_2C, and once it exceeds 150 sccm, carburization is suppressed. This means that N-related species are effective in pulling out C atoms from the W surface. The exact mechanism for avoiding carburization has not been clarified yet; however, the formation of strong C—N bonds may be effective to break C—W bonds.

When H_2 gas is introduced instead of NH_3, there is no clear effect avoidance of carburization; however, when H_2 is added to NH_3 at smaller $FR(NH_3)$, it

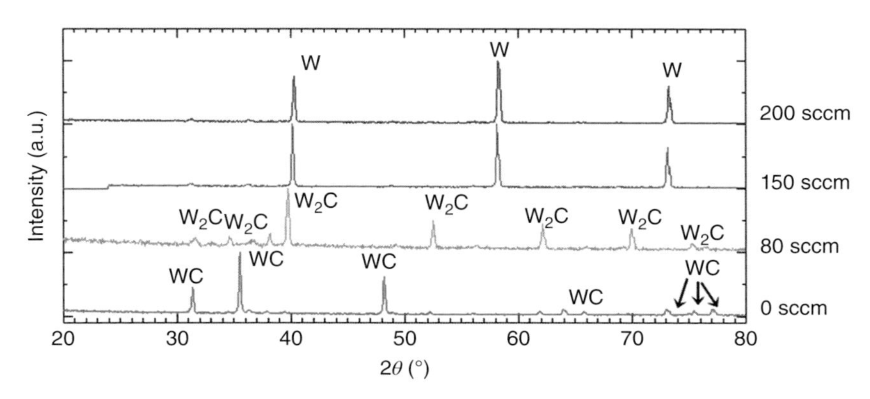

Figure 7.30 XRD spectra of W catalyzer after depositing SiN_x films by HMDS, taking $FR(NH_3)$ as a parameter.

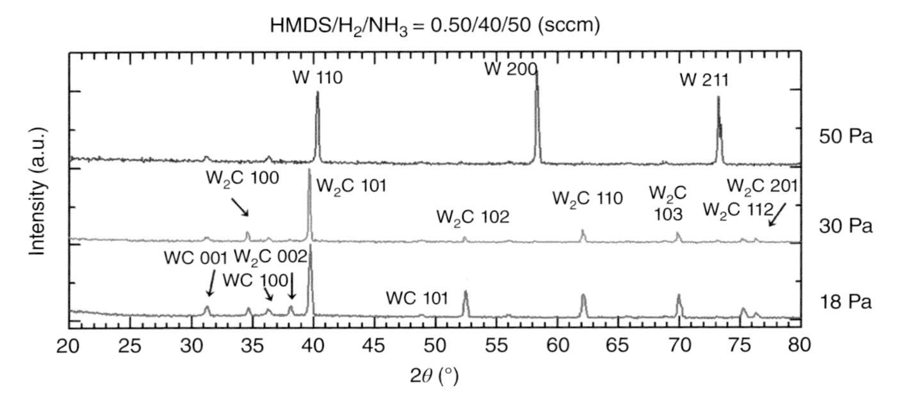

Figure 7.31 XRD spectra of W catalyzer after depositing SiN_x films by HMDS, taking P_g as a parameter.

aids in avoidance of carburization. For instance, when H_2 at a flow rate of $FR(H_2) = 50$ sccm is added to NH_3 at a flow rate of $FR(NH_3) = 80$ sccm under conditions used in Figure 7.30, carburization is suppressed. As explained in Chapter 4, NH_3 is decomposed to $NH_2 + H$ at the surface of the W catalyzer, and subsequent reactions with H in the gas phase are necessary to produce N or N–H. The addition of H_2 may thus help to increase the density of N atoms that are active in pulling out C atoms from the W wire.

Figure 7.31 shows similar XRD patterns when SiN_x films are prepared by various P_g's at FR(HMDS) = 0.5 sccm, $FR(H_2) = 40$ sccm, and $FR(NH_3) = 50$ sccm. As Pg increases, carburization is suppressed more effectively. Summarizing, to avoid carburization of a W catalyzer when C-related source gas is used for SiN_x deposition, a large NH_3 gas flow and a higher gas pressure are effective. This experiment suggests that the extension of lifetimes of catalyzer is sometimes possible by mixing in other gases that prevent modification of the catalyzer by surface reactions.

7.6.9 Long-Life Catalyzer Used in *i*CVD

In *i*CVD, T_{cat} can be reduced to very low values because of the use of reaction initiators to enhance film growth reactions. The low T_{cat} is helpful in extending the lifetime of the catalyzer, as the carbidizing reactions due to decomposition of CH_3 are slowed down significantly. In Chapter 6, we showed that organic films can be deposited at low T_{cat} with the help of reaction initiators. For the decomposition of such initiators, a wire temperature T_{cat} lower than 500 °C can often be chosen. Particularly, when NiCr or other Ni alloy is used as a catalyzer, the decomposition of organic materials is likely to be easier, as indicated in Figure 6.3.

Here, we like to show an example that demonstrates how a Ni-containing alloy is effective in decomposing organic gases. Figure 7.32 shows the Q-mass signal intensity for decomposition of tri-methyl-aluminum (TMA) as a function of T_{cat} for various catalyzing materials. When TMA is decomposed, the fragments of it, such as CH_4, CH_3, and C_2H_4, can be detected, as already shown in Figure 5.59. In Figure 7.32, Inconel, stainless steel (SUS), Iridium (Ir) and W are used as catalyzers. In Figure 5.59, the CH_3 signal intensities are plotted for all

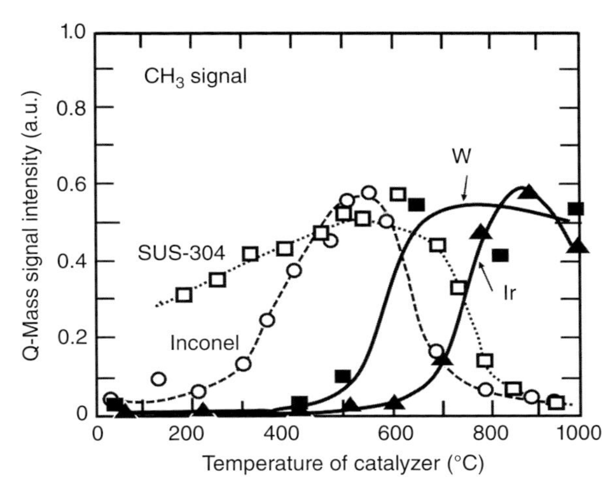

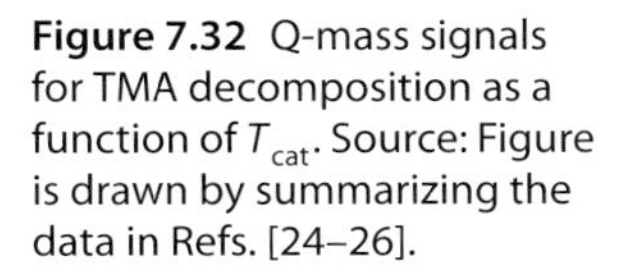
Figure 7.32 Q-mass signals for TMA decomposition as a function of T_{cat}. Source: Figure is drawn by summarizing the data in Refs. [24–26].

the studied catalyzers. The data shown in the figure are taken from references [24–26]. For W and Ir, TMA is decomposed at T_{cat} higher than 500–600 °C; however, for Inconel-600 (72% Ni, 14–17% Cr, and 6–10% Fe) and SUS-304 (8–10% Ni, 18–20% Cr, and about 70% Fe), which are both Ni alloys, TMA is decomposed at T_{cat} lower than 500 °C. This suggests that the reaction initiators can be decomposed at T_{cat} lower than 500 °C, provided that the catalyzer contains Ni, such as NiCr.

The reduction of T_{cat} below 500 °C has a special meaning. Decomposition of $-CH_3$, which is included in organic source gases, becomes quite rare at such low T_{cat}, as the generation of species with a single C atom is suppressed, and thus, carburization of the catalyzer is suppressed. Actually, in *i*CVD, the lifetimes of a catalyzer is extended to very long times when these low T_{cat} values are used.

7.7 Chamber Cleaning

One of the most important issues in CVD systems is cleaning of the chamber wall. In particular, in RF-PECVD, the films deposited on the chamber wall or other parts within the chamber change the matching conditions of the transfer of RF power, so that chamber cleaning is essential to keep the film quality stable. In Cat-CVD, the films deposited on the chamber wall sometimes have influence on the quality of films, as the films previously deposited on the wall react with H atoms, for instance, and additional species are generated other than those formed by the introduced source gases.

For the simplest cleaning, H atoms that are generated by catalytic cracking of H_2 gas at a heated W catalyzer are used to etch Si-related films. In Figure 7.33, the etch rates of c-Si and SiO_2 by H atoms are plotted as a function of H_2 gas pressure P_g. In this case, T_{cat}, FR(H_2), and substrate temperature T_s are kept at 1800 °C, 48 sccm, and room temperature (RT), respectively. From the figure, it can be seen that the Si etch rates exceed over 200 nm/min when P_g is 0.5 Torr (67 Pa), although SiO_2 is not etched at all by a similar density of H atoms. This etch rate may not be large enough for etching or removing Si films such as a-Si

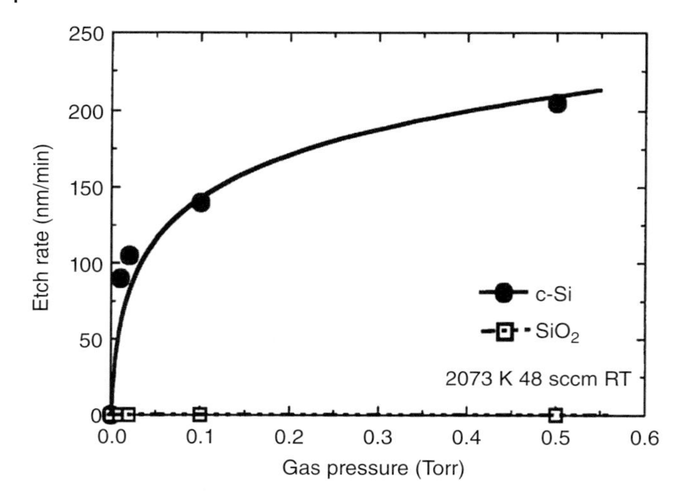

Figure 7.33 Etch rates of c-Si and SiO_2 by H atoms generated by catalytic reactions with a heated W catalyzer from H_2 gas, as a function of gas pressure P_g. Source: Matsumura 2001 [27]. Reprinted with permission of Elsevier.

from the chamber walls, particularly in mass production; however, it is important that halogen gases that are often greenhouse gases are not used in the cleaning process. Etching with H_2 gas can be carried out because a high density of H atoms can be easily produced by catalytic cracking reactions.

To increase the etching or cleaning speed, nitrogen-tri-fluoride (NF_3) gas can also be used, similar to PECVD, although this gas is not recommended in view of environmental issues. When W is used, the surface is easily fluoridated at T_{cat} lower than 2000 °C. Therefore, to avoid fluoridation, T_{cat} is kept as high as 2400 °C. If a W catalyzer is heated in NF_3 ambient and if the surface of the W catalyzer is modified due to fluoridation, the electric currents to heat the W catalyzer, I_{cat}, will change. Figure 7.34 shows I_{cat} as a function of heating times

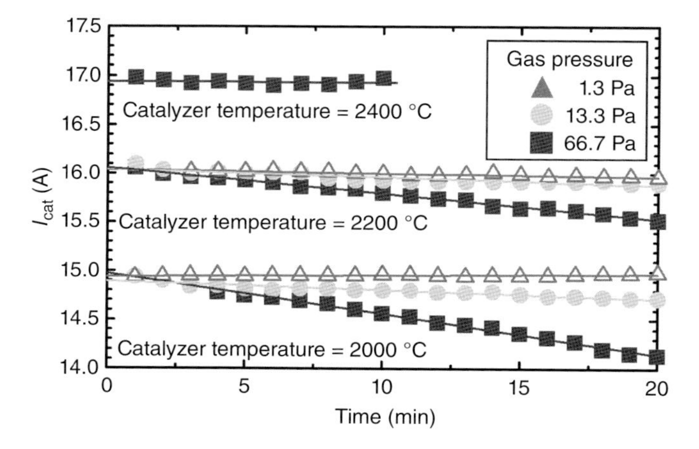

Figure 7.34 Currents for heating a tungsten catalyzer, I_{cat}, as a function of usage times for various T_{cat} and P_g. Source: Matsumura et al. 2003 [28]. Reprinted with permission of Elsevier.

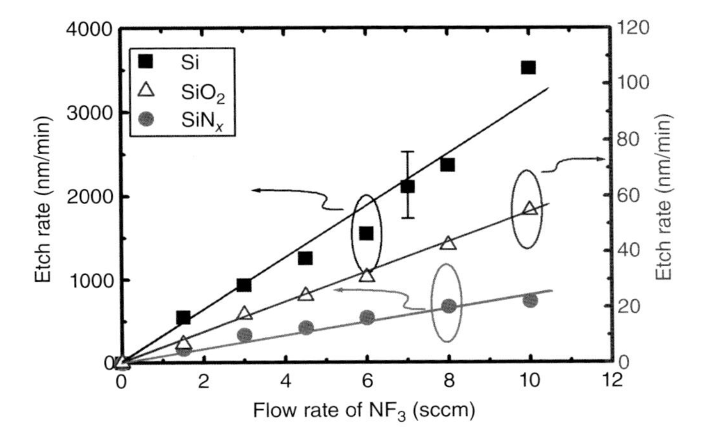

Figure 7.35 Etching rates of c-Si, SiN_x, and SiO_2 as a function of FR(NF_3). Source: Matsumura et al. 2003 [28]. Reprinted with permission of Elsevier.

for various T_{cat} and P_g. The flow rate of NF_3, FR(NF_3), is kept constant at 10 sccm in this case. As shown in the figure, when T_{cat} is 2000 °C, T_{cat} becomes unstable, particularly at higher P_g. However, when T_{cat} is elevated to 2400 °C, it becomes stable even for higher P_g such as 66.7 Pa. This demonstrates that NF_3 cleaning is possible if T_{cat} can be elevated up to 2400 °C.

Next, in Figure 7.35, we show the etch rates of c-Si, SiO_2 and SiN_x, when T_{cat} is kept at 2400 °C, P_g at 66.7 Pa, and T_s at 300 °C. The etching rates are plotted as a function of FR(NF_3). As shown in the figure, the etching rate exceeds 3 μm/min for c-Si. These values are equivalent to the etch rates in an RF-PECVD machine.

7.8 Status of Mass Production Machine

7.8.1 Cat-CVD Mass Production Machine for Applications in Compound Semiconductors

Finally, we like to briefly introduce the status of the development of mass production Cat-CVD machines. As far as the authors know, the first Cat-CVD apparatus sold for mass production was developed by ANELVA Corp., Japan, in 1990s. The apparatus is made for depositing SiN_x passivation films on GaAs devices. As explained in Chapter 8, the formation of passivation films on compound semiconductors such as GaAs is the first successful example from an industrial standpoint. At that time, an ESC was not adapted because the productivity was reduced by it. GaAs wafers are placed on a turning table, and by rotating the table, the elevation of substrate temperature by heat radiation was avoided. Figure 7.36 shows a schematic illustration of the apparatus, and Figure 7.37a shows a photograph of the apparatus and Figure 7.37b shows a photograph of samples during deposition and the rotating table or sample stage. All figures are reproduced from reference [29].

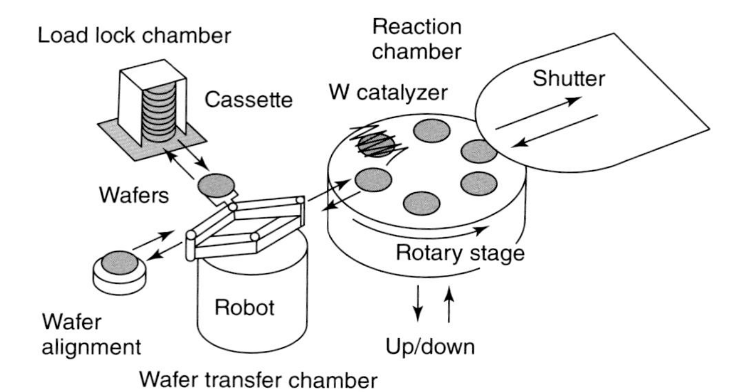

Figure 7.36 Schematic illustration of the first mass production Cat-CVD apparatus used for GaAs passivation. Source: Reprinted with permission from Ref. [29].

Figure 7.37 Photographs of Cat-CVD apparatus used for mass production of GaAs devices. (a) The system and panel. (b) The rotating stage during deposition. Source: Reprinted with permission from Ref. [29].

7.8.2 Cat-CVD Mass Production Apparatus for Large Area Deposition

As mentioned in Chapter 2, in the 1990s, technologies for large area deposition were strongly desired mainly for fabrication of a-Si thin film transistor (TFT) driver plates used in liquid crystal displays (LCDs). At that time, it was believed that large area deposition by RF-PECVD was difficult because of the standing wave problems at RF frequencies. A large area PECVD machine using an array of many hollow cathodes, demonstrated in Figure 2.11, which was developed by Applied Komatsu Technology Corp. or Applied Materials Corp., USA, was not widely known. Thus, the possibility of Cat-CVD for large area deposition attracted strong attention.

There were two approaches to realize large area deposition by Cat-CVD. The challenge was how can we hold the heated W or Ta catalyzer in place over a large area. One idea was that many short W catalyzer wire arrays are mounted on a large plate. As explained already in Section 7.6.2 and Figures 7.15 and 7.16, when the terminals of the catalyzing wires are connected inside a capping system, silicide is not formed at least from the end points of the wires. The structure shown

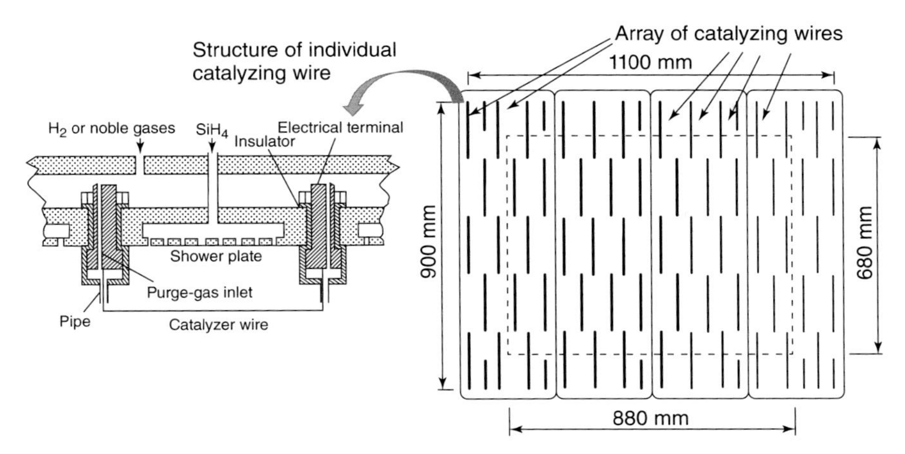

Figure 7.38 Structure of gas showerhead with catalyzing wire arrays for large area Cat-CVD apparatus. Source: Ishibashi et al. 2003 [9]. Reprinted with permission of Elsevier.

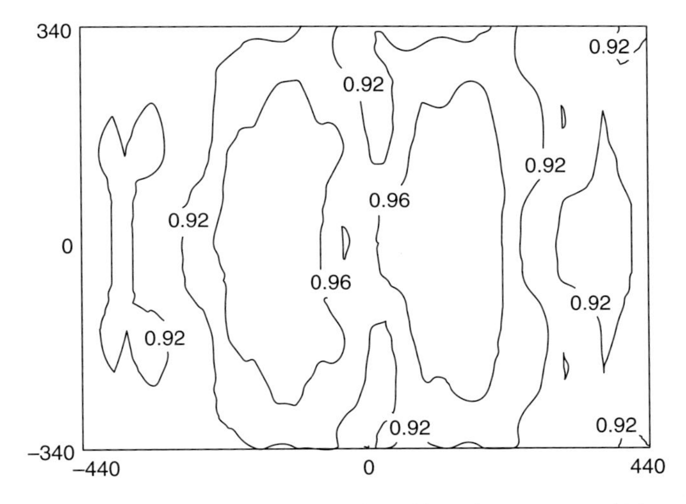

Figure 7.39 Thickness uniformity of a-Si films. The maximum thickness is 1.0 located at the center of contours. Unit of length at vertical and horizontal axes is mm. Source: Matsumura et al. 2003 [28]. Reprinted with permission of Elsevier.

in Figure 7.38 was invented [9]. By extending the area of the array of catalyzing wires, the deposition area could be extended as desired. Figure 7.39 demonstrates the thickness uniformity of a-Si films deposited by the apparatus of Figure 7.38. Good thickness uniformity can be obtained reliably over a wide area. The thickness is normalized with respect to the center of the contours.

This system solved the size problems in Cat-CVD; however, a-Si films are also deposited on the showerhead plate, which is also the supporting plate of the catalyzing wire array. Therefore, the supporting plate with the catalyzers had to be cleaned frequently to avoid the release of Si-flakes from the plate. Particularly, in Cat-CVD, heating and cooling repeatedly take place, depending on the recipe for film deposition. All components surrounding the catalyzing wires suffer from the

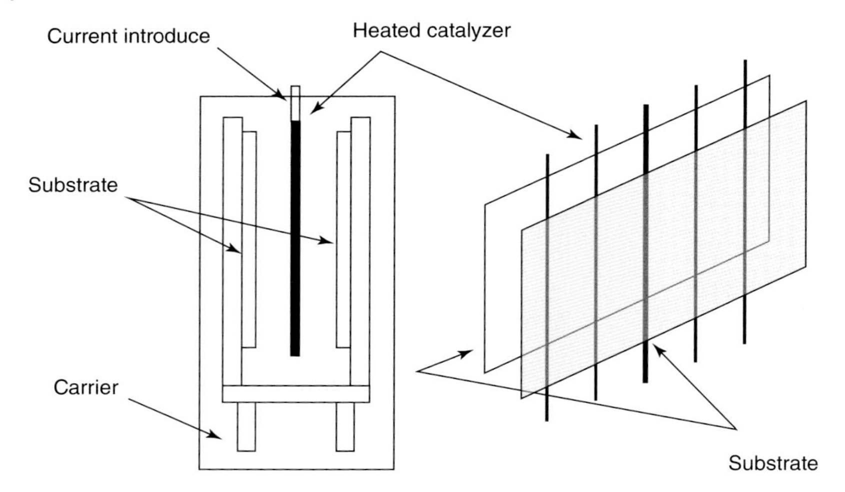

Figure 7.40 Schematic diagram of vertical type large area Cat-CVD apparatus. Source: Osono et al. 2006 [30]. Reprinted with permission of Elsevier.

periodical thermal expansion and shrinkage. This easily leads to flakes of residual films. To avoid this, other approach was pursued.

The catalyzing wires are hung vertically and the substrates are set vertically. This concept was demonstrated in Figure 2.12. In this apparatus, two substrate trays can be placed, one at each side of the catalyzing wires. In this way, the films are deposited on both substrate trays, and the trays facing the catalyzing wires enclose the wire assembly, i.e. all films are deposited on trays instead of on the chamber walls. Therefore, chamber cleaning is not required often. Figure 7.40 shows a schematic diagram of the structure of this vertical type Cat-CVD machine, and Figure 7.41 shows a photograph of the machine [30]. The fundamental concepts are the same as in Figure 2.12.

S. Osono et al. at ULVAC Ltd. systematically studied the a-Si film uniformity and distribution of a-Si quality using the Cat-CVD apparatus shown in the figures [30]. a-Si films were prepared at $T_{cat} = 1700\,°C$, $P_g = 0.1–1\,Pa$, $FR(SiH_4) = 100\,sccm$, and $D_{cs} = 7\,cm$ using a Ta catalyzer. The uniformity of a-Si thickness is shown in Figure 7.42 and the photoconductivity σ_{ph} (or σ_p) and the photosensitivity, which is the ratio of the photoconductivity to the dark conductivity σ_d, at various positions of the deposited a-Si film are described in Figure 7.43. The thickness uniformity of this prototype machine was not sufficiently good; however, the uniformity of the optoelectronic quality was excellent, as the photoconductive properties of a-Si prepared by this large area Cat-CVD apparatus at any position were as high as the best quality film. Various efforts have been undertaken, and finally, they succeeded in developing a large area mass production apparatus with sufficient uniformity both in thickness and quality, based on the machine shown here and such apparatuses have been sold in a market.

Figure 7.41 A photograph of vertical type large area Cat-CVD apparatus. Source: Reproduced with permission from Osono et al. [31]. Copyright 2004, The Japan Society of Applied Physics.

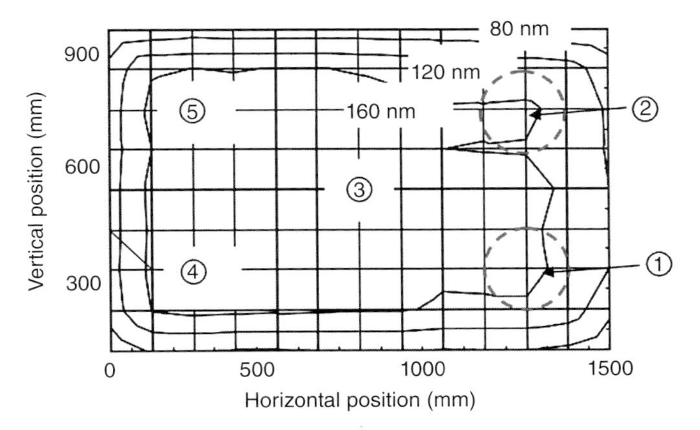

Figure 7.42 Thickness uniformity of a-Si films. The numbers, ① to ⑤, show the position where the properties are measured, but the measured properties are shown in the next figure. Source: Osono et al. 2006 [30]. Reprinted with permission of Elsevier.

7.8.3 Cat-CVD Apparatus for Coating of PET Bottles

Cat-CVD is also applied to prepare gas barrier films as explained later, in Chapter 8. For instance, if PET bottles are coated with gas barrier films, the time during which the best taste of the drink in the PET bottles is preserved can be extended. A Japanese beer and soft drinks selling company, Kirin, has developed a Cat-CVD mass production bottling machine [32]. With this machine, Cat-CVD barrier

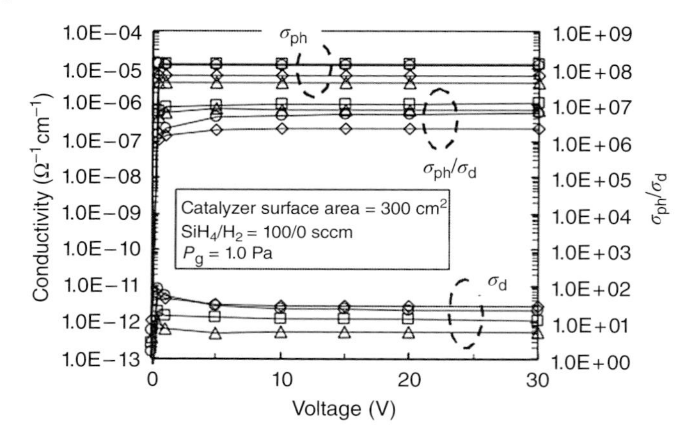

Figure 7.43 Photoconductive properties of a-Si deposited by large area Cat-CVD apparatus. The conductivities are plotted as a function of the applied voltage during measurements. Source: Osono et al. 2006 [30]. Reprinted with permission of Elsevier.

Figure 7.44 A photograph of Cat-CVD barrier film coating machine for PET bottles. Source: Photograph is provided by Courtesy of Kirin Corp.

films are deposited on the inside of many PET bottles at the same time. Figure 7.44 shows a photograph of the Cat-CVD machine.

7.8.4 Prototypes for Any Other Mass Production Machine

Various prototype mass production Cat-CVD machines have been developed. Photographs of some of them are summarized at the end of this chapter in

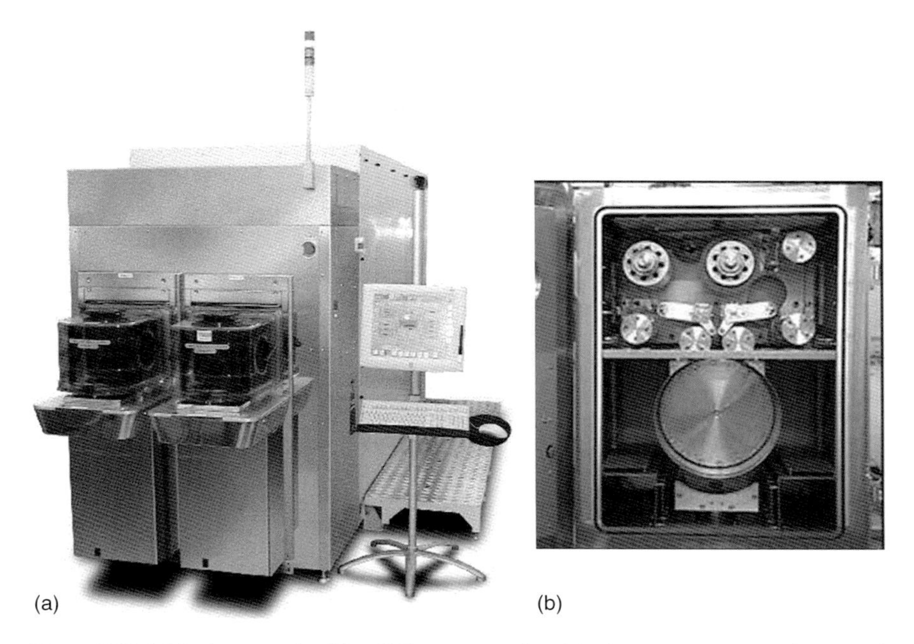

Figure 7.45 (a) A photograph of Cat-CVD apparatus for 12 in. Si process. Source: Photograph is reproduced from Ref. [33] with permission. (b) For roll-to-roll barrier film coating on plastic substrates. Some of them are available in the market. Source: Photograph is provided by Courtesy of Ishikawa Seisakusyo Co.

Figure 7.45. Figure 7.45a is for a 12-in. Si process and Figure 7.45b is for film coating on plastic substrates. Apart from these apparatuses, also coating of poly-tetra-fluoro-ethylene (PTFE) films on razors has been commercialized and mass production machines were manufactured, although this has not been published in the open literature. The possibility to form coatings on various materials is one of the useful advantages of Cat-CVD, and the apparatuses for realizing it have been supplied by industrial companies. However, in most cases, this is not made public.

It has taken some time since the start of the study of Cat-CVD; however, mass production of products using this technology is gradually being realized.

References

1 Honda, N., Masuda, A., and Matsumura, H. (2000). Transport mechanism of deposition precursors in catalytic chemical vapor deposition studied using a reactor tube. *J. Non-Cryst. Solids* 266–269: 100–104.

2 van der Werf, C.H.M., Goldbach, H.D., Loffler, J. et al. (2006). Silicon nitride at high deposition rate by hot wire chemical vapor deposition as passivating and antireflection layer on multi-crystalline silicon solar cells. *Thin Solid Films* 501: 51–54.

3 Asari, S., Fujinaga, T., Takagi, M. et al. (2008). ULVAC research and development of Cat-CVD application. *Thin Solid Films* 516: 541–544.

4 Zheng, W. and Gallagher, A. (2006). Hot wire radicals and reactions. *Thin Solid Films* 501: 21–25.
5 Doyle, J.R., Xu, Y., Reedy, R. et al. (2008). Film stoichiometry and gas dissociation kinetics in hot-wire chemical vapor deposition of a-SiGe:H. *Thin Solid Films* 516: 526–528.
6 Martin, I.T., Teplin, C.W., Stradins, P. et al. (2011). High rate hot-wire chemical vapor deposition of silicon thin films using a stable TaC covered graphite filament. *Thin Solid Films* 519: 4585–4588.
7 Ledermann, A., Weber, U., Mukherjee, C., and Schroeder, B. (2001). Influence of gas supply and filament geometry on large-area deposition of amorphous silicon by hot-wire CVD. *Thin Solid Films* 395: 61–65.
8 Zhang, Q., Zhu, M., Wang, L., and Liu, F. (2003). Influence of heated catalyzer on thermal distribution of substrate in HWCVD system. *Thin Solid Films* 430: 50–53.
9 Ishibashi, K., Karasawa, M., Xu, G. et al. (2003). Development of Cat-CVD apparatus for 1-m-size large-area deposition. *Thin Solid Films* 430: 58–62.
10 Far Associates. Tungsten filament emissivity behavior. http://pyrometry.com/farassociates_tungstenfilaments.pdf (accessed 02 November 2018).
11 Karasawa, M., Masuda, A., Ishibashi, K., and Matsumura, H. (2001). Development of Cat-CVD apparatus – a method to control wafer temperatures under thermal influence of heated catalyzer. *Thin Solid Films* 395: 71–74.
12 Santos, M.T., Muterlle, P.V., and de Carvalho, G.C. (2015)). Emissivity characterization in stainless steels alloys for application in hydroelectric turbines. *Appl. Mech. Mater.* 719–720: 3–12.
13 Matsumura, H., Hayakawa, T., Ohta, T. et al. (2014). Cat-doping: novel method for phosphorus and boron shallow doping in crystalline silicon at 80 °C. *J. Appl. Phys.* 116: 114502–114501, 10.
14 Heintze, M., Zedliz, R., Wanka, H.N., and Schubert, M.B. (1996). Amorphous and microcrystalline silicon by hot wire chemical vapor deposition. *J. Appl. Phys.* 79 (5): 2699–2706.
15 Hobach, C., Beyer, W., and Wagner, H. (1991). Investigation of the precursors of a-Si*H films produced by decomposition of silane on hot tungsten surface. *J. Non-Cryst. Solids* 137–138: 661–664.
16 Ishibashi, K. (2001). Development of the Cat-CVD apparatus and its feasibility for mass production. *Thin Solid Films* 395: 55–60.
17 van der Werf, C.H.M., van Veenendaal, P.A.T.T., van Veen, M.K. et al. (2003). The influence of the filament temperature on the structure of hot-wire deposited silicon. *Thin Solid Films* 430: 46–49.
18 Honda, K., Ohdaira, K., and Matsumura, H. (2008). Study on silicidation process of tungsten catalyzer during silicon film deposition in catalytic chemical vapor deposition. *Jpn. J. Appl. Phys.* 47: 3692–3698.
19 Honda, K. (2008). Change of catalyzing wires used in Cat-CVD system and suppression of its changes. JAIST PhD thesis, March 2008. Chapter 5 (in Japanese).
20 van der Werf, C.H.M., Li, H., Verlaan, V. et al. (2009). Reversibility of silicidation of Ta filaments in HWCVD of thin film silicon. *Thin Solid Films* 517: 3431–3434.

21 Kniffler, N., Pflueger, A., Scheller, D., and Schroeder, B. (2009). Degradation and silicidation of Ta and W-filaments for different filament temperatures. *Thin Solid Films* 517: 3424–3426.

22 Knoesen, D., Arendse, C., Halindintwali, S., and Muller, T. (2008). Extension of lifetime of tantalum filaments in the hot-wire (Cat) chemical vapor deposition. *Thin Solid Films* 516: 822–825.

23 Osono, S., Hashimoto, M., and Asari, S. (2008). "Cat-CVD apparatus", applied from ULVAC. Japanese Patent, Open Number P.2008-300793A, Application Number 2007-148306.

24 Ogita, Y.-I. and Tomita, T. (2006). The mechanism of alumina formation from TMA and molecular oxygen using catalytic-CVD with an iridium catalyzer. *Thin Solid Films* 501: 35–38.

25 Ogita, Y.-I. and Tomita, T. (2006). The mechanism of alumina formation from TMA and molecular oxygen using catalytic-CVD with a tungsten Catalyzer. *Thin Solid Films* 501: 39–42.

26 Ogita, Y.-I., Kudoh, T., and Iwai, R. (2009). Low temperature decomposition of large molecules of TMA using catalyzers with resistance to oxidation in catalytic-CVD. *Thin Solid Films* 517: 3439–3442.

27 Matsumura, H. (2001). Summary of research in NEDO Cat-CVD project in Japan. *Thin Solid Films* 395: 1–11.

28 Matsumura, H., Umemoto, H., Izumi, A., and Masuda, A. (2003). Recent progress of Cat-CVD research in Japan – bridging between the first and second Cat-CVD conferences. *Thin Solid Films* 430: 7–14.

29 Oku, Y., Usuji, T., Totsuka, M. et al. (2004). Cat-CVD technology for improvement of performance of high frequency devices. *Technical Report of Mitsubishi Electric Corp.*, vol. 78, pp. 46 (in Japanese).

30 Osono, S., Kitazoe, M., Tsuboi, H. et al. (2006). Development of catalytic chemical vapor deposition apparatus for larger size substrates. *Thin Solid Films* 501: 61–64.

31 Osono, S., Kitazoe, M., Tsuboi, H. et al. (2004). Development of catalytic chemical vapor deposition equipment for large substrate. *Oyo Buturi* 73: 935–938. (in Japanese).

32 Nakaya, M., Kodama, K., Yasuhara, S., and Hotta, A. (2016). Novel gas barrier SiOC coating to PET bottles through a hot wire CVD method. *J. Polym.* 2016: 7. https://doi.org/10.1155/2016/4657193.

33 Fujinaga, T., Kitazoe, M., Yamamoto, Y. et al. (2007). Development of Cat-CVD apparatus. *ULVAC Tech. J.* 67: 30–34. (in Japanese).

8

Application of Cat-CVD Technologies

The Cat-chemical vapor deposition (CVD) technology has found application and has pursued applications in various fields, not only in the field of electronic devices but also in the field of functional films such as gas barrier films. In this chapter, various examples of application of Cat-CVD technology are presented. Mainly, applications of thin films prepared by Cat-CVD are discussed. A number of applications involving surface treatment using radicals generated in a Cat-CVD apparatus will be introduced in Chapter 9. Many people engaged in the application of Cat-CVD, particularly to solar cells, are using the term, hot-wire chemical vapor deposition (HWCVD), instead of Cat-CVD, and there are many papers using the term HWCVD. However, in this book, for consistency, we use the term Cat-CVD.

8.1 Introduction: Summarized History of Cat-CVD Research and Application

The history of Cat-CVD research has been described in Chapter 1. Since the birth of Cat-CVD technology, various applications have been pursued, and these applications have in turn encouraged fundamental research in this area. The steps in research and the development of the applications are summarized briefly in Figure 8.1.

In the figure, the movement toward industrial implementation is shown with arrows surrounded by thick solid lines and the search for practical applications is shown with arrows of dashed lines. The first successful application is the preparation of passivation films on gallium arsenide (GaAs) and compound semiconductors. Devices using Cat-CVD technology are now available in the market. The application of organic films is also a relatively early successful example. Razors coated with Cat-CVD films are commercially available. Large-scale mass production of solar cells using Cat-CVD films has started in the 2000s, and the market volume of this area appears to be growing rapidly. However, it is not easy to precisely indicate in which solar modules Cat-CVD films are industrially implemented, as most companies do not disclose their production methods. In spite of the lack of public information on implementations of Cat-CVD, the technology has gradually been accepted and adopted in many factories.

Catalytic Chemical Vapor Deposition: Technology and Applications of Cat-CVD,
First Edition. Hideki Matsumura, Hironobu Umemoto, Karen K. Gleason, and Ruud E.I. Schropp.

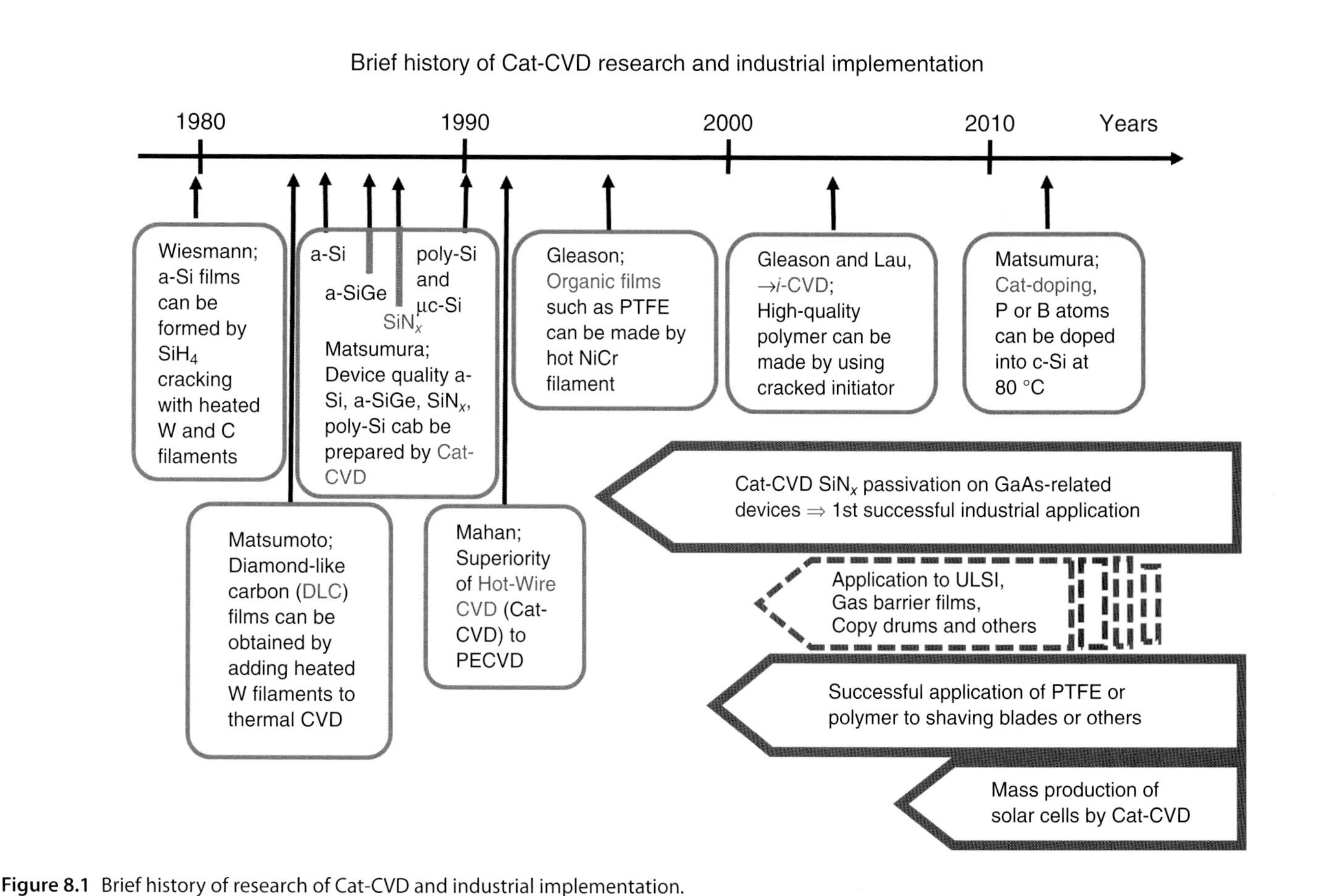

Figure 8.1 Brief history of research of Cat-CVD and industrial implementation.

The research on practical application in the silicon industry making ultralarge-scale-integrated (ULSI) circuits has shown attractive results. The application of Cat-CVD for the preparation of gas barrier films and copier drums appears to be ongoing, but in some cases, it has already been used in mass production. In this chapter, we show various research activities toward application. However, in some cases, we cannot say exactly what has or has not been industrially implemented.

8.2 Application to Solar Cells

8.2.1 Silicon and Silicon Alloy Thin Film Solar Cells

8.2.1.1 Introduction

In 2016, 5.9% of all solar cells produced worldwide were thin film cells, see Figure 8.2. [1]. Only a fraction of this market segment is thin film silicon cells. This fraction is 9.8%. Yet, this represents a market of 500 MWp (megawatts at the peak of sunlight) annually.

Thin film solar cells can be produced at lower rates per square meter. Therefore, even though the efficiency of thin film cells is lower than that of crystalline silicon (c-Si) cells, the cost per Watt peak is lower. At some point in the future, thin film solar cells, including installation costs, may become more cost effective for sustainable electric power generation than conventional c-Si solar cells. For a number of applications, the added advantages are already so unique that thin film methods are the only available production techniques. This is the case

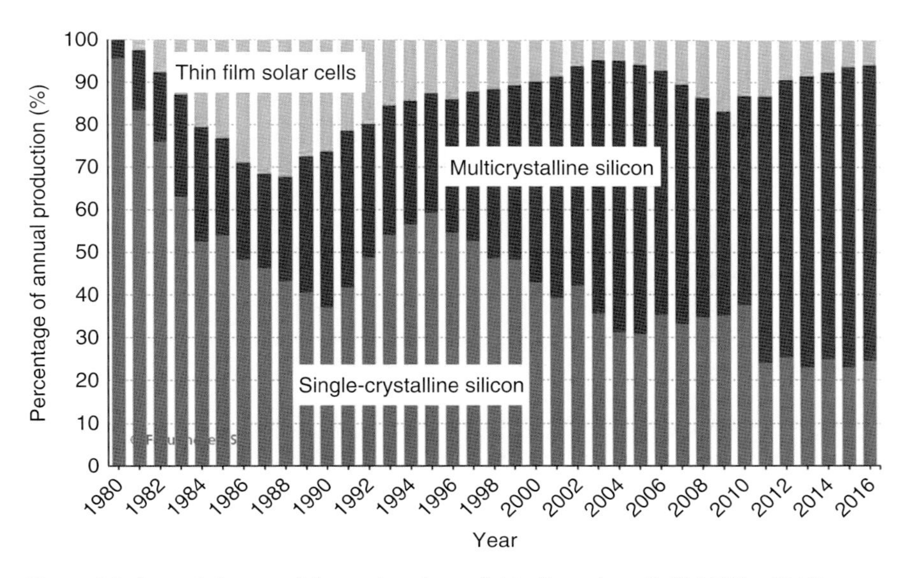

Figure 8.2 Annual change of the market share of thin film solar cells (4.9 GW at 2016), multicrystalline Si (57.5 GW at 2016), and single-crystalline Si (20.2 GW at 2016) solar cells. Source: Reprinted with permission from Ref. [1].

for lightweight and rollable, bendable, or stretchable photovoltaics (PV) on foil. These features may prove crucial for application of PV on consumer articles, vehicles, facades of buildings, and rooftops of residential and industrial buildings that were originally not constructed for carrying the high weight of conventional glass panels.

Another area where thin film solar cells may help a breakthrough is in the high-efficiency segment. Here, complete semitransparent thin film solar cells are added to high-efficiency solar panels to produce a jump in performance through the principle of tandem cell enhancement. Tandem cells or tandem junctions (two or more stacked solar cells with different photoresponse – ratio of photoconductivity and dark conductivity – characteristics) can provide a genuine quantum leap in efficiency over their single-junction counterparts.

The majority of thin film silicon solar cells is made by plasma enhanced chemical vapor deposition (PECVD). The creation of reactive radicals for thin film growth in Cat-CVD is fundamentally different (see Chapter 4). In the Cat-CVD process, the feedstock gases are very efficiently cracked into atomic radicals at the surface of a hot filament (usually tungsten or tantalum) [2], which is held at a temperature higher than 1600 °C.

The reactive species are subsequently transported to the substrate in a low-pressure ambient (typically only 2 Pa for amorphous silicon). Deposition of the silicon layer by Cat-CVD is economically interesting because a high deposition rate is feasible without gas-phase particle formation [3]. It has been shown that ultrahigh deposition rates (more than 100 times that of PECVD) can be achieved [4]. These rates are achieved in a laboratory reactor with only two tungsten filaments located at 3.2 cm from the substrate. The saturated defect density of the hydrogenated amorphous-silicon (a-Si:H) is low, at a value of $2\times10^{16}\,\text{cm}^{-3}$ as measured by the constant photocurrent measurement (CPM) [5]. Even for high deposition rate samples, it is typically $(2–4)\times10^{16}\,\text{cm}^{-3}$ and independent of the deposition rate, up to 13 nm/s, although the void density increases by a factor ~100 [6]. The low defect density means that silicon hydride bonds effectively passivate these voids.

8.2.1.2 Amorphous Silicon Solar Cells

As mentioned briefly in Chapter 1, the first Cat-CVD thin film silicon solar cells were made in 1993 at the University of Kaiserslautern and NREL. In the year 2000, Kaiserslautern developed Cat-CVD-deposited p- and n-type layers for incorporation in p-i-n structures. These cells had an (initial) efficiency of 8.8% for a cell area of $0.08\,\text{cm}^2$ and 8.0% for an area of $0.8\,\text{cm}^2$ [7]. These cells showed a partially irreversible degradation (i.e. not recoverable by annealing), probably because of the high void density of the p-type a-SiC:H layer made at low substrate temperature. A microcrystalline p-type layer, also deposited by Cat-CVD, was effective in increasing the V_{oc} to 900 mV. Later, Kaiserslautern researchers have produced a 10.4% efficient p-i-n cells on SnO_2:F-coated glass [8], which, however, suffered from ~30% efficiency loss because of light-induced degradation. The first a-Si:H/a-Si:H-stacked cells, with the same band gap in the two components, were also made at Kaiserslautern and had an initial efficiency of 7.0%

(V_{oc} = 1.63 V, FF = 0.57). The pin/pin structure was deposited on Asahi U-type SnO_2:F and had a highly reflecting indium–tin oxide (ITO)/Ag back contact. For this structure, a tunnel recombination junction with n- and p-type hydrogenated microcrystalline silicon (μc-Si:H) was developed, deposited entirely with the Cat-CVD technique. NREL reported initial efficiencies of 5.7% in 2001 for n-i-p cells on stainless steel (SS) in a simple structure without back reflector at a deposition rate of up to 50 Å/s [9]. Their stabilized efficiency was 4.8%.

At Utrecht University, researchers have chosen to optimize n-i-p (substrate-type) solar cells [10] because substrate-type cells allow for the use of higher deposition temperatures. This is because n-type layers are more stable in a reactive atmosphere than p-type layers, as was also noted at Kaiserslautern. Thin p-type layers are more prone to deterioration because of the high hydrogen flux in Cat-CVD than n-type layers, because of their higher compositional disorder and concomitant porosity. Moreover, cells in p-i-n configuration (with the intrinsic absorber layers deposited at lower temperature) showed higher sensitivity to light-induced defect creation. Reference [10] presents a materials toolbox for thin film silicon solar cell, which for convenience is reproduced here (Table 8.1).

Later, it was found that n-i-p solar cells with amorphous silicon absorber layers (i-layers) deposited from undiluted silane at 250 °C were remarkably stable, despite the fact that these i-layers were as thick as 400 nm. Their degradation during 1500 hours of 1-sunlight soaking was less than 10%. More remarkably, the i-layers were deposited at 1 nm/s deposition rate. In comparison, to achieve similar performance stability using PECVD i-layers the deposition rate should be reduced to less than 0.3 nm/s. The high stability was attributed to the higher medium range order present in the Cat-CVD absorber layers [11], as evidenced by X-ray diffraction measurements. The linewidth of the first scattering peak in X-ray diffraction measurements was as low as $5.10° \pm 0.09°$, indicating the high medium-range order. Therefore, these films are also sometimes considered to have "proto-crystalline" network structure [12], i.e. at the edge of becoming (micro/nano)crystalline. Further, these films have a high void fraction, consisting of isolated, elongated small voids. It was speculated by Ref. [11] that the void nature provides a higher degree of topological freedom to the amorphous network, which allows the bulk of the film to possess an enhanced medium-range order.

8.2.1.3 Amorphous Silicon–Germanium Alloy Solar Cells

Amorphous semiconductor alloys, such as a-SiGe:H, can also be obtained with high electronic quality [13, 14] using Cat-CVD. Because of the reduced optical band gap, such layers are of interest for application in tandem cells [15]. Early results on a-SiGe:H reported by NREL include the achievement of material with a band gap even close to that of μc-Si:H (1.2 eV) [16].

The amorphous alloys possess a direct optical band gap and the photoresponse (ratio of photoconductivity and dark conductivity) in excess of 2 orders of magnitude at 1.2 eV and 3 orders of magnitude at 1.4 eV [17]. Their work has led to 8.64% single-junction cells, obtained without any band gap profiling in the absorber layer [18].

Table 8.1 HT a-Si:H – amorphous films deposited from pure SiH_4 at 430 °C; LT a-Si:H – amorphous films deposited from a 1 : 1 SiH_4/H_2 mixture at 250 °C; poly1 – polycrystalline films deposited from a highly diluted (1 : 214) SiH_4/H_2 mixture at 500 °C; Poly2 – polycrystalline films deposited at a low SiH_4/H_2 ratio of 1 : 15 at 500 °C.

Properties of various thin film silicon intrinsic absorber materials that have been developed for hot-wire CVD					
Materials property	**Reference (PECVD)**	**HT a-si:H (HWCVD)**	**LT a-Si:H (HWCVD)**	**Poly1 (HWCVD)**	**Poly2 (HWCVD)**
Crystalline fraction (%); Raman				80	95
Average grain size (nm); XRD				28	70
Deposition rate (Å/s)	1.2	18	10	1	5.5
Hydrogen content (at.%); IR	10.5	8	12–13	0.7	0.47
Microstructure parameter	<0.1	<0.1	0.25	n/a	n/a
Ambipolar diffusion length (nm); SSPG	200	260	160	37	568
Photoconductivity $(\Omega\ \text{cm})^{-1}$	1×10^{-5}	4×10^{-6}	3×10^{-5}	1×10^{-6}	2×10^{-5}
Dark conductivity $(\Omega\ \text{cm})^{-1}$	4×10^{-11}	2×10^{-10}	6×10^{-12}	5.6×10^{-7}	1.5×10^{-7}
Band gap (eV)	1.78	1.70	1.80	1.1	1.1
Activation energy (eV)	0.85	0.8	0.9	0.57	0.54
Density of states (cm^{-3}); CPM	5×10^{15}	7×10^{15}			
Density of states (cm^{-3}); ESR				2.9×10^{18}	7.8×10^{16}

CPM, Constant photocurrent measurement; ESR, Electron spin resonance; HT, High temperature; IR, Infra red; LT, Low temperature; SSPG, Steady state photocarrier grating; XRD, X-ray diffraction. In this table, Cat-CVD is described in HWCVD.
Source: Schropp 2002 [10]. Reproduced with permission from Elsevier.

Building on this result, Eindhoven University of Technology has undertaken further optimization, including band gap profiling and multijunction cell optimization. Furthermore, there is a need for faster production of low-bandgap thin film semiconductors because μc-Si:H absorber layers need to have more than 1 μm thickness while it is hard to maintain high electronic quality at deposition rates above 0.5 nm/s. Deposition of a-SiGe:H solar cell absorber layers by Cat-CVD takes three to four times less time, and thus, the vacuum reactor space can be reduced by roughly the same factor, thus reducing the cost of ownership. A thickness of 150 nm is sufficient for the absorber layer, and this can be deposited within five minutes. Moreover, a band gap less than 1.4 eV still yields a good electronic quality material. This could (thus far) not be achieved by PECVD.

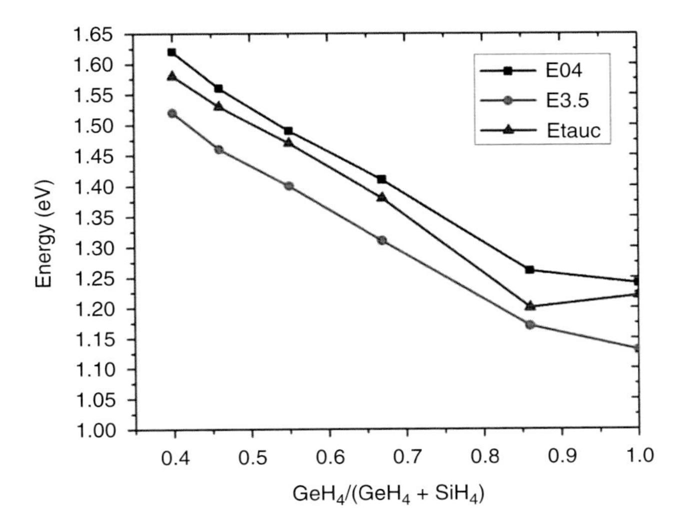

Figure 8.3 Band gap as a function of the $GeH_4/(GeH_4 + SiH_4)$ flow ratio.

Figure 8.3 shows the band gap of the Cat-CVD a-SiGe:H layers (deduced in three different ways from the absorption spectrum) as a function of the $GeH_4/(GeH_4 + SiH_4)$ gas flow ratio, for various definitions of the band gap. It is clear that the $E_{3.5}$ band gap, which is defined as the photon energy at which the absorption coefficient is $10^{3.5}$ cm^{-1}, can be varied over a large range.

By optimizing the n- and p-type layers and the contacts of cells without band gap profiling with E_{gap} = 1.65 eV, an efficiency of 8.94% was obtained (21.31 mA/cm^2, 0.727 V, FF of 0.577, absorber 120 nm thick, measured without light mask). The optimized band gap profiling for cells with E_{gap} as low as 1.37 eV led to an efficiency of 5.73% (27.37 mA/cm^2, 0.458 V, FF of 0.457, absorber 90 nm thick, measured without light mask). The absorber layer of this cell is deposited in just three minutes. Figure 8.4 shows the external quantum efficiency (EQE) spectra for three different band gaps of the absorber layer [19].

The external collection efficiency shows a remarkably high value of 40% at long wavelength, λ = 900 nm. This is at least as high as for μc-Si:H cells with ~2 μm thickness and optimized light trapping. All optimized features were incorporated in an a-SiGe:H/a-Si:H tandem cell, where the bottom cell was a Cat-CVD cell and the top cell a PECVD cell. For this cell, an efficiency of 11.07% was reached (13.0 mA/cm^2, 1.31 V, FF of 0.652, measured without light mask). Later, the current matching was improved and the a-SiGe:H/a-Si:H tandem cells with a 1.37 eV band gap for the bottom cell reached 11.89% efficiency (15.41 mA/cm^2, 1.189 V, FF of 0.6493, without light mask). This is quite remarkable, as the absorber of the bottom cell is only 120 nm (only three minutes and nine seconds deposition time) and that of the top cell is 290 nm.

Finally, an "all Cat-CVD (all hot-wire)" triple junction cell was developed. Here, the three absorber layers (1.86 eV a-Si:H (90 nm), 1.54 eV a-SiGe:H (163 nm), and 1.37 eV a-SiGe:H (90 nm)) were all made by Cat-CVD [20]. The low band gap and high refractive index of the a-SiGe:H materials allow for a

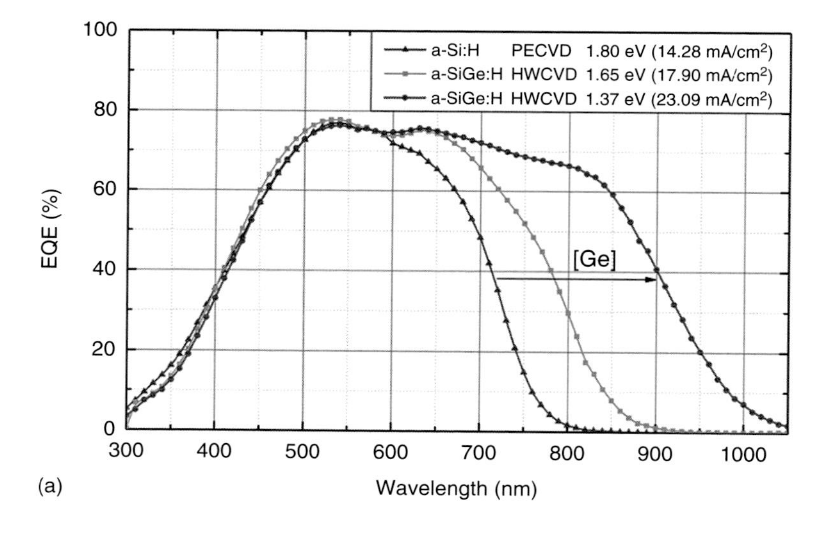

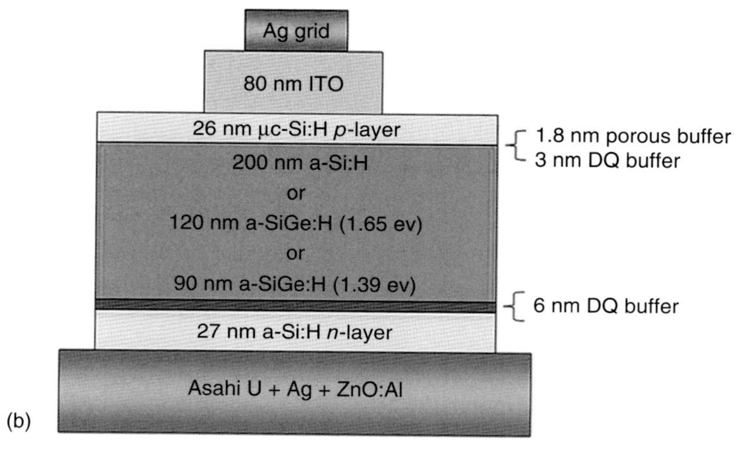

Figure 8.4 (a) The external quantum efficiency (EQE) spectra of three cells with different band gaps. Source: Veldhuizen et al. 2015 [19]. Reprinted with permission from Elsevier. (b) The response extends to the near-infrared spectrum for absorber layers with a decreasing band gap, without sacrificing the response in the blue/green part of the spectrum. Note that the EQE of the 1.37 eV band gap cell is still 40% at 900 nm, while the absorber layer is only 60 nm thick. This thickness is deposited in only three minutes.

thin absorber layer thickness in solar cells, which furthermore preserves the effect of light-trapping textures at subsequent interfaces throughout the device as compared to the devices with the thicker μc-Si:H layers.

Figure 8.5 shows a schematic cross section and schematic band diagram of the cell. The resulting efficiency of 10.2% (8.1 mA/cm^2, 1.84 V, FF of 0.684) was remarkably stable. After 1000 hours of light soaking at AM1.5, 100 mW/cm^2, under V_{oc} conditions, and with artificial cooling to 50°C, the relative reduction of the efficiency is limited to 6%, so that an efficiency of 9.6% remains. The $J-V$ curves are shown in Figure 8.6.

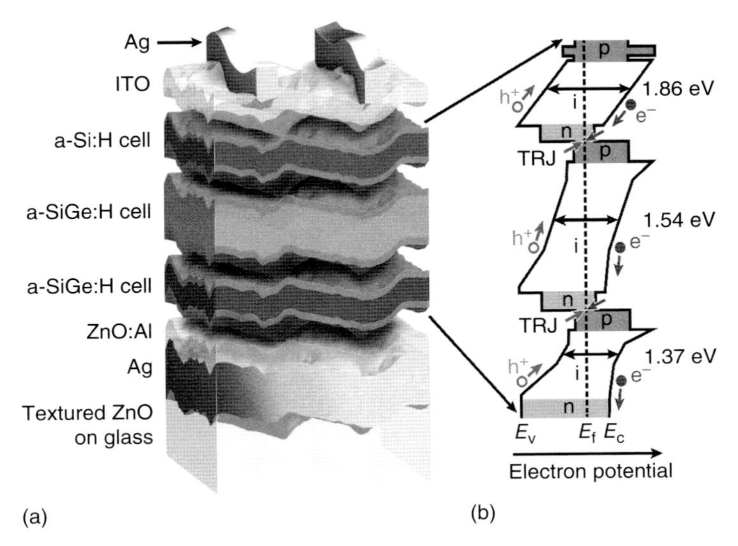

Figure 8.5 (a) Schematic cross section of the triple junction solar cell. The thicknesses of the layers are all to scale, except for the substrate (ZnO on glass). (b) The schematic energy band diagram, where TRJs are the tunnel recombination junctions between the subcells. Source: Veldhuizen and Schropp 2016 [20]. Reprinted with permission from AIP Publishing.

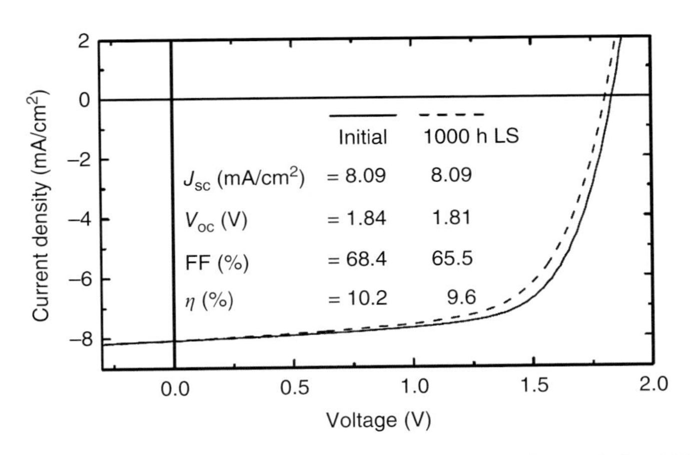

Figure 8.6 Current density–voltage characteristics before and after 1000 hours of light soaking. Source: Veldhuizen and Schropp 2016 [20]. Reprinted with permission from AIP Publishing.

The deposition rate of the silicon alloy semiconductor layers of this cell is 0.4–0.5 nm/s. The deposition time of all photoactive layers together was under as little as 15 minutes. The short deposition time, combined with the stabilized efficiency of the cells, makes Cat-CVD a viable cost-competitive thin film solar cell production method.

8.2.1.4 Microcrystalline Silicon Solar Cells and Tandem Cells

Microcrystalline silicon is a broad category of silicon thin films that have at least some degree of crystallinity. There is no unambiguous definition of microcrystalline silicon: its crystals can be nano- to micrometer sized and it can have a widely varying crystalline volume fraction, even varying over the thickness of the sample itself [21]. It can be a bi- or triphase material, where besides the crystalline phase also an amorphous and/or an intergrain phase is present, usually containing voids. Microcrystalline silicon is often called nanocrystalline silicon; the two terms are roughly interchangeable.

Hydrogenated microcrystalline silicon (μc-Si:H) is of interest as light-absorbing low-bandgap (1.0–1.1 eV) semiconductor layer in thin film silicon tandem cells. Because the nanocrystalline content has an indirect band gap, the absorption coefficient is small, but in most cases, there is internal light scattering, which slightly enhances the total absorption. Enhanced light scattering is often designed into the cell structure by the use of textured surfaces and interfaces. Nevertheless, the μc-Si:H absorber layers need to be considerably thicker than 1 μm, which brings about new challenges, such as producing material that is homogeneous along the growth direction, while the textured (rough) interfaces can lead to cracks in the deposited layer – at the risk of forming shunt paths through the solar cell [22]. Figure 8.7 illustrates the occurrence of cracks, as observed in transmission electron microscope (TEM) images (Figure 8.7a,b) [23], as well as a proper texture in which cracks are avoided (Figure 8.7c).

Optimal films are obtained in a deposition regime that secures complete filling of the volume between nanocrystals with amorphous tissue, thus avoiding long interconnected voids [24]. In contrast to the growth evolution during PECVD, where the μc-Si:H material tends to become increasingly crystalline and void rich as it grows thicker, in Cat-CVD, the material tends to gradually lose its crystallinity. In both deposition techniques, this brings about a defect-rich zone within the film, severely limiting the energy conversion efficiency in the respective cells. In PECVD, this issue is addressed by varying the H_2 dilution of the silane feedstock gas, effectively tapering it down to achieve a constant crystalline volume fraction over the thickness of the film. In contrast, in Cat-CVD, the H_2 dilution needs to be *increased* during deposition. For this reason, the Cat-CVD profiling technique was called *reverse profiling* [25]. The $I-V$ curve can thus be optimized, as shown in Figure 8.8. Cells in the n-i-p configuration on stainless steel coated with a textured Ag/ZnO back reflector, with a 2-μm thick μc-Si:H absorber layer, had a short-circuit current density (under AM1.5100 mW/cm^2 illumination) of 23.4 mA/cm^2, leading to a stable efficiency of 8.52% (0.545 V, FF of 0.668) [26], only slightly lower than that of state-of-the-art PECVD n-i-p cells at that time with a similar device structure [27], which had 8.9% efficiency.

Such layers and cells were used in a-Si:H/μc-Si:H/μc-Si:H and a-Si:H/a-SiGe:H/μc-Si:H n-i-p triple junction cells, with an efficiency of 10.9% (8.35 mA/cm^2, 1.98 V, FF of 0.66) and 9.6% (8.15 mA/cm^2, 1.89 V, FF of 0.62), respectively. The cells with two μc-Si:H cells require special care to achieve current matching: they have rather thick i-layers with a thicknesses of 180, 2400, and 3700 nm, respectively. The cells with a-SiGe:H middle cells have thicknesses 180, 250, and 2000 nm, respectively. The a-Si:H/μc-Si:H/μc-Si:H triple junction cells have

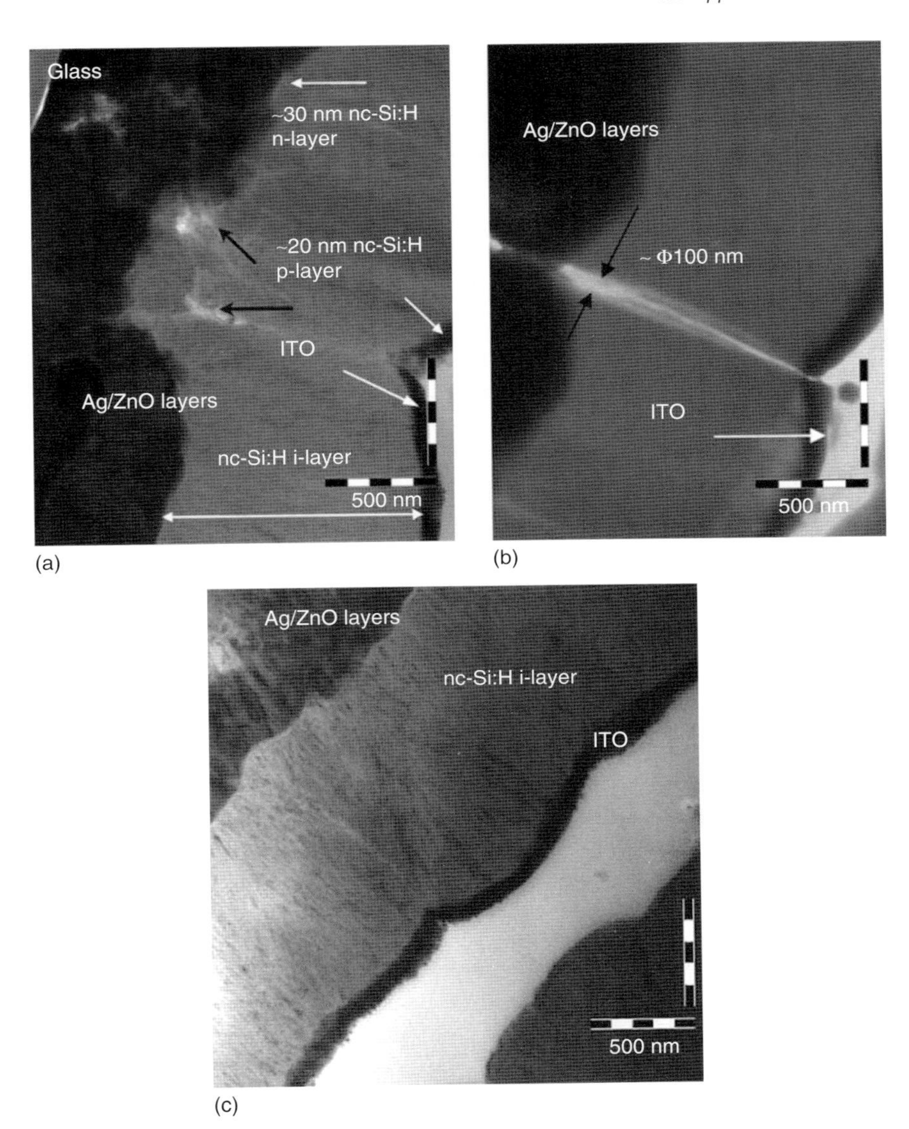

Figure 8.7 (a, b) Two cross-sectional transmission electron microscope (TEM) pictures of a nc-Si:H n-i-p solar cell deposited on a rough Ag/ZnO coating. A Corning glass substrate is used instead of stainless steel to facilitate preparation of the cross section of the sample for TEM. The dark arrows point to the cavities, which are not completely filled with silicon. Source: Schropp et al. 2008 [23]. Figures (a) and (b) are reprinted with permission from John Wiley & Sons. (c) A well-behaved rough back reflector (rms roughness 73 nm). The 1.3-μm-thick μc-Si:H layer has no cracks or cavities.

been exposed to prolonged illumination with 1-sun intensity under open-circuit conditions and a temperature of 50 °C. The efficiency stabilized after 150 hours, and the efficiency degradation was 3.5% relative, after more than 500 hours of light soaking.

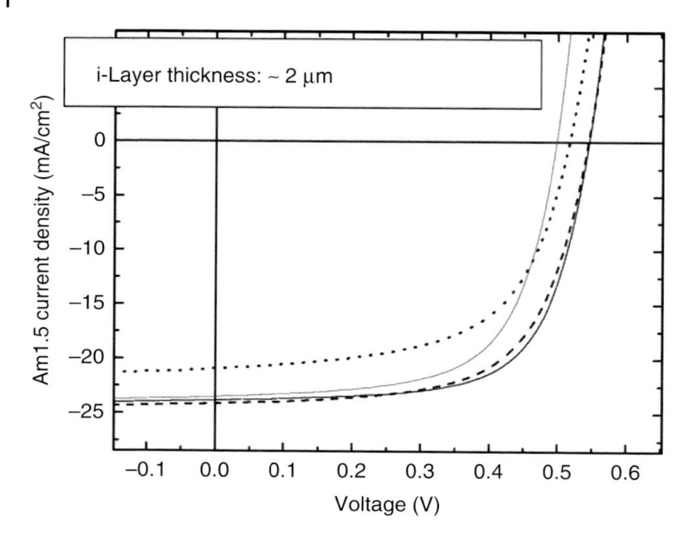

Figure 8.8 AM1.5 *J–V* characteristics of μc-Si:H n-i-p cells on SS with a Ag/ZnO-textured back reflector. Gray line: constant H_2 dilution at $R_H = flow(H_2)/flow\ (H_2 + SiH_4) = 0.953$; dotted line: constant H_2 dilution at $R_H = 0.948$; dashed line: two-step profiling, first half at $R_H = 0.948$, second half at $R_H = 0.952$; solid black line: with the same H_2 profile as two-step profiling, but with a cold start (filaments are deliberately at lower temperature initially). Source: Li et al. 2008 [25]. Reprinted with permission from Elsevier.

8.2.1.5 Nanostructured Solar Cells

In PECVD, nonuniformities in thickness can be caused by variation of the plasma potential over the area of the substrate. This occurs at the scale of the substrate size, as a result of edge effects and due to the size of the wavelength of the high-frequency excitation in comparison to the substrate size (the finite wavelength effect, as discussed in Chapter 2). Therefore, it is very important that the substrate is an equipotential plane. In Cat-CVD, the finite wavelength effect is not an issue because no high-frequency electric fields are used.

A further source of local nonuniformities, taking place at the micrometer and submicrometer scale, is the steepness of local slopes and the presence of protrusions in the substrate surface. In PECVD, these cause variations in local electric field strengths and concomitant variations in impinging ion densities on the (nanoscale-curved) growing surface. As ions contribute to the growth rate in PECVD, the local variations in discharge strength cause local (nanoscale) thickness variations. This is an adverse effect, specifically when using nanostructured substrates, e.g. for light-trapping enhancement, in solar cells. A solution to this issue in PECVD is often found in using strongly decreased deposition rates, as the ion contribution to growth is smaller and longer surface diffusion times for the precursors are available.

This aspect plays virtually no role in Cat-CVD, as in the absence of a plasma, there is no plasma sheath in which ions are accelerated. Likewise, even if a low density of charged particles is present [28], these charged particles do not undergo strong acceleration given the low voltages used in the Cat-CVD assembly (wire and substrate) system. As there is no impact ionization, any contribution of individual ions to film growth, as is the case in PECVD [29], is absent.

In Cat-CVD, as finite wavelength effects, local discharge strength variations, and ion contributions to growth do not play a role, it is easier to achieve conformal coatings at a relatively high deposition rate on high aspect ratio features and nano-sized textures. This is of importance when depositing passivation layers on microelectronic devices with features that have high aspect ratios. Excellent conformity by Cat-CVD deposition in vertical trenches was observed [30].

This advantage of Cat-CVD was also used in nanostructured silicon thin film solar cell development. Examples of these developments are (i) the deposition of a-Si:H on nanostructured plasmonic metallic surfaces [31], (ii) the formation of radial junctions over vertical nanorods [32], and (iii) the deposition of passivation layers on textured and micropillar silicon heterojunction (SHJ) cells [33]. This most recent example is discussed later, in Section 8.2.2.

In the first example for thin film solar cells, 90 nm of a-Si:H is deposited over ZnO-coated silver nanoparticles formed by nanoimprint lithography (Figure 8.9) [34]. A J_{sc} close to 17 mA/cm^2 and an efficiency of 9.4–9.6% was obtained for cells with this thickness of a-Si:H on optimized size distributions of silver nanoparticles [35].

The second example shows conformal coverage of n-i-p solar cells over ZnO/Ag nanorods (Figure 8.10). Here, the ZnO nanorods were grown using a glass substrate with a 1-mm thick sputtered ZnO seed layer submerged in a 0.5-mM solution consisting of equal amounts of zincacetate dihydrate and hexamethylenetetramine at a temperature of 80 °C. The nanorods were over-coated with a thermally evaporated layer of 200 nm Ag and a sputtered layer of 80 nm ZnO:Al as a back contact [33]. The difference in the conformity of thin film layers is easily seen for Cat-CVD (Figure 8.10) and PECVD (Figure 8.11), respectively.

Also, Cat-CVD a-SiGe:H solar cells were made on these ZnO/Ag nanorod substrates [36]. Here, the deposition of the a-SiGe:H absorber layer was performed by using two parallel 0.3-mm Ta filaments at a distance of 35 mm from the substrate. The band gap of a-SiGe:H was as low as 1.37 eV. Note that the thickness of the absorber layer was only 35 nm and that it was deposited within 90 seconds. The J_{sc} generated in this thin absorber layer was 19.1 mA/cm^2. The deposition time of 90 seconds implies a strong reduction with respect to the ~2000 second deposition of an equivalently absorbing μc-Si:H layer by PECVD.

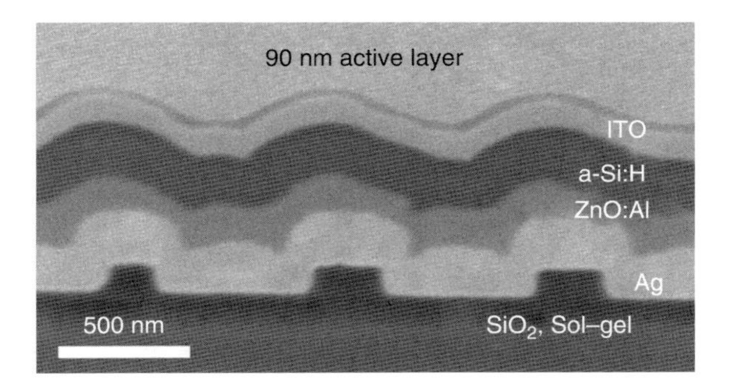

Figure 8.9 Scanning electron microscope (SEM) image of a cross section of a fabricated cell, cut using focused ion beam milling. The 90-nm a-Si:H layer is the dark layer between the ZnO:Al and the ITO layer. Source: Ferry et al. 2011 [34]. Reprinted with permission from OSA Publishing.

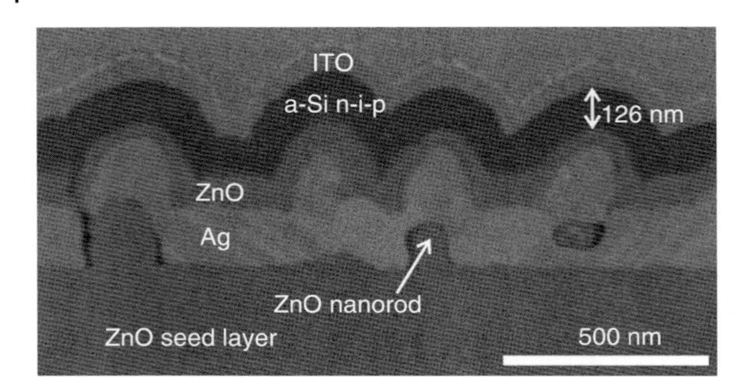

Figure 8.10 A SEM image of an n-i-p solar cell with 126 nm intrinsic a-Si:H layer, deposited by Cat-CVD, over Ag-coated ZnO nanorods.

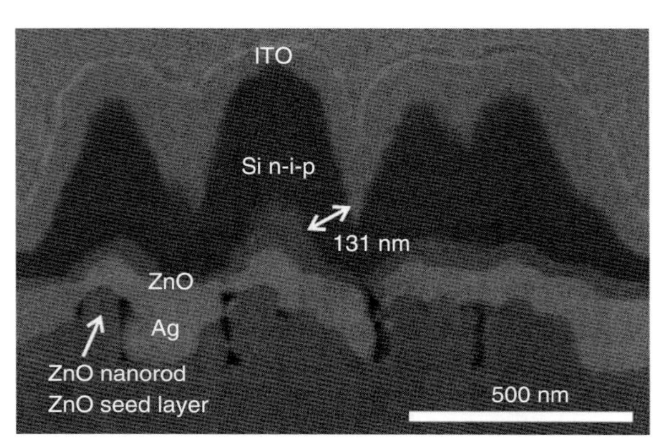

Figure 8.11 A SEM image of an n-i-p solar cell with a nominally 131 nm intrinsic a-Si:H layer, deposited by 13.56 MHz PECVD, over Ag-coated ZnO nanorods. The local thickness on the tips of the rods is larger than that on the sidewalls and in the valleys.

8.2.2 Application to Crystalline Silicon (c-Si) Solar Cells

8.2.2.1 Introduction

Many efforts have been done in the research of silicon thin film solar cells as mentioned above for making low-cost solar cells. During this period, the price of c-Si has also decreased enormously. For instance, the world market price of purified polycrystalline silicon, which is a starting material for making device-grade multicrystalline and single-crystalline Si wafers, was over US$ 500 per kg in 2008, but it rapidly dropped to about US$ 30 per kg in 2012 and it reached about US$ 15 per kg in 2018 [37]. Along with this decreasing price of c-Si materials, the market share of the sum of the multicrystalline Si and the single-crystalline Si solar cells increases as shown in Figure 8.2.

However, even with the reduced price of c-Si, still the Si material cost takes up a high fraction of the price structure of c-Si solar cells. Therefore, the thinning of c-Si wafers used for solar cells is an inevitable movement for lowering the Si material cost. For such a thin single c-Si, the surface passivation becomes highly crucial. The carriers generated in c-Si by sunlight diffuse to the electrodes for power generation. During this movement of carriers, some of the carriers disappear by recombination in the bulk of the c-Si wafer and at the surface of the wafer. When the thickness of c-Si is reduced, the probability that carriers are eliminated by the recombination at the surface increases, as the carriers easily reach the upper and lower surfaces of the wafer. Thus, reducing the surface recombination of carriers by providing excellent surface passivation becomes very important.

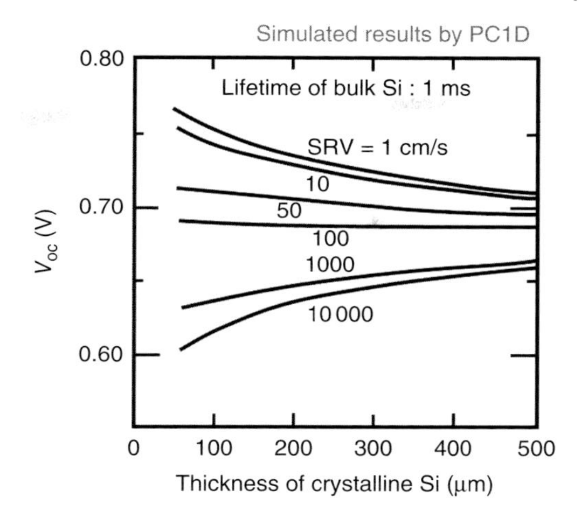

Figure 8.12 Results simulated by using PC1D program for the relationship between open-circuit voltage (V_{oc}) and thickness of crystalline silicon (Si) substrates. The surface recombination velocity (SRV) of carriers is taken as a parameter.

Figure 8.12 shows our simulated results for the relationship between open-circuit voltage (V_{oc}) and the thickness of c-Si wafers used for making solar cells. The simulation was carried out by using *PC1D* program. V_{oc} is linked with the quality of c-Si solar cells and larger is usually better. In the figure, the surface recombination velocity (SRV) is taken as a parameter. Smaller SRV corresponds to lower surface recombination. The figure demonstrates that if SRV is small enough, the use of thinner c-Si results into larger V_{oc}, but that if SRV is larger than 100 cm/s, thinning of c-Si wafer leads to degradation of V_{oc}, and thus to a decrease of overall solar cell quality. The development of high-quality passivation films is, thus, the most important issue in the development of thin c-Si solar cells. In the simulation, the carrier lifetime in bulk Si wafers was assumed as 1 ms. If the value is improved to a larger one, the effect of surface recombination becomes more enhanced and the necessity of high-quality passivation becomes even more serious.

Cat-CVD is expected to be a suitable technique to produce such passivation films, as the films can be deposited at low temperatures without any plasma damage to the surface of substrates. In the following sections, we show some results of passivation of c-Si.

8.2.2.2 Cat-CVD Silicon–Nitride (SiN_x)/Amorphous–Silicon (a-Si)-Stacked Passivation

As mentioned in Section 8.2.3, a hydrogenated amorphous silicon (a-Si) film works as a good passivation material in c-Si. In addition, the interface between silicon–nitride (SiN_x) and a-Si also shows a good interface as verified in thin film transistors (TFTs) with SiN_x gate insulator, as discussed in Section 8.3.1. Utilizing good interface qualities both of a-Si/c-Si and SiN_x/a-Si, we made a SiN_x/a-Si-stacked passivation by Cat-CVD. Before this work, similar structures made by PECVD had been reported, but the quality at that time was not sufficient, until Cat-CVD-stacked passivation was reported [38].

Figure 8.13a shows a schematic image of a sample structure for measuring the carrier lifetime. Both sides of a single c-Si wafer grown by float zone (FZ) method

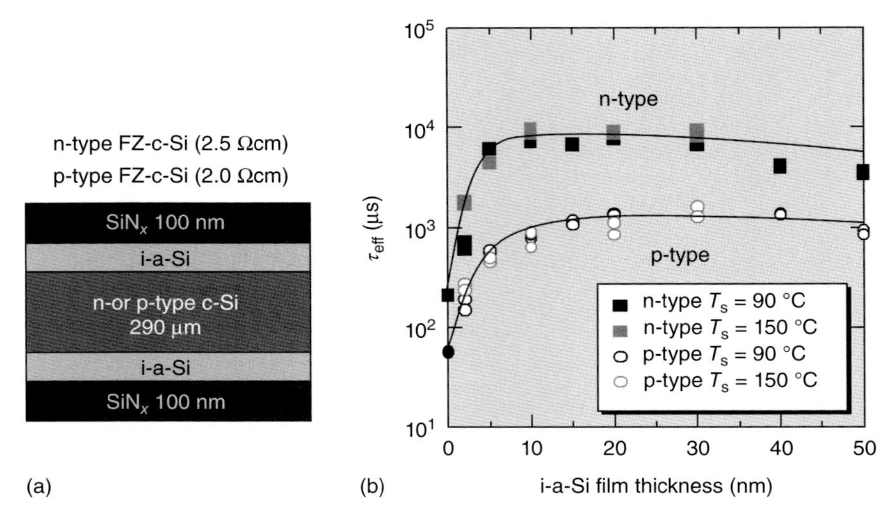

Figure 8.13 Passivation of crystalline silicon (c-Si) by silicon nitride (SiN_x)/intrinsic amorphous silicon (i-a-Si)-stacked layers prepared by Cat-CVD. (a) Device structure and (b) measure carrier lifetime (τ_{eff}) as a function of thickness of inserted i-a-Si layers. Source: Koyama et al. 2010 [38]. Reprinted with permission from IEEE.

were coated by SiN_x/a-Si-stacked layers prepared by Cat-CVD. The deposition parameters of typical SiN_x and a-Si films by Cat-CVD used for this purpose are summarized in Table 8.2. The thickness of the c-Si wafer was 290 μm, and the same stacked layers were deposited on mirror-polished surfaces of both sides of the wafer. The cleaning before the film deposition is mentioned later. The carrier lifetime of the samples was measured by a microwave photoconductive decay (μ-PCD) method using excitation by laser light of a wavelength of 904 nm. μ-PCD is a method to measure the decay of photoexcited carriers by utilizing the reflection of microwaves from the sample surface and the decay rate of the carriers results in a value for the carrier lifetime. Figure 8.13b shows the measured carrier

Table 8.2 Deposition parameters of Cat-CVD i-a-Si and SiN_x films used in stacked passivation on crystalline silicon wafers.

Deposition parameters	i-a-Si	SiN_x
Catalyzing material	W wire	W wire
Surface area of catalyzer, S_{cat} (cm^2)	31	31
Temperature of catalyzer, T_{cat} (°C)	1800	1800
Catalyzer-Substrate distance, D_{cs} (cm)	12	8
Flow rate of SiH_4, FR(SiH_4) (sccm)	10	8.5
Flow rate of NH_3, FR(NH_3) (sccm)	–	200
Gas pressure, P_g (Pa)	0.55	10
Substrate temperature, T_s (°C)	90 and 150	250

FR, Flow rate.

lifetimes of the SiN_x/a-Si-coated n-type c-Si and p-type c-Si, as a function of the thickness of the a-Si insertion layer. The resistivity of the n-type c-Si was 2.5 Ω cm and of the p-type it was 2.0 Ω cm. The resistivities of these n-type and p-type c-Si wafers are suitable for use in solar cell production. The results at $T_s = 90$ and 150 °C are given in the figure as a function of the thickness of i-a-Si layers. Compared with p-type c-Si, the carrier lifetime of n-type c-Si is much higher and it reaches 10 ms when the thickness of the inserted a-Si layer is 10 nm. As n-type c-Si is commonly used in the heterojunction solar cells described below, the results for n-type c-Si are very encouraging.

Here, if we assume that the carriers recombine only at the surface, that is, if we neglect carrier recombination in the bulk of the c-Si, the maximum possible SRV (SRV_{max}) can be estimated from the following Eq. (8.1).

$$\frac{1}{\tau_{eff}} = \frac{1}{\tau_{bulk}} + \frac{2S}{W} \tag{8.1}$$

where τ_{eff}, τ_{bulk}, S, and W refer to the effective carrier lifetime determined by measurement, the carrier lifetime determined by the carrier recombination inside c-Si, SRV of this case, and the thickness of the c-Si wafer, respectively. Here, if we assume that τ_{bulk} is infinite, S becomes SRV_{max} and is derived as $(W/2\tau_{eff})$. In this example, SRV_{max} is derived to be 1.45 cm/s. The value is already small enough. However, if we measure τ_{eff} for various thicknesses of c-Si wafers, we can derive a real SRV.

Figure 8.14 shows the relationship between $(1/t_{eff})$ and $(1/W)$. From Eq. (8.1), it is known that the gradient of plots for $(1/t_{eff})$ vs. $(1/W)$ relation is equivalent to the real SRV and the intercept of an extrapolated line with a $(1/t_{eff})$ axis is equivalent to τ_{bulk}. From this, the real SRV by Cat-CVD SiN_x/a-Si-stacked passivation is 0.18 cm/s. We do not know whether PECVD SiN_x/a-Si-stacked passivation can obtain a similarly low SRV value. At least, by using Cat-CVD

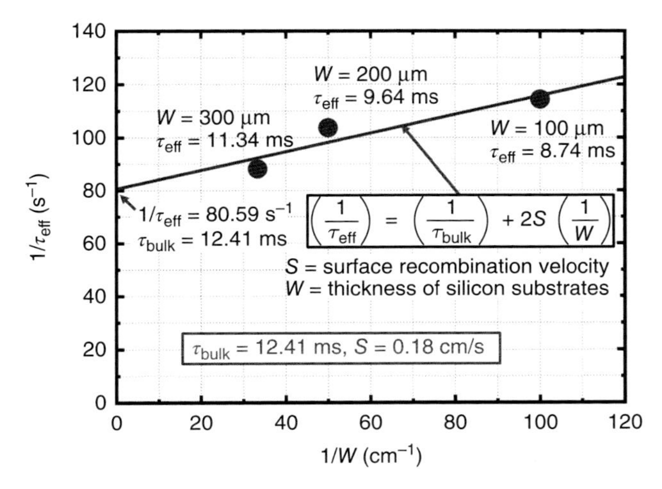

Figure 8.14 The reciprocal of carrier lifetime (t_{eff}) vs. the reciprocal of thickness (W) of crystalline silicon wafers. Source: Nguyen et al. 2017 [39]. Reprinted with permission from Japanese Journal of Applied Physics.

SiN_x/a-Si, we can obtain extremely low SRV, *lower than* ever reported. The value leads to the expectation of even higher V_{oc}'s for thin c-Si solar cells, as can be inferred from Figure 8.12.

8.2.2.3 Cat-CVD SiN_x/a-Si-Stacked Passivation on Textured c-Si Substrates

The above results were obtained with passivation films deposited on mirror-polished c-Si wafers. However, in practical c-Si solar cells, textured structures are formed at the surface of c-Si substrates to confine sunlight effectively. Figure 8.15 shows a cross-sectional scanning electron microscopic (SEM) image of such a textured structure, as an example. The passivation films have to be deposited on such a complicated surface. For obtaining good passivation, the deposited films have to have good coverage feature and also the surface has to be cleaned before film deposition. Apart from excellent film deposition, excellent surface cleaning is required before film deposition for realizing high-quality passivation. In PECVD, as shown in Figure 2.9, Si atoms within a depth of about 2 nm from the surface are reshuffled by the plasma attack and cause a rough surface, because of which any contaminants adhering to the surface may also be removed. However, in Cat-CVD, films are deposited softly and therefore removal of contaminants is not expected to take place in the absence of any special treatment before Cat-CVD deposition. Atomic H cleaning, described in Chapter 9, may be an answer in this case. As we have to keep the optical reflectivity of the textured structure low, the choice of cleaning methods is limited. To benefit from the advantages of Cat-CVD, we have to develop the proper cleaning process for the textured structures, which does not change the textured structure.

We have investigated several cleaning methods for the textured surface and found that the viscosity or the surface tension of the chemical solution used for cleaning is an important factor in preserving the textured surface. The optimal cleaning process appears to be dependent on the chemicals used in making the textures. We have to develop a cleaning chemical process that fits the specific texture process. For our textured surfaces, which are usual for commercial solar cells, a cleaning treatment using a 96–98% concentrated sulfuric acid (H_2SO_4)

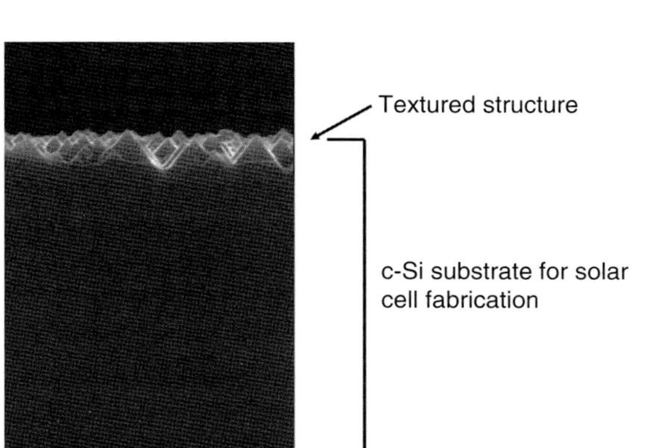

Figure 8.15 Scanning electron microscopic (SEM) cross-sectional view of textured surface of crystalline silicon substrate.

solution was effective when it was heated to lower the viscosity and allow the solution to penetrate into narrow textured valleys [39].

Figure 8.16 shows the carrier lifetime of c-Si substrates coated with Cat-CVD SiN_x/a-Si-stacked passivation layers on both sides, as a function of the temperature of the H_2SO_4 solution for cleaning the c-Si substrates. The results for both flat and textured c-Si substrates are shown in the figure. In the figure, the results by cleaning using a mixture of H_2SO_4 and hydrogen peroxide (H_2O_2) solution, which is known as the "piranha" solution and also known as a strong cleaning solution, are plotted by X-marks together with these results for comparison. For the flat substrates, changing the viscosity does not make any difference, but for textured substrates, the elevation of the temperature induces a remarkable improvement. The carrier lifetime reaches 7–8 ms for the textured substrates. This appears better than the best value ever reported for PECVD passivation, as indicated in the figure. After cleaning, the shape of the texture had not changed at all, and the reflectivity had not changed. The reflectivity after deposition of stacked passivation layers was kept within 3% in the range of wavelength from 400 to 1050 nm. The passivation on practical textured surface by Cat-CVD again appears to add to the advantages of Cat-CVD.

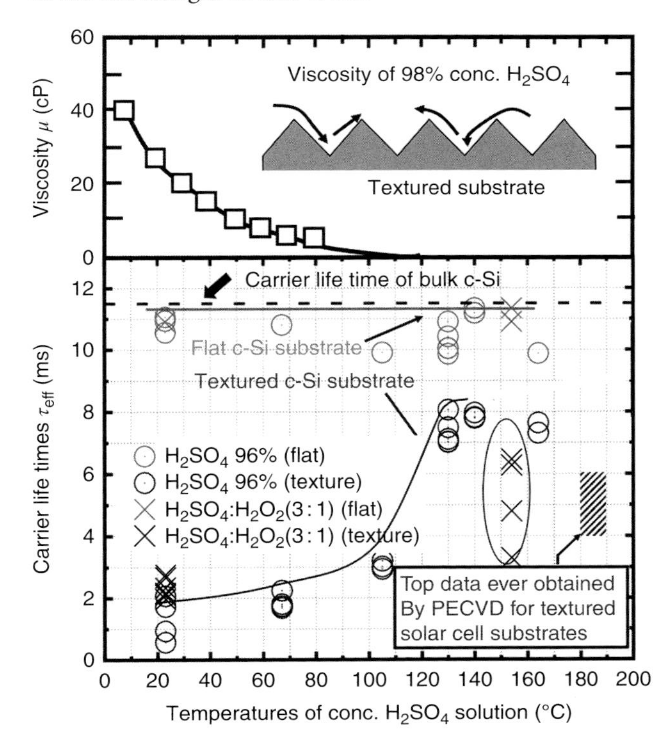

Figure 8.16 Viscosity of concentrated sulfuric acid (H_2SO_4) solution used in cleaning process and carrier lifetime of crystalline silicon wafers coated by Cat-CVD SiN_x/a-Si-stacked passivation layers after cleaning process using H_2SO_4 solution, as a function of its temperature. Source: Nguyen et al. 2017 [39]. Reprinted with permission from Japanese Journal of Applied Physics.

8.2.3 a-Si and c-Si Heterojunction Solar Cells

8.2.3.1 Introduction

As a useful application of Cat-CVD passivation, a-Si/c-Si heterojunction (SHJ) solar cells should be first considered. As mentioned above, for reducing the cost of materials, the thickness of the c-Si substrates has to be reduced. When the substrates are thinned, it is very hard to use high temperatures around 800 °C, which are used in the conventional fabrication process of c-Si solar cells. The substrates are easily damaged by thermal distortion and wafer handling becomes very delicate, making thin c-Si substrates unsuitable for low-cost mass production. Thus, if high-temperature diffusion for the fabrication of p-type layers and n-layers could be replaced by doped a-Si layers deposited on c-Si at low temperature, c-Si solar cell production can be free from temperature problems. In addition, it is recognized that the efficiency of SHJ solar cells is usually higher than that of the conventional c-Si solar cells using thermal diffusion process. The SHJ cell is expected to be a major solar cell type in future, if a low cost fabrication method is developed.

8.2.3.2 Surface Passivation on c-Si Solar Cells

Surface passivation of the a-Si:H/c-Si interface is a key feature in optimizing the performance of SHJ solar cells. To obtain open-circuit voltage (V_{oc}) values reaching to or exceeding 750 mV for such cells, excellent surface passivation is required. The company Sanyo [40] has first introduced the concept of depositing a thin intrinsic layer of a-Si:H between the base and the emitter of SHJ solar cells. Continued research has resulted in a world record level efficiency of 24.7% for c-Si solar cells with 101 cm^2 area, using such heterojunction interfaces in the conventional bifacial structure [41] and of 26.6% for c-Si solar cells with a 180 cm^2 area in a fully back-contacted design at Kaneka Corp. in 2017 [42].

The a-Si:H thin layer is used to render the surface dangling bonds of the wafers electrically inactive by saturating these bonds with atomic hydrogen that is present in ultrathin films of such hydrogenated materials. These films must be made very thin (about 4 to 7 nm) and conformal as they bring about parasitic absorption. PECVD has been often used for the fabrication of a-Si:H for SHJ devices. However, Cat-CVD is an interesting alternative due to the almost complete absence of ions or electric field during deposition, which eliminates the risk of ion bombardment damage. The efficiency of gas use in Cat-CVD is much higher than that in PECVD as explained in Chapter 2, the use of halogen gas is avoided for chamber cleaning by using the system shown in Figure 2.12, and also the passivation quality of Cat-CVD films is usually better than that of PECVD to realize higher solar cell efficiency. All these factors contribute the reduction of fabrication cost per watt, and thus, Cat-CVD is expected to make SHJ solar cells a major c-Si solar cell type in future. In addition, as Cat-CVD passivation films do not need additional annealing, the process can be simplified [43]. Moreover, Cat-CVD is a suitable in-line deposition technique and can lead to lower cost-of-ownership in the mass fabrication of SHJ cells [3, 44].

The advantage of conformal coverage appeared in Cat-CVD films as discussed in Section 8.2.1 can be utilized to cover small size textures with passivation films. As small textures will be essentially used in thin c-Si cells, this is also a useful advantage of Cat-CVD. Contrary to it, when using PECVD for the deposition of a-Si:H, randomly textured surfaces of SHJ wafers sometimes require a smoothening treatment that is needed to reduce the risk of localized recombinative paths situated at the pyramid valleys [45, 46].

Although the efficiency of SHJ solar cells fabricated by Cat-CVD exceeds 25%, as the companies do not make open their data, here, we show our preliminary results as an example. SHJ cells were made using 160-μm-thick, KOH-textured, n-type Czochralski wafers with ⟨100⟩ orientation. Intrinsic a-Si:H films of thickness 4 nm were deposited by Cat-CVD on both sides of the wafer at a rate of ~0.4 nm/s using two heated (1700 °C) parallel 0.3-mm tantalum filaments at a distance of 35 mm from the substrate. Before deposition, only a cleaning treatment for two minutes in a 1% HF solution was applied. The deposition was performed at a substrate temperature of 130 °C and at a pressure of 2 Pa, from undiluted silane. The doped layers on the front and back side were made by PECVD. Indium tin oxide layers of 80 nm thickness were deposited on the front and back side by magnetron sputtering. The back contact mirror layer and front contact grid fingers were made of Ag by thermal evaporation. After fabrication, the cells were annealed in a nitrogen atmosphere at 190 °C for three hours. The average V_{oc} of the finished devices was 704 mV, which is exemplary for the excellent passivation capabilities of the 4-nm-thick Cat-CVD a-Si:H passivation layer. Figure 8.17 shows both the J–V and EQE characteristics of the highest efficiency solar cell. The efficiency is 19.4%, for cells with an active area of 0.69 cm^2 [33].

As mentioned above, the efficiency of SHJ solar cells fabricated by Cat-CVD has exceeded 25% in laboratory level, and the fabrication cost by Cat-CVD is

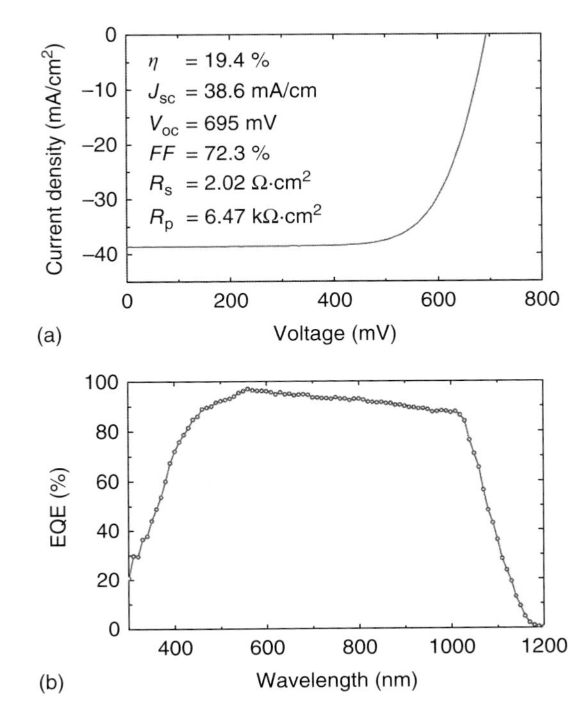

Figure 8.17 (a) Current density–voltage characteristics, which were scaled to match the short-circuit current density (J_{sc}) value that is extracted from the EQE measurements (the integrated J_{sc} from EQE was only 5% lower than the J_{sc} value as measured with the solar simulator) and (b) external quantum efficiency (EQE) results of a silicon heterojunction solar cell with a-Si:H passivation layers deposited by Cat-CVD. The series resistance (R_s) and the parallel or shunt resistance (R_p) are derived from the tangent to the current density at the V_{oc} and J_{sc} positions, respectively. Source: Veldhuizen et al. 2017 [33]. Reprinted with permission from Elsevier.

apparently lower than that of PECVD in mass-production experience. Therefore, several solar cell companies have installed Cat-CVD mass-production machines for fabricating low cost SHJ solar cells and achieved 23% efficiency in mass-production cells in 2017 [47]. Application of Cat-CVD to SHJ solar cells is expected to be an area of the greatest impact.

8.3 Application to Thin Film Transistors (TFT)

8.3.1 Amorphous Silicon (a-Si) TFT

8.3.1.1 General Features of a-Si TFT

As discussed in Chapter 5, the Cat-CVD technique is very suitable for the fast, large area deposition of hydrogenated amorphous silicon. A high-quality semiconductor material, such as amorphous and polycrystalline thin film silicon, is currently applied in drivers for image scanners, in active matrices for liquid crystal displays (LCDs), and for organic light-emitting diode (OLED) displays.

A drawback of amorphous silicon TFTs is the appearance of new dangling bond defects after prolonged stress, e.g. after long durations of light exposure and/or free charge accumulation (as is the case in TFTs operation by the application of a gate bias). The additional defects give rise to a shift in the threshold voltage (V_t), which demarcates the transition between the *off* and the *on* state of the transistor [48]. In fact, any disturbance of the equilibrium Fermi level gives rise to defect creation. The effect is reversible and the thus created metastable dangling bonds can be annealed out by heating the material to 150 °C for several hours in the dark [49].

As the excess hydrogen in a-Si:H is thought to play an important role in the metastable behavior of a-Si:H [50], much research has been performed to reduce the hydrogen content in a-Si:H to an amount not much higher than strictly needed for passivation of disorder-induced dangling bonds. In 1991, Cat-CVD appeared to be the first promising new technique to prepare device quality a-Si:H with a low hydrogen content [51]. Moreover, a high deposition rate, roughly 10 times of that obtained in PECVD, could be obtained. This triggered the research on Cat-CVD TFTs in 1996. Although the first TFTs had low field-effect mobility [52], their performance was improved by optimizing the interface state density, and a state-of-the-art mobility of 0.8 cm^2/V s was obtained, sufficient for matrix switching operation in LCDs [53]. Moreover, upon prolonged gate bias stressing, the threshold voltage instability (ΔV_t) was limited to only 0.3 V. For comparison, the ΔV_t for typical PECVD TFTs is larger than 3 V [54].

The intrinsic stability of Cat-CVD TFTs is an asset in driver matrices, in particular for OLED displays because the TFTs in such displays are in the charge accumulation state as long as the pixel is on, which means that the amorphous silicon is in a nonequilibrium state a much larger fraction of the time than in LCD panels. In 2002, this stability was completely eliminated so that the threshold voltage shift was $\Delta V_t = 0.0$ V [55], by using hydrogen dilution of silane during growth of the channel layer. The semiconductor channel layer was called heterogeneous silicon because of a small fraction of nanocrystals present in the layer. A comparison of threshold voltage shifts in PECVD (glow discharge, GD a-Si),

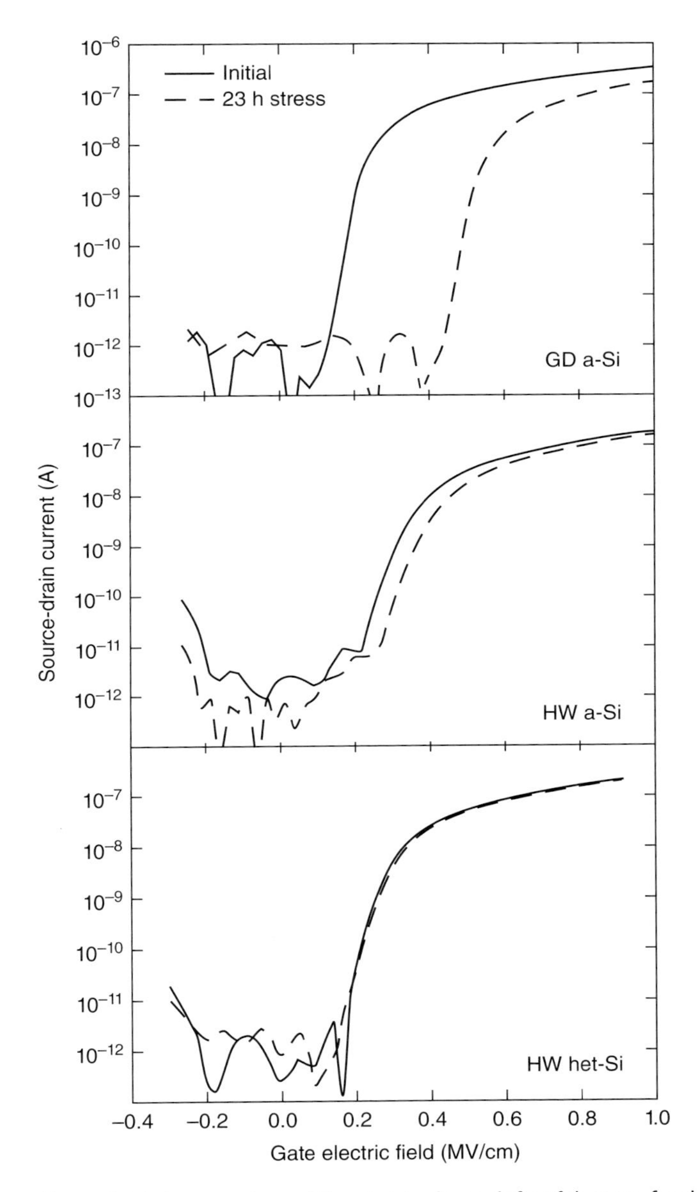

Figure 8.18 A comparison of threshold voltage shifts of the transfer characteristics after 23 hours of continuous bias voltage stress in PECVD (GD a-Si), Cat-CVD (HW a-Si), and heterogeneous silicon Cat-CVD (HW het-Si) TFTs. The solid curves are the initial characteristics and the dashed curves are taken after the stress treatment. Source: Schropp et al. 2002 [55]. Reprinted with permission from Elsevier.

Cat-CVD (Cat a-Si), and heterogeneous silicon Cat-CVD (Cat het-Si) is shown in Figure 8.18.

Although the above first demonstrations of feasible Cat-CVD a-Si:H TFTs were typical laboratory examples using thermally grown SiO_2 as the gate insulator, effort was undertaken to develop "all-Cat-CVD" TFTs using SiN_x deposited by

Cat-CVD as the gate insulator to establish a true thin film device and to facilitate continuous deposition of the entire device at a low processing temperature.

To that end, in Chapter 5, the preparation of SiN_x by Cat-CVD is discussed. In a feasibility study of ultrahigh deposition rates, 7 nm/s was obtained, primarily by increasing the process pressure to 0.8 mbar (80 Pa) and recalibrating the SiH_4/NH_3 gas mixture to obtain a dielectric with a N/Si ratio of 1.2 [56]. Layers made at a high rate of 3 nm/s were used in novel TFTs [57]. Along with the deposition rate of 1 nm/s for a-Si:H, this should lead to a deposition time for the SiN_x/a-Si:H stack of less than four minutes. The n^+-doped μc-Si:H regions under the source/drain contacts are 20–30 -nm thick.

In addition to the high deposition rate of SiN_x, an added advantage of the Cat-CVD process is that the feedstock SiH_4 gas is efficiently used. The utilization rate of SiH_4 for the layers made at 3 nm/s is as high as 77%. The H content in the films is still only 9 at.%, despite the high deposition rate. SiN_x is very dense and has a mass density of 3.0 g/cm^3. This also leads to high etch resistance as judged by the 16 buffered hydrofluoric acid (BHF) etch rate of only 7 nm/min.

The test structure is depicted in Figure 8.19. The devices were postannealed in forming gas and the device characteristics showed a threshold voltage of 1.7–2.4 V, a mobility of 0.24–0.38 cm^2/V s and ON/OFF ratios of 10^6.

8.3.1.2 Cat-CVD a-Si TFT: Differences from PECVD a-Si TFT

Apart from the practical advantages of Cat-CVD a-Si TFT, there are also fundamental differences in Cat-CVD a-Si TFT from the conventional PECVD a-Si TFT. In Figure 8.18, we already showed the drain current–gate voltage characteristics and revealed that Cat-CVD a-Si TFT was more stable than the PECVD one. If we measure the characteristics of Cat-CVD a-Si TFT carefully in the off current region, although the measurement of such low currents is not carried out so often because the value of the off currents is already low enough for LCD uses, the difference between Cat-CVD a-Si TFT and PECVD a-Si TFT becomes clear.

Figure 8.20a,b shows the drain currents I_D–gate voltage V_G characteristics for Cat-CVD a-Si TFT and PECVD a-Si TFT. The channel width (W) and the channel length (L) are written in the figure for both TFTs [58]. The structure of the a-Si TFT is drawn schematically in Figure 8.20c. The process temperature was 320 °C for Figure 8.20a and 180 °C for Figure 8.20b. All TFTs were fabricated

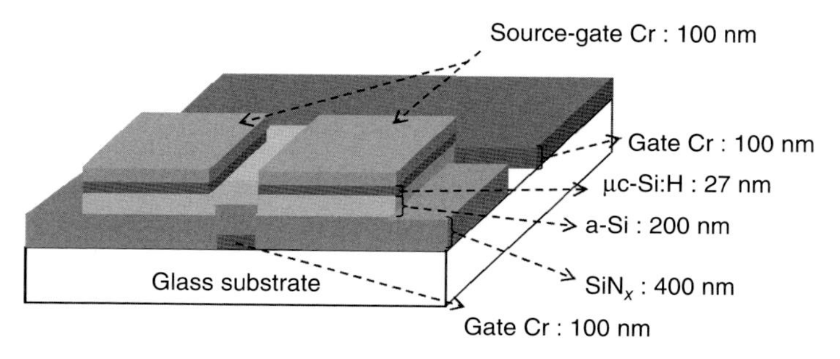

Figure 8.19 Schematic layout of TFT. Source: Schropp et al. 2008 [57]. Reprinted with permission from Elsevier.

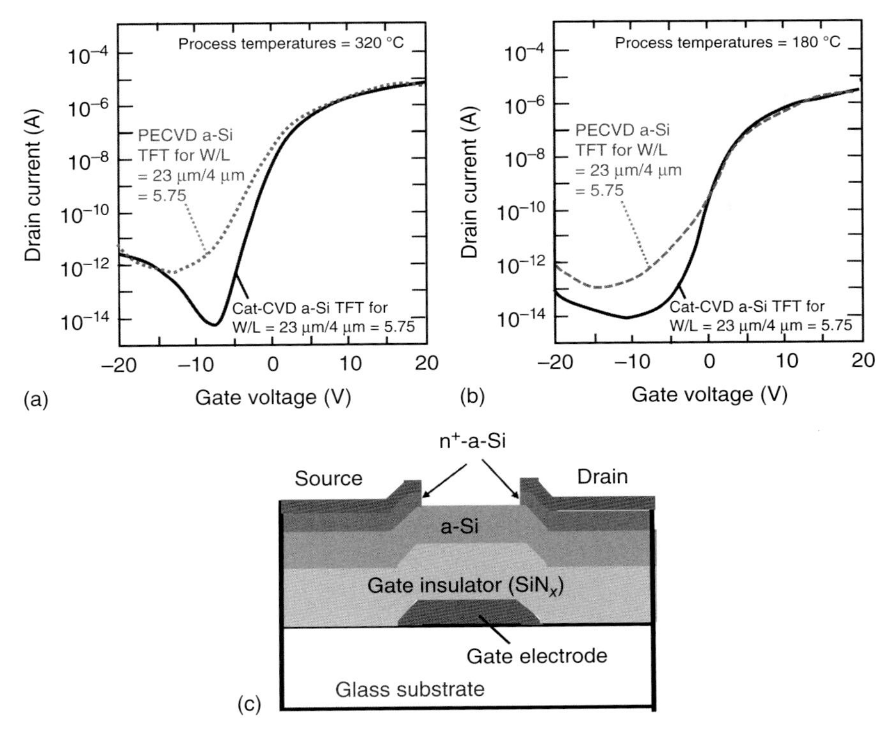

Figure 8.20 (a) Drain currents–gate voltage characteristics of Cat-CVD a-Si TFT and PECVD a-Si TFT. The process temperatures including deposition temperatures were 320 °C. (b) Similar characteristics for the process temperatures of 180 °C. The channel width W and the channel length *L* for these TFT are all written in the figure. (c) Schematic diagram of the TFT structure. Source: Nishizaki et al. 2009 [58]. Reprinted with permission from Elsevier.

in the conventional test line of an LCD company, except for the deposition of Cat-CVD films. The data of PECVD TFTs is provided by them. The off currents of both PECVD and Cat-CVD TFTs reach to values less than 10^{-12} A. However, the lowest off currents approach values less than 10^{-14} A in the case of Cat-CVD a-Si TFT fabricated both at 320 and 180 °C. Probably, this occurs because the defect density at the middle of the band gap in Cat-CVD a-Si is lower than that in PECVD a-Si even if the process temperatures are lowered to 180 °C. It is believed that lowering the process temperatures to 180 °C or less makes the selection of transparent substrates easier to reduce the cost of LCD.

In a PECVD TFT, a-Si films are deposited on the gate SiN_x layers. If plasma deposition of a-Si introduces defects in SiN_x near the interface of a-Si/SiN_x, the existence of such defects may create another conduction path through defects and the off current may be increased and the TFT performance would be degraded. In addition to the stability of a-Si film itself, the quality of SiN_x at the interface between a-Si and SiN_x is also very important. If the defects are created by plasma in a region of the SiN_x near the interface, for instance, some of the electrons moving in the channel region of a-Si may be trapped at such defects in the SiN_x and induce an extra potential that changes the threshold voltage.

This may also be a reason of instability of TFT operation [59]. Actually, this type of threshold voltage instability (by trapped charges in the gate insulator) was sometimes observed, as the threshold voltage could return to the original value by applying voltage of opposite polarity.

Here, we show the effect of plasma exposure during device fabrication in later TFT operation of the device. To determine the plasma effect, the TFT was exposed to Ar plasma or H_2 plasma for 30 or 180 seconds [60]. The experimental procedure is illustrated in Figure 8.21a. The TFT is exposed to plasma from the surface side of a-Si, not at SiN_x directly. The results are not direct evidence of the effect of the plasma damage to the gate insulator; however, the effect of plasma exposure is clearly observed. As the surface of a-Si is damaged, the current leakage path through defects was made at the surface of a-Si and it increased the off currents as shown in Figure 8.21b. Similarly, it can be expected that the SiN_x gate insulator is as easily damaged during the initial step of a-Si deposition. By using Cat-CVD a-Si, we can make TFTs with extremely low off currents in some cases, keeping the on currents the same as for PECVD TFTs.

If the W/L ratio is increased, both on and off currents are increased by the same fraction. As the off currents are much smaller than those of PECVD TFT, the on currents can be increased in this way, while limiting the off currents to the off current level of PECVD a-Si TFT. The drive currents of a-Si TFT can thus

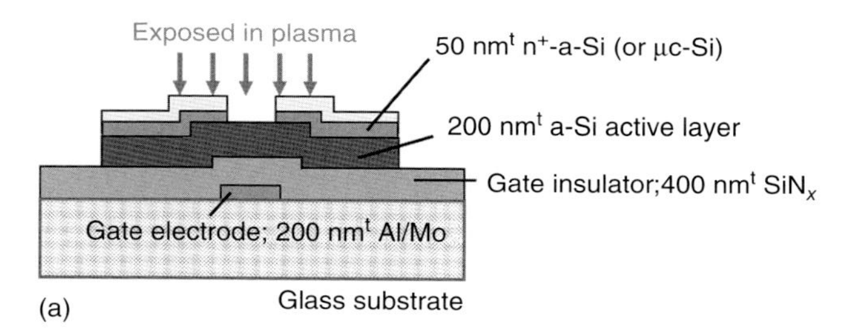

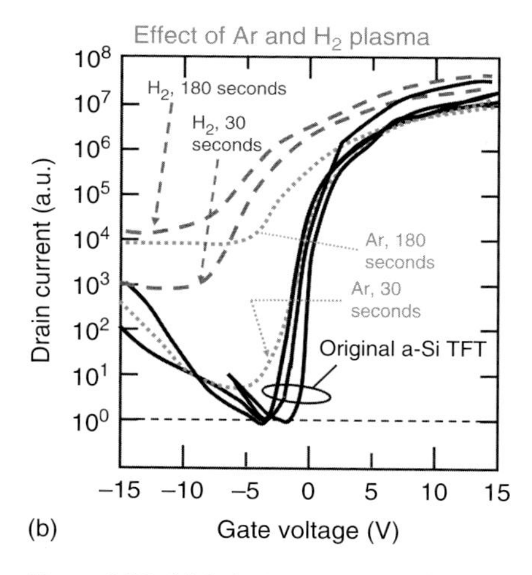

Figure 8.21 (a) A device structure of the experiment to determine the effect of plasma exposure. (b) Drain current–gate voltage characteristics of Cat-CVD a-Si TFT before and after exposure to H_2 plasma or Ar plasma, for 30 or 180 seconds. Source: Matsumura et al. 2011 [60]. Reprinted with permission from Elsevier.

be improved to almost the same level of poly-Si TFTs mentioned below [61]. The high current drivability becomes important for larger and high definition displays such as a 8 K definition display, which is expected to be the next-generation television format. This means that Cat-CVD is also useful in the future.

8.3.2 Poly-Si TFT

To increase the size of active-matrix LCDs and to relax the design rules, a-Si:H pixel TFTs are often replaced by polysilicon TFTs. Poly-Si TFTs have a higher field-effect mobility and allow larger drive currents, so that the pixel aperture ratio can be increased and bright, low-power LCDs can be obtained even when they have a large panel size. Preferably, poly-Si should be obtained as deposited – by direct deposition – so that extra crystallization steps (e.g. by laser or flash lamp) can be eliminated from the processing sequence. Formation of polycrystalline silicon by Cat-CVD was reported in 1991 [62]; however, in 1997, it was also shown that Cat-CVD can deliver high-quality intrinsic poly-Si layers usable for TFT without such post-treatments [63]. Using a low H_2 dilution of the SiH_4 gas, typically 10 : 1, the crystals expand conically during growth and reach diameters in the order of 1 μm within 3–4 μm of deposition thickness. At the top of the film, the crystal grains with (220) crystal orientation form a highly compact layer and form a pyramid-like surface texture. However, the seed density under these low dilution conditions is low and the incubation time to fully crystallized growth is rather long. Therefore, a seed layer is used that is made under high hydrogen dilution conditions, typically 100 : 1, to avoid amorphous material at the onset of deposition [64].

Princeton University and Utrecht University jointly prepared TFTs using this profiled polysilicon film in the top-gate configuration, to make use of the larger compact grain structure at the top of the film as the channel material [65]. A

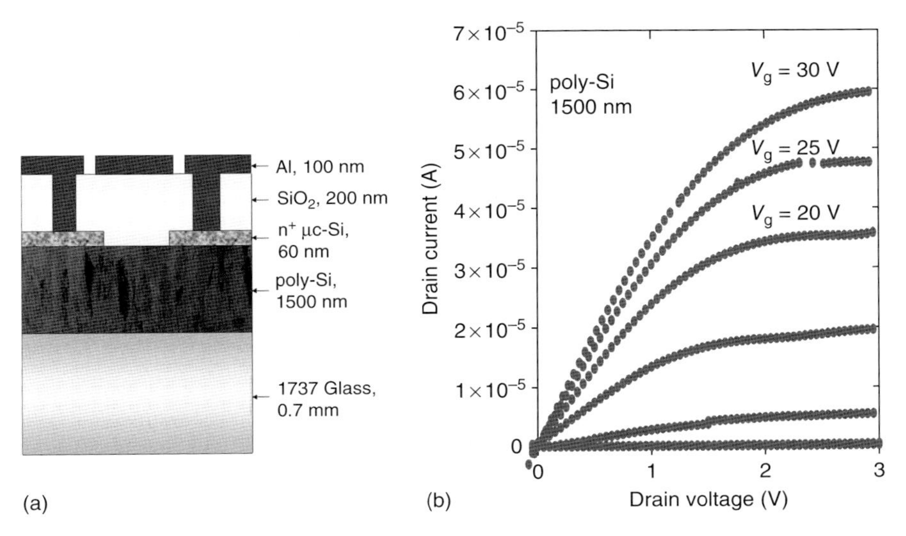

Figure 8.22 (a) Schematic cross section of the top-gate TFT structure. (b) Drain current I_D vs. drain voltage V_{SD} characteristics of the Cat-CVD poly-Si TFT. Source: Schropp et al. 2000 [65]. Reprinted with permission from Materials Research Society.

schematic cross section is shown in Figure 8.22a and the output characteristics are shown in Figure 8.22b.

Later, other groups also produced directly deposited poly-Si TFTs. For instance, an electron mobility of 25 cm^2/V s was obtained using Cat-CVD top-gate microcrystalline silicon TFTs, by the Barcelona group, in collaboration with Rennes [66]. All processing steps were performed at temperatures under 200 °C, and the only post-treatment of the microcrystalline Si structure is a forming gas treatment at 200 °C for one hour. This allows for integration of complementary metal oxide semiconductor (CMOS) circuitry with microelectromechanical systems or chemical sensors on plastic flexible substrates.

As early efforts for making Cat-CVD poly-Si TFT, finally, we introduce the work by SONY group, although they presented their work only in Japanese, and thus it has not been known widely [67]. The work was partially introduced in the related international conference by H. Matsumura et al. later in 2003 [68]. They noted the structural difference of poly-Si films along the growth direction when it was grown on glass substrates by Cat-CVD. The grains are small in an initial growth region near the substrate, followed by a gradual increase of the size of the grains, but when the grains grow bigger, the grains are not connected to each other and empty regions are born in some parts of the grain boundary. Such low-density regions of grain boundaries are easily oxidized. This phenomenon has been mentioned in Section 5.2.2. TFT is a device in which the currents should flow in lateral direction but not along the growth direction. Therefore, the existence of oxidized boundary regions is not acceptable for obtaining large mobility. Thus, they etched back the oxidized surface of the poly-Si to bring out the highly packed poly-Si grain region below oxidized region and made a top-gate TFT using the highly packed poly-Si region. The structure of their TFT is shown already in Figure 5.29, and the depth distribution of the electron mobility is shown in Figure 5.30. The peak mobility was about 40 cm^2/V s in TFTs made under selected conditions.

Another development in the field of poly-Si TFTs is the fabrication of ambipolar poly-Si (or microcrystalline) TFT by B.R. Wu et al. [69]. This may show the possibility to make a complementary TFT system in which n-channel poly-Si TFT and p-channel TFT are working complementary to realize low power consumption devices. Poly-Si TFT made by Cat-CVD has a big feasibility as a key device for future large area electronics.

8.4 Surface Passivation on Compound Semiconductor Devices

8.4.1 Passivation for Gallium–Arsenide (GaAs) High Electron Mobility Transistor (HEMT)

Gallium–arsenide (GaAs) is a high-performance semiconductor material. Their electron mobility is several times larger than that of c-Si at room temperature RT, and in addition, if a special structure in which the region of carrier transport is separated from the carrier generation regions (the impurity-doped regions) is adopted, the electron mobility is much enhanced over that of conventional GaAs devices, particularly at low temperatures. The high electron mobility transistor

(HEMT), or modulation-doped transistor [70], is a specially designed device for obtaining extremely high electron mobility. The electron mobility is determined by both the scattering by phonons and the scattering by impurities generating carriers. The suppression of phonon scattering is possible by lowering the operation temperatures, and the suppression of impurity scattering is carried out by the separation of electrons moving in the channel from the highly impurity-doped region by which electrons are provided. However, the flow of electrons in the channel has to be controlled by the electric field from the gate electrodes.

First, in 1997, R. Hattori et al. presented a paper about Cat-CVD SiN_x passivation on GaAs-based HEMT at the GaAs integrated circuit (IC) Symposium held at Anaheim, California, USA [71]. In their HEMT, a n-type aluminum–gallium arsenide (AlGaAs) layer was used as a carrier generation layer and an intrinsic indium–gallium arsenide (InGaAs) layer on an intrinsic GaAs substrate was used as the channel. This is the first successful report on the application of Cat-CVD to GaAs-based devices. In these GaAs-based HEMTs, the source, gate, and drain electrodes were separately formed and the gap between the gate and source or between the gate and drain was exposed to air. To suppress the leakage currents at this gap due to moisture in air, the surface was passivated. They compared the device characteristics after passivating the gap by Cat-CVD SiN_x films with those of devices passivated by PECVD SiN_x films, as illustrated in Figure 8.23a. In

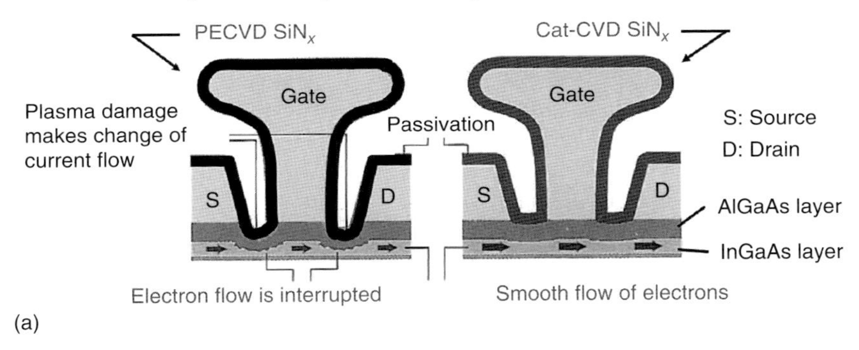

(a)

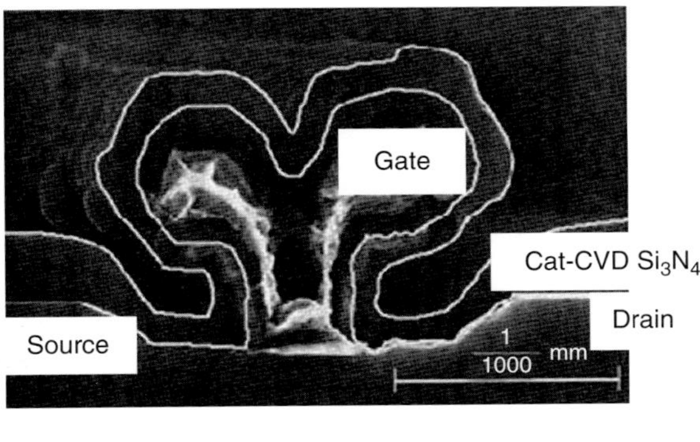

(b)

Figure 8.23 (a) Cross-sectional views of device structures of GaAs-related high-electron mobility transistors (HEMTs) coated by PECVD silicon nitride (SiN_x) and Cat-CVD SiN_x layers. (b) Cross-sectional scanning electron microscopic (SEM) image of HEMT. Source: Figures are drawn based on data shown in Ref. [71].

Table 8.3 Device characteristics GaAs-based high-electron mobility transistor (HEMT) and comparison between Cat-CVD and PECVD.

Device parameters	PECVD SiN_x used	Cat-CVD SiN_x used
Noise Figure (NF) at 12 GHz (dB)	0.53	0.44
Mutual conductance, g_m (mS)	70.7	75.2

Figure 8.23b, the SEM cross-sectional image of a GaAs-based HEMT, passivated with a Cat-CVD SiN_x film, is shown. The Cat-CVD film covers the electrodes conformally. The results of the device performance are summarized in Table 8.3. In the table, the noise figure (NF) during 12 GHz operation and the mutual conductance (g_m) are shown together. Smaller NF and larger g_m are a measure of better performance of these devices. The table demonstrates the superiority of Cat-CVD SiN_x passivation over PECVD SiN_x passivation.

They also succeeded in making a good performance GaAs field-effect transistor (FET) with a gate width of 100 μm and a gate length of 0.7 μm by using Cat-CVD SiN_x passivation [72]. In the device, g_m was 22 mS for Cat-CVD-passivated devices and 15 mS for the PECVD-passivated ones. The results encouraged companies to use Cat-CVD films on GaAs-based devices. Since then, a number of other groups started the research of Cat-CVD.

8.4.2 Passivation for Ultrahigh-Frequency Transistors

In 2006, M. Higashiwaki et al. succeeded in fabricating high-performance aluminum–gallium nitride (AlGaN)/gallium nitride (GaN) heterostructure field-effect transistor (HFET) by using Cat-CVD SiN_x films as gate insulators and as passivation layers [73]. The device consists of a surface AlGaN barrier layer and a GaN-active layer, and on the surface of the AlGaN barrier, the passivation layer is deposited. As Cat-CVD is a suitable method for depositing high-quality thin SiN_x without any serious influence on the AlGaN, the thickness of the AlGaN barrier layer can be chosen thinner, and thus, the drive currents of the transistors can be improved. Before their work, the current gain cutoff frequency (f_T) had reached plateaus of 110–120 GHz. However, by introducing Cat-CVD SiN_x films and using thin AlGaN barrier layers, it drastically increased to 163 GHz. The results opened the era of millimeter wave signal processing.

Their group had also further improved f_T, to 181 GHz in 2008. The device structure shown in their paper is presented in Figure 8.24 [74]. It is shown that only a 2-nm-thick SiN_x layer functions as the gate insulator and at the same time as a passivation layer. This work encouraged many researchers on GaAs- or GaN-related devices to pay attention to Cat-CVD technology. In fabrication of compound semiconductor devices, the advantage of using Cat-CVD SiN_x has become well known.

8.4.3 Passivation for Semiconductor Lasers

A semiconductor laser is another important device often made from compound semiconductors. Again, the superiority in forming the passivation layer by

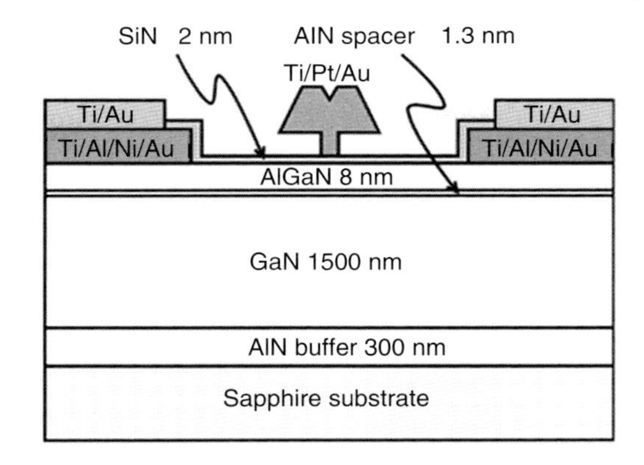

Figure 8.24 A schematic cross-sectional view of aluminum–gallium nitride (AlGaN)/gallium nitride (GaN) field-effect transistors. Between GaN and sapphire substrate, an aluminum nitride (AlN) buffer layer is inserted. Source: Higashiwaki et al. 2008 [74]. Reprinted with permission from IEEE.

Cat-CVD SiN_x has been proven to be attractive. The experimental results are very promising. The operation lifetime appears extended. Presently, some of the semiconductor lasers that are already commercially used in optical communication are fabricated with the help of Cat-CVD technology.

In many instances, Cat-CVD technology is actually used as the method for preparation of passivation on compound semiconductor devices, such as GaAs-related devices as already shown in Figure 8.18. It is now often used for the fabrication of monolithic microwave-integrated circuit (MMICs) devices with ultrahigh-frequency transistors. Such devices are used for the operation of radars that are now one of the key components for safety devices on automobiles detecting obstacles, for instance. The longtime reliability of devices coated by Cat-CVD films is now generally recognized, and thus, high-frequency devices using Cat-CVD films are being launched in communication satellites. The use of Cat-CVD technology is expanding silently, without any public news releases from companies.

8.5 Application for ULSI Industry

Application to ULSI fabrication has been pursued by a number of research groups. As mentioned in Chapter 5, conformal step coverage has been realized with SiN_x films deposited by Cat-CVD and excellent insulating performance has been obtained for SiN_x with low H contents in the film. Here, we introduce an example of an application in this field.

Figure 8.25 shows an illustrative cross-sectional view of a metal oxide silicon (MOS) transistor used in ULSI. The gate electrode has side walls that are used for making lightly doped regions near to the source (S) and drain (D) electrodes. All these structures are covered with liner SiN_x films for avoiding moisture penetration. The conformal step coverage and low H contents are both required for SiN_x films prepared at temperatures lower than 400 °C. If the insulators of the side walls and/or the liner insulating films include a lot of H atoms, these H atoms are penetrating into the channel region of MOS transistors and change the operation after long-term operation. Particularly, when the gate voltage is applied, ionized H atoms can easily move to the channel regions. It is known that H atoms are

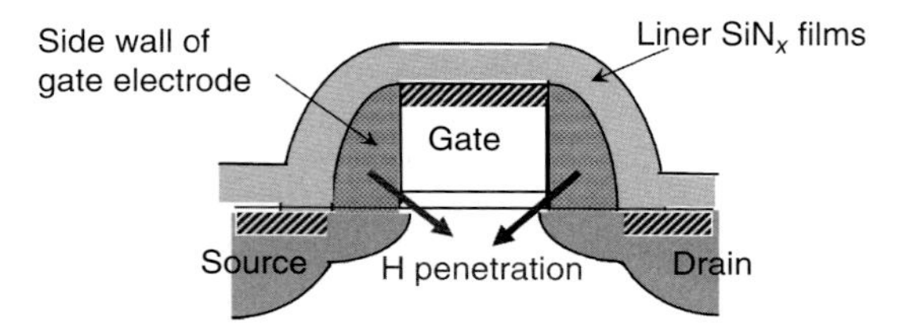

Figure 8.25 A schematic cross-sectional view of metal oxide silicon transistor with side walls of a gate electrode and a liner silicon nitride (SiN_x) layer. Source: Figure is provided by the courtesy of Professor Yoichi Akasaka at Osaka University.

active in changing the carrier concentration by making bonds with doping impurities. By this, the threshold voltage, which is one of the key parameters for MOS operation, starts to shift. This phenomenon is more clearly observed when the operation temperature is elevated. Therefore, it is sometimes called the negative bias temperature instability (NBTI) or NBT instability.

The magnitude of the threshold voltage shift (ΔV_{th}) in a certain voltage range is often used as a measure expressing the instability. Figure 8.26 shows the lifetimes within which ΔV_{th} is limited to 30 mV as a function of the reciprocal of the gate voltage (V_g). To observe the phenomenon within a short time, often, V_g is set at a much larger voltage than during usual operation. Therefore, the relation of the gate voltage is plotted as the reciprocal of it.

Some combinations with the method to prepare SiN_x films are described in the literature. The side wall (S) and liner films (L) are both prepared by Cat-CVD and compared with the use of films prepared by conventional low-temperature chemical vapor deposition (LPCVD). The Cat-CVD SiN_x films used as both the side wall and the liner films are prepared with T_{cat} = 2000 °C and T_s = 100 °C. The side wall LPCVD SiN_x is prepared at 600 °C and liner LPCVD SiN_x at 450 °C.

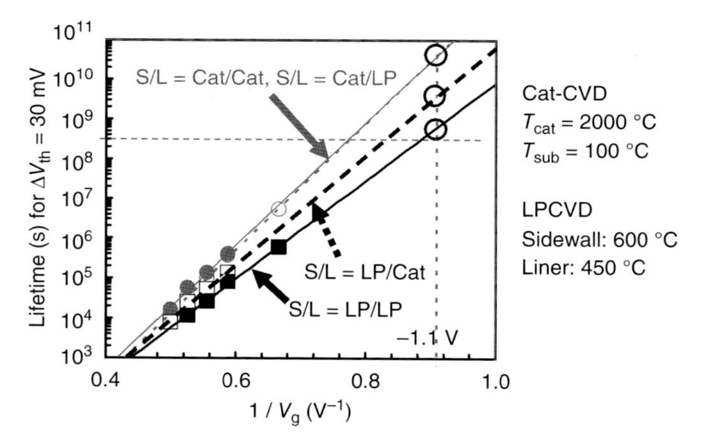

Figure 8.26 Lifetimes for restricting the threshold voltage shift (V_{th}) to less than 30 mV as a function of the reciprocal of the gate voltage (V_g). S and L refer to side wall and liner insulator, respectively. Source: Figure is provided by the courtesy of Professor Yoichi Akasaka at Osaka University.

When the side wall and liner films are prepared by LPCVD, the lifetime of the device, operated at $V_g = 1.1$ V, is expected to be about 10^8–10^9 seconds, which is equivalent to 3.2–32 years. However, when they both are made by Cat-CVD, the expected lifetime is extended to 10^{10}–10^{11} seconds, equivalent to 320–3200 years [75]. Herewith, in addition to compound semiconductor devices, the longtime reliability of devices using Cat-CVD films is again confirmed in ULSI.

8.6 Gas Barrier Films for Various Devices Such as Organic Devices

8.6.1 Inorganic Gas Barrier Films, SiN_x/SiO_xN_y, for OLED

As explained in Section 5.3.4, dense and stable films are obtained at temperatures lower than 100 °C by Cat-CVD. Particularly, Cat-CVD SiN_x films are resistive to moisture penetration even if the films are obtained at temperatures lower than 100 °C and samples are set in 100% humidity of 2 atm pressure water vapor at 120 °C for 24 hours. This property is applied in gas barrier films used for OLED. However, a simple experiment reveals that it is not so simple in practice. An OLED covered with a 1000-nm-thick Cat-CVD SiN_x single layer showed dark spots after operation under accelerated stress circumstances of 90% humidity at 60 °C for 40 hours. When the positions of the dark spots are carefully observed by scanning electron microscopy (SEM), we find growth of hillocks or cracks in the Cat-CVD SiN_x films starting from the substrate surface. The Cat-CVD SiN_x films are dense and hard and therefore not flexible or soft enough when the substrate surface is not smooth. Figure 8.27a shows the dark spots appearing on the surface of an OLED and Figure 8.27b shows the cross-sectional SEM view at a similar position, showing the dark spots.

To avoid the growth of hillocks or cracks in the SiN_x films, we deposited a SiN_x layer first, then paused for a five-minute interval, and again started the deposition of the SiN_x film at the same growth conditions. The result is shown in the SEM cross-sectional view of Figure 8.28a. In this experiment, SiN_x films were

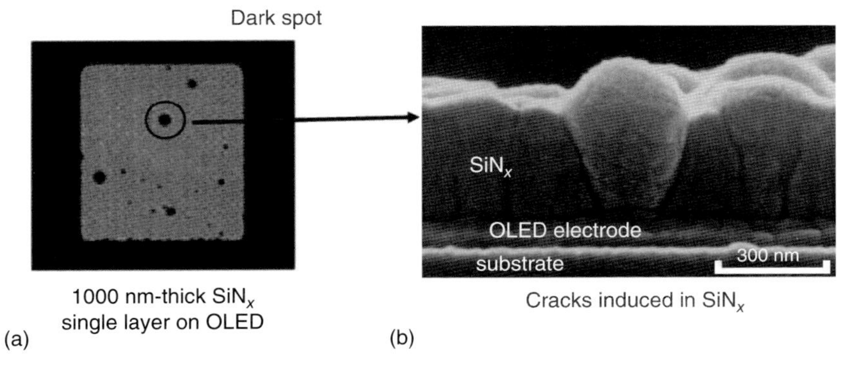

Figure 8.27 (a) OLED and dark spots appeared on it. OLED is coated by a 1000-nm-thick Cat-CVD single SiN_x layer. (b) Cross-sectional SEM image of a SiN_x single layer coated on OLED.

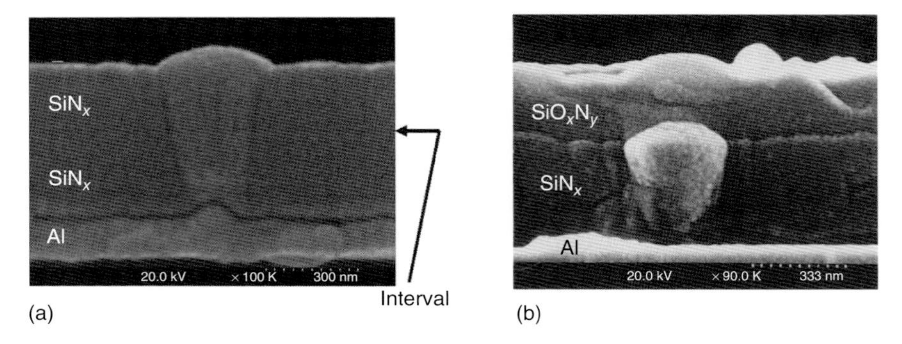

Figure 8.28 (a) Cross-sectional SEM image of Cat-CVD SiN_x/SiN_x double layers deposited on rough aluminum (Al) electrodes. (b) Cross-sectional SEM image of Cat-CVD SiO_xN_y/SiN_x-stacked layers deposited on rough Al electrodes.

deposited on rough aluminum (Al) electrodes in order to be able to see the cracks clearly. From this experiment, we concluded that even if we interrupt growth for a certain time interval, the growth of cracks just resumes. We therefore needed other methods to stop the growth of cracks. As an alternative, we started to use different types of films on the SiN_x layer.

We investigated stacked layers of SiN_x and silicon-oxy-nitiride (SiO_xN_y). The SiO_xN_y films are described in Section 5.4. The effect of these stacked layer structures was remarkable. Figure 8.28b shows a SEM cross-sectional view of the Cat-CVD SiN_x/SiO_xN_y-stacked layers deposited on a rough surface [76]. The cracks in SiN_x that started from the surface of the substrates are ending at the stacked SiO_xN_y layer deposited on top of it. The second SiO_xN_y layer does not suffer from any propagation of cracks. Building on this success, we next attempted to cover actual OLEDs provided from an OLED production company.

We prepared three types of gas barrier structures. The first one is that of a 100-nm-thick SiO_xN_y layer deposited on a 100-nm-thick SiN_x layer to form a 200-nm-thick stacked bilayer in total. The second structure is that of a 100-nm-thick SiO_xN_y/100-nm-thick SiN_x-stacked layer, repeated to form seven layers in total. The total thickness thus becomes 700 nm. The third one is that a single SiN_x film of a thickness of 1000 nm. Figure 8.29 shows the results of OLEDs after a few hours of exposure to accelerated circumstances of 90% humidity at 60 °C. Figure 8.29a–c shows the first structure, the second structure, and the third structure of the gas barrier layers, respectively. It is found that by making stacked layers, the dark spots do not occur at all; however, the corners of the OLED is gradually shrunk. When we carefully observed one of the shrinking corners, we discovered that there were specific origins from where the gas barrier ability of the films was broken. There was a particle of a size of about 100 mm at a corner and there was a step of about 700 nm height in total at the other corner of OLED. The 700-nm-thick stacked layers were not thick enough to cover all OLEDs perfectly. This result is demonstrated in Figure 8.30.

Thus, finally, we decided to deposit seven stacked layers of 300-nm-thick SiN_x/300-nm-thick SiO_xN_y layers (the total thickness was 2100 nm) on the OLED, and they were again placed in 90% humidity at 65 °C circumstances. The

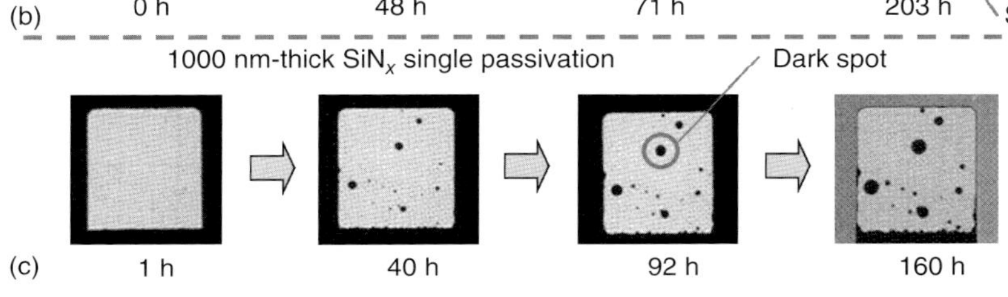

Figure 8.29 Time dependence of light emission patterns of OLED in accelerated circumstances of 90% humidity at 60 °C. The OLED is covered with (a) 100-nm-thick SiO_xN_y/100-nm-thick SiN_x-stacked passivation layer (layers in total 200 nm thick) and (b) 100-nm-thick SiO_xN_y/100-nm-thick SiN_x multistacked layers (totally seven layers and thickness 700 nm in total), and (c) 1000-nm-thick SiN_x single layer. All films are prepared by Cat-CVD.

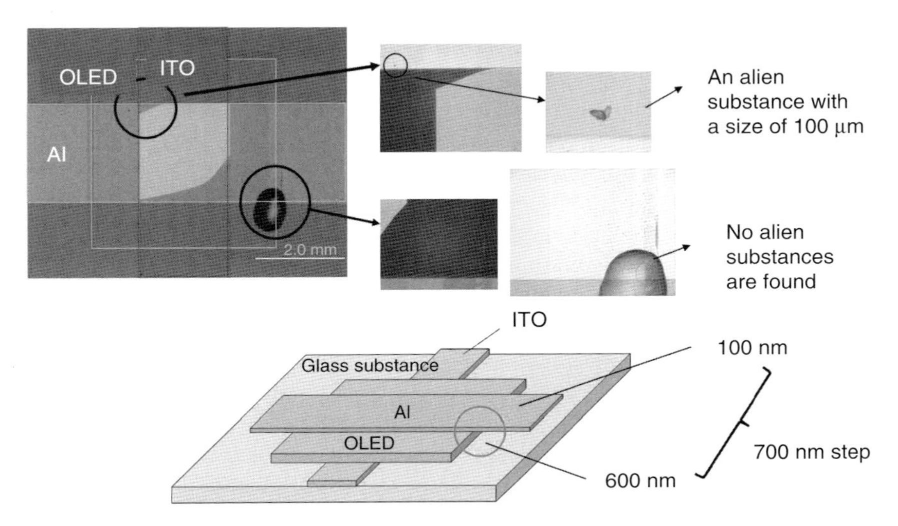

Figure 8.30 OLED covered with Cat-CVD 100-nm-thick SiO_xN_y/100-nm-thick SiN_x multistacked layers (totally seven layers and thickness 700 nm in total), after 203 hours under accelerated circumstances of 90% humidity at 60 °C.

Lifetime test in accelerated circumstance;
at 60 °C in 90% humidity

SiO_xN_y/SiN_x stacked passivation
(thickness of each layer is 300 nm, totally 7 layers, 2100 nm)

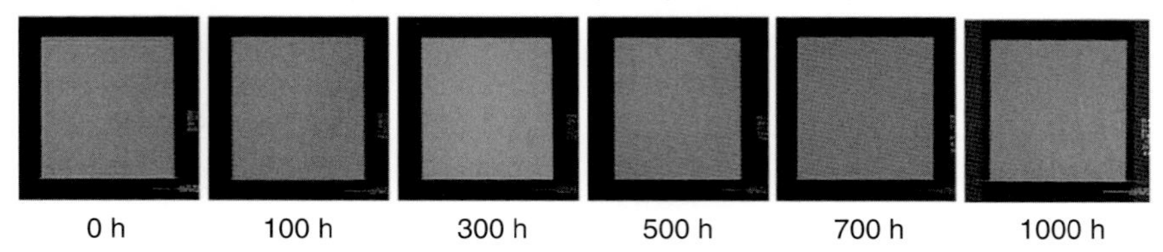

Figure 8.31 Time dependence of light emission patterns of OLED under accelerated circumstances of 90% humidity at 60 °C. The OLED is covered with Cat-CVD 300-nm-thick SiO_xN_y/300-nm-thick SiN_x multistacked layers (totally seven layers and the total thickness is 2100 nm). Source: Ogawa et al. 2008 [76]. Reprinted with permission from Elsevier.

results are demonstrated in Figure 8.31. Even after 1000 hours of exposure, no degradation was observed [77]. In conclusion, Cat-CVD SiN_x films are dense and suitable as gas barrier films, but for actual passivation of OLEDs, the films have to overcoat structures with tall steps, unintentional particles, or contaminants. Therefore, sufficient thickness and flexibility are required for these gas barrier films. To achieve this, other stacked structures, including inorganic/organic stacked layers, started to collect attention.

Finally, we emphasize one more advantage of the use of Cat-CVD films, based on the plasma damage-free nature of the deposition method. The deposition of passivation films on OLED itself sometimes induces initial degradation of OLED performance, even if the stability of OLED performance is guaranteed after film deposition. In the case of deposition of SiN_x/SiO_xN_y passivation films by Cat-CVD, no initial degradation is observed. As shown in Figure 8.32, the luminescence characteristics vs. applied voltage shows no degradation after deposition of gas barrier films by Cat-CVD.

8.6.2 Inorganic/Organic Stacked Gas Barrier Films

Electronic devices such as solar cells and displays, in particular, halide perovskite solar cells (PSCs) and OLEDs, can be very sensitive to damage by oxidizing ambient conditions, due to atmospheric oxygen and water vapor. This is particularly the case if these devices are made on flexible plastic substrates because these substrates are generally highly permeable to oxygen and water vapor. Therefore, these devices need to be encapsulated to protect them against atmospheric influences. Preferably, this is done by thin film encapsulation as this preserves flexibility and rollability of the modules or panels. However, even devices built on glass substrates that are by nature highly protecting substrates against moisture and oxygen still require hermetic encapsulation of the top surfaces. Also here, thin film encapsulation can be useful, as it can be uniformly and conformally applied on micro- and nanostructured devices over large areas.

In many cases, a single organic or inorganic layer cannot provide the required low water vapor transmission rate (WVTR). In practice, OLEDs and most likely

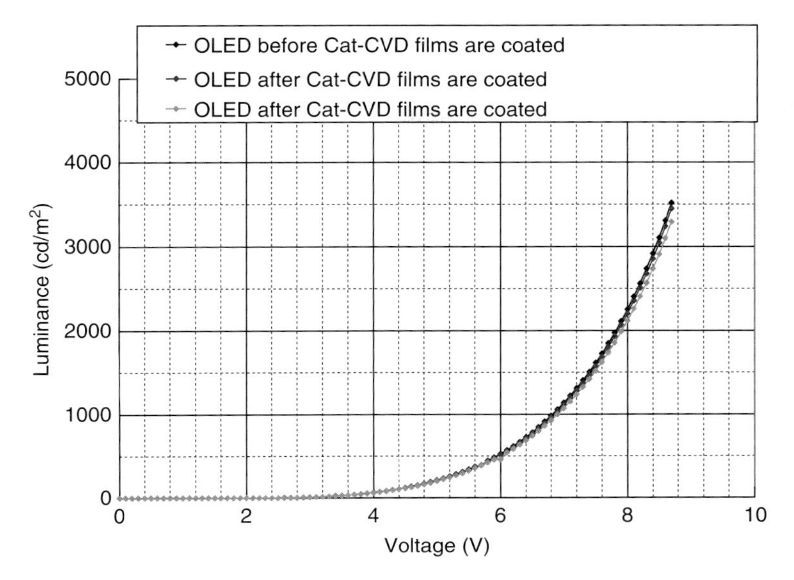

Figure 8.32 Light emission characteristics of OLED coated by Cat-CVD films. Luminescence intensity is plotted as a function of applied voltage to the OLED. The characteristics before and after Cat-CVD film coating are shown. Source: Ogawa et al. 2008 [76]. Reprinted with permission from Elsevier.

also PSCs require an encapsulation with a WVTR as low as $<10^{-6}$ g/m^2/day [78]. The requirement for the barrier coating on PSCs has not been established exactly at this moment, but given their high sensitivity to humidity [79], it is likely that a WVTR of at least lower than 10^{-6} g/m^2/day is needed for both water and oxygen. There is as yet no individual thin film that can deliver such WVTR. Therefore, thin film encapsulation often requires a multilayer structure. Various multilayer structures have been proposed as a gas and water vapor barrier layer [80–82]. Many of these multilayer structures contain multiple dyads of inorganic films and polymer films, in which the organic layers decouple defects (such as pinholes and cracks) that are inevitably present in consecutive inorganic layers.

In this manner, chances of the existence of permeation paths like pinholes or cracks propagating through the entire multilayer stack are decreased. In many cases, the combination of different layers in the stack leads to the necessity to combine widely varying deposition techniques, such as sputtering, PECVD, atomic layer deposition (ALD), or evaporation for the inorganic layer and spray deposition, spin coating, or *i*CVD for the organic (polymeric) layers.

The required multiprocessing sequence leads to high cost of the barrier stacks. Therefore, in principle, it is desirable to deposit the inorganic/organic stack in a single process using a single deposition technology. For this reason, we have investigated the use of Cat-CVD SiN_x layers in combination with *i*CVD polymer layers. Both deposition techniques are Cat-CVD techniques involving heated wires in a vacuum system and can in principle be combined easily in a continuous in-line or roll-to-roll deposition system. A combination of silicon nitride (SiN_x)

and polymer is very suitable to create such a multilayer. As both types of layers can be made highly transparent, the multilayer encapsulation can be used for displays, sensors, and solar cells. Further, Cat-CVD avoids the impact of ionic species that are present in sputtering or PECVD, thus eliminating potential damage to sensitive active layers.

We start our multilayer barrier deposition sequence with a poly-glycidyl methacrylate (PGMA) polymer layer. The PGMA layers were made in an *i*CVD chamber made after a design originating from MIT labs [83]. The monomer, GMA (97%, Aldrich), and the initiator, *tert*-butyl peroxide (TBPO) (98%, Aldrich), were fed into the reactor through a gas mixing stage. TBPO was thermally decomposed at a parallel array of nichrome wires, 3 cm above the substrate and heated to 220 °C. The substrate holder was water cooled to keep the substrate at 17 °C.

The low-temperature SiN_x layer is deposited directly on top of the PGMA layer. It was shown earlier, by Alpuim et al. 2009, that transparent, dense SiN_x layers can be made at temperatures as low as 100 °C by Cat-CVD [84]. In our experiments, to control the substrate temperature more reliably, a large substrate to filament distance is used in our experiments to prevent gradual increase of the substrate temperature by radiative heating. Alternatively, an artificially cooled substrate susceptor can be used to keep the substrate below 110 °C. Pure silane (SiH_4) and ammonia (NH_3) were used as source gasses. No hydrogen dilution was used. The ammonia flow was 150 sccm and the silane flow was 5 sccm. The deposition pressure was set at a value of 0.04 mbar (4 Pa). The source gasses were catalytically decomposed at two tantalum filaments with a diameter of 0.125 mm, held at 2100 °C. The wires are placed 20 cm from the substrate. To control the duration of the deposition, a shutter was situated between the sample and the filaments.

The N/Si ratio was close to stoichiometry, at 1.2. The etch rate in 16BHF was 25 nm/min, indicating a density of 2.5 g/cm^3 [85]. The ratio of N—H to Si—H bonds is around 1, as determined by Fourier transform infrared spectroscopy. The layers are highly transparent, with an extinction coefficient of only $k = 0.002$ at 400 nm wavelength.

During deposition of SiN_x on top of PGMA, deposited by *i*CVD, initially an oxygen-rich SiO_xN_y layer is formed. This is a result of oxygen atoms being released from the PGMA material, which reacts with atomic H and (or) amine species [86]. This mechanism allows for the formation of SiO_x material without using oxidative source gasses, which is beneficial for the performance of moisture barriers as SiO_x like materials can seal off defects in the polymer layer and provide an extra adhesive layer bonding through the epoxide ring-opening reactions between the organic and inorganic layers. The oxide layer is visible in the cross-sectional TEM micrograph of Figure 8.33 between the dark SiN_x regions and the light PGMA regions, just underneath the SiN_x layers.

A dyad layer on a Corning Eagle XG borosilicate glass substrate was inspected to confirm its integrity and its optical transparency. A SiN_x layer of 70 nm was deposited on 150 nm PGMA. Figure 8.34 shows the optical transmission and reflection spectra in the visible/NIR (near infra red) spectrum and Figure 8.35 the fourier transform infra red (FTIR) spectra of the individual layers as well as

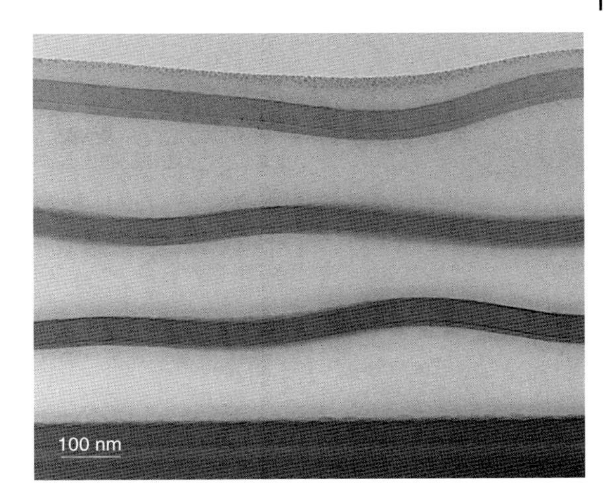

Figure 8.33 Cross-sectional TEM micrographs, using bright-field mode, of a SiN_x/PGMA/SiN_x/PGMA /SiN_x/PGMA/SiN_x/PGMA multilayer. The oxide layer is clearly visible in cross-sectional TEM micrographs between the dark SiN_x regions and the light PGMA regions, just underneath the SiN_x layers. Source: Spee et al. 2014 [86]. Reprinted with permission from NRC Research Press.

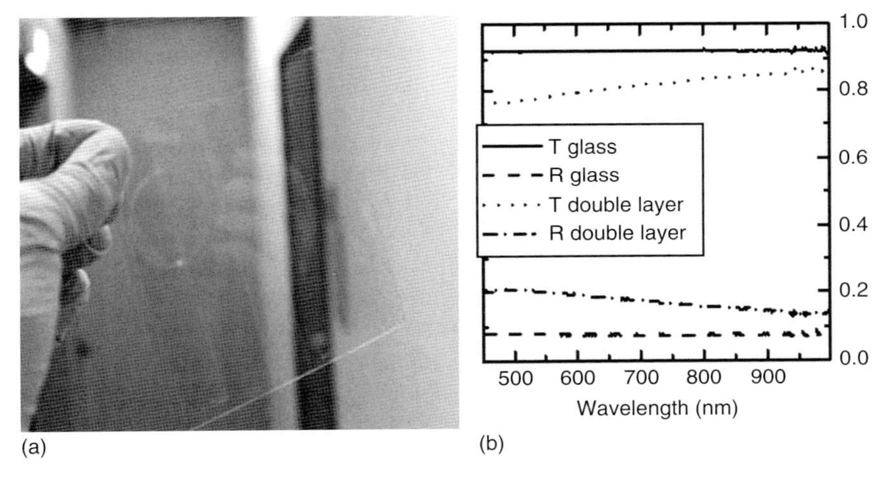

Figure 8.34 (a) Photograph of a dyad consisting of 70 nm SiN_x deposited on 150 nm PGMA on glass and (b) reflection and transmission spectra of the dyad on glass, compared to bare Corning Eagle XG borosilicate glass. Source: Spee et al. 2011 [85]. Reprinted with permission from American Scientific Publishers.

the dyad. It is shown that the optical transmission is very suitable for optoelectronic applications and that the SiN_x deposition process does not deteriorate the PGMA, as all polymer FTIR absorption peaks are unaltered in the dyad.

A layer with excellent barrier function was made simply by using a configuration of two SiN_x layers with a PGMA interlayer. The triple stack could be obtained pinhole free, with a WVTR of 5×10^{-6} g/m^2/day, as determined by the accurate Ca test method [87]. This WVTR value, established using a 60 °C and 90% relative humidity (RH) atmosphere, implies a barrier function that is sufficient for even the most sensitive device [88]. The optical transmission of 5 mm × 5 mm Ca squares covered by three types of samples after various incremental exposure durations is shown in Figure 8.36. The top row shows the results for an individual SiN_x layer, deposited at 100 °C, and the middle row for an individual SiN_x layer

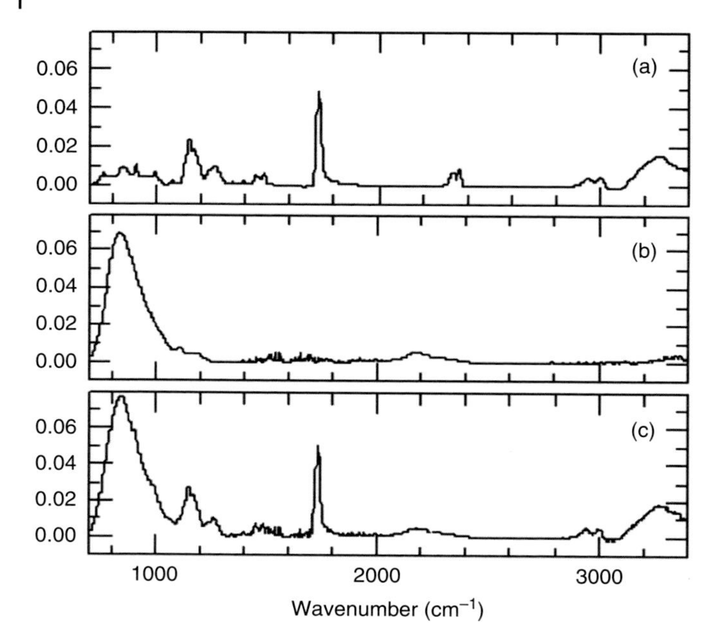

Figure 8.35 FTIR spectra of (a) 150 nm PGMA, (b) 70 nm SiN_x, and (c) double layer. The peaks due to the epoxy ring and the C=O bond are still clearly present. The small peak in (c) around 2300 cm^{-1} is caused by CO_2 gas in the FTIR setup. Source: Spee et al. 2011 [85]. Reprinted with permission from American Scientific Publishers.

deposited at high temperature (not suitable for use with soft polymers). The third row shows the results for the SiN_x/PGMA/SiN_x layer stack. Here, the SiN_x layers were 100 nm and the PGMA layer was 200 nm. Even after 190 days in this accelerated humidity test, no significant degradation of the Ca layer occurs.

Similar SiN_x/organic films such as PGMA or hexamethyldisiloxane (HMDSO) multistacked layers are investigated as gas barrier films for the next-generation flexible OLED [89]. They will be used in portable smart phones. The gas barrier films for coating OLED are still an active research area for future-generation devices.

8.6.3 Gas Barrier Films for Food Packages

The world market of consumer goods requiring gas barrier films is very large. Most food packaging bottles made by polyethylene terephthalate (PET) and other packaging materials require gas barrier films to extend storage times. Thus, various methods have been attempted to make high-quality gas barrier films with low cost. Among them, in some cases, gas barrier films made by Cat-CVD have also been investigated. Some results on gas barrier ability were introduced already in Section 5.4.2 for SiO_xN_y films made by hexamethyldisilazane (HMDS).

M. Nakaya et al. also succeeded in developing an inner coating system for PET bottles by silicon-oxy-carbide (SiO_xC_y) films prepared by Cat-CVD using vinylsilane (SiH_3–CH–CH_2) as the source gas [90]. The system is schematically

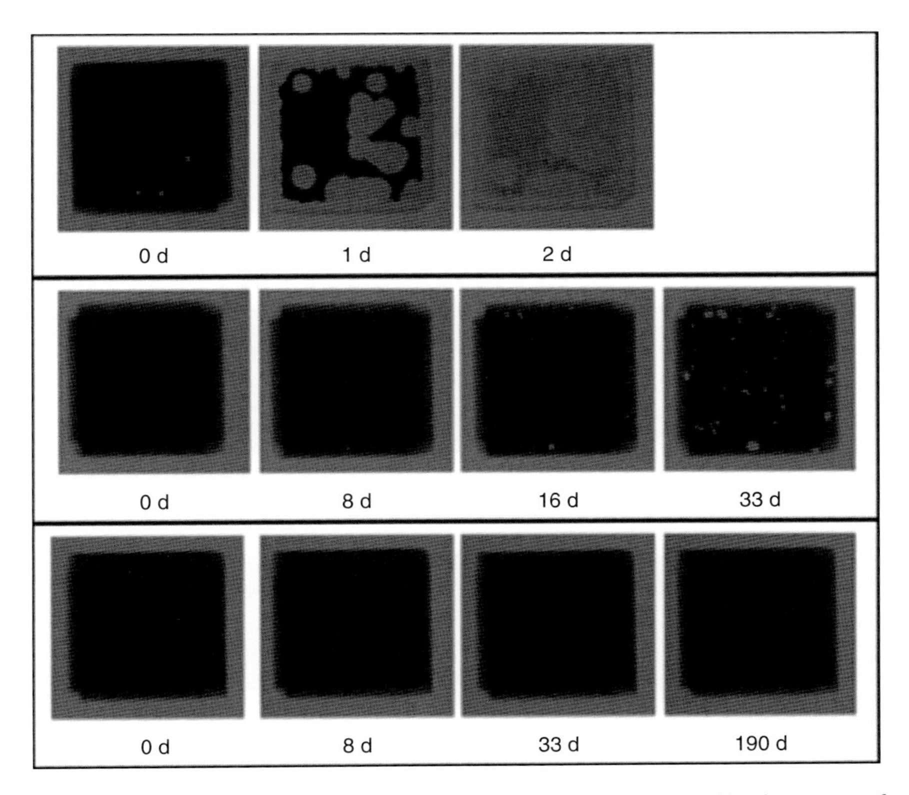

Figure 8.36 The optical transmission of 5 mm × 5 mm Ca squares covered by three types of samples after various incremental humidity exposure durations. The top row shows the results for an individual SiN_x layer, deposited at 100 °C. The middle row for an individual SiN_x layer deposited at high temperature (not suitable for use with soft polymers). The third row shows the results for SiN_x/PGMA/SiN_x layer. Here, the SiN_x layers were 100 nm and the PGMA layer was 200 nm. Source: Spee et al. 2012 [88]. Reprinted with permission from John Wiley and Sons.

illustrated in Figure 8.37. Catalyzing wires are put into an array of many PET bottles, each individually through the neck of the PET bottle. Figure 8.38 shows the water vapor loss from sealed PET bottles, as a function of time. Initially, PET bottles were filled with 500 g distilled water, but the weight of PET bottles gradually decreased by water vapor permeation through the walls of PET bottles. The result for a noncoated PET bottle is compared with that coated by Cat-CVD SiO_xC_y films with a thickness of about a few tens of nanometers. Figures 8.37 and 8.38 are cited from the reference [90]. The deposition time for the SiO_xC_y film is only on the order of seconds. It is confirmed that by depositing Cat-CVD films, the water vapor penetration can be suppressed.

Compared with the gas barrier ability required in OLED, the requirement to the inner coating of PET bottles is much looser; however, with this Cat-CVD coating, a beverage in a PET bottle can be kept for more than six months, whereas a conventional PET bottle without inner coating can keep the beverage only for three months.

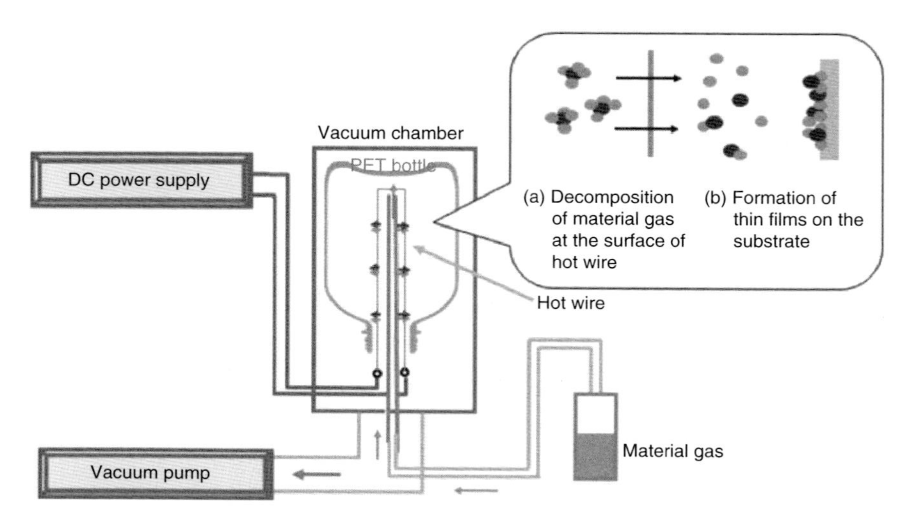

Figure 8.37 Schematic illustration of PET bottle inner coating system by Cat-CVD. Catalyzing wires are put inside each PET bottle through the bottle neck. Source: Nakaya et al. 2016 [90]. Reprinted with permission from Journal of Polymers.

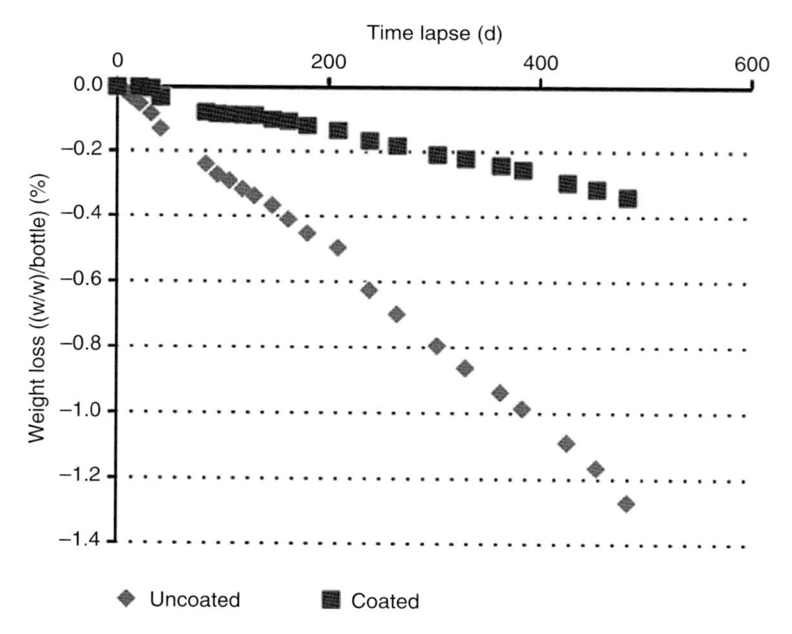

Figure 8.38 Variation of total weight of PET bottles as a function of time. Initially, the PET bottle is filled with 500 g distilled water, but as water vapor migrates through the wall of PET bottle, the total weight is gradually decreasing. For a PET bottle coated on the inside by Cat-CVD films, the rate of decrease of weight is smaller than that for noncoated one. Source: Nakaya et al. 2016 [90]. Reprinted with permission from Journal of Polymers.

In mass production, similar structures with catalyzing wires are arrayed and the coating on the inside of many PET bottles is performed at the same time. When PECVD and related technology are used, it is not easy to generate completely the same plasma in an array of many bottles within a short time. Considering, for instance, that an array of 500 PET bottles should be coated simultaneously using 500 RF electrodes for generating plasma, it is easily understood that this is not straightforward to do this in the 500 small volumes. In PECVD, there may be more fluctuation of the quality of the gas barrier films than in the case of Cat-CVD. Deposition without using plasma is one of the bright advantages in Cat-CVD.

8.7 Other Application and Summary of Present Cat-CVD Application

As mentioned above, Cat-CVD has many applications. The area in which Cat-CVD is used is still growing and expanding. The application is not limited to the area mentioned here. For instance, gas barrier films are also used in the back sheet of solar cells. Another application area of Cat-CVD is the use of radicals generated in the Cat-CVD apparatus. This topic will be discussed in Chapter 9.

Finally, we summarize the situation of industrial implementation of Cat-CVD technology briefly, based on Figure 8.1.

(1) *Solar cell application*: One of the successful areas of application. Mass production machine used in the fabrication of solar cells have been sold since 2007. The number of machines installed in the factory is increasing. Solar cells using Cat-CVD in part of the manufacturing steps can be bought in the market.
(2) *Application to passivation of compound semiconductor devices*: Mass production has been started in the early 2000s or before. Cat-CVD, as used in part of the processing of high-frequency transistors and semiconductor lasers, has been installed in the production lines.
(3) *Application to consumer goods*: Some of the present-day high-grade razor blades use Cat-CVD or Cat-CVD-related technology (*i*CVD). Some companies are selling razor blades coated by Cat-CVD films. This will be expanding.
(4) *Application to passivation of OLEDs*: Inorganic/organic multistacked barrier layers are useful candidates as the passivation structures of future OLEDs. Industrial implementation will be expected.
(5) *Application to copy drum production*: Small size markets have been created. Future expansion is expected.
(6) *Application to food packaging*: Various efforts are ongoing. The future will be bright.
(7) *Application to ULSI or Si industry*: There has been competition with ALD technique. There may be still feasibility for Cat-CVD to be installed in this area.

The number of applications of Cat-CVD technology is still growing. We can rely on a bright future for this technology.

References

1 Fraunhofer ISE and PSE AG (2017). Photovoltaics Report (July 2017). https://www.ise.fraunhofer.de/content/dam/ise/de/documents/publications/studies/Photovoltaics-Report.pdf. (accessed 26 February 2018).

2 Umemoto, H. (2014). Gas-phase diagnoses in catalytic chemical vapor deposition (hot-wire CVD) processes. *Thin Solid Films* 575: 3–8.

3 Schropp, R.E.I. (2015). Industrialization of hot wire chemical vapor deposition for thin film applications. *Thin Solid Films* 595: 272–283. https://doi.org/10.1016/j.tsf.2015.07.054.

4 Nelson, B., Iwaniczko, E., Mahan, A.H. et al. (2001). High-deposition rate a-Si:H n-i-p solar cells grown by HWCVD. *Thin Solid Films* 395: 292–297.

5 Mahan, A.H. and Vanecek, M. (1991). A reduction in the Staebler-Wronski effect observed in low H content a-Si:H films deposited by the hot wire technique. *AIP Conf. Proc.* 234: 195–202.

6 Mahan, A.H., Xu, Y., Williamson, D.L. et al. (2001). Structural properties of hot wire a-Si:H films deposited at rates in excess of 100 Å/s. *J. Appl. Phys.* 90: 5038–5047.

7 Weber, U., Koob, M., Dusane, R.O. et al. (2000). a-Si-:H based solar cells entirely deposited by hot-wire CVD. In: *Proceedings of the 16th European Photovoltaic Solar Energy Conference*, Glasgow (1–5 May 2000), 286–291. London: James & James.

8 Bauer, S., Schröder, B., Herbst, W., and Lill, M. (1998). A significant step towards fabrication of high efficient, more stable a-Si:H solar cells by thermo-catalytic CVD. In: *Proceedings of the Second World Conference on Photovoltaic Solar Energy Conversion*, Vienna (6–10 July 1998), 363. Ispra: Joint Research Centre (European Commission).

9 Nelson, B., Iwaniczko, E., Mahan, A.H. et al. (2001). High-deposition rate a-Si:H n-i-p solar cells grown by HWCVD. *Thin Solid Films* 395: 292–297.

10 Schropp, R.E.I. (2002). Advances in solar cells made with hot wire CVD: superior films and devices at low equipment cost. *Thin Solid Films* 403–404: 17–25.

11 Schropp, R.E.I., van Veen, M.K., van der Werf, C.H.M. et al. (2004). Protocrystalline silicon at high rate from undiluted silane. *Mater. Res. Soc. Symp. Proc.* 808: A8.4.1.

12 Koh, J., Lee, Y., Fujiwara, H. et al. (1998). Optimization of hydrogenated amorphous silicon p-i-n solar cells with two-step i-layers guided by real time spectroscopic ellipsometry. *Appl. Phys. Lett.* 73: 1526–1528.

13 Matsumura, H. (1987). High-quality amorphous silicon germanium produced by catalytic chemical vapor deposition. *Appl. Phys. Lett.* 51: 804–805.

14 Datta, S., Xu, Y., Mahan, A.H. et al. (2006). Superior structural and electronic properties for amorphous silicon–germanium alloys deposited by a low temperature hot wire chemical vapor deposition process. *J. Non-Cryst. Solids* 352: 1250–1254.

15 Guha, S., Cohen, D., Schiff, E. et al. (2011). Industry–academia partnership helps drive commercialization of new thin-film silicon technology. *Photovoltaics International* 13: 134–140.
16 Nelson, B.P., Xu, Y., Williamson, D.L. et al. (1998). Hydrogenated amorphous silicon germanium alloys grown by the hot-wire chemical vapor deposition technique. *Mater. Res. Soc. Symp. Proc.* 507: 447–452.
17 Xu, Y., Mahan, A.H., Gedvilas, L.M. et al. (2006). Deposition of photosensitive hydrogenated amorphous silicon–germanium films with a tantalum hot wire. *Thin Solid Films* 501: 198–201.
18 Mahan, A.H., Xu, Y., Gedvilas, L.M., and Williamson, D.L. (2009). A direct correlation between film structure and solar cell efficiency for HWCVD amorphous silicon germanium alloys. *Thin Solid Films* 517: 3532–3535.
19 Veldhuizen, L.W., van der Werf, C.H.M., Kuang, Y. et al. (2015). Optimization of hydrogenated amorphous silicon germanium thin films and solar cells deposited by hot wire chemical vapor deposition. *Thin Solid Films* 595: 226–230. https://doi.org/10.1016/j.tsf.2015.05.055.
20 Veldhuizen, L.W. and Schropp, R.E.I. (2016). Very thin and stable thin-film silicon alloy triple junction solar cells by hot wire chemical vapor deposition. *Appl. Phys. Lett.* 109: 093902. https://doi.org/10.1063/1.4961937.
21 Stuckelberger, M., Biron, R., Wyrsch, N. et al. (2017). Review: Progress in solar cells from hydrogenated amorphous silicon. *Renew. Sust. Energ. Rev.* 76: 1497–1523. https://doi.org/10.1016/j.rser.2016.11.190.
22 Li, H., Franken, R.H., Stolk, R.L. et al. (2008). Mechanism of shunting of nanocrystalline solar cells deposited on rough Ag/ZnO substrates. *Solid State Phenom.* 131–133: 27–32.
23 Schropp, R.E.I., Li, H., Rath, J.K., and van der Werf, C.H.M. (2008). Thin film nanocrystalline silicon and nanostructured interfaces for multibandgap triple junction solar cells. *Surf. Interface Anal.* 40: 970–973. https://doi.org/10.1002/sia.2816.
24 Vetterl, O., Finger, F., Carius, R. et al. (2000). Intrinsic microcrystalline silicon: A new material for photovoltaics. *Sol. Energy Mater. Sol. Cells* 62: 97–108.
25 Li, H., Franken, R.H., Stolk, R.L. et al. (2008). Improvement of μc-Si:H n–i–p cell efficiency with an i-layer made by hot-wire CVD by reverse H_2-profiling. *Thin Solid Films* 516 (5): 755–757.
26 Schropp, R.E.I., Li, H., Franken, R.H. et al. (2008). Nanostructured thin films for multibandgap silicon triple junction solar cells. *Thin Solid Films* 516: 6818–6823.
27 Yan, B., Yue, G., Owens, J.M. et al. (2006). Over 15% efficient hydrogenated amorphous silicon based triple-junction solar cells incorporating nanocrystalline silicon. In: *4th World Conference on Photovoltaic Energy Conversion*, Waikoloa Village, Hawaii, USA (7–12 May 2006), 1477–1480. IEEE. https:/dx.doi.org/10.1109/WCPEC.2006.279748
28 Hong, J.-S., Kim, C.-S., Yoo, S.-W. et al. (2012). In-situ measurements of charged nanoparticles generated during hot wire chemical vapor deposition of silicon using particle beam mass spectrometer. *Aerosol Sci. Technol.* 47 (1): 46–51.

29 Hamers, E.A.G., Fontcuberta i Morral, A., Niikura, C. et al. (2000). Contribution of ions to the growth of amorphous, polymorphous, and microcrystalline silicon thin films. *J. Appl. Phys.* 88 (6): 3674–3688.

30 Wang, Q., Ward, S., Gedvilas, L. et al. (2004). Conformal thin-film silicon nitride deposited by hot-wire chemical vapor deposition. *Appl. Phys. Lett.* 84: 338–340.

31 Ferry, V.E., Verschuuren, M.A., Li, H.B.T. et al. (2009). Improved red-response in thin film a-Si:H solar cells with soft-imprinted plasmonic back reflectors. *Appl. Phys. Lett.* 95: 183503.

32 Kuang, Y., van der Werf, C.H.M., Houweling, Z.S., and Schropp, R.E.I. (2011). Nanorod solar cell with an ultrathin a-Si:H absorber layer. *Appl. Phys. Lett.* 98: 113111.

33 Veldhuizen, L.W., Vijselaar, W.J.C., Gatz, H.A. et al. (2017). Textured and micropillar silicon heterojunction solar cells with hot-wire deposited passivation layers. *Thin Solid Films* 635: 66–72.

34 Ferry, V.E., Verschuuren, M.A., Li, H.B.T. et al. (2011). Light trapping in ultrathin plasmonic solar cells. *Opt. Express* 18 (S2): A237–A245. https://doi.org/10.1364/OE.18.00A237.

35 Ferry, V.E., Verschuuren, M.A., Claire van Lare, M. et al. (2011). Optimized spatial correlations for broadband light trapping nanopatterns in high efficiency ultrathin film a-Si:H solar cells. *Nano Lett.* 11 (10): 4239–4245.

36 Veldhuizen, L.W., Kuang, Y., and Schropp, R.E.I. (2016). Ultrathin tandem solar cells on nanorod morphology with 35-nm thick hydrogenated amorphous silicon germanium bottom cell absorber layer. *Sol. Energy Mater. Sol. Cells* 158 (part 2): 209–213.

37 PV insights. Grid the World. PV Poly Silicon Weekly Spot Price. http://pvinsights.com/ (accessed 31 October 2018).

38 Koyama, K., Ohdaira, K., and Matsumura, H. (2010). Extremely low surface recombination velocities on crystalline silicon wafers realized by catalytic chemical vapor deposited SiNx/a-Si stacked passivation layers. *Appl. Phys. Letters* 97: 082108-1-3.

39 Nguyen, C.T., Koyama, K., Higashimine, K. et al. (2017). Novel chemical cleaning of textured crystalline silicon for realizing surface recombination velocity <0.2 cm/s using passivation catalytic CVD SiN_x/amorphous silicon stacked layers. *Jpn. J. Appl. Phys.* 56: 056502-1-7.

40 Tanaka, M., Taguchi, M., Matsuyama, T. et al. (1992). Development of new a-Si/c-Si heterojunction solar cells: ACJ-HIT (artificially constructed junction-heterojunction with intrinsic thin-layer). *Jpn. J. Appl. Phys.* 31: 3518.

41 Taguchi, M., Yano, A., Tohoda, S. et al. (2014). 24.7% record efficiency HIT solar cell on thin silicon wafer. *IEEE J. Photovoltaics* 4: 96–99. https://doi.org/10.1109/JPHOTOV.2013.2282737.

42 Yoshikawa, K., Yoshida, W., Irie, T. et al. (2017). Exceeding conversion efficiency of 26% by heterojunction interdigitated back contact solar cell with thin film Si technology. *Sol. Energy Mater. Sol. Cells* 173: 37–42. https://doi.org/10.1016/j.solmat.2017.06.024.

43 Schüttauf, J.W.A., van der Werf, C.H.M., van Bommel, C.O. et al. (2010). Crystalline silicon surface passivation by hydrogenated amorphous silicon

layers deposited by HWCVD, RF PECVD and VHF PECVD: the influence of thermal annealing on minority carrier lifetime. In: *Proceedings of the 25th European Photovoltaic Solar Energy Conference*, Valencia (6–9 September 2010), 1114–1117. https://doi.org/10.4229/25thEUPVSEC2010-2AO.1.2.
44 Schäfer, L., Harig, T., Höfer, M. et al. (2013). Inline deposition of silicon-based films by hot-wire chemical vapor deposition. *Surf. Coat. Technol.* 215: 141–147.
45 Edwards, M., Bowden, S., Das, U., and Burrows, M. (2008). Effect of texturing and surface preparation on lifetime and cell performance in heterojunction silicon solar cells. *Sol. Energy Mater. Sol. Cells* 92: 1373–1377. https://doi.org/10.1016/j.solmat.2008.05.011.
46 Fesquet, L., Olibet, S., Damon-Lacoste, J. et al. (2009). Modification of textured silicon wafer surface morphology for fabrication of heterojunction solar cell with open circuit voltage over 700 mV. In: *Proceedings of the 34th IEEE PVSC Conference*, Philadelphia, PA, USA (7–12 June 2009), 754–758. IEEE https://doi.org/10.1109/pvsc.2009.5411173.
47 Wang, Q. (2018). HW/CAT-CVD for the high performance crystalline silicon heterojunction solar cells. Conference Program and Abstract book of 10th Int. Conf. on Hot-Wire (Cat) & Initiated Chemical Vapor Deposition, held at Kitakyushu, Sept. 3–6, 2018, p. 85.
48 Powell, M.J., Van Berkel, C., Franklin, A.R. et al. (1992). Defect pool in amorphous-silicon thin-film transistors. *Phys. Rev. B* 45: 4160. https://doi.org/10.1103/PhysRevB.45.4160.
49 Staebler, D.L. and Wronski, C.R. (1977). Reversible conductivity changes in discharge-produced amorphous Si. *Appl. Phys. Lett.* 31: 292. https://doi.org/10.1063/1.89674.
50 Stutzmann, M., Jackson, W.B., and Tsai, C.C. (1985). Light-induced metastable defects in hydrogenated amorphous silicon: a systematic study. *Phys. Rev. B* 32: 23. https://doi.org/10.1103/PhysRevB.32.23.
51 Mahan, A.H., Carapella, J., Nelson, B.P. et al. (1991). Deposition of device quality, low H content amorphous silicon. *J. Appl. Phys.* 69: 6728. https://doi.org/10.1063/1.348897.
52 Meiling, H. and Schropp, R.E.I. (1996). Stability of hot-wire deposited amorphous-silicon thin-film transistors. *Appl. Phys. Lett.* 69: 1062. https://doi.org/10.1063/1.116931.
53 Meiling, H. and Schropp, R.E.I. (1997). Stable amorphous silicon thin-film transistors. *Appl. Phys. Lett.* 70 (20): 2681. https://doi.org/10.1063/1.118992.
54 van Berkel, C. and Powell, M.J. (1987). Resolution of amorphous silicon thin-film transistor instability mechanisms using ambipolar transistors. *Appl. Phys. Lett.* 51: 1094. https://doi.org/10.1063/1.98751.
55 Schropp, R.E.I., Stannowski, B., and Rath, J.K. (2002). New challenges in thin film transistor (TFT) research. *J. Non-Cryst. Solids* 299–302: 1304–1310. https://doi.org/10.1016/S0022-3093(01)01095-X.
56 Verlaan, V., Houweling, Z.S., van der Werf, C.H.M. et al. (2008). Deposition of device quality silicon nitride with ultra high deposition rate (> 7 nm/s) using hot-wire CVD. *Thin Solid Films* 516 (5): 533–536.

57 Schropp, R.E.I., Nishizaki, S., Houweling, Z.S. et al. (2008). All hot wire CVD TFTs with high deposition rate silicon nitride (3 nm/s). *Solid State Electron.* 52: 427–431.
58 Nishizaki, S., Ohdaira, K., and Matsumura, H. (2009). Comparison of a-Si TFTs fabricated by Cat-CVD and PECVD methods. *Thin Solid Films* 517: 3581–3583.
59 Nishizaki, S., Ohdaira, K., and Matsumura, H. (2008). Study on stability of amorphous silicon thin-film transistors prepared by catalytic chemical vapor deposition. *Jpn. J. Appl. Phys.* 47: 8700–8706.
60 Matsumura, H., Hasagawa, T., Nishizaki, S., and Ohdaira, K. (2011). Advantage of plasma-less deposition in Cat-CVD to the performance of electronic devices. *Thin Solid Films* 519: 4568–4570.
61 Nishizaki, S., Ohdaira, K., and Matsumura, H. (2009). A-Si TFT with current drivability equivalent to poly-Si TFTs. In: *Technical Digest of International Thin Film Transistor Conference (ITC09)* (5–6 March 2009). Paris: Ecole-Polytechnique.
62 Matsumura, H. (1991). Formation of polysilicon films by catalytic chemical vapor deposition (cat-CVD) method. *Jpn. J. Appl. Phys.* 30: L1522–L1524.
63 Rath, J.K., Meiling, H., and Schropp, R.E.I. (1997). Low-temperature deposition of polycrystalline silicon thin films by hot-wire CVD. *Sol. Energy Mater. Sol. Cells* 48: 269–277.
64 Rath, J.K., Tichelaar, F.D., Meiling, H., and Schropp, R.E.I. (1998). Hot-Wire CVD poly-silicon films for thin film devices. *Mater. Res. Soc. Symp. Proc.* 507: 879–890.
65 Schropp, R.E.I., Stannowski, B., Rath, J.K. et al. (2000). Low temperature poly-Si layers deposited by Hot Wire CVD yielding a mobility of 4.0 cm^2/Vs in top gate Thin Film Transistors. *Mater. Res. Soc. Symp. Proc.* 609: A31.3.
66 Saboundji, A., Colon, N., Gorin, A. et al. (2005). Top-gate microcrystalline silicon TFTs processed at low temperature (200°C). *Thin Solid Films* 487: 227–231.
67 Kasai, H, Kusumoto, N., Yamanaka, H. and Yamoto, H. (2001). Fabrication of high mobility poly-Si TFT by cat-CVD method. *Technical Report of the Institute of Electronics*. Information and Communication Engineering of Japan, ED2001–4 and SDM2001–4, pp. 19–25 [in Japanese].
68 Matsumura, H., Umemoto, H., Izumi, A., and Masuda, A. (2003). Recent progress of cat-CVD research in Japan – bridging between the first and second Cat-CVD conferences. *Thin Solid Films* 430: 7–14.
69 Wu, Bing-Rui, Tsalm, T.-H., Wuu, D.S. (2014). Ambipolar micro-crystalline silicon thin film transistors prepared by hot-wire chemical vapor deposition. *Abstract of HWCVD 8* (13–16 October 2014), Brunswick, Germany.
70 Sze, S.M. (2002). *Semiconductor Devices – Physics and Technology*, 2nde), Section 7.3 in Chapter 7. Wiley.
71 Hattori, R., Nakamura, G., Nomura, S., Ichise, T., Masuda, A., and Matsumura, H. (1997). Noise Reduction of pHEMTs with Plasmaless SiN Passivation by Catalytic-CVD. *Technical Digest of 19th Annual IEEE GaAs IC Symposium*, Anaheim, California, USA (12–15 October 1997) pp. 78–80.

72 Oku, T., Totsuka, M. and Hattori, R. (2000). Application of cat-CVD to wafer fabrication of GaAs FET. *Extended Abstract of the 1st International Conference on Cat-CVD (Hot-Wire CVD) Process*, Kanazawa, Japan (14–17 November 2000), pp. 249–252.

73 Higashiwaki, M., Matsui, T., and Mimura, T. (2006). AlGaN/GaN MIS-HFETs With f_T of 163 GHz Using Cat-CVD SiN Gate-Insulating and Passivation Layers. *IEEE Electr. Device Lett.* 27: 16–18.

74 Higashiwaki, M., Mimura, T., and Matsui, T. (2008). GaN-based FETs using Cat-CVD SiN passivation for millimeter-wave application. *Thin Solid Films* 516: 548–552.

75 Yoichi Akasaka (2007). Data of NBTI (negative bias temperauure instability) of MOS transistors was provided from Yoichi Akasaka at Osaka University, Osaka, Japan.

76 Ogawa, Y., Ohdaira, K., Oyaidu, T., and Matsumura, H. (2008). Protection of organic light-emitting diodes over 50,000 hours by Cat-CVD SiN_x/SiOxNy stacked thin films. *Thin Solid Films* 516: 611–614.

77 Matsumura, H. and Ohdaira, K. (2009). New application of Cat-CVD technology and recent status of industrial implementation. *Thin Solid Films* 517: 3420–3423.

78 Burrows, P.E., Graff, G.L., Gross, M.E. et al. (2001). Ultra barrier flexible substrates for flat panel displays. *Displays* 22: 65–69.

79 Huang, J., Tan, S., Lund, P.D., and Zhou, H. (2017). Impact of H_2O on organic-inorganic hybrid perovskite solar cells. *Energy Environ. Sci.* 10: 2284–2311.

80 Nakayama, H. and Ito, M. (2011). Super H_2O-barrier film using Cat-CVD (HWCVD)-grown SiCN for film-based electronics. *Thin Solid Films* 519: 4483–4486. https://doi.org/10.1016/j.tsf.2011.01.311.

81 Coclite, A.M., Ozaydin-Ince, G., Palumbo, F. et al. (2010). Single-chamber deposition of multilayer barriers by plasma enhanced and initiated chemical vapor deposition of organosilicones. *Plasma Process. Polym.* 7: 561–570.

82 Kim, S.Y., Kim, B.J., Kim, D.H., and Im, S.G. (2015). A monolithic integration of robust, water-/oil-repellent layer onto multilayer encapsulation films for organic electronic devices. *RSC Adv.* 5: 68485–68492.

83 Lau, K.K.S. and Gleason, K.K. (2006). Initiated chemical vapor deposition (*i*CVD) of poly(alkyl acrylates): an experimental study. *Macromolecules* 39: 3688–3694. https://doi.org/10.1021/ma0601619.

84 Alpuim, P., Goncalves, L.M., Marins, E.S. et al. (2009). Deposition of silicon nitride thin films by hot-wire CVD at 100 C and 250 C. *Thin Solid Films* 517: 3503–3506. https://doi.org/10.1016/j.tsf.2009.01.077.

85 Spee, D.A., van der Werf, C.H.M., Rath, J.K., and Schropp, R.E.I. (2011). Low temperature silicon nitride by hot wire chemical vapour deposition for the use in impermeable thin film encapsulation on flexible substrates. *J. Nanosci. Nanotechnol.* 11: 8202–8205. https://doi.org/10.1166/jnn.2011.5100.

86 Spee, D.A., van der Werf, C.H.M., Rath, J.K., and Schropp, R.E.I. (2014). Moisture barrier enhancement by spontaneous formation of silicon oxide interlayers in hot wire chemical vapor deposition of silicon nitride on

poly(glycidyl methacrylate). *Can. J. Phys.* 92: 593–596. https://doi.org/10.1139/cjp-2013-0581.

87 Carcia, P.F., McLean, R.S., and Reilly, M.H. (2010). Permeation measurements and modeling of highly defective Al_2O_3 thin films grown by atomic layer deposition on polymers. *Appl. Phys. Lett.* 97: 221901. https://doi.org/10.1063/1.3519476.

88 Spee, D.A., van der Werf, C.H.M., Rath, J.K., and Schropp, R.E.I. (2012). Excellent organic/inorganic transparent thin film moisture barrier entirely made by hot wire CVD at 100°C. *Phys. Status Solidi (RRL)* 6 (4): 151. https://doi.org/10.1002/pssr.201206035.

89 Robert J. Visser, (2016). "HWCVD for Creating Polymer Based Optical and Barrier Films on Large Area R2R Equipment", Extended Abstract of 9th International Conference on Hot-Wire (Cat) and Initiated Chemical vapor Deposition, Philadelphia, USA, Sept., 6–9, (2016).

90 Nakaya, M., Kodama, K., Yasuhara, S., and Hotta, A. (2016). Novel gas barrier SiOC coating to PET bottles though a hot wire CVD method. *J. Polym.* 2016: 4657193-1-7. http://dx.doi.org/10.1155/2016/4657193.

9

Radicals Generated in Cat-CVD Apparatus and Their Application

The catalytic chemical vapor deposition (CVD) technology is not limited to the deposition of films. Many kinds of radicals can be generated in Cat-CVD chambers. In this chapter, radical generation and application of such radicals are discussed. This is not strictly "CVD" anymore. However, we treat this as a family technology of Cat-CVD.

9.1 Generation of High-Density Hydrogen (H) Atoms

9.1.1 Generation of High-Density H Atoms

Catalytic decomposition of hydrogen gas (H_2) is almost the oldest study of the reactions on a tungsten (W) catalyzer. It started in the 1910s. It has been known that the catalytic cracking of H_2 on heated W is the simplest method to obtain hydrogen (H) atoms. However, the absolute density measurements as well as the application studies are not so old, and some of them are in a scope of this book. We already showed H density as a function of the catalyzer temperature, T_{cat}, for an Ir catalyzer in Figure 4.7.

Figure 9.1 demonstrates H density as a function of temperature of the catalyzer (T_{cat}) when a W wire is used as a catalyzer. The H density was measured by two-photon laser induced fluorescence (LIF) and vacuum ultraviolet (VUV) laser absorption. In this experiment, the flow rate of H_2, FR(H_2), was 150 sccm and the gas pressure, P_g, was 5.6 Pa [1]. The inner diameter of a stainless steel (SUS) cylindrical chamber was 45 cm and the height was 40 cm. The measuring point was set at the center of the chamber and at only 10 cm apart from the catalyzer, to make the influence of the chamber wall small. As already mentioned in Section 4.5.1, concerned with H generation, the difference of the H density generated by various different metal catalyzers appears small.

Figure 9.2 also shows H density as a function of P_g, for fixed T_{cat} at 2200 K. FR(H_2) was 150 sccm at $P_g = 5.6$ Pa [1]. The H density, [H], is likely to increase in proportion to the square root of the H_2 pressure. This can be simply understood, as H_2 is decomposed to two H atoms by $H_2 \rightarrow H + H$ and H density is roughly proportional to the square root of the H_2 density, [H_2], that is, H_2 pressure; although strictly speaking, this holds when both H and H_2 are in equilibrium conditions.

Catalytic Chemical Vapor Deposition: Technology and Applications of Cat-CVD,
First Edition. Hideki Matsumura, Hironobu Umemoto, Karen K. Gleason, and Ruud E.I. Schropp.

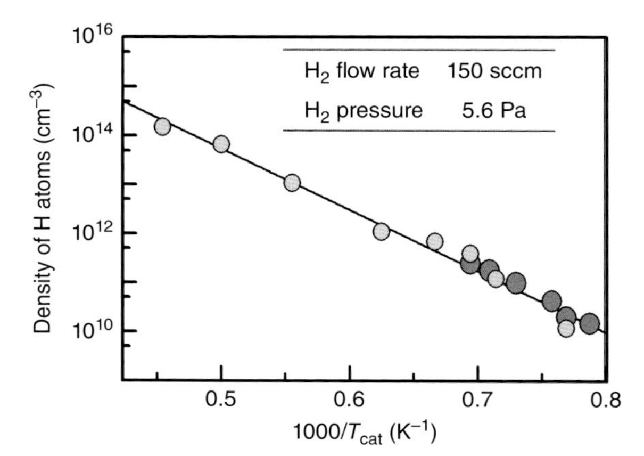

Figure 9.1 Atomic hydrogen (H) density as a function of reciprocal of the temperature of the catalyzer, T_{cat}. Dark closed circles show the results by VUV absorption measurement and open circles by two-photon LIF measurement. Source: Umemoto et al. 2002 [1]. Copyright 2002. Reprinted with permission from The Japan Society of Applied Physics.

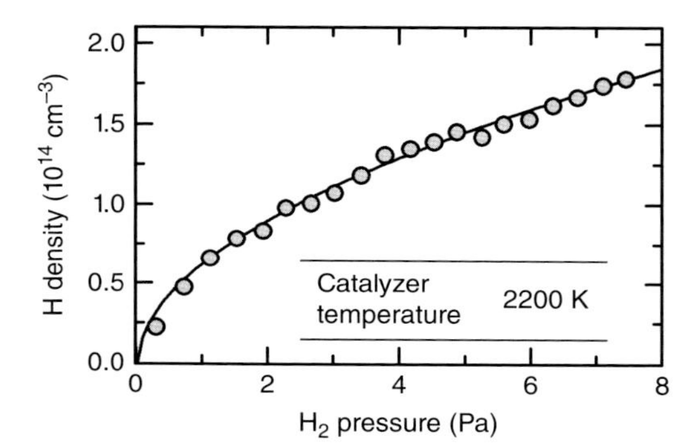

Figure 9.2 H-atom density as a function of gas pressure of H_2, P_g. Source: Umemoto et al. 2002 [1]. Copyright 2002. Reprinted with permission from The Japan Society of Applied Physics.

These two figures demonstrate that the H density around 10^{14} cm^{-3} can be easily obtained just by placing heated W wires in a chamber. A similar density of H atoms can also be obtained by plasma decomposition such as microwave plasma [2]. However, an important point in Cat-CVD technology is that we can obtain high-density H atoms in a very simple apparatus without coproducing other active species, such as ions and electronically excited species. In addition, in catalytic cracking, we need not mind the gas pressure for decomposition of molecules, where a plasma can only be generated at certain limited gas pressures. For instance, J. Larjo et al. have succeeded in producing H atoms of 1.2×10^{17} cm^{-3} by decomposing a mixture of 3.9 kPa of H_2 and 80 Pa of CH_4 by using a TaC filament at 2700 °C [3]. This catalytic cracking system has many advantages for generating H atoms.

As mentioned later in this chapter, H atoms are used in various applications. They can be used for cleaning or may be used even for sterilization processes. If such H atoms can be obtained in atmospheric pressure or in a simple vacuum system, it may be very useful.

In principle, H atoms can be generated even at atmospheric pressure if we use the catalytic cracking method. The development of an H generator operating at atmospheric pressure is desired in many occasions. For this purpose, we have to find the way to lengthen the lifetimes of H atoms. In the presence of gases, H atoms may easily be lost in reactions in the gas phase. For example, H atoms react with SiH_4 to produce H_2 and SiH_3. However, in a pure H_2 system, the only removal process in the gas phase is recombination, $H + H \rightarrow H_2$.

As recombination reactions are always exothermic, a third body, M, which accepts excess energy, is necessary for recombination reactions of two atoms, such as

$$H + H + M \rightarrow H_2 + M \tag{9.1}$$

M may be a H_2 molecule itself or an added gas molecule. The lifetime of H atoms, that is, how long H atoms can survive in the system, can be calculated from the rate constant. D.L. Baulch et al. reported the reaction rate constant of Eq. (9.1), k_{H_2}, to be 8.8×10^{-33} cm^6/s at 300 K when the third body M is H_2 [4]. Then, the lifetime of H atoms, τ_H, which is decided by the reaction rate of Eq. (9.1), can be calculated as follows.

First, (i) a single H atom recombines with another H atom to produce a H_2 molecule, when the H atom collides with another H atom and a third body M at the same time, i.e. by three-body collisions. (ii) The number of collisions producing H_2 per unit time should be proportional to the product of [H] and M density, [M], and k_{H_2}. Thus, (iii) as τ_H should be expressed by the inverse of the number of reactive collisions per unit time, τ_H is derived as Eq. (9.2):

$$\begin{aligned}\tau_H &= 1/(k_{H_2}[H]\,[M]) \\ &= 1/\{8.8 \times 10^{-33}(\text{cm}^6/\text{s})\,[H]\,(\text{cm}^{-3})\,[M]\,(\text{cm}^{-3})\}\ (\text{sec})\end{aligned} \tag{9.2}$$

Here, for simplicity, we assume that M is only H_2 gas. The ratio of [H] to $[H_2]$ is estimated from the results shown in Figures 9.1 and 9.2 to be about 0.1. For instance, when P_g = 6 Pa, T_{cat} = 2200 K (1927 °C), and T_g = 300 K (27 °C), $[H_2]$ is 1.45×10^{15} cm^{-3} as shown in Table 2.1, whereas [H] is approximately 1.5×10^{14} cm^{-3} as shown in Figure 9.2. In this case, we approximate that the gas pressure is decided only by that of the H_2 gas.

Figure 9.3 shows the simulated results of the lifetimes of H atoms as a function of P_g of H_2 gas, when only H_2 molecules exist in a chamber with an infinite size. In fact, H atoms may recombine on the chamber walls, but this effect is not taken into account in this figure. The recombination processes on the walls will be discussed in Section 9.1.2. The gas temperature is assumed to be 300 K (27 °C) or 523 K (250 °C), and the various ratios of the density of H atoms to that of H_2 molecules ($[H]/[H_2]$) from 0.01 to 1.0 are taken as a parameter. However, a practical value of this ratio may be around 0.1 as already mentioned above. In this calculation, as the gas pressure of H_2, the values before decomposition of H_2 by cracking are simply used. The reaction rate constant, $k_{H_2} = 6.3 \times 10^{-33}$cm^6/s, is used in the calculation for 523 K (250 °C). These results show that the lifetime is longer than the typical residence time of gas molecules in the chamber, say about one second, and that the recombination in the gas phase can be ignored when the gas pressure is less than 100 Pa.

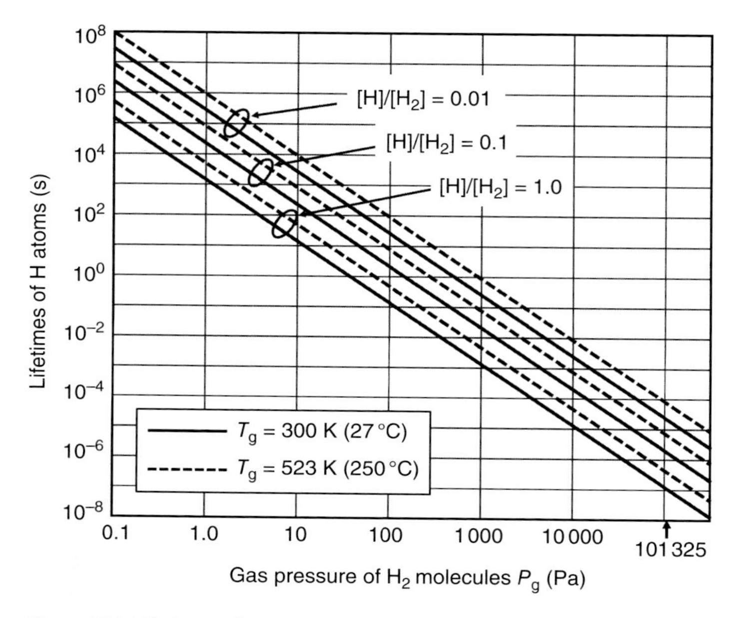

Figure 9.3 Lifetimes of H atoms in infinite size chamber filled with H_2 molecules vs. gas pressure of H_2, neglecting partial pressure of H atoms.

9.1.2 Transportation of H Atoms

The above calculation on the lifetime of H atoms is based only on gas-phase reactions. However, annihilation of H atoms on chamber walls is more important under typical CVD conditions, as the chamber size is not infinite. The recombination rate of H atoms on solid surfaces has been estimated by many researchers. The recombination rate depends not only on the material of the wall surface but also on their surface roughness and temperature. For example, the recombination rate on silicon dioxide (SiO_2) is 1 order of magnitude smaller than that on SUS [5], and the loss probability is high at high temperatures [6, 7]. Time-resolved measurements of H-atom density after the termination of plasma discharge show that the lifetime of H atoms is typically in the order of millisecond [5, 8, 9], much shorter than the typical residence time of molecules in a chamber, in the order of second as has been mentioned in Section 2.1.4. Then, when we wish to transport H atoms, the recombination loss of H atoms on chamber walls becomes a large problem.

S.G. Ansari et al. have shown that SiO_2 or polytetrafluoroethylene (PTFE, also commercially known as Teflon) coating of SUS is effective to lengthen the lifetime of H atoms [10]. They used a water-cooled jacket with an inner diameter of 7.2 cm and a length of 28 cm, installed in a cylindrical chamber. They prepared three identical jackets made of SUS. One was used without coating. The others were used after coating with SiO_2 or PTFE. The SiO_2 coating was carried out by natural oxidation of perhydropolysilazane, an inorganic polymer consisting of cyclic $(H_2Si{-}NH)_n$. After applying a xylene solution of perhydropolysilazane, the jacket was baked at 420 K in air for three hours. PTFE films can be deposited

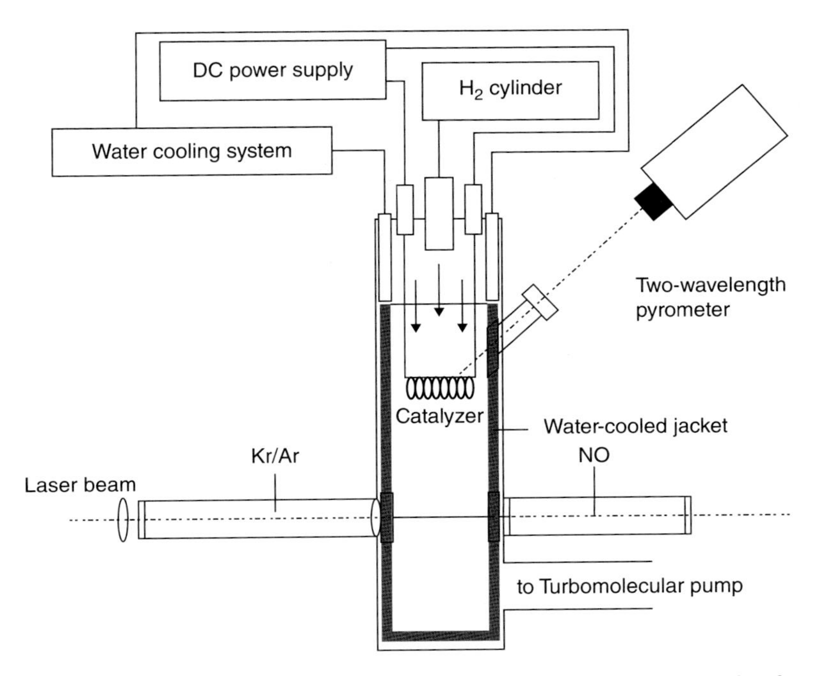

Figure 9.4 A schematic diagram of experimental setup to know the effect of surface conditions of a chamber for H-atom transport. Source: Ansari et al. 2005 [10]. Reprinted with permission from IEEE.

by Cat-CVD as has been discussed in Chapter 6, but in their work, PTFE films, 30 μm in thickness, were coated by spraying and annealing. In addition, H_3PO_4 coating was also tried on the SiO_2 layer. The experimental setup is schematically shown in Figure 9.4. The length and the diameter of the W catalyzing wire was 20 cm and 0.5 mm, respectively. The H-atom density was monitored by VUV absorption at 10 cm downstream from the wire. Figure 9.5 shows the measured H-atom density as a function of the reciprocal of the catalyzer temperature T_{cat} for four coating conditions: noncoating, SiO_2 coating, PTFE coating, and phosphoric acid coating. FR(H_2) was 150 sccm and P_g was 8 Pa. It can be seen that the H-atom density increases drastically using these coatings. In other words, it is possible to lengthen the lifetime of H atoms and to improve the transportation efficiency by coating. At high temperatures, the H-atom density saturates and the absolute density is less than that shown in Figure 9.1. The main cause of this difference can be attributed to the chamber size. The data of Figure 9.1 were taken using a chamber with an inner diameter of 45 cm, whereas a much smaller jacket was used to obtain the data in Figure 9.5. Besides the increase in the surface/volume ratio, the surface temperature of the jacket must have been elevated.

Concluding, the lifetime of H atoms is mainly decided by the recombination probability on the chamber walls. To lengthen the lifetime of H atoms, as well as to increase the H-atom density, SiO_2 or PTFE coating on smooth surfaces appears effective.

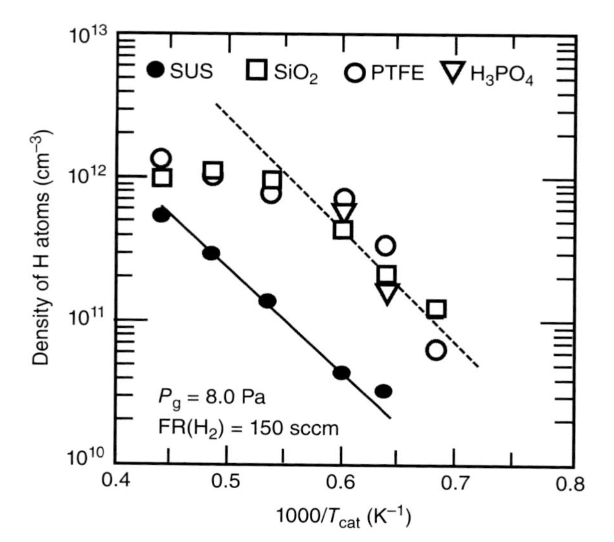

Figure 9.5 H-atom density vs. reciprocal of temperature of catalyzer T_{cat}, for various surface conditions of the chamber wall.

9.2 Cleaning and Etching by H Atoms Generated in Cat-CVD Apparatus

9.2.1 Etching of Crystalline Silicon

H atoms are very active species, enhancing the reactions forming hydrogenated materials. By this way, some elements or some contaminants on a solid surface can be removed by forming volatile hydrogenated species. This property is utilized for surface cleaning of substrates and the etching of some kinds of solids. Here, as an example, we like to show the etch rate of c-Si, although an experimental result for c-Si etch has been already shown in Figure 7.33 for demonstrating the feasibility of chamber cleaning by H atoms.

Figure 9.6 shows the relationship between the etch rate of c-Si and T_{cat} for generating H atoms. The gas pressure P_g, flow rate of H_2 gas FR(H_2), and substrate temperature T_s are 1.3 Pa, 48 sccm, and RT, respectively. It is confirmed that as T_{cat} increases, that is, as the density of generated H atoms increases, the etch rate of c-Si increases. In this case, we have to pay attention to the surface conditions of c-Si. As shown in Figure 7.33 or Figure 9.7 below, c-Si can be etched by H atoms, but SiO_2 cannot be etched by the exposure to H atoms. If the surface of c-Si is oxidized, etching by H atoms becomes quite slow or difficult.

Figure 9.7 shows the etch rate of various materials such as c-Si, poly-Si, a-Si, and SiO_2 as a function of substrate temperatures T_s. The data in the figure are cited from two references [11, 12] and two groups of data are summarized. In the figure, the results for two groups of process parameters are demonstrated for c-Si etching. In the first group, P_g, FR(H_2), and T_{cat} are kept at 13 Pa, 10 sccm, and 1650 °C, respectively. In the other one, they are 1.3 Pa, 48 sccm, and 1800 °C, respectively. For both groups, the same tendency is observed in the etch rates of c-Si. The c-Si etch rate has strong dependence on T_s. Q-mass measurement

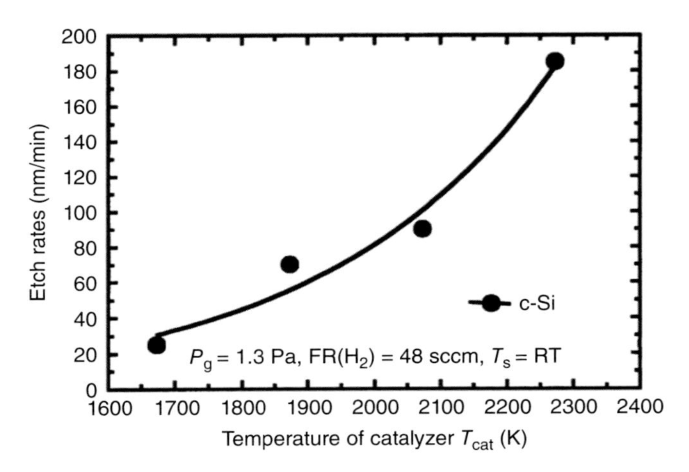

Figure 9.6 Etch rates of c-Si as a function of temperature of catalyzer, T_{cat}. Source: Uchida et al. 2001 [11]. Reprinted with permission from Elsevier.

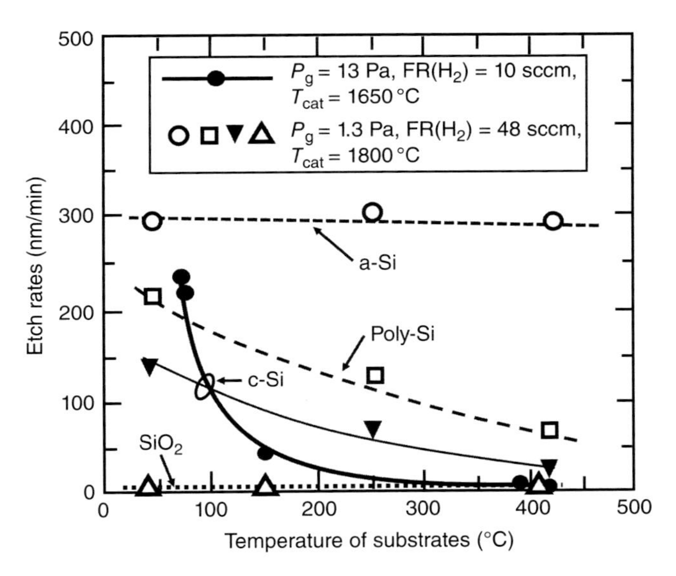

Figure 9.7 Etch rates of various materials such as c-Si, poly-Si, a-Si, and SiO_2 as a function of substrate temperature. The results for two sets of process parameters are summarized for c-Si etching.

for c-Si etching detects SiH_4 as a product. This etching process is also a forming process of SiH_4 from Si and H atoms. When T_s is high, H atoms cannot stay on the c-Si surface long enough to form volatile SiH_4. When T_s is low, however, the time that H atoms are staying on the c-Si surface becomes sufficient for forming SiH_4, and thus, the surface is etched. This tendency is observed even in poly-Si etching. This figure also shows that substrate temperature dependence is more significant at high gas pressures.

The etch rate of a-Si appears different. It is seen that the etch rate of a-Si by H atoms is much larger than that of c-Si and poly-Si. The etch rate of a-Si at 250 °C is almost four times larger than that of c-Si and even about two times larger than that of poly-Si. This is probably because H atoms can penetrate into a-Si easier than c-Si and poly-Si. In the case of a-Si, the etch rate is almost constant against the variation of substrate temperatures.

As shown in the figure, the etch rate of SiO_2 is extremely small. There are not enough data and reports about H etching, and it is not easy to explain the exact etching process of a-Si or SiO_2. However, we can say that H atoms generated in a Cat-CVD apparatus can be used for etching of materials in many cases.

9.2.2 Cleaning of Carbon-Contaminated Surface

H atoms can also etch carbon (C) by forming CH_4. By utilizing this phenomenon, there have been some attempts to remove C atoms from the surface of materials. As already shown in Section 7.6.8, the C atoms that are alloyed with W can be removed by cracked species of NH_3. NH_3 may be a better candidate to enhance C etching compared to H_2. However, as N atoms may sometimes induce other effects such as nitridation, as will be mentioned in Section 9.7, here, we concentrate on explaining the results of etching by H atoms.

As an example, we show the results of cleaning of multilayered mirrors used as reflectors installed in the extreme ultraviolet (EUV) lithography [13, 14]. For making small size semiconductor devices, the patterning accuracy of a dimension of several nanometers is required. There have been many attempts to develop the technologies satisfying such requirements. As one attempt, photolithography using EUV light with a wavelength of 10–15 nm is expected for the next-generation mass production system. At such a short wavelength, optical lenses cannot be used because of the large absorption in optical components. The reflection mirror system is the only possible solution for focusing optical light and build optical confinement system. In such a case, the reflection mirror itself is not simple. To increase reflectivity, the mirror is often made by multilayered materials. For instance, the multilayer mirrors, made by piling up 50 thin bilayers of molybdenum (Mo) and Si on (100) c-Si substrates, are used. The top of the multilayered mirror is capped by a 2.7-nm-thick ruthenium (Ru) layer.

During the patterning process, if the surface of the mirror is contaminated by C atomic layers, the reflectivity is easily degraded. The C contamination may come from the very small amount of vapor of the photodecomposition products of resist or the residual solvent used to dissolve the photoresist and certain components of vacuum equipment. The EUV light may work to make a very thin contamination layer by some kind of photo-CVD process. Anyway, the surface of the mirrors is contaminated by C layers quite often. Thus, we have to remove them without damaging the mirror beneath. As a method to do it, cleaning by H atoms produced in a Cat-CVD apparatus has been attempted, as high-density H atoms are generated without plasma and there is no need to worry about plasma-induced damage.

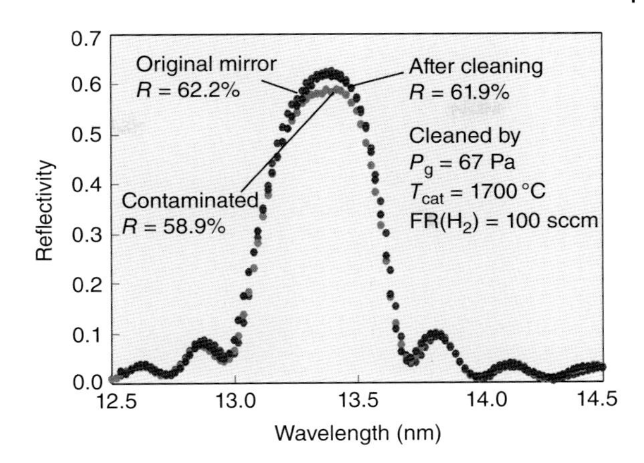

Figure 9.8 Spectra of reflectivity of multilayered mirrors of original, after contaminated, and after cleaned by H atoms. Source: Nishiyama et al. 2005. Data are provided by courtesy of authors of Ref. [13].

Figure 9.8 shows the reflectivity of the EUV mirror before and after H-atom cleaning [13]. In this case, H-cleaning was carried out under the following conditions: T_{cat} = 1700 °C, P_g = 67 Pa, FR(H_2) = 100 sccm, and T_s = RT. Just after fabrication of such a multilayered mirror by a sputtering method, the reflectivity R was 62.2%, but after use, the surface was contaminated by C atoms and R degraded to 58.9%. In practical use, the mirrors are required to be used under EUV irradiation for more than 30 000 hours, keeping the reflectivity loss less than 1.6% from their original reflectivity. That is, R should be kept at values more than 61.3%. After H-cleaning, R recovered to 61.9%. If this H-cleaning system is employed, the requirement is satisfied by periodic cleaning.

As will be mentioned later in Section 9.4, H atoms are also effective in reducing a thin oxide layer formed on metal layers. The surface of the Ru cap layer sometimes suffers from uncontrolled oxidation and is covered with a thin oxide layer. Even if such an oxide layer exists in addition to the C contaminated layer, the H treatment is still effective for reducing it and recovering total reflection.

9.3 Photoresist Removal by Hydrogen Atoms

Above results encourage to use H atoms for removing organic materials. Then, we attempted to remove the photoresists that are widely used for the patterning process of electronic devices or other industrial components.

In the fabrication of semiconductor devices, the photoresist process is quite often carried out for patterning. After patterning, the patterned photoresists have to be removed completely. Oxygen (O) plasma is widely used to ash the photoresists by oxidation. The ash process is one of the successful technologies in the semiconductor process; however, it is sometimes hard to apply for removal of photoresists that are ion implanted. Some parts of photoresists show hardening, and it becomes difficult to remove them by ashing. In addition, the use of O plasma also brings about plasma damage to the samples.

A. Izumi and H. Matsumura reported in 2002 a new approach to remove the photoresists by using high-density H atoms instead of using O plasma [15]. They

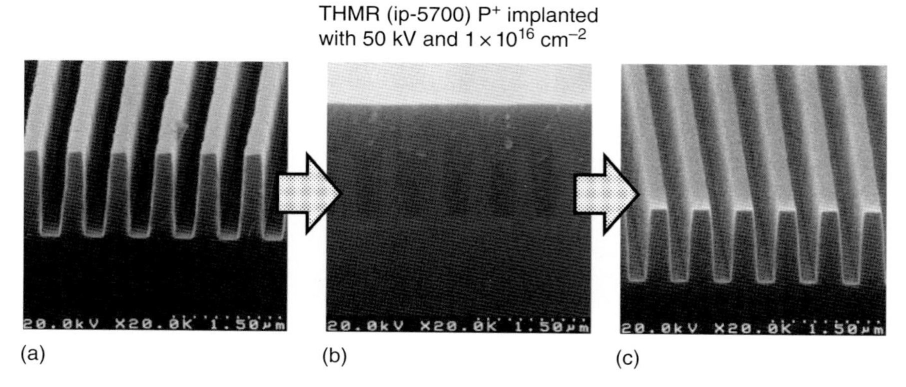

Figure 9.9 Photographs of line and patterns: (a) before the photoresist coating, (b) after the photoresist coating and ion implantation, and (c) after removal of such resist by H atoms.

succeeded in removing even the ion-implanted photoresists. Since then, various studies have been reported concerned with this topic and removal rates over 1 μm/min have been achieved, which is a measure of adoption in mass production. For example, the removal rate of positive-tone Novolak can be as large as 2.4 μm/min when the catalyzer temperature is 2000 °C and the hydrogen pressure is 4.5 Pa [16].

In the case of O-plasma ashing, the ashes remain sometimes on the sample surface after the process and another extra cleaning process is required for removing these ashes. However, in the resist removal process by H atoms, the photoresists are vaporized as volatile hydrocarbon gases, and thus, there is nothing remaining on the surface of the samples. One point we may notice is that P atoms in photoresist become volatile PH_3 by the reaction with H atoms but that they are converted to a residual solid material, P_4O_{10}, by the reaction with O atoms.

Figure 9.9 shows the photographs of line-and-space patters of Si substrates covered with photoresists and after removal of them. The width of a line is 0.25 μm and the pitch of line and space patterns is 0.5 μm. Figure 9.9a shows the original line-and-space patters, panel (b) shows them after being covered with photoresists and then after P ion implantation of 50 kV with a dose of 1×10^{16} cm^{-2}, and panel (c) shows the patters after the removal of photoresists by H atoms. The ion implantation dose is heavy, and usually such a high-dose implantation of P atoms causes photoresist hardening and makes the removal difficult. In addition, the removal of such high-dose implanted photoresists by O-plasma ashing often leaves many residues on the patterns. In Figure 9.9c, we cannot see any residues in the patterns. In this case, the photoresists are positive-type ones for i-line lithography, THMR (ip-5700) made by Tokyo-Ohka Kogyo (TOK) Co. Ltd. In the processing, T_{cat}, P_g, FR(H_2), and T_s are set at 1700 °C, 2.7 Pa, 100 sccm, and RT, respectively.

It is confirmed that the photoresists are completely removed even after ion implantation of a dose of 10^{16} cm^{-2}. The photoresists removal after ion implantation was also studied by H. Horibe et al. [17]. They used positive tone Novolak resist, AZ6112, made by AZ-Electronic Materials Co. for their studies. After

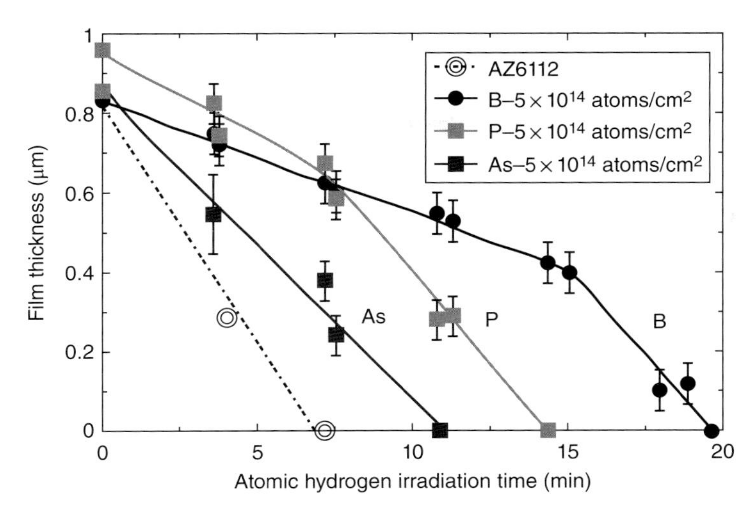

Figure 9.10 The residual thickness of photoresists on B, P, or As ion-implanted samples as a function of H treatment time. Source: Horibe et al. 2011 [17]. Reprinted with permission from Elsevier.

100 °C prebaking of the photoresists, B, P, and As ions of 70 keV were implanted with a dose of 5×10^{12}–5×10^{15} cm^{-2}. The temperature before implantation was room temperature. The photoresists were removed by H atoms generated from the H_2 gas diluted to 10% by N_2 gas for safety. Total gas pressure, a net gas flow of H_2, T_{cat}, and initial substrate temperature were set at 21 Pa, 30 sccm, 2420 °C, and 25 °C, respectively. The original thickness of the photoresists after prebaking was about 0.8–1.0 μm. The residual thickness was measured as a function of atomic hydrogen treatment time, as shown in Figure 9.10.

In the figure, it is shown that the photoresist removal rate after ion implantation is low initially; however, after etching the hardened region, it is enhanced to 0.1 μm/min. The initial region with low etch rates corresponds to the ion-implanted hardened region. After removal of the ion-implanted region, the etch rates start to increase. The removal rates of the photoresists depend on the photoresist itself and on the process parameters for etching. Particularly, some of the photoresists have a strong etch rate dependence on substrate temperature.

Figure 9.11 shows the removal rates of photoresists as a function of substrate temperature. The photoresist was ip-3650 made by TOK, and in the photoresist removal, T_{cat}, P_g, FR(H_2), the distance between the sample and the catalyzer D_{cs}, and surface area of the catalyzer S_{cat} were kept at 1800 or 2000 °C, 67 Pa, 300 sccm, 3.5 cm, and 18 cm^2, respectively [18]. The photoresists with a simple flat surface and with line and space patterns were formed on c-Si substrates. The photoresists were spin-coated on c-Si substrates and then they were prebaked at 90 °C for 90 seconds. After making line and space patterns if required, they were postbaked at 110 °C for 90 seconds. Probably as the number of H atoms attacking the sample surface is considered sufficient, to enhance the reaction of the photoresists with H atoms, the substrate temperatures were elevated. As expected, it is demonstrated in the figure that the photoresist removal rates can be increased

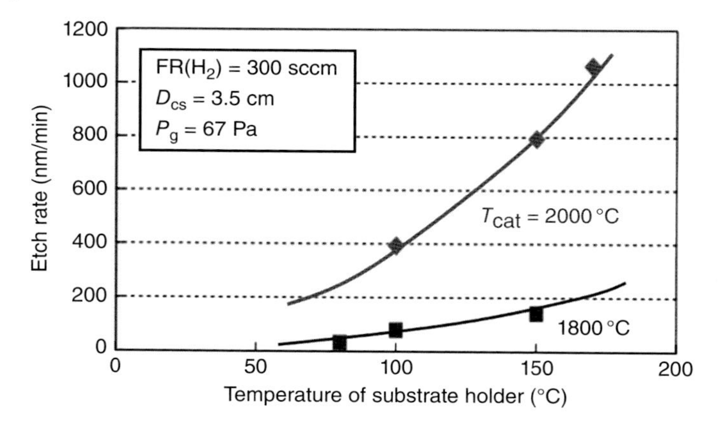

Figure 9.11 The photoresist removal rates vs. temperatures of substrate holder, taking catalyzer temperature T_{cat} a parameter. The real substrate temperatures are usually 20–30 °C lower than the holder temperatures. The flow rate of H_2, FR(H_2), distance between catalyzer and substrates, D_{cs}, and gas pressure, P_g, are all described in the figure. Source: Hashimoto et al. 2006 [18]. Reprinted with permission from Elsevier.

to values over 1 μm/min, which satisfy the industrial requirement for mass production. Actually, model machines of a photoresist remover using H atoms have been made by two companies at least.

Recently, photoresist removal by using a H_2/O_2 mixture has been reported [19]. Of course, when W is used as a catalyzer, the $[H_2]/[O_2]$ ratio must be large enough not to oxidize the catalyzer. By the addition of a small amount of O_2, not only the removal rate increased but also the removal uniformity improved [20].

In the case of photoresist removal, the surface of the samples is always exposed to contaminants that come from the W catalyzer together with H atoms. In the case of film deposition by Cat-CVD, such contaminants are embedded inside the films and thus distributed throughout the whole film. However, in the case of radical treatment, the contaminants emitted from the catalyzer may be piled up at the surface of the samples. The effect of contaminants may be more serious than the deposition. Therefore, the density of W atoms on the sample surface was monitored by the total reflection X-ray fluorescence (TRXF).

Figure 9.12 shows the sheet W concentration at the surface of a 15-nm-thick SiO_2 layer on c-Si substrates, as a function of the catalyzer temperatures T_{cat} [18]. The SiO_2-coated c-Si was used as SiO_2 protects the etching by H atoms. In this experiment, P_g, T_s, FR(H_2), D_{cs}, and S_{cat} were 67 Pa, 150 °C, 300 sccm, 3.5 cm, and 21 cm^2, respectively. The exposure time was two minutes in this case. When the removal rates of photoresists exceed 1 μm/min, usually two minutes is enough for the resist removal process. The W sheet concentration increases as T_{cat} increases. However, even at T_{cat} = 2000 °C, it is kept at lower than 5×10^{10} atoms/cm^2. This value is usually acceptable as a semiconductor process.

Figure 9.13 shows a similar W sheet concentration on SiO_2-coated c-Si substrates as a function of exposure time to H atoms. In this experiment, T_{cat} was fixed at 2000 °C, but other parameters were the same as those of Figure 9.12. If the process is finished within five minutes, the W contamination can be suppressed to

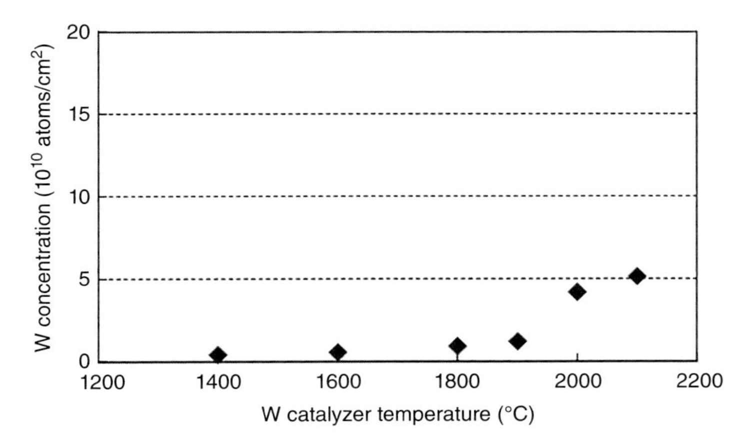

Figure 9.12 W concentration on SiO_2-coated Si substrates vs. catalyzer temperatures T_{cat}. Source: Hashimoto et al. 2006 [18]. Reprinted with permission from Elsevier.

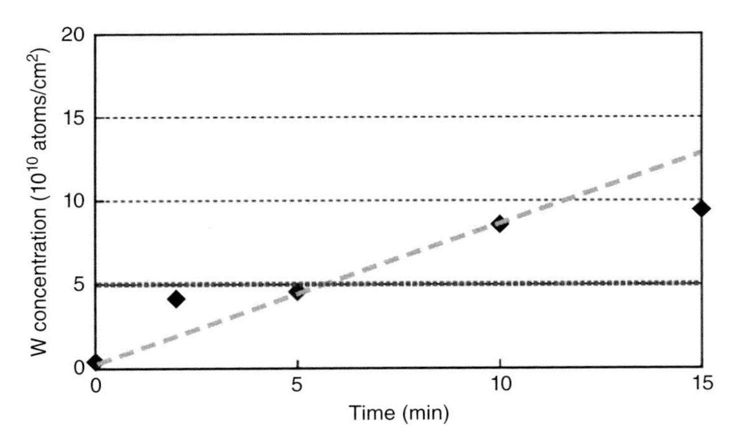

Figure 9.13 W concentration on SiO_2-coated Si substrates vs. H treatment time for T_{cat} = 2000 °C. Source: Hashimoto et al. 2006 [18]. Reprinted with permission from Elsevier.

less than 5×10^{10} atoms/cm^2. The flux density of the W contaminants estimated from a broken line in Figure 9.13 is 8×10^9 atoms/cm^2 min.

When H_2 molecules were decomposed to H atoms at 2000 °C, the density of H atoms can be 10^{14} cm^{-3} as shown in Figure 9.1, although the H density was measured at the center of a relatively large chamber with an inner diameter of 45 cm and the effect of the chamber wall was minimized. In this case, the thermal velocity of H atoms with temperatures of 250–2000 °C is about 3.3×10^5–6.9×10^5 cm/s as shown in Table 2.2. Here, roughly, we approximate 4×10^5 cm/s for the thermal velocity. If we assume that the actual H density is only 10^{11} cm^{-3}, considering the effects of consuming H atoms by the materials surrounding samples, much less than the value shown in Figure 9.1, the number of H atoms colliding with the photoresists should be 10^{16} atoms/cm^2 s or 6×10^{17} atoms/cm^2 min, as evaluated from Eq. (2.6). From these values and the results of Figure 9.13, we can say that the flux of H atoms attacking the photoresist includes a fraction of about 1.3×10^{-8} as W contaminants. This value appears much smaller than the estimated density of W contaminants in the flux of deposition species. This may be an effect of the high-density H atoms. Although we do not know the exact reason, we can say that W contamination is not a serious problem in photoresist removal.

9.4 Reduction of Metal Oxide by H atoms

9.4.1 Reduction of Various Metal Oxides

As already shown in Figure 9.7, SiO_2, which is made by thermal oxidation of Si, is not etched quickly by H atoms. However, removing O atoms or reduction of metal oxide is not so difficult. A. Izumi et al. reported the results of reduction of a surface oxide layer on various metals by H atoms [21]. They revealed that oxides of copper (Cu), Ru, niobium (Nb), Mo, rhodium (Rh), palladium (Pd), iridium (Ir), and platinum (Pt) were all reduced by H-atom treatments.

In their case, T_{cat}, S_{cat}, D_{cs}, and FR(H_2) were fixed at 1700 °C, 12.6 cm^2, 6 cm, and 50 sccm, respectively, but T_s varied from 40 to 180 °C. The time for H treatment was one minute. Various metals were deposited on (100) c-Si substrates by the sputtering method, and then, they were oxidized in O_2 plasma generated by electron cyclotron resonance (ECR). The thickness of oxidized layers depends on the metals; however, roughly, it varied in a range of a few nanometers.

Figure 9.14 shows the observed results by X-ray photoelectron spectroscopy (XPS) for oxidized Pd, taking T_s as a parameter. The peak due to Pd–O disappeared after just one minute of H treatment, even at low substrate temperatures. This demonstrates that an oxidized layer can be easily removed by H treatments of only one minute.

They also reported the reduction rates of various metal oxides and their activation energies. The results are summarized in Table 9.1. It is known that the

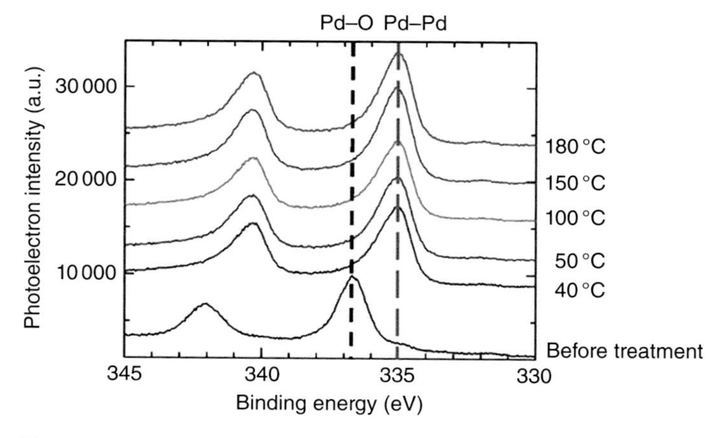

Figure 9.14 XPS spectra of oxidized Pd and those after one minute H-atom treatment. Source: Izumi et al. 2008 [21]. Reprinted with permission from Elsevier.

Table 9.1 Rates of reduction of metal oxide layers and their activation energies, when they are reduced by H atoms.

Metals forming oxides	Reduction rates (nm/min)	Activation energy of reduction by H atoms (eV)
Cu	1.5	5.7×10^{-3}
Ru	0.13	7.6×10^{-4}
Nb	0.01	8.5×10^{-2}
Mo	0.14	1.5×10^{-2}
Rh	2.0	2.4×10^{-2}
Pd	0.30	6.4×10^{-2}
Ir	0.34	7.7×10^{-3}
Pt	2.1	3.7×10^{-3}

activation energy is quite low and reduction of metal oxide by H atoms is a very easy process [21]. As mentioned in Chapter 5, Ir is used as a catalyzer for SiO_2 deposition. Even if it is slightly oxidized, the surface of it may be reduced and returned to original Ir.

9.4.2 Characteristic Control of Metal Oxide Semiconductors by H Atoms

The H treatment of metal oxides is also attempted for many purposes. As an example, here, we introduce the results of H treatment of cuprous oxide Cu_2O, which is a well-known semiconductor. The conductivity or carrier concentration of Cu_2O is decided by defects formed with excess O atoms. If the Cu_2O with such defects is exposed to H atoms, some O atoms are bonded with H and the carrier concentration and the carrier mobility can be adjusted.

Figure 9.15 shows the resistivity, mobility, and carrier concentration of Cu_2O as a function of gas pressure after the exposure to H atoms [22]. The density of H atoms depends on the gas pressure P_g during the H treatment. In this case, Cu_2O films were prepared by thermal oxidation of sputtered Cu films on c-Si substrates. The oxidation temperature was 300 °C. Thickness of Cu_2O films is about 100 nm. H atoms were generated by catalytic cracking reaction of H_2 gas with a W catalyzer of $T_{cat} = 1050$ °C. The substrate temperature T_s during the process was fixed at 200 or 300 °C. As already shown in Figure 7.33 or Figure 9.2, as P_g increases, the density of generated H atoms increases. When P_g increases from 0.19 mTorr (0.025 Pa) to 1.0 mTorr (0.13 Pa), the density of excess O atoms may decrease due to pulling out by H atoms and the carrier concentration also decreases, and instead, the mobility increases. In this case, for P_g over 1.0 mTorr, the surface of Cu_2O started to be reduced and Cu appears. Just below the condition for the reduction of Cu_2O, the carrier concentration and the mobility are well controlled by using the conditions generating low density of H atoms.

This is one of example of various utilizations of H atoms.

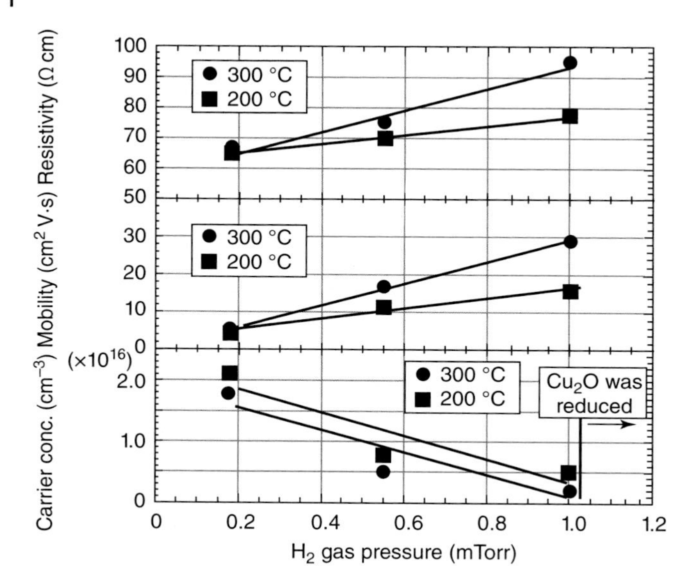

Figure 9.15 Resistivity, mobility, and carrier concentration of cuprous oxide (Cu_2O) as a function of P_g after the H-atom treatment for 30 minutes. Source: Tabuchi and Matsumura 2002 [22]. Copyright (2002). Reprinted with permission from The Japan Society of Applied Physics.

9.5 Low-Temperature Formation of Low-Resistivity Metal Lines from Liquid Ink by H Atoms

As another application of H atoms generated in Cat-CVD apparatus, we can introduce the formation of low-resistivity metal lines by exposure to H atoms. In electronic devices, there are many metal lines for interconnection of various components. One of the features of integrated circuits (ICs) is the formation of metal lines by dry processes such as vacuum evaporation or sputtering. However, there are other approaches for making metal lines from liquid using conductive paste such as a silver (Ag) paste. The liquid process is also important for fabrication of other devices. For instance, metal lines in solar cells are formed by screen printing using metal paste. Metal lines made by metal paste are useful for the interconnection of three-dimensional lamination of several ICs. Such a liquid process fits to a new process using an inkjet or to printing electronics in which all device components are formed by using liquid.

Such a metal paste or a metal ink consists of metal nanoparticles and organic binders. After painting the metal paste or metal ink, usually annealing at a temperature around 200–300 °C is required to make low-resistivity metal lines. The annealing treatment is necessary for evaporation of organic binders and connecting metal nanoparticles for electric conduction. However, sometimes, the annealing over 100 °C is severe for substrates, e.g. when they are plastic. In addition, even after annealing at 300 °C, the resistivity is sometimes not as low as

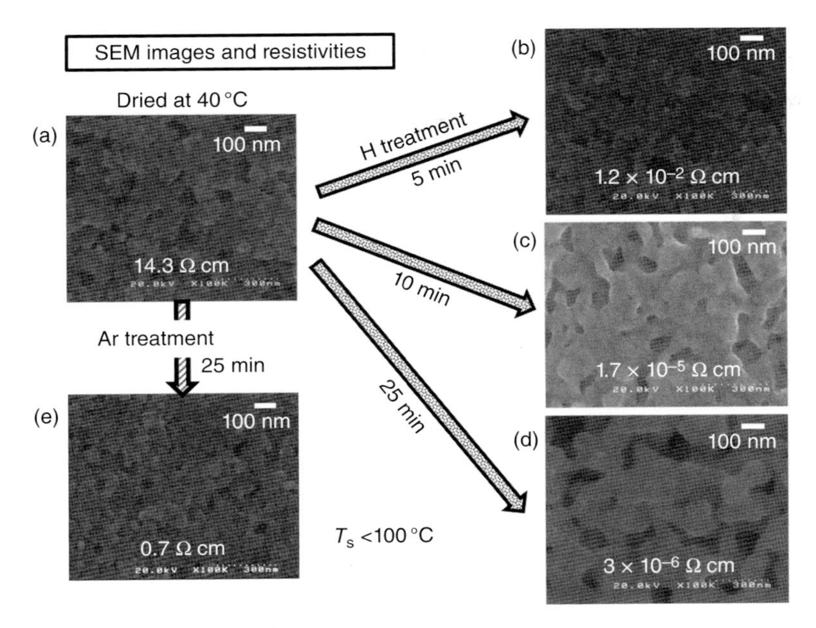

Figure 9.16 Summary of scanning electron microscopy (SEM) images and resistivities of Ag lines after H treatments, (a) original Ag line just after dried at 40 °C, (b) after H treatment for 5 minutes, (c) for 10 minutes, and (d) for 25 minutes. (e) The SEM image and resistivity of the same Ag line after Ar treatment for 25 minutes for comparison.

metal lines formed by vacuum evaporation. By using H-atom treatment, these problems can be solved at the same time.

As explained above, H atoms are effective in removing carbon (C) atoms or photoresists. Organic binders are considered almost similar to photoresist. In addition, even if the surface of metal nanoparticles is oxidized slightly and covered with thin-oxidized layer, such oxide layers can also be reduced to make real contacts among metal particles. By this reduction, the resistance after connection of metal particles can be lowered. This effect cannot be expected with the conventional simple annealing process. By the conventional annealing treatments, it has been hard to obtain low-resistivity metal lines comparable to the resistivity of evaporated metal lines.

There are some reports concerned with the formation of low-resistivity metal lines such as Cu lines using liquid metal ink with successive H-atom treatments [23]. Here, we introduce the results of formation of low-resistivity Ag lines using a well-known Ag paste [24].

In Figure 9.16, the images of scanning electron microscopy (SEM) and the resistivities of Ag lines after H treatments are demonstrated. In addition, similar results after Ar treatment are demonstrated for comparison.

At first, metallic functional liquid including 40-nm-size Ag nanoparticles, the so-called Ag nanoink or Ag paste, are dropped into a trench with a width of 10 μm made on the surface of plastic substrates to form straight metal lines. The Ag nanoink was diluted by 1,3-propanediol [$HO(CH_2)_3OH$] and pure water with volume ratios of 2 : 5 : 5, respectively. The samples were annealed at 40 °C in air for drying and solidifying the ink. The thickness of metal lines was about 1.6 μm but becomes 1 μm after H treatment. The sample was set in a Cat-CVD chamber and treated by H atoms generated there. T_{cat}, P_g, T_s, and FR(H_2) were kept at 1350 °C, 70–100 Pa, initially at RT and finally lower than 100 °C, and 100 sccm, respectively. The resistivity was measured by the four-probe method.

Figure 9.16a shows the SEM image and the resistivity of Ag metal line just after drying at 40 °C, panel (b) shows the SEM image and the resistivity after a 5-minute H treatment, panel (c) shows those after 10 minutes, and panel (d) shows those after 25 minutes. For comparison, panel (e) shows the SEM image and the resistivity after a 25-minute treatment with Ar gas. It is clear that the Ar treatment does not show any significant effect of lowering the resistivity and the growth of size of metal nanoparticles. Contrary to the Ar treatment, the effect of H treatment is apparent. By increasing the treatment time, the resistivity is lowered and the particle size increases. By reduction of the surface of the metal nanoparticles, the particles can grow and easily combine with neighbor particles. The resistivity of 3×10^{-6} Ω cm obtained by H treatment appears comparable with, or only slightly higher, that obtained by vacuum evaporation of Ag. At least, the resistivity of H-treated samples is lower than that of Ag lines made by conventional annealing at temperatures higher than 200 °C. By the conventional annealing method, it is hard to obtain a resistivity less than 1×10^{-5} Ω cm. The usefulness of the H treatment for lowering the resistivity of liquid-based process is apparent.

9.6 Low-Temperature Surface Oxidation – "Cat-Oxidation"

We have explained various applications of the H-atom treatment. The use of radicals generated in a Cat-CVD apparatus is not limited to H atoms. For instance, when we produce oxygen (O)-related radicals, such as OH, we can oxidize the surface of c-Si at a temperature as low as 250 °C, whereas the conventional thermal oxidation for making excellent insulating silicon dioxide (SiO_2) layers usually requires temperatures over 900 °C. The oxidation by O-related radicals generated in the Cat-CVD apparatus is called "Cat-oxidation" in this book. Here, at first, this low-temperature Cat-oxidation is introduced by following the references by A. Izumi [25] and the works with his colleagues.

The formation of SiO_2 is one of the key technologies for Si devices. The high-temperature process in conventional thermal oxidation sometimes causes the change of impurity profiles formed before oxidation. If we can form the SiO_2 layer at low temperatures, keeping the same quality as that of thermal SiO_2, it may open another possibility for Si process.

As already mentioned, O atoms easily oxidize the surface of metal catalyzers such as a W catalyzer. Once W-oxide is formed, as the vapor pressure of such W-oxides is high, the samples are easily contaminated by W atoms. If W is used without forming oxidized layers, the contamination is kept at an acceptable level for device fabrication as shown in Chapter 7. Therefore, for Cat-oxidation, at first, the oxidation of W should be avoided.

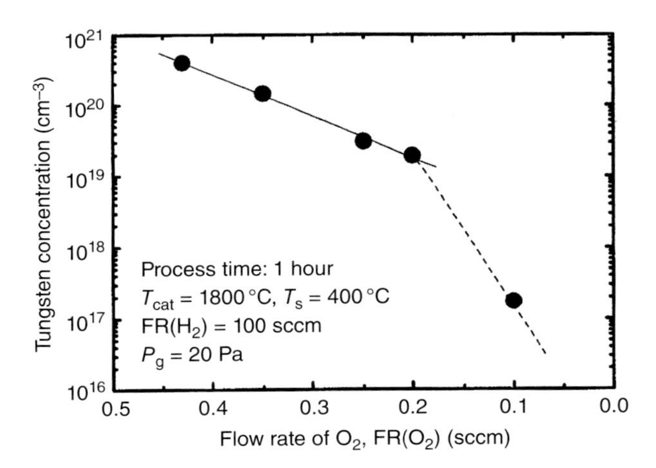

Figure 9.17 Density of tungsten (W) atoms detected by the Rutherford backscattering (RBS) method as a function of the net flow rates of oxygen (O_2) gas, FR(O_2).

The simplest way to avoid the oxidation of W is to dilute O_2 by reduction gas such as H_2 instead of using pure O_2. Figure 9.17 shows our data for the W concentration in the SiO_2 layers after one-hour Cat-oxidation as a function of net FR(O_2). The data were taken by the Rutherford backscattering (RBS) technique using 2.0 MeV He ions as probe ions. As already explained, the RBS method is highly sensitive to heavy atoms such as W atoms in light materials such as SiO_2 or Si. In this experiment, FR(H_2), P_g, T_{cat}, and T_s were 100 sccm, 20 Pa, 1800 °C, and 400 °C, respectively. O_2 gas was diluted to 0.5% by He gas. From this figure, it is found that when the net FR(O_2) exceeds 0.2 sccm, the W concentration jumps up to an order of magnitude over 10^{19} per cm^3; however, it enormously drops down to 2×10^{17} per cm^3 when the net FR(O_2) decreases to 0.1 sccm. If we assume that the thickness of the SiO_2 layer is about 5 nm, the flux density of W atoms is evaluated to be 1×10^{11} atoms/cm^2 for one hour and 1.7×10^9 atoms/cm^2 min.

The flux of W contaminants in the H-atom treatments is about 8×10^9 atoms/cm^2 min for T_{cat} = 2000 °C as shown in Figure 9.13. When T_{cat} is decreased to 1800 °C, the W flux density goes down to 1/100 of the value at T_{cat} = 2000 °C, considering the data shown in Figure 7.11. Hence, the W flux density is approximated as about 8×10^7 atoms/cm^2 min, although the data were obtained at different parameter values and such facts are neglected, here, to obtain a rough image.

The W contaminant density obtained in Cat-oxidation appears by 1 order of magnitude larger than that obtained in H-atom treatments. To reduce the incorporation of W atoms, in the following experiments, the net FR(O_2) is reduced to 0.05 sccm and T_{cat} is also lowered to 1650 °C as explained later. Thus, we are expecting that the W concentration inside the Cat-oxidized SiO_2 is less than 10^{15} per cm^3, considering both the extrapolation of the data in Figure 9.17 to FR(O_2) = 0.05 sccm and the decrease of T_{cat} following a trend shown in Figure 7.11.

When H_2 gas is used as the major source gas, we have to worry about the surface etching of c-Si during oxidation. Thus, the second important issue is how to

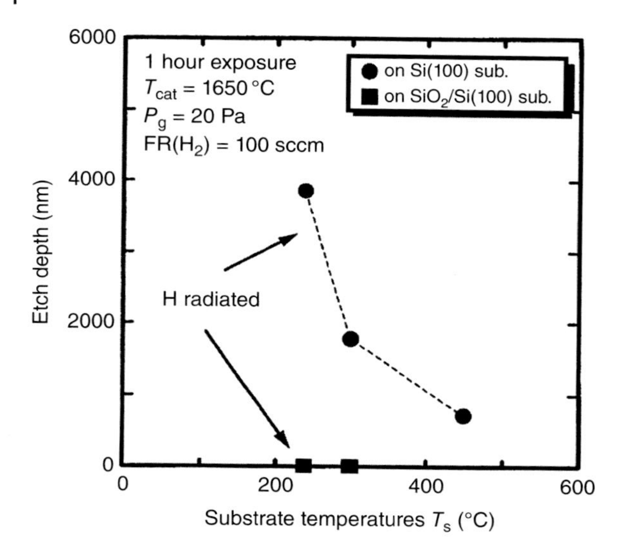

Figure 9.18 Etch depth of bare crystalline silicon (c-Si) and c-Si coated with chemically formed thin silicon dioxide (SiO_2) layer, as a function of substrate temperature T_s, when the samples are exposed to H atoms for one hour.

avoid it and how to carry out the Cat-oxidation properly. For this purpose, the chemical coating process of c-Si substrates before Cat-oxidation becomes very important. After the conventional cleaning process of Si substrates such as *RCA cleaning* process [26], the surface oxidation layer is once removed by 1% diluted hydrofluoric acid (HF) solution, and then, the surface of c-Si substrates is again chemically oxidized by 10-minute dipping in 90 °C solution of H_2SO_4 and H_2O_2 mixture with a mixing ratio of 4 : 1, respectively. The thickness of such a chemically formed preoxidized layer is about 0.6 nm.

Figure 9.18 shows the etch depth of c-Si with and without preoxide layers, after one-hour exposure to H atoms. The conditions generating H atoms are the same as those for Cat-oxidation, except for FR(O_2) = 0 sccm. The conditions of Cat-oxidation are summarized in Table 9.2. Bare c-Si is easily etched by the exposure to H atoms. However, the formation of preoxide layers by chemical process can avoid etching of c-Si. In many applications using radicals generated in the

Table 9.2 Process parameters for Cat-oxidation.

Process parameters	Setting conditions
Flow rate of H_2, FR(H_2)	100 sccm
Flow rate of He containing 0.5% O_2, FR(He)	10 sccm
Net flow rate of O_2, FR(O_2)	0.05 sccm
Temperature of catalyzer, T_{cat}	1650 °C
Substrate temperature, T_s	RT to 250 °C
Gas pressure during process, P_g	20 Pa
Surface area of catalyzer, S_{cat}	30 cm^2
Distance from substrate to catalyzer, D_{cs}	5 cm

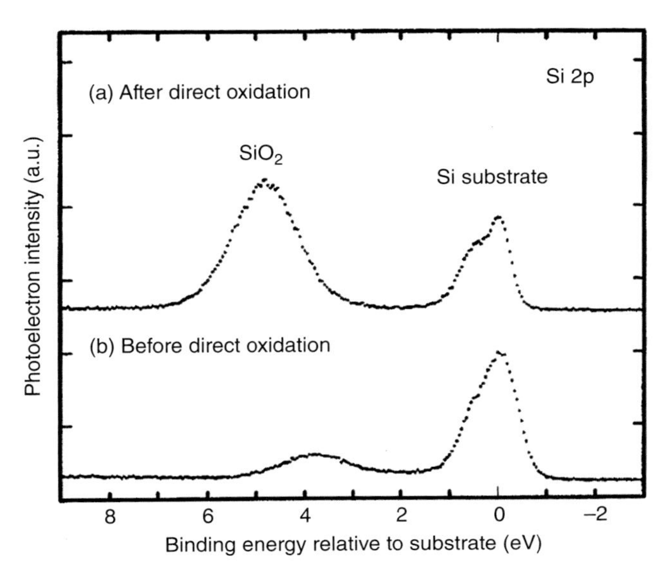

Figure 9.19 Spectra of X-ray photoelectron spectroscopy (XPS) for silicon (Si) 2p-orbital of crystalline silicon (c-Si), (a) after and (b) before Cat-oxidation. The signal from bulk c-Si substrates and the signal when Si is bonded with oxygen atom are indicated. Source: Izumi 2001 [25]. Reprinted with permission from Elsevier.

Cat-CVD apparatus, avoiding surface etching is a key point to obtain satisfactory results showing the effect of radicals.

Figure 9.19 shows the XPS spectra of Si(2p) signals from the surface of (100) c-Si substrates, panel (a) after 60-minute oxidation and panel (b) before oxidation [25]. The chemical shift to higher binding energy of Si(2p) is clearly observed after oxidation. A new big spectral peak appeared after Cat-oxidation at the relative binding energy of 5 eV, which is measured from the original binding energy of a bulk c-Si. After formation of an oxide layer, the number of emitted photoelectrons from c-Si substrates decreases as they have to pass through the oxide layer. Therefore, the signal intensity from the c-Si substrates depends on the thickness of the SiO_2 layer on c-Si, and the thickness of an oxide layer can be estimated from this ratio of the signal from the c-Si substrate to that from SiO_2. The estimated thickness is 3.2 nm in this case.

Figure 9.20 shows the current density (J)–voltage (V) characteristics for Cat-oxidized SiO_2 and thermally oxidized SiO_2, taking the thickness of SiO_2 layers as a parameter. The J–V characteristics were measured by making a structure of metal electrode/Cat-oxidized layer/c-Si/metal electrode. The 2.4-nm-thick Cat-oxidized SiO_2 sample in the figure was made at a relatively high substrate temperature over 300 °C. In most cases, the Cat-oxidized SiO_2 appears more leaky than thermally oxidized SiO_2; however, when Cat-oxidation is carried out at high temperatures around 300 °C or higher, the current density is likely to approach the value of thermal SiO_2 and becomes almost equivalent to them.

Figure 9.21 also shows the interface state density at the Cat-oxidized SiO_2/c-Si and thermally grown SiO_2/c-Si interfaces, derived from the capacitance

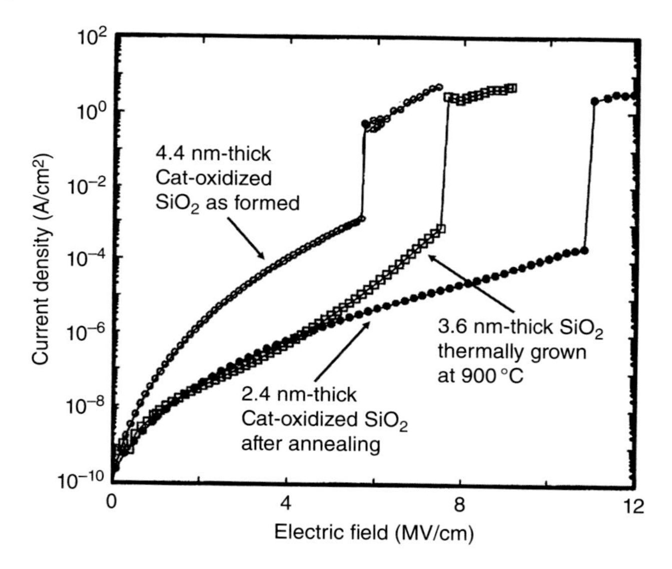

Figure 9.20 Current density (*J*) vs. voltage (*V*) characteristics of Cat-oxidized film, thermally oxidized film, and Cat-oxidized film after annealing at 300 °C.

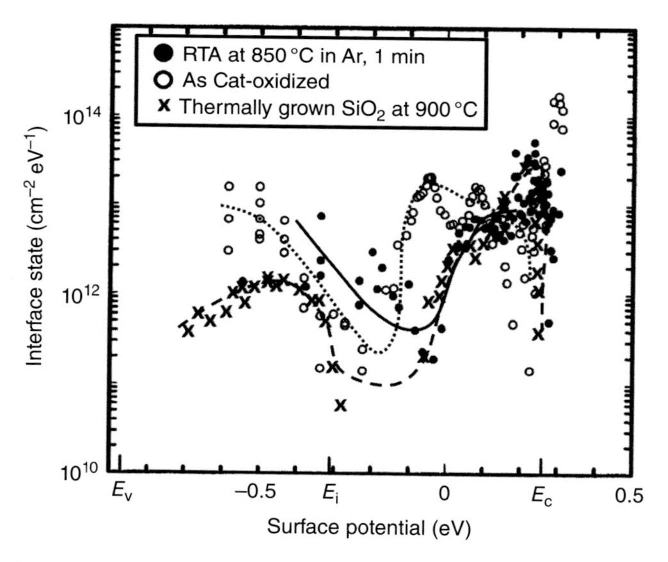

Figure 9.21 Interface states of various energies for 3.3-nm-thick Cat-oxidized layer, Cat-oxidized layer with postrapid thermal annealing (RTA) at 800 °C for one minute, and thermally oxidized layer at 900 °C.

(*C*)–voltage (*V*) measurements for the same samples used in *J*–*V* measurements. For Cat-oxidized SiO_2, the results of the sample as grown and after one-minute rapid thermal annealing (RTA) at 850 °C are shown. The thermal SiO_2 was prepared by a dry oxidation at 900 °C. The SiO_2/c-Si interface state density at the middle of the band gap of c-Si is known to be 10^{10}–10^{12} cm^{-2} eV^{-1}

for the thermally oxidized SiO_2, and in the order of 10^{12} cm^{-2} eV^{-1} for the SiO_2 made by plasma oxidation. The interface state density of Cat-oxidized SiO_2 appears to be at the order of 10^{12} cm^{-2} eV^{-1} and similar to that of plasma-oxidized SiO_2. This value is inferior to that of the thermally grown SiO_2. However, it also appears to decrease to the order of 10^{11} cm^{-2} eV^{-1} and is likely to approach the value of thermal SiO_2 when RTA at 850 °C for one minute is applied on the Cat-oxidized SiO_2.

In summary, it is known that (i) oxidation at a low temperature around 250 °C is possible by using Cat-oxidation, (ii) contamination of catalyzer W is negligible if special care is taken to highly dilute the O_2 gas with H_2 gas, (iii) to avoid surface etching by H atoms, the formation of preoxide coating layer by simple chemical process is essential before Cat-oxidation, and (iv) the electrical properties are inferior to those of the conventional thermal SiO_2; however, by applying RTA at 850 °C for only one minute, they are much improved and approach the quality of thermal SiO_2. The Cat-oxidation will be a useful technique to obtain oxide layers by a simple apparatus.

9.7 Low-Temperature Surface Nitridation – "Cat-Nitridation" of Si and GaAs

Silicon nitride (Si_3N_4 in stoichiometric, SiN_x in a general form) films or silicon oxynitride (SiN_xO_y) films are important materials for Si industry as mentioned in Chapter 5. Such SiN_x or SiN_xO_y films have been obtained by various methods. For instance, the direct thermal nitridation of c-Si has been studied for obtaining thin Si_3N_4 layers on c-Si by using dry NH_3 gas, similar to the direct thermal oxidation of c-Si by dry O_2 gas. However, the conventional thermal process requires high temperatures from 800 to 1200 °C. The temperatures appear even higher than those of thermal oxidation. Such a high-temperature process may induce thermal diffusion of various materials inside the devices and restricts the fabrication processes.

As low-temperature methods, PECVD of SiN_x and plasma nitridation have been studied. However, these processes always suffer from the surface damage due to plasma. Thus, again, the study on the low-temperature nitridation of c-Si by using cracked species of NH_3 generated in Cat-CVD apparatus, "Cat-nitridation," appears to attract expectation. In this section, we discuss the low-temperature Cat-nitridation of c-Si to demonstrate that the surface of c-Si can be converted to a SiN_x layer at low temperatures, similar to Cat-oxidation. In this section, we also introduce the low-temperature nitridation of GaAs to make GaN at surface.

A. Izumi et al. reported in 1997 that the surface of c-Si could be converted to Si_3N_4 at temperatures as low as 250 °C by exposing c-Si to species generated by cracking reaction of NH_3 molecules with a heated W catalyzer [27]. They also found that the surface of GaAs was converted to gallium nitride (GaN) at low temperatures by exposing GaAs to NH_3-decomposed species and verified that Cat-nitridation is not limited to Si [28]. In Table 9.3, typical parameters

Table 9.3 Various parameters for Cat-nitridation of c-Si and GaAs.

Parameters	Cat-nitridation for c-Si	Cat-nitridation for GaAs
Materials for catalyzers	W	W
Substrates used in experiment	(100) Si	(100) GaAs
Temperature of catalyzer, T_{cat} (°C)	1000–1700	1220–1700
Surface area of catalyzer, S_{cat} (cm^2)	20	20
Distance from catalyzer to substrate, D_{cs} (cm)	4–5	4–5
Gas pressure, P_g (Pa)	0.1–4	0.67–1.0
Flow rate of NH_3, FR(NH_3) (sccm)	50–200	50
Flow rate of H_2, FR(H_2) (for cleaning) (sccm)	—	50
Substrate temperature, T_s (°C)	200–250	150–280
References	[27]	[28, 29]

of Cat-nitridation are summarized for c-Si and GaAs. Generally speaking, the adjustment of the catalyzer temperature T_{cat} is an important factor to obtain good results for Cat-nitridation and also for Cat-oxidation. Nitridation process is always competing with etching process by H atoms. If the etching is dominant, we cannot observe nitridation or oxidation. Although it depends on other parameters such as FR(NH_3) or P_g, generally, lower T_{cat} is likely to show better results.

At first, they checked the W contamination for Cat-nitridation of c-Si. Figure 9.22 shows a RBS spectrum of c-Si, which was exposed to NH_3-decomposed species for 30 minutes. RBS was carried out by using 2.8 MeV He ions as probe ions. In the figure, two c-Si RBS spectra observed from the random direction and from the direction parallel to the crystalline axis are shown. The observation along the crystalline axis is to lower the detection limits of W atoms by reducing the piling up noise in signals from Si atoms. In the figure, the RBS signals of W atoms, which should come out at the channel of 450, cannot be seen. As already explained in this book, the RBS is one of the most sensitive techniques to detect heavy atoms embedded in light materials. The detection limit of W atoms in c-Si can be less than 1 ppm.

If the W concentration in the nitride layer and the thickness of it are assumed to be less than 1×10^{16} cm^{-3} and 3 nm, respectively, the flux density of W contaminants is roughly evaluated as 1×10^{8} atoms/cm^2 min. This value is by 1 order of magnitude smaller than in the case of Cat-oxidation. In both cases, the W contamination is not a serious problem.

Figure 9.23 shows the XPS spectra from N (1s), O (1s), and Si (2p) for the surface of (100) c-Si before Cat-nitridation and after 30-minutes of Cat-nitridation at T_{cat} = 1700 °C, T_s = 200 °C, and P_g = 1 Pa. The peak from N (1s) and the chemical shift to the higher binding energy of Si (2p) are clearly observed after Cat-nitridation. The peak of the chemical shift of Si (2p) is deconvoluted to two major spectra with a peak at 103.3 eV and another peak at 101.9 eV as shown in

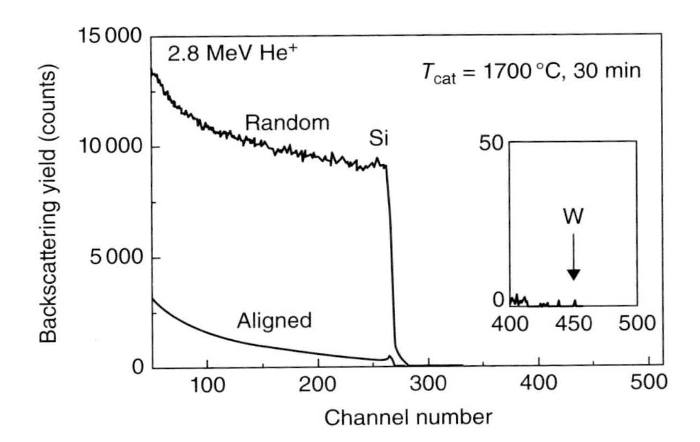

Figure 9.22 Spectra of the Rutherford backscattering (RBS) of Cat-nitirided crystalline silicon (c-Si). c-Si is observed from random direction and along aligned direction of crystalline axis. The signal position for tungsten (W) atoms is indicated by an arrow. Source: Izumi and Matsumura 1997 [27]. Reprinted with permission from AIP Publishing.

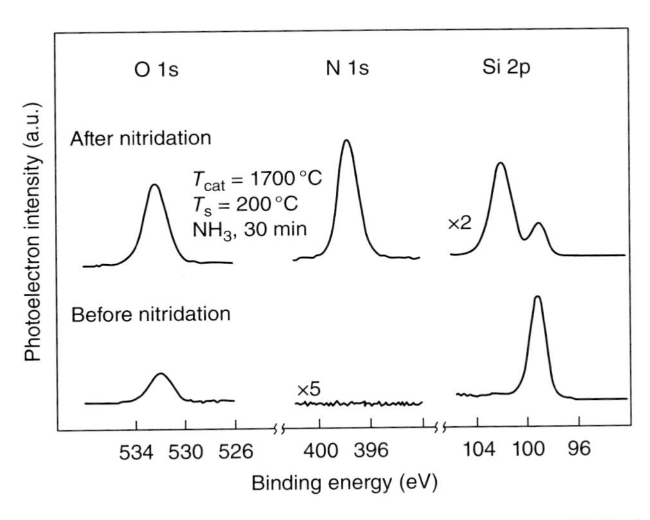

Figure 9.23 Spectra of X-ray photoelectron spectroscopy (XPS) of oxygen (O) 1s-orboital, nitrogen (N) 1s-orbital, and silicon (Si) 2p-orbital for crystalline silicon (c-Si) substrates before and after Cat-nitridation. Source: Izumi and Matsumura 1997 [27]. Reprinted with permission from AIP Publishing.

Figure 9.24. The peak positions are assigned as Si—O and Si—N bonds. In this case, the stoichiometry and the thickness of the Cat-nitride layer are estimated as Si:N:O = 1 : 0.9 : 0.3 and 3.8 nm, respectively. The origin of O atoms is not clear, but often O atoms are observed. There was no drying system of NH_3 in the experiment system used. This may be one of the possible reasons. As another possibility, because the Cat-nitrided layer is very thin, it may be naturally oxidized in air after Cat-nitridation, similar to the formation of natural oxide at the surface of c-Si.

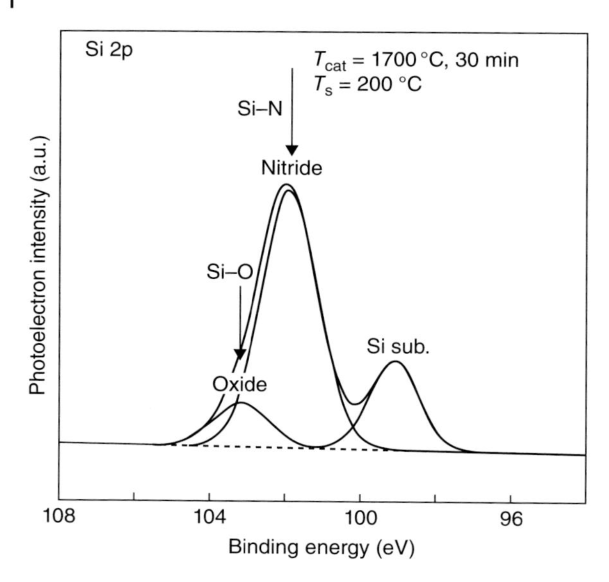

Figure 9.24 Spectrum of X-ray photoelectron spectroscopy (XPS) of silicon (Si) 2p-orbital and its deconvoluted profiles showing the existence of bonds with oxygen (O) and nitrogen (N). Source: Izumi and Matsumura 1997 [27]. Reprinted with permission from AIP Publishing.

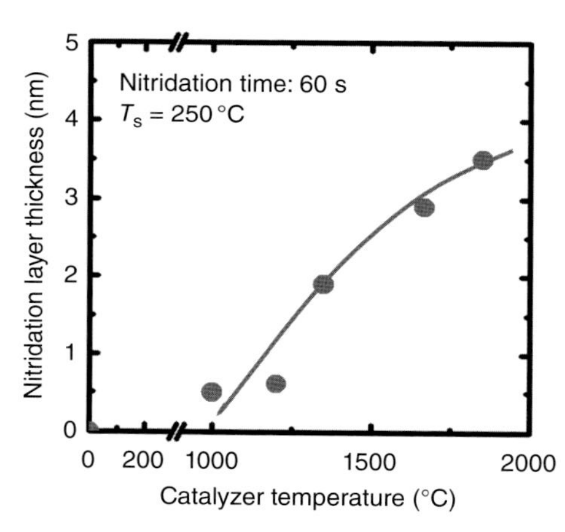

Figure 9.25 Thickness of Cat-nitridation layer on a crystalline silicon substrate as a function of catalyzer temperature.

Figure 9.25 shows the thickness of the Cat-nitrided layer as a function of T_{cat}. In this case, T_s, P_g, and FR(NH_3) were 250 °C, 4 Pa, and 200 sccm, respectively. The process time was set at only 60 seconds. As shown in the figure, as T_{cat} increases, the thickness increases. Figure 9.26 also shows the thickness of nitride layer as a function of the process time for $T_{cat} = 1350\,°C$, $P_g = 4$ Pa, and $T_s = 250\,°C$. As the process time increases, the thickness of the nitride layer increases approximately in proportion to the square root of the process time. This relation implies that Cat-nitridation appears to follow the Grove–Deal model, which was considered for the growth of thermal oxide layer [30]. The model tells us that the nitride layer grows by the nitridation species passing through the nitride layer already formed. According to this model, if we extend the process time, the thickness should increase slowly. However, at the moment, even when changing various parameters, growth over 4.8 nm was not realized [27]. The electrical properties were

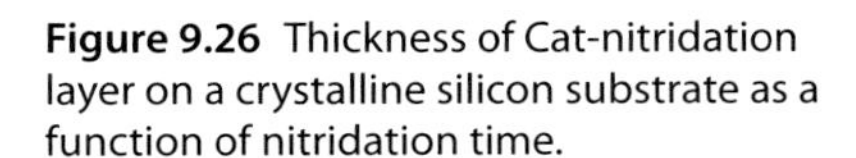
Figure 9.26 Thickness of Cat-nitridation layer on a crystalline silicon substrate as a function of nitridation time.

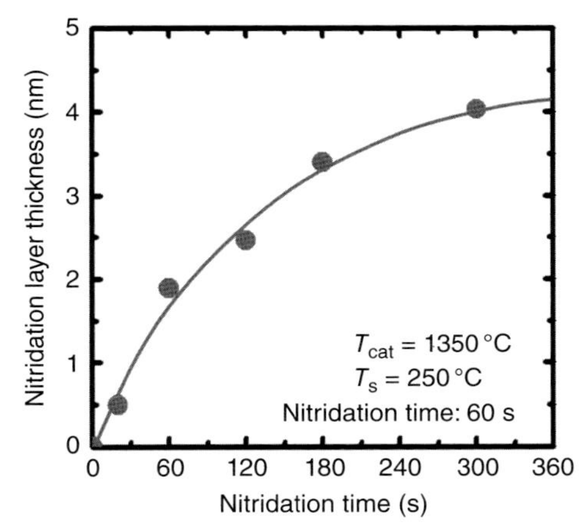

also studied, and it was revealed that the Cat-nitridation before the deposition of Cat-CVD SiN_x improved the interface quality. However, at the moment, the data do not appear enough.

Surface cleaning and surface nitridation of GaAs were also reported by A. Izumi et al. [28, 29]. Usually, the surface of GaAs is cleaned chemically using hydrochloride (HCl)-based chemicals; however, it is not easy to clean the surface by a dry process before film deposition. As the surface of many compound semiconductors such as GaAs is sensitive to plasma damage, the surface treatment using plasma is not effective. Thus, it is desired to develop a surface cleaning method of compound semiconductors by using the plasma-less Cat-CVD technology. In addition, contrary to c-Si, SiO_2 is not a good passivation material for GaAs, and thus, it is another important issue to obtain the stable surface of GaAs by a new technology without using plasma.

Figure 9.27 shows XPS spectra of As (3d) and Ga (3d) from the surface of (100) GaAs (panel (a)) before processing, (panel (b)) after HCl chemical cleaning, (panel (c)) after dry cleaning by H atoms generated in the Cat-CVD apparatus, and (panel (d)) after processing by NH_3-decomposed species generated in the Cat-CVD apparatus [28]. All spectra presented here were observed at the photoelectron takeoff angle of 35°. T_{cat}, T_s, and P_g were kept at 1700 °C, 280 °C, and 0.67 Pa, respectively, and the process time was 30 minutes in this experiment. In the figure, the peaks originating from O-related bonds are indicated by closed inverted triangles and those originating from GaAs surface by closed circles. It is apparent that the intensity of O-related signals decreases after any form of cleaning process. The surface oxidation layer can be removed by Cat process similar to the HCl chemical process. The intensity of O-related As (3d) peak is much reduced by atomic H process compared to other cleaning methods. If we focus on the variation of the intensity of O-related As (3d), the cleaning process by H atoms is best and that by NH_3 is the second best, but still better than HCl process.

Surface nitridation could not be observed clearly when T_{cat} was 1700 °C. Probably at such a high temperature, the cleaning process was dominant. When T_{cat} was reduced down to 1220 °C, the surface of GaAs appears to become nitrided, and GaN can be seen [29].

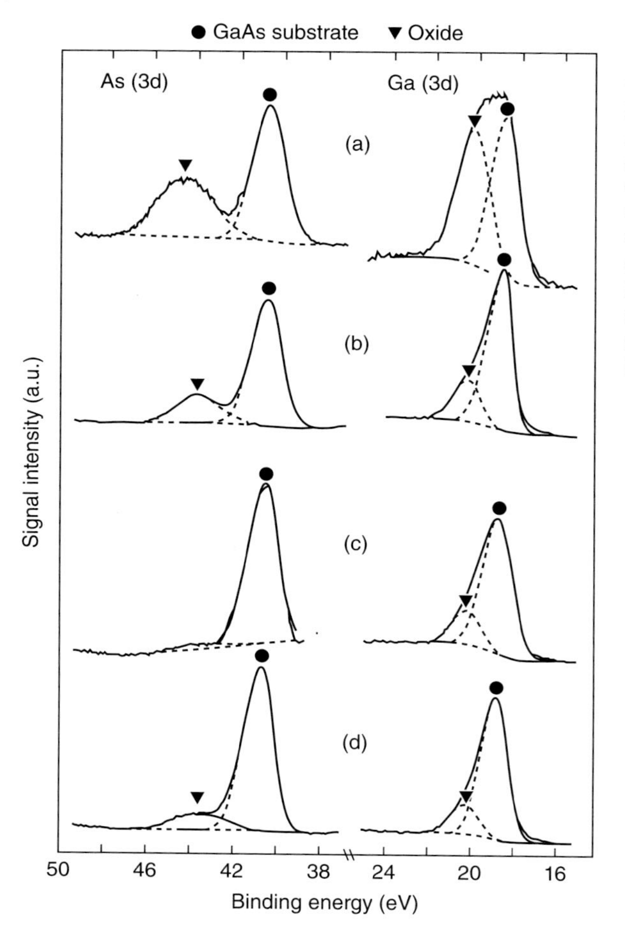

Figure 9.27 Spectra of X-ray photoelectron spectroscopy (XPS) of gallium (Ga) 3d-orboital and arsenic (As) 3d-orbital for gallium arsenide (GaAs) substrates with Cat-nitridation layer on them. (a) Before Cat-nitridation, (b) after chemical cleaning using thin hydrochloride (HCl) solution, (c) after dry cleaning by H atoms, and (d) after cleaning by NH_3-cracked species.

Figure 9.28 shows the XPS spectra of N (1s), As (3d), and Ga (3d) from the surface of (100) GaAs. In this case, in addition to the change of T_{cat} from 1700 to 1220 °C, T_s and P_g were also changed to 150 °C and 1.0 Pa, respectively. Only NH_3 gas with FR(NH_3) = 50 sccm was introduced into the chamber [29]. The peaks related with O bonds are indicated by closed inverse triangles, and in the figure, the position of Ga LMM Auger signal is also indicated. The expression of Ga LMM means the series of electron transition to generate Auger electrons finally. The series of the transition process is (i) after the emission of electrons at the inner L-orbital of a Ga atom by X-ray, (ii) an electron at the M-orbital drops down to the L-orbital by giving the energy to another electron at the M-orbital, and then (iii) such an electron receiving the energy at the M-orbital is emitted as an Auger electron. Thus, to show the history of generating Auger electrons, it is called LMM Auger electron.

From the figure, it is found that the O atoms at the surface of GaAs are completely removed after only three minutes, as seen in the spectral peaks of As (3d)

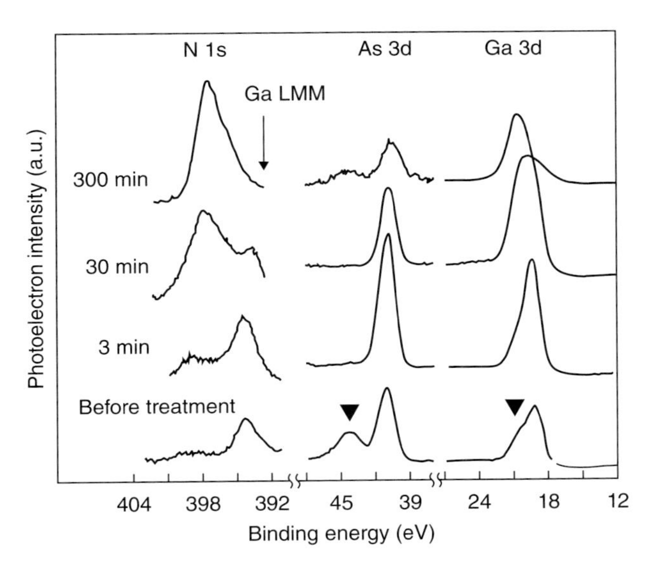

Figure 9.28 Spectra of X-ray photoelectron spectroscopy (XPS) of nitrogen (N) 1s-orbital, arsenic (As) 3d-orbital, and gallium (Ga) 3d-orboital for gallium arsenide (GaAs) substrates with Cat-nitridation layer on them. The spectra before Cat-nitridation and after 3, 30, and 300-minute Cat-nitridation are demonstrated. Source: Izumi et al. 1999 [29]. Reprinted with permission from Elsevier.

and Ga (3d) and that a N (1s) peak is clearly observed separately from the Ga LMM Auger signal after the three-minute process together with the shift of Ga (3d) spectrum to higher energy. Such changes of the peak spectra were observed only when the catalyzer was heated in a NH_3 flow. This is the evidence of the effects of NH_3-decomposed species of cleaning and of the formation of GaN over a three-minute process time. The component in the nitride layer at the GaAs surface was estimated from the spectra as Ga:N = 1 : 1 for both the 3-minute treated sample and the 30-minute treated sample. The thickness of the nitride layer was a few nanometers.

The GaAs surface with ultrathin GaN layer becomes very stable to ambient. Figure 9.29 shows the change of XPS spectra of original GaAs, 30-minute treated GaAs, and 30-minute treated GaAs after 60 days of exposure to air. Even after keeping the sample in air, no oxidation can be observed at the surface of GaAs, This Cat-nitridation process is also useful to make a stable surface of GaAs. As already mentioned, Cat-CVD technology including surface nitridation improves remarkably the fabrication and the performance of GaAs devices.

In addition, attention should also be paid to the flatness of GaAs surface after Cat-nitridation. According to the observation with an atomic force microscope (AFM), the root mean square (RMS) of the surface roughness of GaAs surface is less than only 0.28 nm after 30-minute Cat-nitridation [27]. If the plasma process is adapted, the roughness easily exceeds a few nanometers as demonstrated in Chapter 2. Advantage of Cat-processes, particularly for GaAs, is apparent.

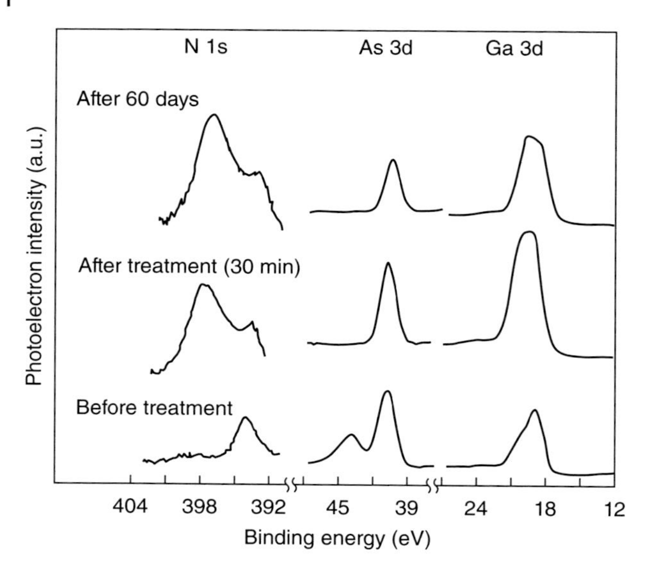

Figure 9.29 Spectra of X-ray photoelectron spectroscopy (XPS) of nitrogen (N) 1s-orbital, arsenic (As) 3d-orbital, and gallium (Ga) 3d-orboital for gallium arsenide (GaAs) substrates with Cat-nitridation layer on them before and after Cat-nitridation. Additionally, the spectra after 60-day exposure to air after Cat-nitridation is demonstrated. Source: Izumi et al. 1999 [29]. Reprinted with permission from Elsevier.

9.8 "Cat-Chemical Sputtering": A New Thin Film Deposition Method Utilizing Radicals

One of the unique points concerned with the application of radicals generated in Cat-CVD is an invention of a new Si film deposition method. As already mentioned, H atoms can etch solid Si if the surface is not covered with oxide layers. In this etching process, SiH_4 molecules are generated. If we use the newly generated SiH_4 as a source for next deposition, Si films can be deposited without supplying dangerous SiH_4 gas from the outside of the deposition chamber.

Figure 9.30 shows a schematic diagram of this new deposition system of Si films [12]. In the system, a Si wafer is placed on the target holder cooled with cooling water, and substrates are placed on the substrate holder heated at 200–300 °C. Between the substrates and the Si wafer, catalyzing wires are spanned keeping their spanned area parallel to the Si wafer and the substrate. As shown in Figure 9.30, the etch rates of Si by H atoms depend on the temperature of Si. When the substrates are heated at about 300 °C, for instance, the etch rate of Si is low. However, when the Si wafer is cooled to near RT, Si is etched efficiently and SiH_4 is generated and Si films are deposited by species generated by the cracking reaction on the heated catalyzer from SiH_4 generated at the Si wafer. That is, just like a sputtering apparatus, the Si wafer is working as a target of sputtering and Si films are deposited without supplying any SiH_4 gas from outside and generating ions.

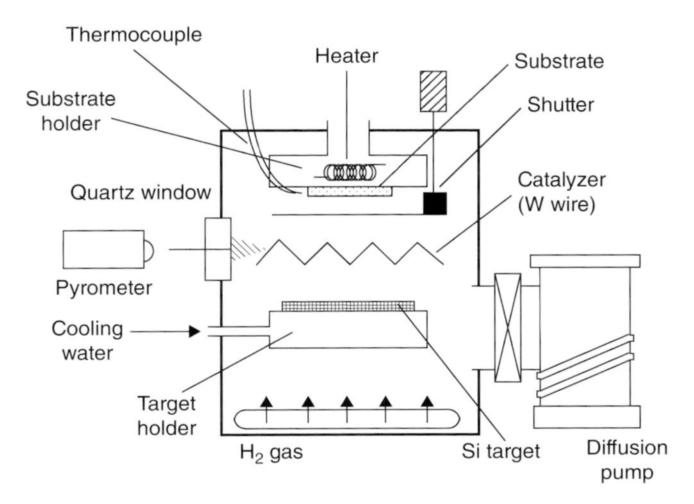

Figure 9.30 Schematic diagram of experimental apparatus for Cat-sputtering. Source: Matsumura et al. 2001 [12]. Copyright (2001). Reprinted with permission from The Japan Society of Applied Physics.

Figure 9.31 An image of scanning electron microscope (SEM) of a polycrystalline silicon film prepared by Cat-sputtering. Source: Matsumura et al. 2001 [12]. Copyright (2001). Reprinted with permission from The Japan Society of Applied Physics.

Tentatively, we call this as "Catalytic Chemical Sputtering (Cat-sputtering)," as a nickname [12]. Of course, this is completely different from the "sputtering" as physical meaning. The phenomena utilize the chemical transport of species.

As generated SiH_4 is automatically highly diluted by H_2, the deposited Si films become poly-Si. Figure 9.31 shows a photograph of the SEM image of such a poly-Si film. The thickness of the film is 0.9 μm. We can see that a poly-Si film with a grain size over 1 μm is obtained.

This attempt is a trial for utilization of radicals generated in the Cat-CVD system. The concept utilizing radicals to make other species will widely expand to other areas.

References

1 Umemoto, H., Ohara, K., Morita, D. et al. (2002). Direct detection of H atoms in the catalytic chemical vapor deposition of the SiH_4/H_2 system. *J. Appl. Phys.* 91 (3): 1650–1656.
2 Yamada, T., Ohmi, H., Kakiuchi, H., and Yasutake, K. (2017). Hydrogen atoms density in narrow-gap microwave hydrogen plasma determined by calorimetry. *J. Appl. Phys.* 119:063301/1–063301/7.
3 Larjo, J., Koivikko, H., Lahtonen, K., and Hernberg, R. (2002). Two-dimensional atomic hydrogen concentration maps in hot-filament diamond-deposition environment. *Appl. Phys. B Lasers Opt.* 74 (6): 583–587.
4 Baulch, D.L., Cobos, C.J., Cox, R.A. et al. (1992). Evaluated kinetic data for combustion modelling. *J. Phys. Chem. Ref. Data* 21: 411–429. https://kinetics.nist.gov/kinetics/index.jsp.
5 Kae-Nune, P., Perrin, J., Jolly, J., and Guillon, J. (1996). Surface recombination probabilities of H on stainless steel, a-Si:H and oxidized silicon determined by threshold ionization mass spectrometry in H_2 RF discharges. *Surf. Sci.* 360: L495–L498.
6 Rousseau, A., Granier, A., Gousset, G., and Leprince, P. (1994). Microwave discharge in H_2: influence of H-atom density on the power balance. *J. Phys. D* 27: 1412–1422.
7 Kim, Y.C. and Boudart, M. (1991). Recombination of O, N, and H atoms on silica: kinetics and mechanism. *Langmuir* 7: 2999–3005.
8 Tserepi, A.D. and Miller, T.A. (1994). Two-photon absorption laser-induced fluorescence of H atoms: a probe for heterogeneous processes in hydrogen plasmas. *J. Appl. Phys.* 75: 7231–7236.
9 Rousseau, A., Cartry, G., and Duten, X. (2001). Surface recombination of hydrogen atoms studies by a pulsed plasma excitation technique. *J. Appl. Phys.* 89: 2074–2078.
10 Ansari, S.G., Umemoto, H., Morimoto, T. et al. (2005). Technique for the production, preservation, and transportation of H atoms in metal chambers for processings. *J. Vac. Sci. Technol. A* 23 (6): 1728–1731.
11 Uchida, K., Izumi, A., and Matsumura, H. (2001). Novel chamber cleaning method using atomic hydrogen generated by hot catalyzer. *Thin Solid Films* 395: 75–77.
12 Matsumura, H., Kamesaki, K., Masuda, A., and Izumi, A. (2001). Catalytic chemical sputtering: a novel method for obtaining large-grain polycrystalline silicon. *Jpn. J. Appl. Phys.* 40, Part 2 (3B): L289–L291.
13 Nishiyama, I., Oizumi, H., Motai, K. et al. (2005). Reduction of oxide layer on Ru surface by atomic-hydrogen treatment. *J. Vac. Sci. Technol. B* 23: 3129–3131.

14 Motai, K., Oizumi, H., Miyagaki, S. et al. (2008). Cleaning technology for EUV multilayer mirror using atomic hydrogen generated with hot wire. *Thin Solid Films* 516: 839–843.

15 Izumi, A. and Matsumura, H. (2002). Photoresist removal using atomic hydrogen generated by heated catalyzer. *Jpn. J. Appl. Phys.* 41 (7A): 4639–4641.

16 Yamamoto, M., Horibe, H., Umemoto, H. et al. (2009). Photoresist removal using atomic hydrogen generated by hot-wire catalyzer and effects on Si-wafer surface. *Jpn. J. Appl. Phys.* 48: 026503/1–026503/7.

17 Horibe, H., Yamamoto, M., Maruoka, T. et al. (2011). Ion-implanted resist removal using atomic hydrogen. *Thin Solid Films* 519: 4578–4581.

18 Hashimoto, K., Masuda, A., Matsumura, H. et al. (2006). Systematic study of photoresist removal using hydrogen atoms generated on heated catalyzer. *Thin Solid Films* 501: 326–328.

19 Yamamoto, M., Umemoto, H., Ohdaira, K. et al. (2016). Oxygen additive amount dependence of the photoresist removal rate by hydrogen radicals generated on a tungsten hot-wire catalyst. *Jpn. J. Appl. Phys.* 55: 076503/1–076503/5.

20 Yamamoto, M., Maejima, K., Umemoto, H. et al. (2016). Enhancement of removal uniformity by oxygen addition for photoresist removal using H radicals generated on a tungsten hot-wire catalyst. *J. Photopolym. Sci. Technol.* 29: 639–642.

21 Izumi, A., Ueno, T., Miyazaki, Y. et al. (2008). Reduction of oxide layer on various metal surfaces by atomic hydrogen treatment. *Thin Solid Films* 516: 853–855.

22 Tabuchi, N. and Matsumura, H. (2002). Control of carrier concentration in thin cuprous oxide Cu_2O films by atomic hydrogen. *Jpn. J. Appl. Phys.* 41: 5060–5063.

23 Kumahira, Y., Nakako, H., Inada, M. et al. (2009). Novel materials for electronic device fabrication using ink-jet printing technology. *Appl. Surf. Sci.* 256: 1019–1022.

24 Kieu, N.T.T., Ohdaira, K., Shimoda, T., and Matsumura, H. (2010). Novel technique for formation of metal lines by functional liquid containing metal nanoparticles and reduction of their resistivity by hydrogen treatment. *J. Vac. Sci. Technol. B* 28 (4): 776–782.

25 Izumi, A. (2001). Surface modification of silicon related materials using a catalytic CVD system for ULSI application. *Thin Solid Films* 395: 260–265.

26 RCA Clean, in Wikipedia. https://en.wikipedia.org/wiki/RCA_Clean.

27 Izumi, A. and Matsumura, H. (1997). Low-temperature nitridation of silicon surface using NH_3-decomposed species in a catalytic chemical vapor system. *Appl. Phys. Lett.* 71 (10): 1371–1372.

28 Izumi, A., Masuda, A., Okada, S., and Matsumura, H. (1996). Novel surface cleaning of GaAs and formation of high quality SiN_x films by Cat-CVD. Institute of Physics Conference Series, No.155, Chapter 3. Paper presented at 23rd International Symposium Compound Semiconductors, St. Petersburg, Russia (22–27 September 1996), pp. 343346.

29 Izumi, A., Masuda, A., and Matsumura, H. (1999). Surface cleaning and nitridation of compound semiconductors using gas-decomposition reaction in Cat-CVD method. *Thin Solid Films* 343–344: 528–531.

30 Grove, A.S. (1967). *Physics and Technology of Semiconductor Devices,* Chapter 2. Wiley.

10

Cat-doping: A Novel Low-Temperature Impurity Doping Technology

Generation of high-density radicals in a catalytic chemical vapor deposition (Cat-CVD) apparatus and their application are explained in the previous chapter. In this chapter, another application of such radicals, directing to impurity doping into semiconductors, is demonstrated. Impurity doping is an essential technology to fabricate semiconductor devices. It is demonstrated that impurity doping into semiconductors are successfully carried out at the temperatures as low as 80 °C when the radicals generated in Cat-CVD apparatus are used.

10.1 Introduction

Doping of impurities such as phosphorus (P) and boron (B) into crystalline silicon (c-Si) is a key to fabricate silicon (Si) devices. Usually, impurity doping is carried out by thermal diffusion or ion implantation, and in some cases by plasma. In the thermal diffusion, usually, the temperatures over 1000 °C are used for the enhancement of diffusion process. On the other hand, in ion implantation, impurities are doped into semiconductors by accelerating impurity ions under 10–100 kV. In ion implantation, temperatures used in the process can be lowered compared with the diffusion method; however, for the activation of impurities and elimination of defects created by ion bombardment, annealing at about 800 °C or higher is always required. On the other hand, plasma doping is known as a method to dope impurities into a shallow region at low temperatures [1]. This technique utilizes the sheath voltage in plasma or sometimes extra applied voltage to implant impurity ions into semiconductors. Therefore, however, this method induces plasma damage or implanted damage, and usually, annealing is required to obtain good doping quality.

In addition, by the most of doping methods, except for plasma doping, the impurities are always introduced into deeper area than near-surface area. For instance, even if low-energy ion implantation is adapted, it is not easy to introduce P or B ions in c-Si only in the depth shallower than a few tens of nanometers.

During the study on Cat-CVD, it was discovered that P or B could be doped into c-Si at temperatures as low as 80 °C by exposing c-Si substrates to the species generated by catalytic cracking of phosphine (PH_3) or diborane (B_2H_6) gas on heated tungsten (W) catalyzers. In spite of no voltages are applied near the substrates,

Catalytic Chemical Vapor Deposition: Technology and Applications of Cat-CVD,
First Edition. Hideki Matsumura, Hironobu Umemoto, Karen K. Gleason, and Ruud E.I. Schropp.

contrary to the method using plasma, the impurities are still doped into semiconductors at low temperatures. This phenomenon itself appears quite new. The history of this research is short, and there are so many unsolved issues. However, for showing the new feasibility of Cat-CVD (not CVD any more) technology, finally, this new method or phenomenon, named "Catalytically cracked impurity doping = Cat-doping," is introduced in this chapter [2].

10.2 Discovery or Invention of Cat-doping

The phenomenon of low-temperature impurity doping was discovered in the way of the study on amorphous silicon (a-Si)/c-Si heterojunction solar cells. One of the students, who was fabricating a-Si/c-Si heterojunction solar cells, attempted to clean the c-Si surface before a-Si deposition by Cat-CVD, by using H atoms generated in a Cat-CVD apparatus. The story began from this H-atom cleaning.

Cleaning by H atoms should be done very carefully. For instance, when the catalyzer temperature T_{cat} for generating H atoms is too high, the surface of c-Si is easily etched and the surface is roughened. Figure 10.1 shows the surface roughness of c-Si after the H-atom cleaning for one minute at a substrate temperature T_s of 150 °C and with a gas pressure P_g of 1 Pa, as a function of T_{cat}. The surface roughness was measured by an atomic force microscope (AFM). The root mean square (RMS) of the surface roughness is shown in a right vertical axis of the figure. The figure demonstrates that up to T_{cat} = 1300 °C, the surface roughness after the H-atom cleaning is negligible. Thus, cleaning was carried out at T_{cat} = 1300 °C.

After the H-atom cleaning, in that experiment, 40-nm-thick intrinsic amorphous-Si (i-a-Si) layers were deposited on both sides of c-Si wafers to know the effect of passivation by a-Si layers. Through the H-atom cleaning and a-Si deposition, the substrate temperature T_s was kept at relatively low, about only 150 °C, to avoid epitaxial growth on c-Si wafers during a-Si deposition. Usually, in such experiments, to avoid cross-contamination, H-atom cleaning and deposition of i-a-Si and P- or B-doped a-Si are all carried out in separate different chambers. The carrier lifetimes of c-Si covered with 40-nm-thick i-a-Si prepared at T_{cat} = 1700 °C were typically around 0.7 ms for this c-Si wafers without annealing of i-a-Si after the deposition at that time. However, one day, because the H-atom process chamber was busily used by other groups, the student used unusually another chamber for H-atom cleaning process to minimize the waiting time. The chamber had been used for P-doped n-a-Si deposition and the chamber wall was covered with P-doped n-a-Si. The result was interesting. The carrier lifetimes of only such samples increased drastically. The value exceeded 2 ms as indicated with an arrow in Figure 10.1 [3].

If the use of the n-a-Si deposition chamber for H-atom cleaning has an influence on the improvement of the carrier lifetime, we should be able to detect P atoms in the sample. The result of observation by the secondary ion mass spectroscopy (SIMS) is shown in Figure 10.2. It was apparent that P atoms were incorporated in c-Si. If we could think that (i) the surface potential of c-Si was bent

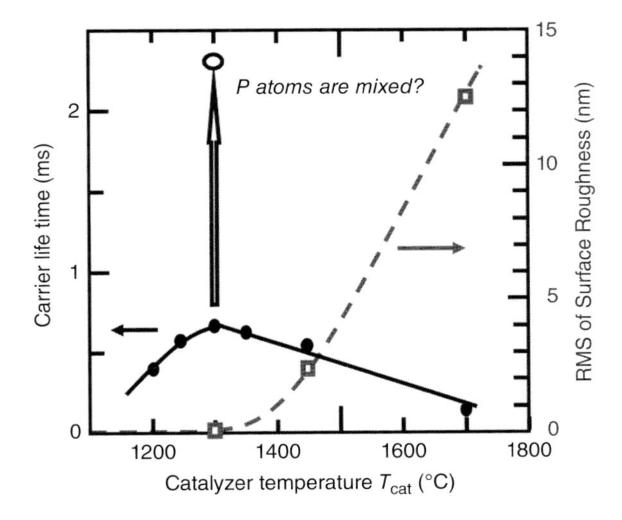

Figure 10.1 Carrier lifetimes and root mean square (RMS) of etched surface roughness, as a function of catalyzer temperature, T_{cat}. Source: Reprinted with permission from Ref. [3].

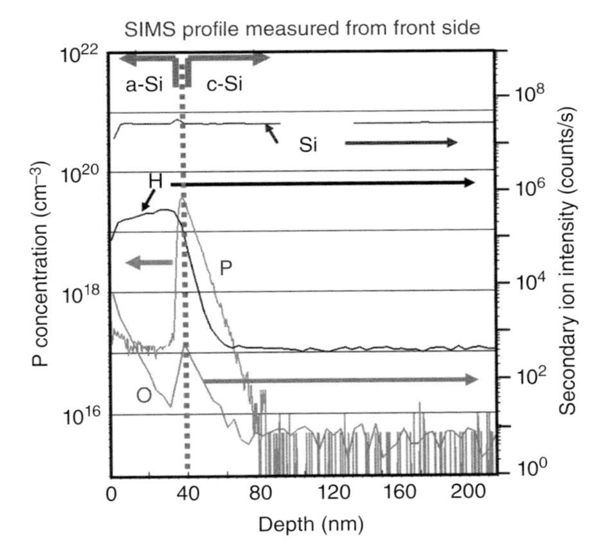

Figure 10.2 Concentration profiles of phosphorus (P) and hydrogen (H) measured by secondary ion mass spectroscopy (SIMS). The crystalline silicon (c-Si) sample is capped with 40-nm-thick amorphous silicon (a-Si) film.

down due to the heavy incorporation of P atoms only at the interface of a-Si/c-Si, (ii) holes were repulsively gone away from the interface, and (iii) the recombination at the interface was reduced, we can explain the improvement of the carrier lifetimes. This speculation also includes an important assumption that P atoms can be doped in c-Si at 150 °C without using any accelerated source of impurities into c-Si.

Is it really possible to dope impurities into c-Si at such low temperatures? In the case of plasma doping, we can think the help of plasma sheath voltage for the explanation of impurity doping, as mentioned above. However, in the present case, no voltage is applied. If so, how can the phenomenon be explained? We started the systematic study on this new phenomenon, named Cat-doping.

10.3 Low-Temperature and Shallow Phosphorus (P) Doping into c-Si

10.3.1 Measurement of Electrical Properties of a Shallow-Doped Layer

As indicated from the data shown in Figure 10.2, even if P atoms are doped, the doping depth may be limited to a shallow region. In the measurement of the electrical properties, such as the carrier concentration and the mobility, for such shallow doping area in c-Si, we have to be careful to eliminate the effect of surface defects in c-Si. As schematically illustrated in Figure 10.3, if the surface potential of c-Si is forced to be bent due to the existence of high-density surface defects, the measured results of doped carrier density may be affected on this band bending. It is important to understand that the proper surface passivation or a cap layer is necessary for exact measurement for such very shallow properties. There have been known various passivation films for c-Si surface such as silicon dioxide (SiO_2) and i-a-Si. Particularly, i-a-Si is an excellent passivation material formed at low temperatures as described in Chapter 8. Actually, the Cat-doping phenomenon was discovered for c-Si sample covered with i-a-Si as mentioned above. Thus, we checked the effect of deposition of i-a-Si on c-Si on the measurement of electrical properties.

The electrical properties were measured by the van der Pauw method, based on the Hall effect measurement [4]. The samples have a rectangular shape with a size of 10 mm × 10 mm, and four electrodes with a diameter of 1 mm were formed at the four corners of the rectangular samples. In the van der Pauw measurements, theoretically, the size of the electrodes has to be as small as possible for obtaining results without errors. However, actually, this size of 1 mm diameter against 10 mm square samples was enough to keep measuring errors less than 10%, considering the results of test measurements of samples with known properties.

Figure 10.4 shows two types of van der Pauw electrodes used in the experiments to check both the effect of passivation coating and the electrode structures. One is n-a-Si/aluminum (Al)-stacked electrode shown in Figure 10.4a and the other is Al single electrodes shown in Figure 10.4b. Usually, below such electrodes, a thin i-a-Si layer is inserted as a capping layer. When the thickness of the i-a-Si layer

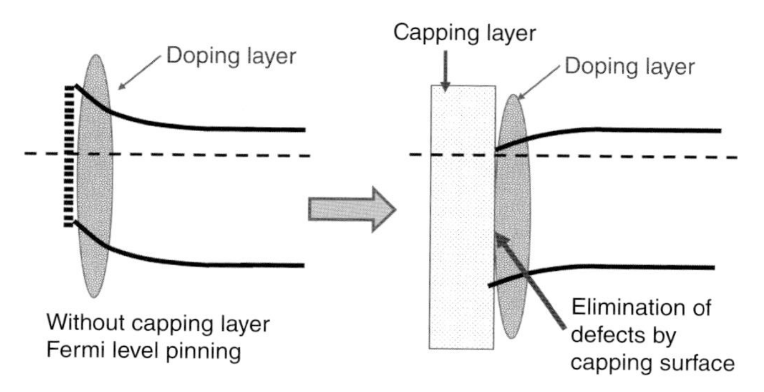

Figure 10.3 Schematic illustration of surface states of crystalline silicon (c-Si) without a capping layer (left) and with a capping layer (right).

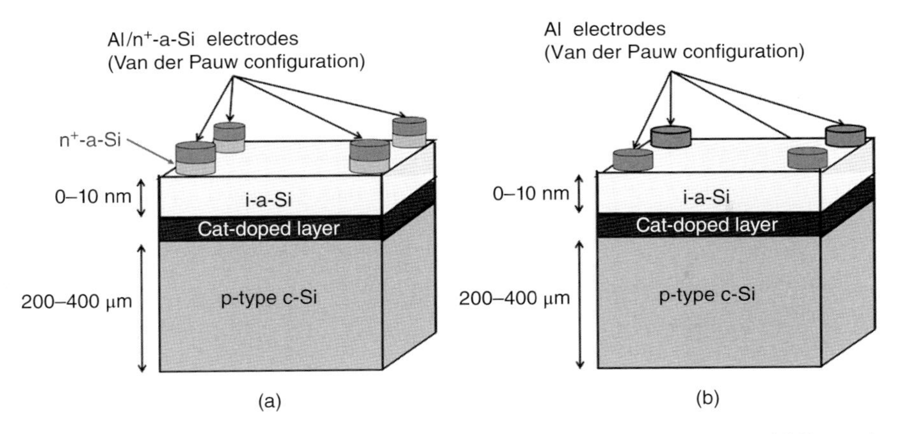

Figure 10.4 Two types of electrode structures for van der Pauw measurements. (a) Electrodes consist of heavily doped n-type amorphous silicon (n^+-a-Si) and successively deposited aluminum (Al), and (b) electrodes consist of only Al single layer. Source: Reprinted from open access Ref. [2] with sincere thanks.

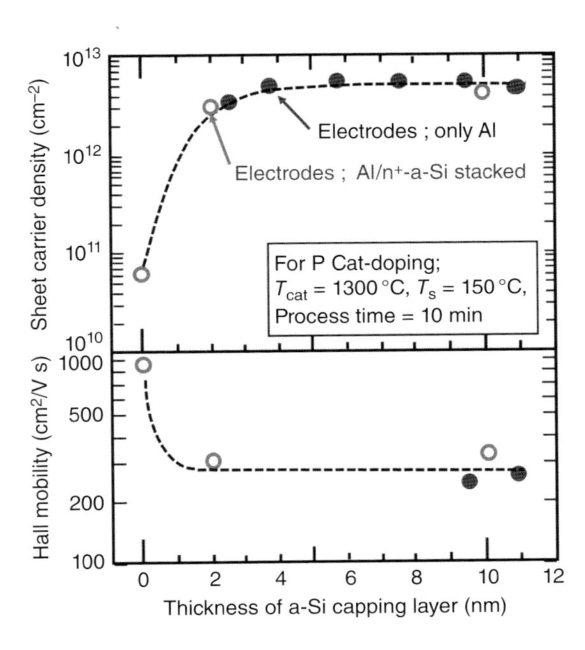

Figure 10.5 Sheet carrier density and Hall mobility of samples, as a function of thickness of capping amorphous silicon (a-Si) layers. The results measured by two types of electrodes are shown together. Source: Reprinted from open access Ref. [2] with sincere thanks.

becomes 0, these electrodes directly contact with Cat-doped layers in c-Si, and we can know the effect of the capping layers.

Figure 10.5 shows the sheet carrier density and the carrier mobility of P Cat-doped c-Si. The typical P Cat-doping conditions are summarized in Table 10.1, together with the conditions for B-doping. However, in this experiment, the parameters were fixed at $T_{cat} = 1300\,°C$, $T_s = 150\,°C$, $P_g = 1$ Pa, and FR(PH_3) = 0.43 sccm. The PH_3 gas was diluted to 2.25% by helium (He) gas and the flow rate of total He gas was 19 sccm. The process time was 10 minutes. When

Table 10.1 Typical phosphorus (P) and boron (B) Cat-doping into crystalline silicon (c-Si).

	P Cat-doping using PH_3	**B Cat-doping using B_2H_6**
Material of catalyzer	Tungsten (W)	Tungsten (W)
Catalyzer temperature, T_{cat}	RT to 1800 °C, mainly kept at 1300 °C	RT to 1800 °C
Substrate temperature, T_s (°C)	50–350	50–350
Net flow rate of doping gas (=PH_3 or B_2H_6), FR(PH_3) or FR(B_2H_6). (Both are diluted to 2.25% by helium) (sccm)	0.43–0.45	0.43–3.38
Flow rate of H_2, FR(H_2) (sccm)	0–20	0–150
Gas pressure during Cat-doping, P_g (Pa)	0.5–3	0.5–3
Catalyzer–substrate distance, D_{cs} (cm)	10–12	10–12

Al single electrodes were used, the measurement of the sample without i-a-Si capping passivation layer was impossible. When n-a-Si/Al-stacked electrodes were used, the measurement of such samples was possible, but even in this case, the data were fluctuated and might include large errors. However, when the i-a-Si layer with a thickness larger than 4 nm was inserted, the measurement of the sheet carrier density became stable. The measured mobility also became stable for the thickness of i-a-Si over 2 nm and was 200–300 cm^2/V s. This means that, by the capping of c-Si by i-a-Si layers thicker than several nanometers, we can avoid the effect of surface defects of c-Si on the measurement of a shallow surface doping region. To insure the measurements, we decided to use the samples capped with 10-nm-thick i-a-Si layers.

We should also notice that the values of mobility cannot be observed if the currents flow inside the capping i-a-Si but can be observed only when the currents flow in c-Si. That was, it was concluded that the measurement of Cat-doped c-Si layer was clearly carried out by this capping process.

The conduction type was confirmed from the polarity of the Hall voltage under the magnetic flux of 0.32 T, applied normally on the sample surface. Figure 10.6 shows the Hall voltage as a function of applied currents for a P Cat-doped sample and p-type bulk c-Si. In the present experiments, unless particularly mentioned, the data for Cat-doping into (100) Si are demonstrated. In the figure, the similar result is also shown for a B Cat-doped sample mentioned later. In this experiment, P atoms were Cat-doped into p-type c-Si with the hole density of 10^{13}–10^{14} cm^{-3}. P Cat-doping was carried out at $T_s = 350$ °C for the samples shown in the figure; however, the results for the samples prepared at $T_s = 80$ °C were not so different. When P atoms are Cat-doped into p-type c-Si, if the P Cat-doped region is converted to n-type, the polarity of the Hall voltage has to be changed. The figure confirms that p-type c-Si is converted to n-type c-Si by P Cat-doping. In addition, the B Cat-doped sample shows p-type conduction. The results again confirm the validity of our capping method to investigate the properties of the shallow Cat-doping layers.

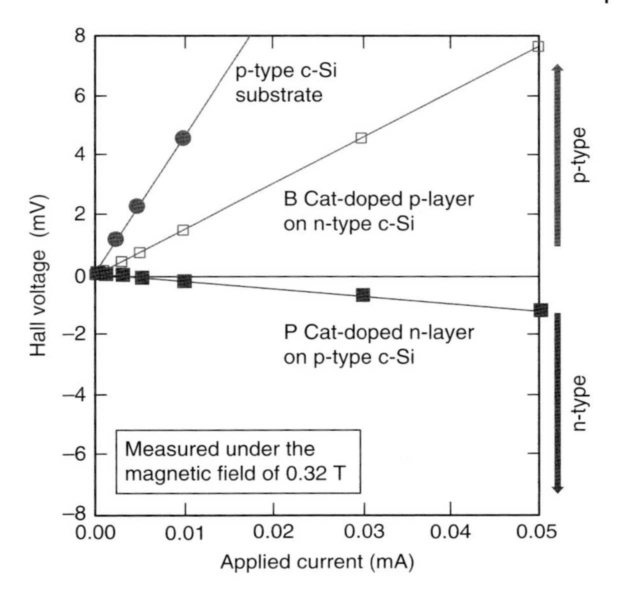

Figure 10.6 The Hall voltages vs. applied currents under the magnetic flux of 0.32 T, for p-type crystalline silicon (c-Si), phosphorus (P) Cat-doped p-type c-Si, and B Cat-doped n-type c-Si. The conduction types of P or B Cat-doped regions are converted to opposite conduction type. Source: Reprinted from open access Ref. [2] with sincere thanks.

Figure 10.7 shows the sheet carrier density of originally p-type c-Si as a function of T_{cat} during P Cat-doping. In this case, T_s was 80 °C and the process time was 10 minutes. When T_{cat} was lower than 800 °C, the sample was kept in p-type; however, when T_{cat} exceeded 1000 °C, c-Si was converted to n-type, and the sheet carrier density increased as T_{cat} increased. As mentioned in Chapter 4, PH_3 molecules are decomposed to P and 3H on the W surface when T_{cat} is heated over 1000 °C and the amount of cracked species increases exponentially as T_{cat} increases. The results shown in the figure clearly demonstrate that the existence of PH_3-cracked species is essentially necessary for this low-temperature doping.

Finally, we confirmed whether P atoms were really incorporated in Si substitutional sites. For it, we measured the measuring temperature dependence of the sheet carrier density of P Cat-doped c-Si and derived the activation energy of the carrier density. The result is shown in Figure 10.8. The activation energy derived from the gradient of the plots was about 0.045 eV, which is equivalent to that obtained in the case when P atoms were incorporated in c-Si substitutional sites and working as the conventional donors. The result confirms that the P atoms are normally doped into Si substitutional sites of c-Si at only 80 °C.

10.3.2 Measurement of Concentration Profiles of Cat-Doped Impurities by SIMS

The concentration profiles of P atoms were observed by SIMS. In the SIMS measurement of the profiles, we have to be careful to two things. One is the miss assignment of the mass number. The mass number of P atoms is 31, and this is the same as the mass number of SiH_3, which is easily created in Si and H systems. To distinguish P atoms from SiH_3 species, the mass resolution has to be improved and also we have to be always careful to check the ratio of the isotopes of Si atoms to judge whether the signals are coming from Si-related species or from P atoms.

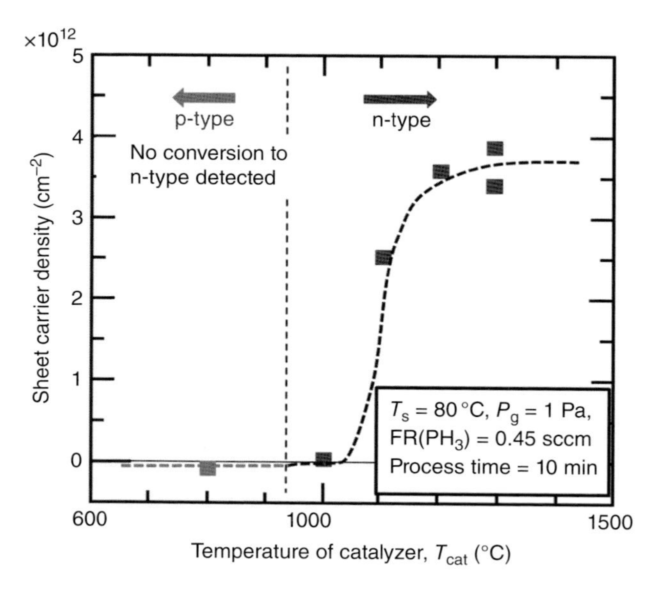

Figure 10.7 Sheet carrier density of P Cat-doped p-type crystalline silicon (c-Si) as a function of temperature of catalyzer T_{cat}. Temperature of substrate, T_s, and flow rate of hydrogen molecules (H_2), FR(H_2), are all described in the inset of the figure. The conduction types are also indicated. Source: Reprinted from open access Ref. [2] with sincere thanks.

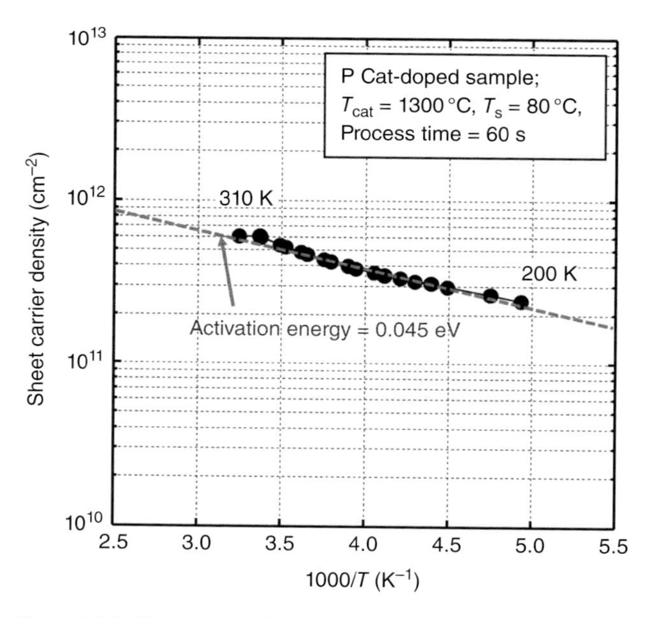

Figure 10.8 Sheet carrier density of phosphorus (P) Cat-doped sample as a function of the reciprocal of the substrate temperature T_s. Conditions of P Cat-doping are described in the figure. Source: Reprinted from open access Ref. [2] with sincere thanks.

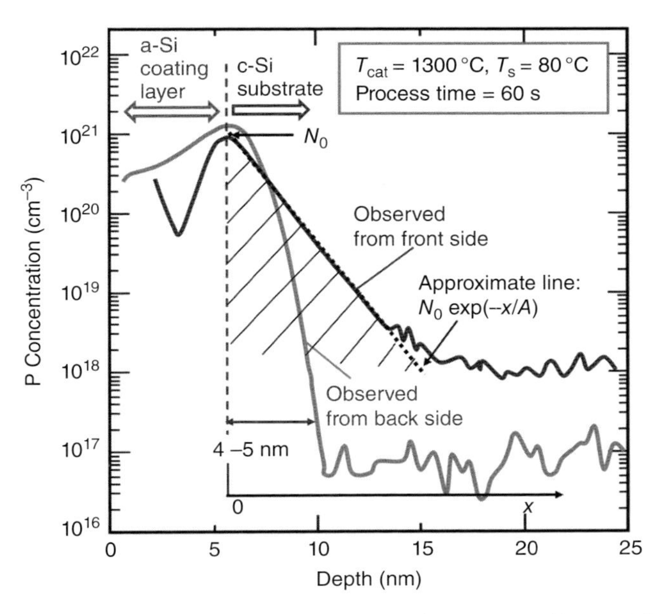

Figure 10.9 Cat-doped phosphorus (P) profiles measured by SIMS, from front side and back side of the sample. The hatched area for the profile measured from front side corresponds to total P atoms incorporated in c-Si. The sample is capped with a thin a-Si layer. The conditions of P Cat-doping are described in the figure. Source: Reprinted from open access Ref. [2] with sincere thanks.

For this reason, usually, relatively high-energy probe ions of 5 keV are used for detecting P atoms. However, this relatively high-energy probe ions are likely to induce a second problem, knock-on effect.

The probe ions collide with impurity atoms during the SIMS measurement and sometimes knock on them to deeper regions. The measured profiles are distorted by this effect. Figure 10.9 shows the SIMS profiles of P atoms measured from both the front side and the back side of the sample. In this experiment, P Cat-doping was carried out under the conditions: $T_{cat} = 1300\,°C$, $T_s = 80\,°C$, and the process time of 60 seconds. After Cat-doping, the surface of c-Si was protected by a 10-nm-thick i-a-Si layer. When the sample was observed from the back side, it was etched from the back side of the sample in advance. The figure demonstrates that two profiles are so different and that when we like to know the precise profile without the effect of knock-on, we have to measure the P profiles from the back side, although the measurement process becomes complicated.

On the other hand, the measurement from the front side has an advantage. As shown in Figure 10.9, the profile observed from the front side can be often approximated by an exponential form following $N_0\exp(-x/A)$, where N_0, x, and A refer to the surface concentration of P atoms, the depth measured from a-Si/c-Si interface, and a characteristic factor expressing the sharpness of the profile, respectively. In this case, the total amount of measured atoms N_{total}

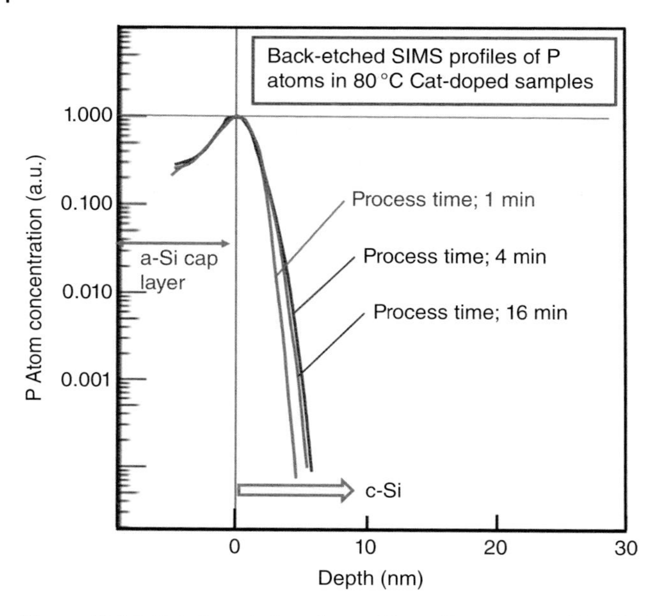

Figure 10.10 Cat-doped phosphorus (P) profiles measured by SIMS from back side of samples. Process times, 1, 4, and 16 minutes, are taken as a parameter. Source: Reprinted from open access Ref. [2] with sincere thanks.

incorporated in c-Si is simply calculated from the area hatched in the figure. The value can be simply calculated as $N_{\text{total}} = N_0 \times A$. This relation is often used in data processing for various purposes.

Figure 10.10 shows the P profiles measured by SIMS from the back side, taking the process time as a parameter. In this Cat-doping experiment, T_{cat} and T_{s} were kept at 1300 °C and 80 °C, respectively, and the c-Si was also protected by a 10-nm-thick capping i-a-Si layer after Cat-doping. Even if the profiles are measured from the back side, the depth resolution still exists because of the spatial fluctuation of etching rates as schematically illustrated in Figure 10.11. Even if the profile is very sharp like a delta function, such a profile is observed as a broadening form due to the fluctuation of definition of the depth. When the etching fluctuation in measuring area exists, the profile is convoluted with a Gaussian distribution showing the resolution of measuring systems. The figure shows that P atoms are doped into c-Si after the process time of only 1 minute and that the profiles appear deeper after 4 minutes process but does not expand any more even after 16 minutes. The value of depth resolution is not clearly known, but it appears smaller than the profile of one minute because the change of the profiles can be observed.

Even if the precise discussion on the profiles is difficult, from the figure, it can be concluded that the P-doped layer by Cat-doping is limited at the depth at around only 2–5 nm and that the doping depth does not appear expanding from that value. The depth at which P atoms can penetrate into c-Si at 80 °C appears to saturate at about 2–5 nm, without any relations to the process time.

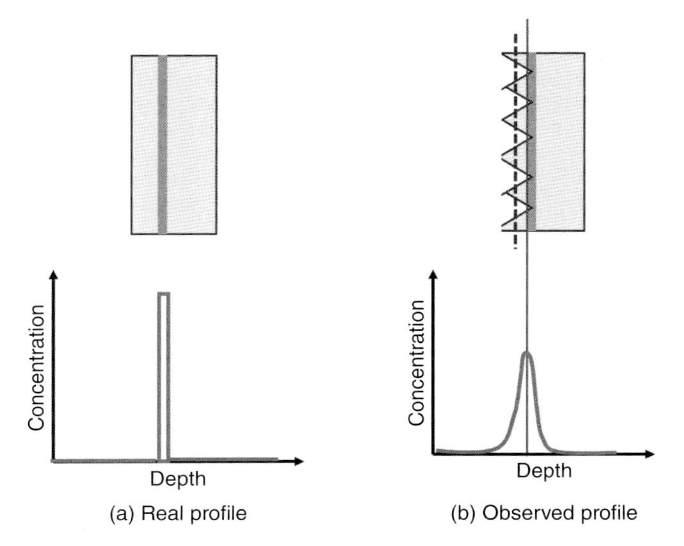

Figure 10.11 Illustration of really observed profile when the surface of the sample is roughened due to non-uniform back-etching. An original delta-function like profile (a) is observed as a Gaussian like profile (b) when roughened. This decides the depth resolution.

The phenomenon does not appear to be the same as the widely known thermal diffusion process. The diffusion constant of P atoms in c-Si at 80 °C, estimated from the extrapolation of known diffusion constants at the temperatures from 900 to 1300 °C, is too small to explain the enough amount of P doping [5].

When P atoms are doped by the thermal diffusion under the constant supply of diffusion sources at the surface, the profile $C(x,t)$ is known to be expressed by a complimentary error function (erfc) as shown in Eq. (10.1). Here, x and t refer to the position in c-Si measured from the surface and the process time for doping, respectively.

$$C(x,t) = C_s \mathrm{erfc}\left(\frac{x - x_0}{2\sqrt{Dt}}\right)$$

$$\mathrm{erfc}(y) = \frac{2}{\sqrt{\pi}} \int_y^{\infty} \exp(-u^2) \mathrm{d}u \tag{10.1}$$

where C_s, x_0, and D express the surface concentration of impurity, initial doping depth which is usually set near 0, and the diffusion constant, respectively.

In Figure 10.12, the profile of P atoms after 60 seconds of Cat-doping, already shown in Figure 10.10, is again replotted. The closed rectangle shows measured SIMS plots and a solid curve shows the erfc fitting to the plots. When x_0 is adjusted to 0.66 nm, the plots clearly fit to erfc. This x_0 value is almost equal to the value of fluctuation observed at the interface between c-Si and a-Si prepared by Cat-CVD, as shown in Figure 2.9. In the deeper region than 0.66 nm, P atoms really exist inside c-Si. At this stage, if we ignore the problem in the magnitude of the diffusion constant, the P profile appears to follow the conventional diffusion theory. This will be discussed in the following sections.

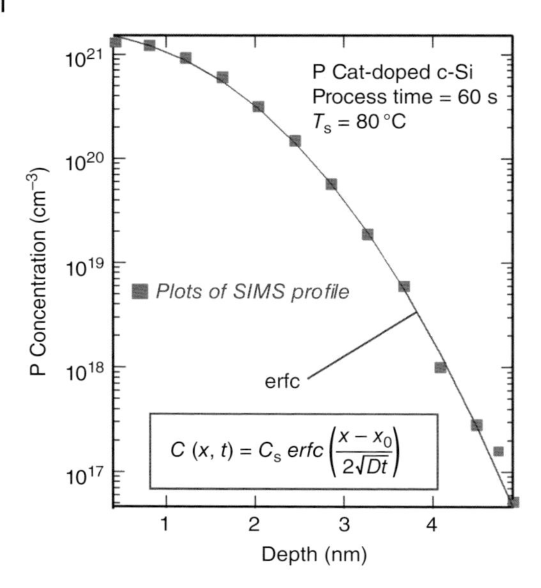

Figure 10.12 Detailed profile of Cat-doped phosphorus (P) atoms, measured by SIMS from back side. Solid curve expresses complimentary error function erfc. The erfc is described in the figure, where $C(x,t)$, C_s, x, x_0, and D refers to P profile at the position x and time t, surface concentration, position, initial position, and diffusion constant, respectively. Source: Reprinted from open access Ref. [2] with sincere thanks.

10.3.3 Estimation of Diffusion Constant

From this fitting and the insertion of $t = 60$ seconds into Eq. (10.1), we can derive the value of D as $D = 7.37 \times 10^{-17}$ cm^2/s $= 2.65 \times 10^{-5}$ μm^2/h. Similar fitting was carried out for the sample prepared at T_s = 350 °C and D was also evaluated as 6.30×10^{-17} cm^2/s $= 2.27 \times 10^{-5}$ μm^2/h. In spite of the elevation of T_s, D did not increase. We derived D values from several samples, and we could obtain a similar order of D values. The estimation from this fitting method appears to include so large errors, probably because of the influence of depth resolution in SIMS profiles.

We also estimated D in Cat-doping from the data shown in Figure 10.9 by a more direct way. The SIMS profile measured from the front side is not accurate as a point of the profile; however, the total amount of P atoms incorporated in c-Si should be correctly expressed. As already mentioned in Section 10.3.2, the total amount of P atoms, N_{total}, is simply expressed by $N_0 \times A$, when the profile is approximately expressed by $\exp(-x/A)$. This N_{total} has a relation with D, followed by Eq. (10.2) [5].

$$N_{total} = \frac{2N_0}{\sqrt{\pi}} \sqrt{Dt} \tag{10.2}$$

From this equation, $D = 2.44 \times 10^{-16}$ cm^2/s $= 8.78 \times 10^{-5}$ μm^2/h was derived from the data of Figure 10.9. The diffusion constant of P atoms in c-Si at temperatures lower than 900 °C is not clear. Thus, here, we just show the simple extrapolated values to the temperatures down to 80 °C in Figure 10.13 from the known values at the temperatures over 900 °C [5]. At 80 °C, as a common sense, P atoms cannot diffuse in c-Si; however, actually, diffusion or diffusion-like phenomenon occurs. If simply D value is considered, in Cat-doping, it is drastically increased by more than 20 orders of magnitude. How should we think this?

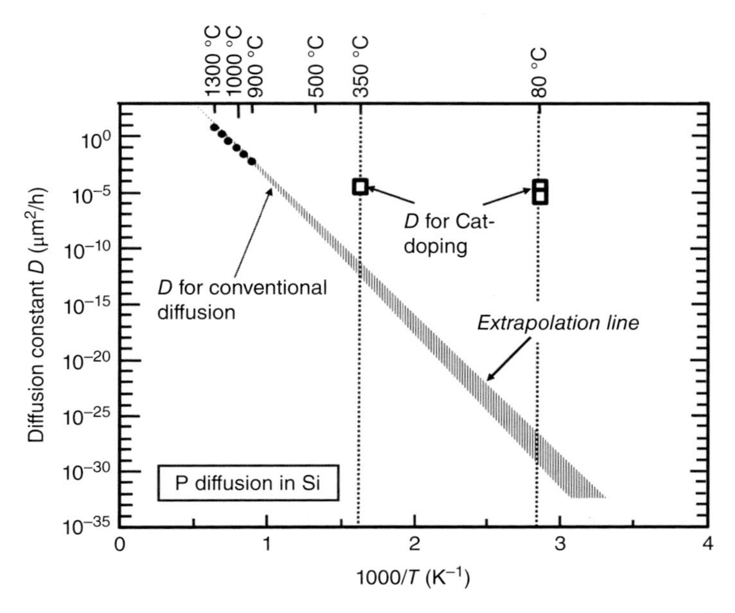

Figure 10.13 An extrapolated line of diffusion constant D from 900 to 80 °C, as a function of reciprocal temperature T. D for Cat-doping is plotted by open squares in the figure.

10.3.4 Properties of Cat-Doped P Atoms

Before we begin to speculate the mechanism of Cat-doping, we collect more information on the Cat-doping phenomenon and confirm the experimental results.

At first, to confirm the doping depth again, we attempted P Cat-doping through very thin SiO_2 layers formed on c-Si. If some species related to P atoms penetrate into c-Si during Cat-doping, the amount of P atoms detected inside c-Si should depend on the thickness of SiO_2 layers. The thickness of SiO_2 layers, which was chemically formed by dipping c-Si samples in 90 °C hydrogen peroxide (H_2O_2) solution, was adjusted to be 0, 1.5, 4, and 7 nm. The thickness was checked by an ellipsometer. P Cat-doping was performed at $T_s = 350$ °C for various process times into p-type c-Si with a hole carrier concentration of 10^{13} to 10^{14} cm^{-3} and with SiO_2 layers of various thickness. After Cat-doping, surface SiO_2 was removed by hydrofluoric acid (HF) solution and immediately 10-nm-thick i-a-Si capping layers were deposited on Cat-doped c-Si substrates. The measured sheet carrier densities are plotted in Figure 10.14 as a function of process time, taking the thickness of SiO_2 layers as a parameter. When c-Si was not covered with SiO_2, just after one-minute process, p-type c-Si originally was converted to n-type. When the thickness of the SiO_2 layer was 1.5 nm, after P cat-doping, the c-Si samples were converted to n-type, but the sheet carrier density became much smaller than that without SiO_2 layers. When the thickness of the SiO_2 layer was 4 nm, still the conversion from p-type to n-type could be observed, but for the SiO_2 thickness of 7 nm, the conversion itself could not be observed any more. Although the penetration of P atoms in SiO_2 must be different from that in c-Si,

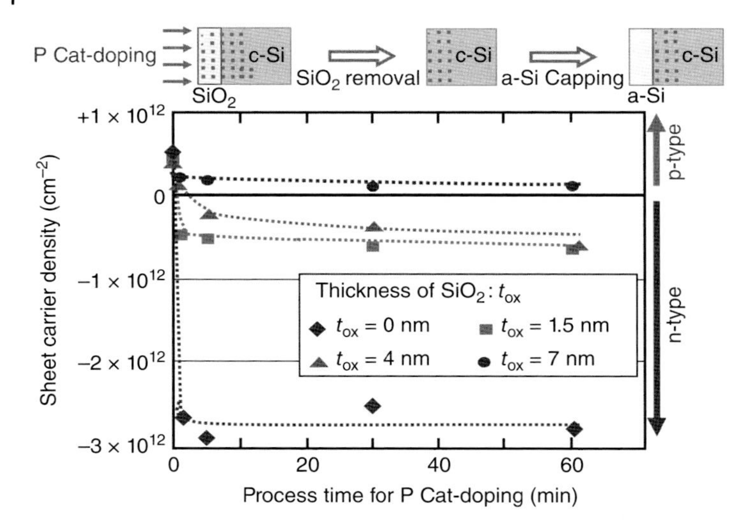

Figure 10.14 Sheet carrier density and conduction type of phosphorus (P) Cat-doped samples through thin SiO_2 layers, as a function of Cat-doping process time. Thicknesses of the SiO_2 layers, t_{ox}, 0, 1.5, 4, and 7 nm, are taken as a parameter. At top, the procedure to prepare the samples for measurement is illustrated. Source: Reprinted from open access Ref. [2] with sincere thanks.

this experiment confirms that P atoms can penetrate into c-Si through a thin SiO_2 layer in Cat-doping process.

From these series of experiments, the penetration of P atoms into c-Si at low temperatures could be understood as doubtless truth. Then, next, we measured the sheet carrier density of P Cat-doped c-Si for various process times. Figure 10.15 shows the sheet carrier density of P Cat-doped c-Si as a function of the Cat-doping process time, taking T_s as a parameter [6]. The sheet carrier density is likely to increase as T_s increases. Within the initial five minutes, the sheet carrier density also appears increasing in proportion to the square root of process time. If the normal diffusion theory holds, the total number of incorporated impurities should increase in proportion to the square root of the time [5]. The figure shows that, within the initial five minutes, the phenomenon appears to follow the simple diffusion mechanism even when T_s was as low as 80 °C; however, that after five minutes, the sheet carrier density saturates. This tendency can also be seen in SIMS profiles of Figure 10.10.

The total sheet densities of P atoms after P Cat-doping in c-Si were also measured by SIMS as a function of the process time in Figure 10.16. In the figure, the results for T_s = 80 and 350 °C are shown. The total number of P atoms incorporated in c-Si is likely to increase as T_s increases, and in both T_s, the number of incorporated P atoms itself saturates after 5–10 minutes of Cat-doping process. It is known that the saturation behavior seen in the sheet carrier density originates from the saturation of total number of incorporated P atoms. At the same time, from the comparison between the sheet carrier density and the number of P atoms measured by SIMS, the electrically activation ratio

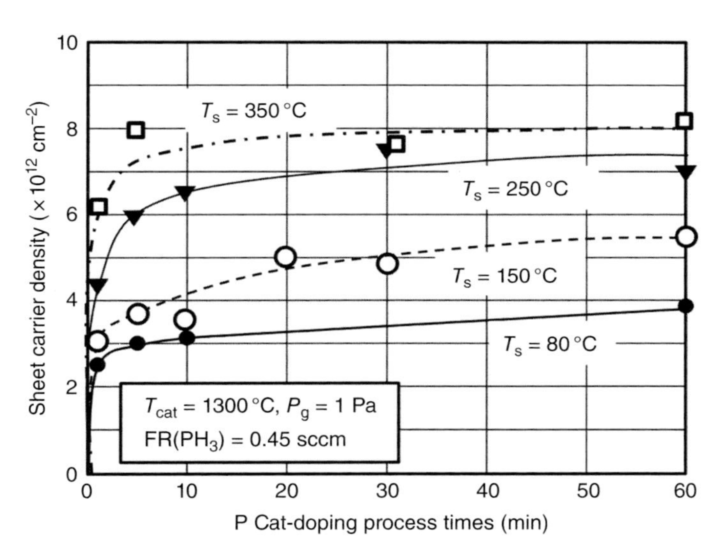

Figure 10.15 Sheet carrier density of phosphorus (P) Cat-doped samples as a function of Cat-doping process time. The substrate temperature T_s during Cat-doping is taken as a parameter. Conditions of Cat-doping are described in the figure. Numbers described at a vertical axis should be multiplied by 10^{12} as indicated in the unit.

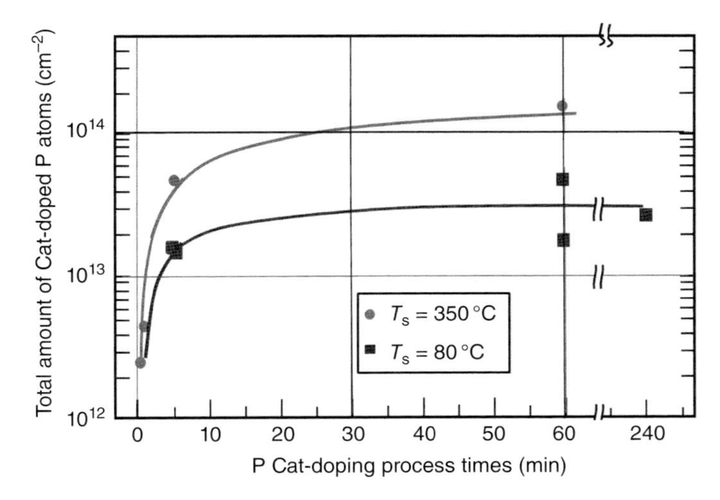

Figure 10.16 Total amount of Cat-doped phosphorus (P) atoms evaluated from SIMS profiles as a function of process time. Results for substrate temperatures T_s of 80 and 350 °C are plotted. Source: Reprinted from open access Ref. [2] with sincere thanks.

of Cat-doped P atoms is estimated as several to 10% for both T_s = 80 and 350 °C. Considering the temperature, these values may be reasonable.

All data mentioned above are concerned with the Cat-doping into (100) Si substrates. If we change the substrates from (100) Si to (111) Si, how do the data change? Next, we show the similar data taken by using (111) Si. Figure 10.17 shows the sheet carrier density as a function of the process time, similar to the

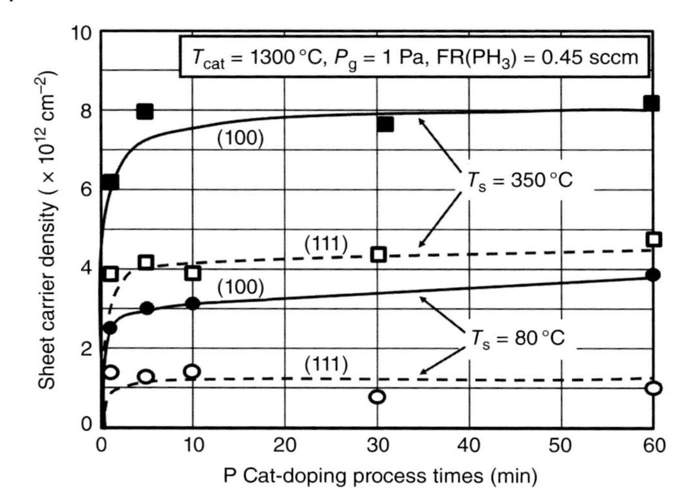

Figure 10.17 Sheet carrier densities of phosphorus (P) Cat-doped samples into (100) c-Si and (111) c-Si as a function of P Cat-doping process time for substrate temperatures T_s of 80 and 350 °C. Conditions for P Cat-doping are described in the figure.Numbers described at a vertical axis should be multiplied by 10^{12} as indicated in the unit.

data shown in Figure 10.15, but in the figure, the results of P Cat-doping into (111) c-Si substrates are demonstrated besides those into (100) substrates [6]. T_s was taken as a parameter. The results of Cat-doping into (111) c-Si show the similar tendency to those of (100) c-Si; however, the sheet carrier density of (111) c-Si appears smaller than that of (100) c-Si. This suggests that the incorporation of P atoms into (100) c-Si is easier than that into (111) c-Si.

The difference of diffusion constants of impurities in (100) c-Si and (111) c-Si substrates is well known in the conventional substitutional diffusion. As the frequency for passing over the barriers during diffusion depends on the direction of impurity movement, this difference is understandable [5]. However, although the conventional diffusion mechanism might not be applied in Cat-doping, the results appear suggestive as some similarities to the conventional diffusion are still remained. The results may give us a hint to consider the mechanism of P cat-doping. In (111) orientation, replacement of Si atoms to P atoms to incorporate P atoms in substitutional sites may be more complicated than the case of (100)-oriented c-Si.

All these data described above suggest that (i) the Cat-doping phenomenon is seen only at the region near to the surface of c-Si and (ii) the phenomenon appears to be related to the c-Si structures.

10.3.5 Mechanism of Cat-doping

10.3.5.1 Possibility of Diffusion Enhancement by H Atoms

The mechanism of the Cat-doping phenomenon has not been clarified. The study on clarifying the mechanism itself has not been enough, and only a few speculative ideas are presented.

At first, we introduce the study explaining the mechanism based on the first-principle calculation. The calculation is carried out for the system in

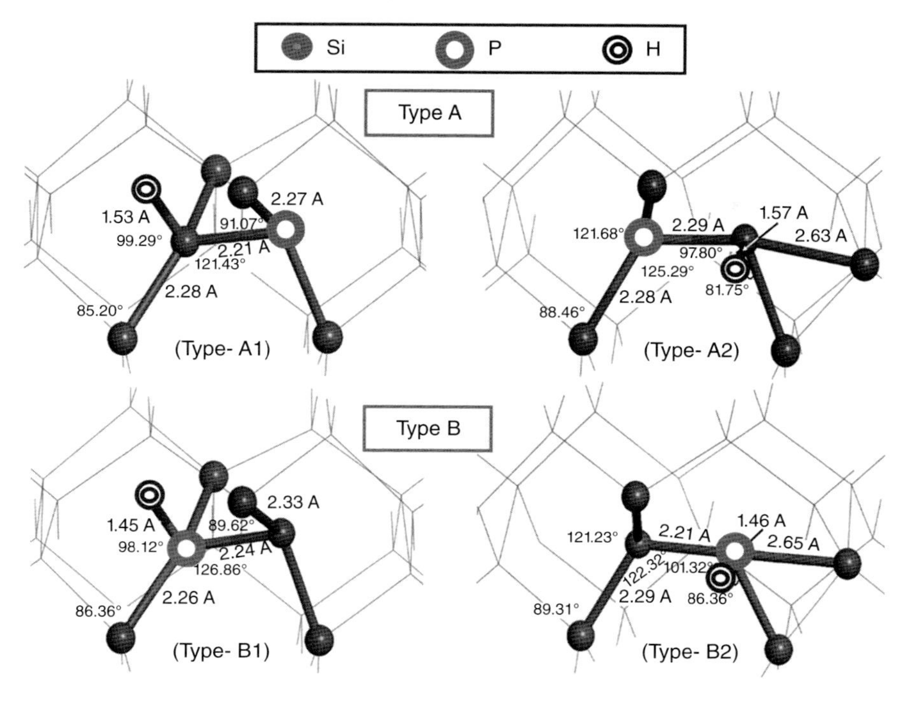

Figure 10.18 Results of the first-principle calculations for the system consisting of 216 Si atoms, 1 P atom, and 1 H atom. Four stable configurations are shown. Two types of configurations are considered. In Type A, H bonds with Si, and in Type B, H bonds with P. For working as a donor, only Type A has a meaning. Source: Reprinted with permission from Ref. [7].

which a P atom moves together with a H atom [7]. In the calculation model, 216 Si atoms are taken into account to construct a crystalline structure of Si, and a single P atom is introduced into the place near a center of the Si crystal model together with a single H atom. In this case, the calculation indicates that there exist only four stable configurations for bonds among H, Si, and P atoms, as schematically illustrated in Figure 10.18. In the stable configurations, two configurations of type A and type B are considered. In type A, a H atom makes a bond directly with a Si atom, and a P atom makes bonds with Si atoms surrounding it. In type B, a H atom bonds with a P atom directly. In this type-B configuration, the P atom may not work as a donor.

Thus, if we may think the movement of a P atom at the configuration (Type-A1) to the next stable configuration (Type-A2) located at the neighboring lattice, we can calculate the diffusion of a P atom in c-Si under the existence of a H atom. The energy level of Type-A2 is about 0.5 eV higher than that of Type-A1 as shown in Figure 10.20 later, as the extra formation energy is required to form this configuration.

Figure 10.19 demonstrates the various configurations of Type-A1 and Type-A2. In the Type-A1 configurations, there are three locations of H, Si, and P atoms, (a–c) in the three-dimensional space, and when the P atom changes its location, it has to move through the distance described in the figure. In the Type-A2

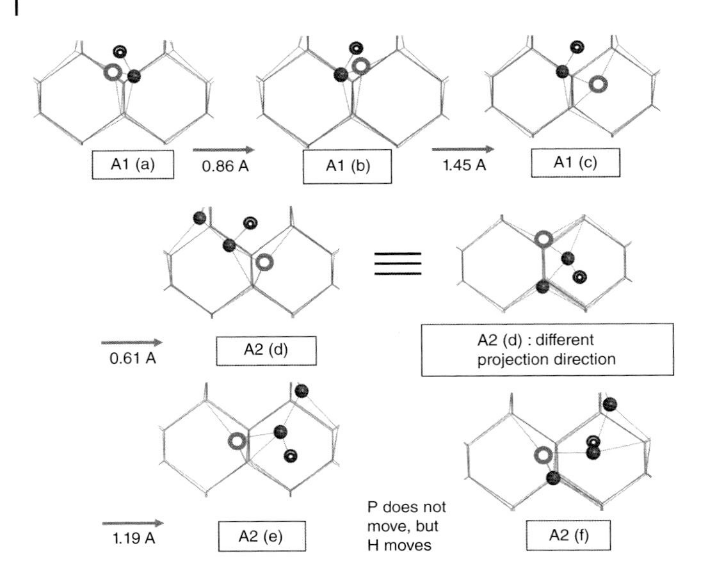

Figure 10.19 Results of the first-principle calculations for the system consisting of 216 Si atoms, 1 P atom, and 1 H atom. The stable locations of H, Si, and P atoms and the length of displacement of P atoms between the two neighboring locations are described. By this way, the movement of P atoms is demonstrated. Source: Reprinted with permission from Ref. [7].

configurations, there are also three locations, (d–f). In this case, when the location (e) changes to the location (f), actually, only the H atom moves. As this is mentioned in the three-dimensional space, it may not be clear in the explanation; the details are explained in the Ref. [7].

Figure 10.20 demonstrates the potential energy for the movement of a P atom from a location of a stable configuration to a neighboring location of another stable configuration, together with the distance that the P atom moves by overcoming the potential barriers formed between two stable locations. The highest potential barrier in the P movement is about 1.29 eV. This energy includes the extra formation energy of the Type-A2 and is much lower than the potential barrier of P substitutional diffusion in c-Si, which is believed about 3.0 eV. Even if such an extra formation energy is considered, the barrier height for the P movement in the existence of the H atom is much smaller than the conventional substitutional diffusion. That is, if a H atom makes a bond with a Si atom neighbor to a P atom, the P diffusion becomes much easier than the conventional substitutional diffusion.

This calculation gives us a hint; however, to explain the saturation in the doping depth, we have to consider another extra mechanism to break stable configuration. This is an unsolved point of this model.

10.3.5.2 Vacancy Transportation Model

Other explanations of the mechanism of Cat-doping do not have concrete calculations or evidences, and we can say just speculation or fantasy.

Another possible explanation is considering the interaction between H atoms in the gas phase during Cat-doping and Si atoms. We illustrate this model by following the process steps shown in Figure 10.21. At first, (1) H atoms attack c-Si

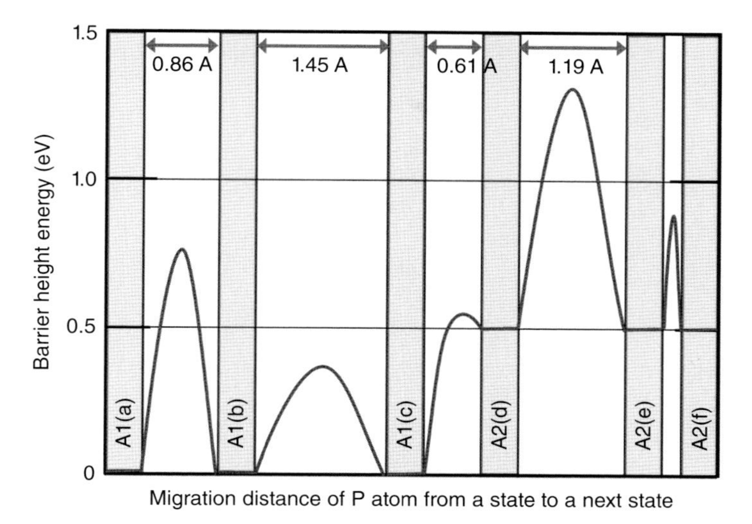

Figure 10.20 Results of the first-principle calculations for the system consisting of 216 Si atoms, 1 P atom, and 1 H atom. The barrier height and the length of displacement of P atoms between the two neighboring locations are described. Source: Reprinted with permission from Ref. [7].

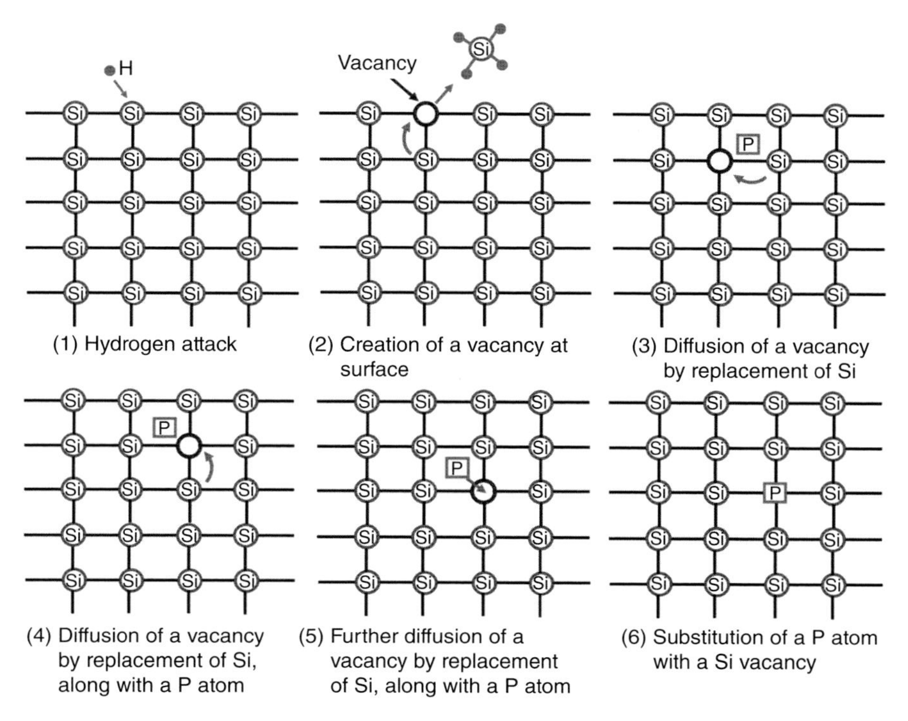

Figure 10.21 A model of P diffusion with vacancy created by H-atom attack on the c-Si surface.

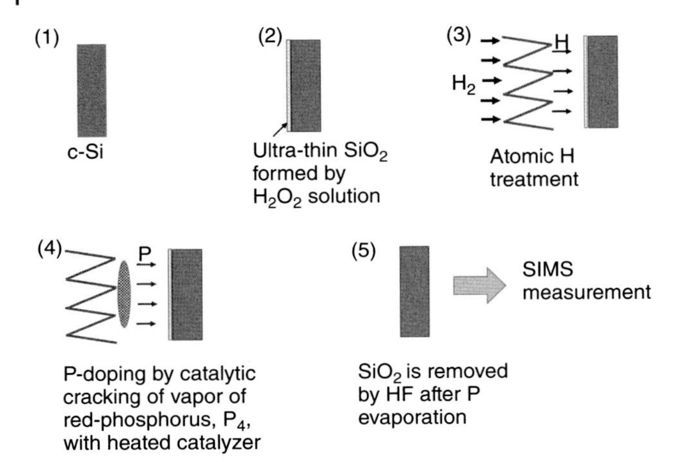

Figure 10.22 Experimental procedure to know the effect of H atoms in Cat-doping process.

surface and (2) pull a Si atom out from the surface of c-Si, and then, the surface Si structure will be locally changed by forming a surface Si vacancy. (3) This Si vacancy is replaced by another Si atom located at deeper regions, and thus, such a vacancy moves to deeper regions. If such a vacancy may make a complex with a single P atom [8], a pair of the vacancy-P atom moves to deeper regions from step (4) to (5). However, the lifetime of such a pair is usually short and finally (6) the P atom combines with the vacancy. That is, the P atom takes a place at the Si substitutional site. However, because of short lifetimes of the vacancy-P pair, the P penetration depth is limited to several nanometers.

There have not been any reports showing the actual and realistic supporting data for this model except for our preliminary experiments. In our experiments, c-Si was exposed to P atoms produced from P_4 in the absence of H atoms, after exposed to H atoms independently. A slight increase in the total amount of incorporated P atoms in c-Si could be observed by this atomic hydrogen pretreatment. Figure 10.22 shows a schematic illustration of the experimental procedures. At first, (1) c-Si substrates are prepared, and (2) the surface of them is covered with ultrathin SiO_2 layers, which are formed by dipping the c-Si in H_2O_2 solution. The SiO_2 layers are used to avoid etching by H atoms. The thickness of the SiO_2 layer is usually less than 2 nm. Then, (3) the c-Si substrates are exposed to H atoms in a Cat-CVD chamber to introduce H atoms into c-Si and also introduce the Si vacancy at the surface or at the adjacent surface of c-Si. The substrate temperature T_s was kept at 350 °C, T_{cat} 1800 °C, P_g 1 Pa, and the flow rate of H_2, FR(H_2), 20 sccm. The time for H treatment varied from 0 to 60 minutes. After H treatment, (4) the c-Si substrates were transferred to another chamber without breaking vacuum to be exposed to P atoms. The P atoms were generated by the catalytic cracking of P_4 vapor, not PH_3, with heated W wires at T_{cat} = 1300 °C [9]. The P_4 vapor was produced by heating red phosphorus powders on tantalum (Ta) boats. The red phosphorus powders are heated only by the radiation from the W catalyzer. After that, (5) the samples are dipped in HF solution to remove all surface oxide layers. By this way, all adsorbed materials on surface SiO_2 layer are

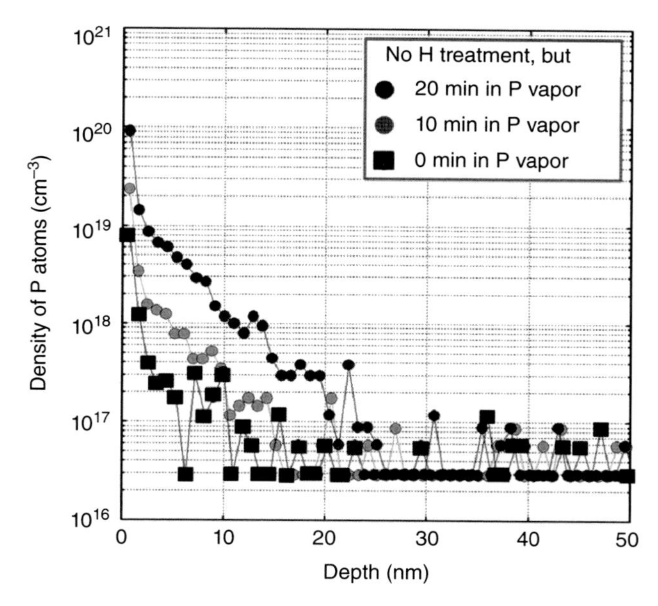

Figure 10.23 Phosphorus (P) SIMS profiles measured from front side of samples. Exposure time to P atoms is taken as a parameter.

removed, and only P atoms incorporated in c-Si can be observed. Then, the total amount of P atoms incorporated by this way was measured by SIMS from the front side, as for knowing the total amount of P atoms, the SIMS measurement from the front side is adequate as already explained.

Figure 10.23 shows the SIMS profiles of P atoms after P_4 cracking treatment for various times, without H treatments. That is, it is known at first that P atoms are incorporated in c-Si without any helps of H atoms. This appears to claim that the existence of single P atoms is the most important factor for Cat-doping. The result may seem to deny the speculation of vacancy model described in Figure 10.21. However, other experimental results appear to support the role of H atoms again.

Figure 10.24 shows the SIMS profiles of P atoms for the samples that had H-atom treatments for various times and then P_4 cracking treatment was performed for 10 minutes. As shown in Figure 10.23, P atoms are incorporated in c-Si without any helps of H atoms; however, if the samples were treated by H atoms in advance, more P atoms are incorporated. The results appear to verify the positive effect of the existence of H atoms in Cat-doping.

The above speculation based on the first-principle calculation and the speculation based on the vacancy model might contain some parts of truth in Cat-doping mechanism; however, we may be allowed to think the phenomenon very simply.

10.3.5.3 Si-Modified Surface Layer Model

It is well known that the surface of Si atoms in c-Si is rearranged when it is free from adsorption of other materials in ultrahigh vacuum [10]. It is believed that the influence of this rearrangement of Si atomic array structure continues to the depth equivalent to about 10 atomic layers. The region of c-Si, shallower than

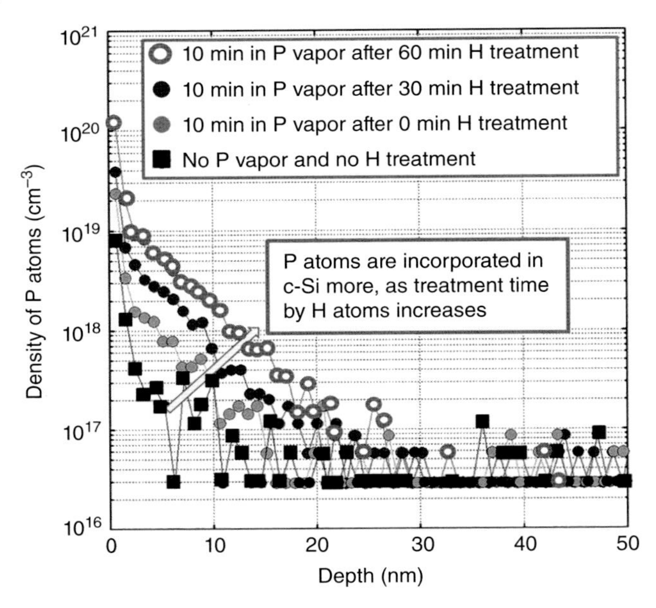

Figure 10.24 Phosphorus (P) SIMS profiles measured from front side of samples. Exposure time to P atoms is fixed at 10 minutes, but time for exposure to H atoms (H treatment time) before exposure to P atoms is taken as a parameter. For comparison, the result for the sample without pretreatment is shown together.

about 1–2 nm, may be affected by the surface rearrangement of solid Si. However, is it really limited to only 1–2 nm depth? If Si surface suffers from stress, for instance, due to the attacking of H atoms, is it not possible that the depth of the surface modified layer extends to 5 nm? As introduced in Chapter 9, actually, c-Si is oxidized by cracked O-related species at only 250 °C, but the depth was limited to about 5 nm. The c-Si can also be converted to nitride by exposing c-Si in N-related species, but the depth of nitride layer was again about 5 nm. Although we cannot provide the exact image of such modified surface layers of c-Si, if we can approve that the c-Si at a region of the depth up to several nanometers is not common c-Si, the most of phenomena concerned with the surface modification such as surface oxidation, surface nitridation, and Cat-doping, introduced in this book, are all understood totally. Considering on many experimental results, it appears natural to imagine that there is a special surface-modified layer at the surface of c-Si, such a surface-modified layer spreads up to about 5 nm in depth, and that all phenomena occurring in such modified layers are often different from the phenomena seen in the bulk c-Si.

10.4 Low-Temperature Boron (B) Doping into c-Si

Compared with P Cat-doping, Cat-doping of boron (B) atoms is a little bit complicated. B atoms easily react with W to make W-boride. The data are sometimes depending on the process time used for consuming B atoms for making W-boride. Figure 10.25 shows the cross-sectional views of the electron probe

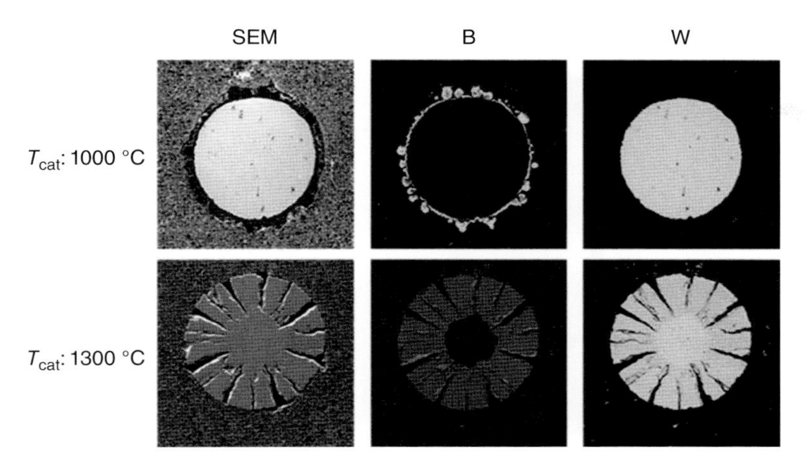

Figure 10.25 Cross-sectional views obtained by electron probe microanalysis (EPMA) for tungsten (W) catalyzer used in boron (B) Cat-doping. Images of scanning electron microscopic (SEM) mode and images of B and W distribution mapping are demonstrated for catalyzer temperature T_{cat} of 1000 and 1300 °C.

microanalyzer (EPMA) for the W catalyzers used in B Cat-doping. In the figure, the simple scanning electron microscope (SEM) images are demonstrated for B Cat-doping at T_{cat} of 1000 and 1300 °C. Similar images of B and W are also shown in the figure. In this experiment, P_g, FR(B_2H_6), and process time were 1 Pa, 3.15 sccm as a net value, and 10 minutes, respectively. It is demonstrated that when T_{cat} = 1000 °C, a major part of W wire is kept as W, and, at only the surface of the W wire, B signals are detected. However, when T_{cat} is elevated to 1300 °C, the situation becomes completely different. In the catalyzing wire, W and B signals are observed. The image of W atoms is fitting to the image of B atoms. That is, the W catalyzer is converted to W-boride. During the formation of this W-boride, the most of B species are wasted to form W-boride and B Cat-doping becomes less efficient.

In addition, diborane (B_2H_6), which is a major source of B Cat-doping, can be decomposed at the temperatures over 300 °C on the surface of substrates, and thus, sometimes, the Cat-doping data depends on the substrate temperatures when it exceeds over 300 °C. By B Cat-doping, n-type c-Si can be converted to p-type c-Si at only 80 °C, similar to P Cat-doping; however, the data become complicated by the above reasons. The result of the Hall effect is already shown in Figure 10.6.

Figure 10.26 shows the sheet carrier density of B Cat-doped n-type (100) c-Si as a function of T_{cat}, taking T_s as a parameter. The other conditions were all the same as the case of Figure 10.25. The carrier concentration of original n-type c-Si was 10^{13}–10^{14} cm^{-3}. When T_s = 80 °C, the n-type c-Si was converted to p-type c-Si only when T_{cat} exceeded 500 °C. However, contrary to P Cat-doping, it behaved complicatedly when T_{cat} increased from RT to 1800 °C. When T_{cat} increased over about 1000 °C, the sheet carrier density started to decrease. This is probably because B atoms started to be used for forming W-boride as explained above. In other words, if we use a complete boride catalyzer, the results may become more stable.

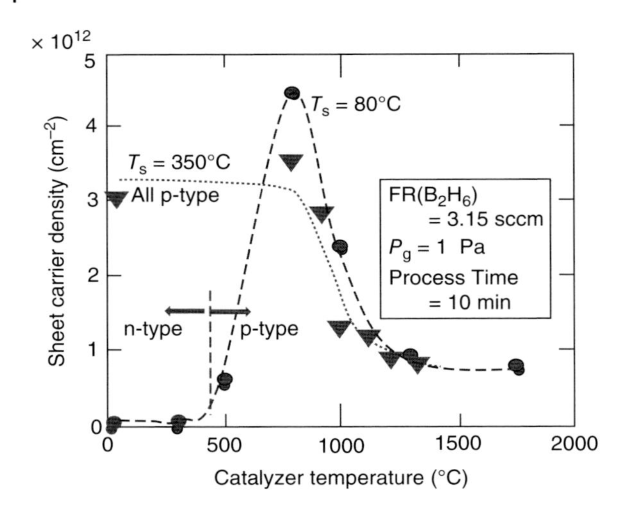

Figure 10.26 Sheet carrier density of c-Si, which is B Cat-doped at substrate temperatures T_s of 80 and 350 °C. Conditions for Cat-doping are described in the figure. Source: Reprinted from open access Ref. [2] with sincere thanks.

In addition, when $T_s = 350\,°C$, the surface of n-type c-Si was always converted to p-type without heating the catalyzer. This is not Cat-doping anymore; however, if the simple B doping at the temperature of 350 °C is successfully carried out, this may also be a very useful technique. In this case, when T_{cat} is elevated over about 1000 °C, the sheet carrier density started to decrease similar to the case of $T_s = 80\,°C$.

To check whether B atoms are incorporated without heating the catalyzer, SIMS profile of B atoms of the sample, which is prepared at $T_s = 350\,°C$, $P_g = 1$ Pa, FR(B_2H_6) = 3.15 sccm, and with the process time = 10 minutes, but T_{cat} = RT, is observed. The SIMS profile is shown in Figure 10.27 [11]. It was

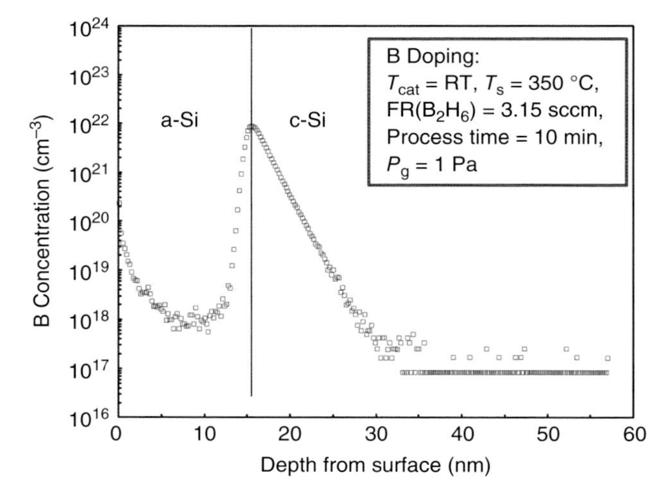

Figure 10.27 SIMS profile of boron (B) atoms Cat-doped into c-Si. The measurement is carried out from the front side. Conditions of Cat-doping are summarized in the figure. Source: Reprinted with permission from Ref. [11].

measured from the front side; however, we can confirm the incorporation of B atoms in c-Si without heating the catalyzer. In this sample, after B doping, immediately, the surface was covered with a 10-nm-thick a-Si layer, which was prepared with $T_{cat} = 1800\,°C$, $T_s = 90\,°C$, $P_g = 0.5\,Pa$, and $FR(SiH_4) = 10\,sccm$.

Total amount of B atoms incorporated is evaluated from this figure as $1.3 \times 10^{15}\,cm^{-2}$. The activation ratio of incorporated B atoms in c-Si, estimated from Figures 10.26 and 10.27, is quite low and less than 1%, although the absolute value of SIMS measurement sometimes includes large errors. The diffusion constant D can also be roughly estimated by using the relation shown in Eq. (10.2). In this case, $D = 2.3 \times 10^{-17}\,cm^2/s = 8.2 \times 10^{-6}\,\mu m^2/h$. The value appears similar to that of P Cat-doping. This again leads a speculation that various materials are easily incorporated in the surface-modified region.

10.5 Cat-Doping into a-Si

Similar to Cat-doping into c-Si, impurities are incorporated into a-Si by Cat-doping. That is, by Cat-doping, i-a-Si can be doped to be n-type a-Si or p-type a-Si. Figure 10.28 shows the sheet resistivity of 20-nm-thick i-a-Si deposited on quartz substrates after P or B Cat-doping as a function of T_{cat}, for $T_s = 50$ and $350\,°C$. Although the Cat-doping conditions are almost the same as those shown in Table 10.1, the present conditions are summarized again in Table 10.2. In the case of P doping, the resistivity starts to increase when T_{cat} exceeds $1000\,°C$, similar to the case of P Cat-doping into c-Si. However, in the case of B Cat-doping, again similar to Cat-doping into c-Si, the results behave complicatedly. The conductivity does not increase at all for $T_s = 50\,°C$, but the sheet conductivity for $T_s = 350\,°C$ samples is likely to decrease for T_{cat} exceeding $1000\,°C$. This is probably due to B consumption for making W-boride.

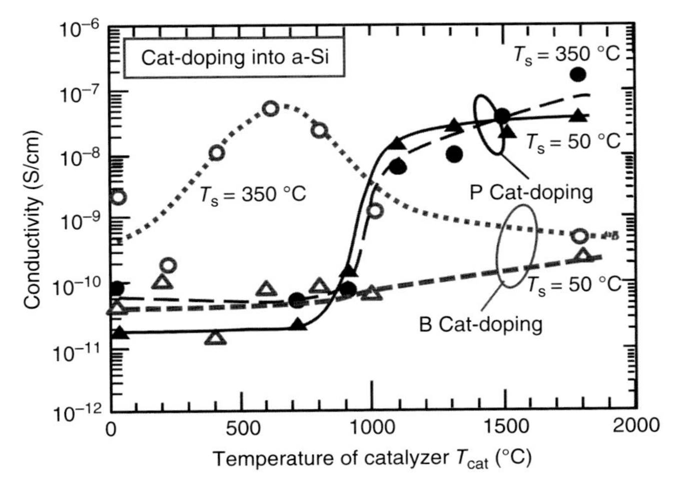

Figure 10.28 Conductivity of phosphorus (P) and boron (B) Cat-doped a-Si films on glass substrates. Substrate temperatures T_s of 50 and 350 °C are taken as a parameter.

Table 10.2 Typical phosphorus (P) and boron (B) Cat-doping into Cat-CVD intrinsic amorphous silicon (i-a-Si) and parameters for the preparation of such i-a-Si films.

	Cat-doping into i-a-Si		
	P Cat-doping by PH_3	**B Cat-doping by B_2H_6**	**i-a-Si deposition**
Material of catalyzer	Tungsten (W)	Tungsten (W)	Tungsten (W)
Catalyzer temperature, T_{cat}	RT to 1800 °C	RT to 1800 °C	1800 °C
Substrate temperature, T_s (°C)	50–350	50–350	160
Flow rate of SiH_4, FR(SiH_4)	–	–	10 sccm
Flow rate of H_2, FR(H_2)	–	0–60 sccm	–
Net flow rate of doping gas (=PH_3 or B_2H_6), FR(PH_3) or FR(B_2H_6). (Both are diluted to 2.25% by helium)	0.45 sccm	0.45–0.68 sccm	–
Gas pressure during Cat-doping, P_g (Pa)	2	2–4	1
Catalyzer–substrate distance, D_{cs} (cm)	10–12	10–12	10–12

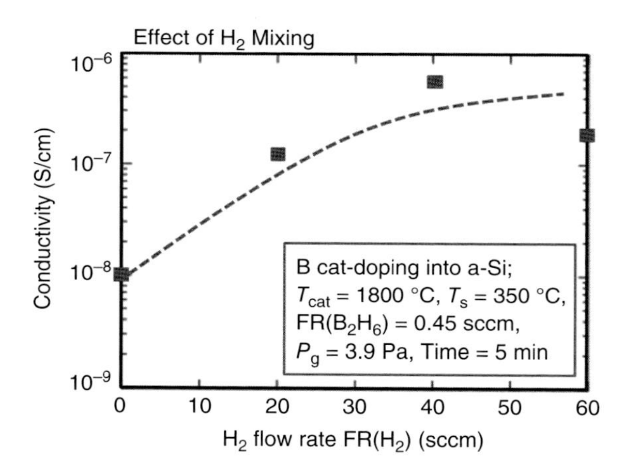

Figure 10.29 Conductivity of boron (B) Cat-doped a-Si as a function of the flow rate of hydrogen (H_2), FR(H_2), mixed with diborane (B_2H_6), which is used for B Cat-doping. Conditions are summarized in the figure. Source: Reprinted with permission from Ref. [12].

Here, when H_2 gas is introduced with B_2H_6, the conductivity starts to increase as shown in Figure 10.29. In the decomposition of B_2H_6, as described in Chapter 4, B_2H_6 is decomposed to $BH_3 + BH_3$ on the W catalyzer, and B atoms are generated by successive reactions with H atoms in the gas phase. The results shown in Figure 10.29 suggests that for Cat-doping, bare B atoms are required instead of B_nH_m (n, m = 1,2,3...) compounds.

Figure 10.30 shows the SIMS profiles of P and B atoms, which are Cat-doped into a-Si. The profiles were measured from the back sides of samples. The Cat-doping conditions are different for B and P Cat-doping. In the case of P atoms, P Cat-doping into c-Si is also shown together for comparison. The P Cat-doping into c-Si was carried out at T_{cat} = 1300 °C for 1 minute, but after the deposition, the sample was annealed at 350 °C for 30 minutes. Some of the key parameters for the preparation of samples are described in the figure.

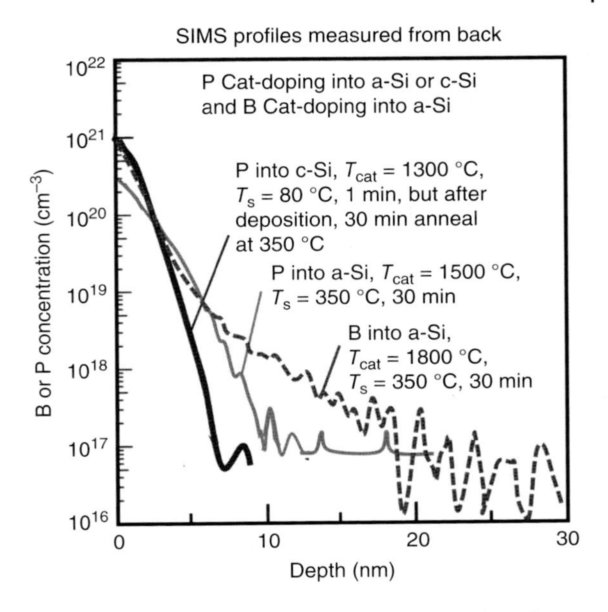

Figure 10.30 Phosphorus (P) and boron (B) SIMS profiles after P or B Cat-doping into amorphous silicon (a-Si). The result for P Cat-doping into crystalline silicon (c-Si) after annealing at 350 °C for 30 minutes is also drawn for comparison. Conditions for P and B Cat-doping are described in the figure. Source: Reprinted with permission from Ref. [12].

Compared with the case of Cat-doping into c-Si, the P and B atoms distribute slightly deeper in a-Si but still shallower than 10 nm or if we say the depth for actually heavily doped region, it would be 5 nm or less. The conductivity data were approximately described in the above figures under the assumption that doping impurities distributed uniformly through all the depth, in 20 nm. As the actual distributions of P and B atoms are concentrated at the vicinity of a-Si surface, the actual conductivity at the doping layer will be four to five times larger than the values shown in Figures 10.28 and 10.29.

10.6 Feasibility of Cat-Doping for Various Applications

10.6.1 Surface Potential Control by Cat-doping Realizing High-Quality Passivation

As described above, the total amount of P and B atoms incorporated in c-Si is not so large and the thickness of the doping layer is very thin. Therefore, it may not be easy to make simple p–n junction devices by Cat-doping technology. However, the impurity existing at only surface area may be useful for controlling the surface potential of c-Si. The control of the surface potential is particularly useful to obtain high-quality surface passivation layers for c-Si solar cells. In the solar cells, carriers excited by sunlight have to reach the electron or hole electrodes without any losses due to recombination. When the thickness of c-Si substrates becomes thinner than about 100 μm to reduce the Si material cost, the recombination at the surface becomes one of the major recombination processes as carriers can move easily across the c-Si substrates to collide with the surfaces of both sides. Here, if the surface potential is bent, one of either electrons or holes is forced to go away from the surface, and thus, the possibility of recombination at the surface

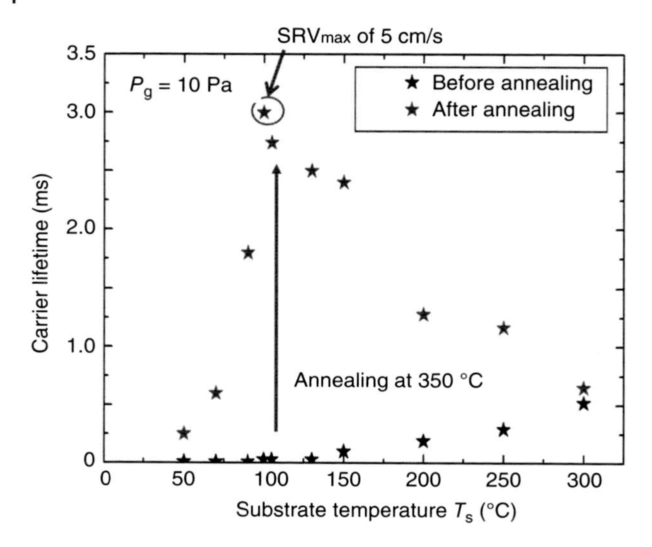

Figure 10.31 Carrier lifetimes of c-Si coated with Cat-CVD SiN_x films as a function of substrate temperature T_s during deposition of SiN_x films. After 350 °C annealing, the carrier lifetimes are drastically improved when T_s is 100–150 °C. Source: Reprinted with permission from Ref. [15].

can be enormously reduced. Cat-doping may be useful to improve the passivation quality of c-Si and also to improve the properties of grain boundary in poly-Si or microcrystalline Si [13].

On the other hand, SiN_x films are widely used as antireflection coating layers on c-Si solar cells. However, in spite of extensive use of SiN_x, the direct deposition of SiN_x films on c-Si often induces the defects related to N atoms [14], as already described in Chapter 8. At an initial stage of SiN_x deposition by using NH_3 and SiH_4 mixture for instance, N atoms supplied from NH_3 are doped by following the similar process to Cat-doping. To avoid this N incorporation inside c-Si, the lowering initial deposition temperature of SiN_x and the change of recipe for SiN_x deposition are effective on reducing the amount of N atoms in c-Si [15]. In the carefully selected recipe, NH_3 gas is not introduced alone without mixing SiH_4 to avoid N Cat-doping.

Figure 10.31 shows the carrier lifetimes of c-Si, which is coated by Cat-CVD SiN_x on both sides, as a function of T_s during deposition of SiN_x films. The carrier lifetime was measured by the microwave photoconductivity decay (μ-PCD) method [16]. The deposition at lower temperatures near 100–150 °C and annealing at 350 °C after film deposition is effective to improve the carrier lifetimes. Probably, low-temperature deposition helps to suppress the N incorporation in c-Si and to increase H contents in the films. Such a large amount of incorporated H atoms may work eliminating the defects or the negative effect of incorporated N atoms during annealing. The carrier lifetime is elevated to 3 ms from the values much less than 1 ms by annealing [15].

Here, if we introduce the Cat-doping to these samples, what happens? Figure 10.32 shows schematic illustration of the phenomenon when P Cat-doping is introduced before deposition of SiN_x films on n-type c-Si. By the P Cat-doping,

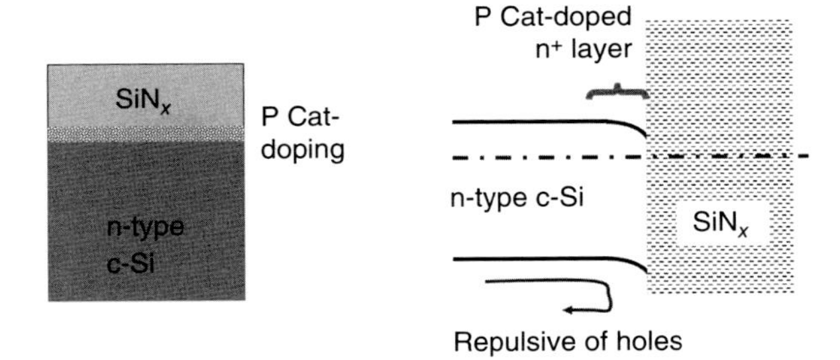

Figure 10.32 Schematic illustration for SiN_x deposition after phosphorus (P) Cat-doping into c-Si is carried out.

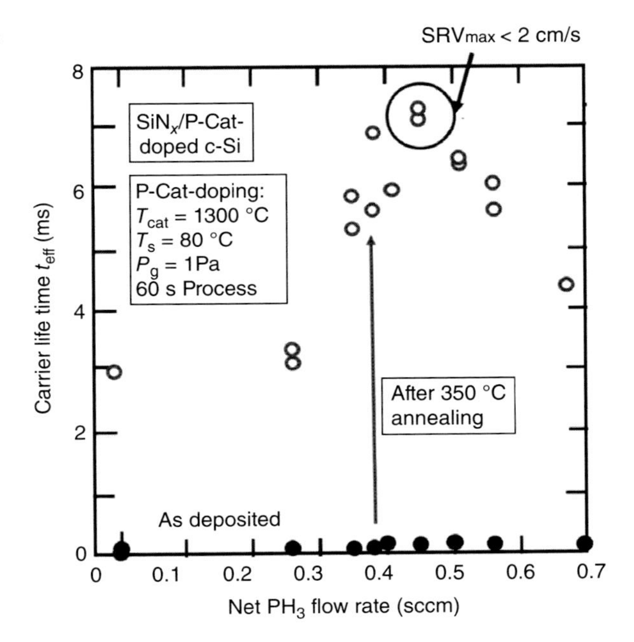

Figure 10.33 Carrier lifetimes of c-Si samples coated by Cat-CVD SiN_x films after phosphorus (P) Cat-doping, as a function of net flow rate of phosphine (PH_3) gas used for P Cat-doping. Conditions for Cat-doping are summarized in the figure. Source: Reprinted with permission from Ref. [15].

the band of c-Si is expected to be bent downward and the holes are forced to leave from the surface to reduce the surface recombination. Figure 10.33 shows the carrier lifetimes of SiN_x-coated c-Si as a function PH_3 flow rate as the source gas of P Cat-doping. T_s for SiN_x deposition was 100 °C and annealed at 350 °C after finishing all processes. It is clearly shown that the carrier lifetimes are improved to 7–8 ms from 3 ms by this P Cat-doping [17]. Only surface doping is possible by Cat-doping, but it is still useful like this.

This is an example of usefulness of Cat-doping. Another example shall be demonstrated below. The Cat-doping will find further applications.

10.6.2 Cat-doping into a-Si and Its Application to Heterojunction Solar Cells

Finally, we introduce another example of application of Cat-doping.

As already introduced in Chapter 8, the Si heterojunction (SHJ) solar cell is one of the most expectable c-Si solar cells for obtaining high-energy conversion

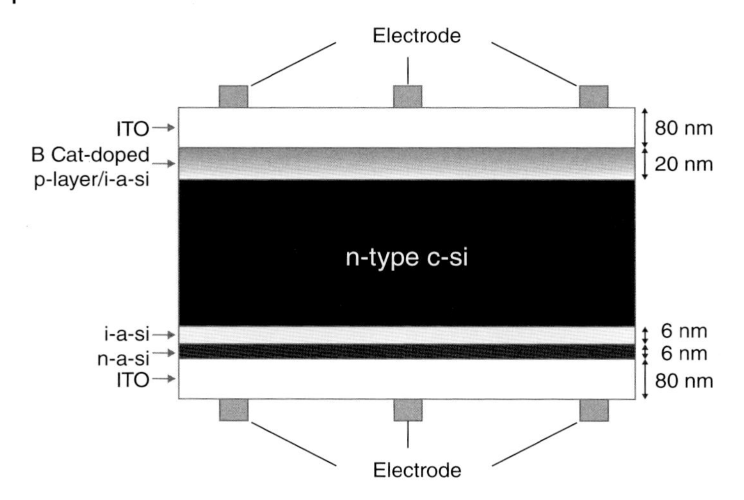

Figure 10.34 Structure of a-Si/c-Si heterojunction solar cells, which include a boron (B) Cat-doped p-type a-Si layer. Source: Reprinted with permission from Ref. [19]. Copyright 2017. The Japan Society of Applied Physics.

efficiency. In the SHJ solar cells, instead of forming p–n junction inside c-Si, p-a-Si is deposited on n-type c-Si for making p–n junction, although usually, to improve the surface passivation quality, i-a-Si layers are inserted between c-Si and doped a-Si layer. As all processes to fabricate SHJ cells are performed at low temperatures around 200 °C, SHJ cells are more suitable for making solar cells with thin c-Si substrates due to reduction of thermal stress.

In such SHJ solar cells, particularly, back contact cells in which all electrodes for electrons and holes are placed only at the back of c-Si solar cells are investigated. By this way, the shadowing loss of the front electrodes can be eliminated and also the metal wiring lines can be simplified. Actually, solar cells with the efficiency over 26.7% have been successfully produced [18] by this structure.

However, it is not easy to make patterned back electrodes with commercially acceptable cost. If n-a-Si or p-a-Si is formed after deposition of i-a-Si layers by conversion of conduction type, we may simplify the fabrication steps to reduce the production cost. Here, we introduce a preliminary experiment by K. Ohdaira et al. [19]. They made SHJ solar cells with a structure illustrated in Figure 10.34. An n-type c-Si was used as a substrate, and a 20-nm-thick i-a-Si layer was deposited on the front side of the c-Si substrate. At the back side, 6-nm-thick i-a-Si and 6-nm-thick n-a-Si layers were successively deposited. After that, just for checking the feasibility of Cat-doped a-Si, i-a-Si at the front is exposed to B-related species and B was Cat-doped to make p-a-Si on i-a-Si layers. Then, on both front and back sides, 80-nm-thick indium tin oxide (ITO) layers were deposited and finally metal electrodes were formed by the screen printing.

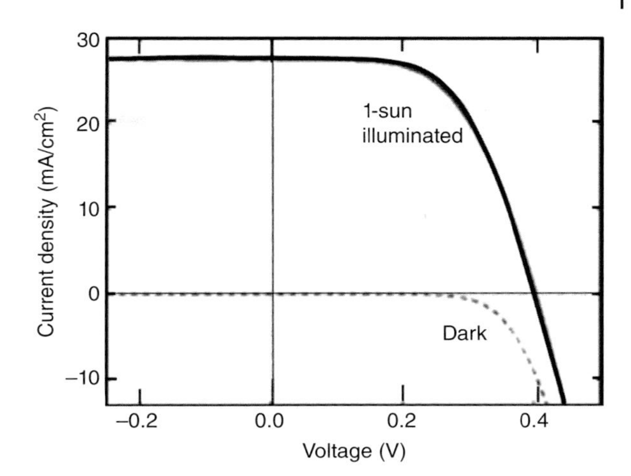

Figure 10.35 Characteristics of a-Si/c-Si heterojunction solar cells fabricated by using B Cat-doping. Source: Reprinted with permission from Ref. [19]. Copyright 2017. The Japan Society of Applied Physics.

Figure 10.35 shows the characteristics of fabricated solar cells. At first, we can confirm that the solar cells are operating properly. The short-circuit currents and the open-circuit voltage, which correspond to the values at the intercepts with current axis and with the voltage axis, are only 28 mA/cm^2 and 0.43 V, respectively. The values were not so good; however, according to other experiments, the carrier lifetime of Cat-doped a-Si/i-a-Si/c-Si structure was quite large and reached 5 ms [18]. The results in the figure probably suffered from other parts such as ITO deposition or metal formation processes. Important things are the fact that the Cat-doping layer can work as p-a-Si layer.

There have been some other attempts to apply this technology to various materials or various fields. For instance, N atoms were introduced into silicon–carbide (SiC) at temperatures as low as 300 °C to control the surface potential of SiC devices, although the low temperature doping into SiC was not so easy by the conventional other known technologies.

The Cat-doping is a new technology, and the attempts of application are not so widely known. However, this technology will be used gradually, although it will take time as Cat-CVD required the time for industrial implementation.

References

1 Strack, H. (1963). Ion bombardment of silicon in glow discharge. *J. Appl. Phys.* 34: 2405–2409.
2 Matsumura, H., Hayakawa, T., Ohta, T. et al. (2014). Cat-doping: novel method for phosphorus and boron shallow doping in crystalline silicon at 80 °C. *J. Appl. Phys.* 116: 114502-1–114502-10.
3 Matsumura, H., Miyamoto, M., Koyama, K., and Ohdaira, K. (2011). Drastic reduction in surface recombination velocity of crystalline silicon by surface

treatment using catalytically-generated radicals. *Sol. Energy Mater. Sol. Cells* 95: 797–799.

4 van der Pauw, L.J. (1958). A method of measuring specific resistivity and Hall effect of discs of arbitrary shape. *Philips Res. Rep.* 13: 1–9.

5 Grove, A.S. (1967). *Physics and Technology of Semiconductor Devices*, Chapter 3. Wiley.

6 Hayakawa, T., Nakashima, Y., Koyama, K. et al. (2012). Distribution of phosphorus atoms and carrier concentration in single-crystal silicon doped by catalytically generated phosphorus radicals. *Jpn. J. Appl. Phys.* 51: 061301-1–061301-9.

7 Anh, L.T., Cuong, N.T., Lam, P.T. et al. (2016). First-principles study of hydrogen-enhanced phosphorus diffusion in silicon. *J. Appl. Phys.* 119: 045703-1–045703-7. https://doi.org/10.1063/1.4940738.

8 Fair, R.B. and Tsai, J.C.C. (1977). A quantitative model for the diffusion of phosphorus in silicon and emitter dip effect. *J. Electrochem. Soc.* 124: 1107–1118.

9 Umemoto, H., Kanemitsu, T., and Kuroda, Y. (2014). Catalytic decomposition of phosphorus compounds to produce phosphorus atoms. *Jpn. J. Appl. Phys.* 53: 05FM02-1–05FM02-4.

10 Smeu, M., Guo, H., Ji, W., and Wolkow, R.A. (2012). Electronic properties of Si(111)-7 × 7 and related reconstructions: Density functional theory calculation. *Phys. Rev. B* 85: 195315-1–195315-9.

11 Ohta, T., Koyama, K., Ohdaira, K., and Matsumura, H. (2015). Low temperature boron dopiung into crystalline silicon by boron-containing species generated in Cat-CVD apparatus. *Thin Solid Films* 575: 92–95.

12 Seto, J., Ohdaira, K., and Matsumura, H. (2016). Catalytic doping of phosphorus and boron atoms on hydrogenated amorphous silicon films. *Jpn. J. Appl. Phys.* 55: 04ES05-1-04ES05-4.

13 Liu, Y., Kim, D.Y., Lambertz, A., and Ding, K. (2017). Post-deposition catalytic-doping of microcrystalline silicon thin layer for application in silicon heterojunction solar cell. *Thin Solid Films* 635: 63–65. https://doi.org/10.1016/j.tsf,2017.02.003.

14 Higashimine, K., Koyama, K., Ohdaira, K. et al. (2012). Scanning transmission electron microscope analysis of amorphous-Si insertion layers prepared by catalytic chemical vapor deposition, causing low surface recombination velocities on crystalline silicon wafers. *J. Vac. Sci. Technol. B* 30: 031208-1–031208-6. https://doi.org/10.1116/1.4706894.

15 Thi, T.C., Koyama, K., Ohdaira, K., and Matsumura, H. (2014). Passivation quality of stoichiometric SiN_x single passivation layer on crystalline silicon prepared by catalytic chemical vapor deposition and successive annealing. *Jpn. J. Appl. Phys.* 53: 022301-1–022301-6. https://doi.org/10.7567/JJAP.53.022301.

16 Deb, S. and Nag, B.R. (1962). Measurement of lifetime of carriers in semiconductors through microwave reflection (letters to editor). *J. Appl. Phys.* 33: 1604–1604.

17 Thi, T.C., Koyama, K., Ohdaira, K., and Matsumura, H. (2014). Drastic reduction in the surface recombination velocity of crystalline silicon passivated

with catalytic chemical vapor deposited SiN_x films by introducing phosphorous catalytic-doped layer. *J. Appl. Phys.* 116: 044510-1–044510-7. https://doi.org/10.1063/1.4891237.

18 Yoshikwa, K., Kawasaki, H., Yoshida, W. et al. (2017). Silicon heterojunction solar cell with interdigitated back contacts for a photoconversion efficiency over 26%. *Nat. Energy* 2: 17032-1–17032-8. https://doi.org/10.1038/energy.2017.32.

19 Ohdaira, K., Seto, J., and Matsumura, H. (2017). Catalytic phosphorus and boron doping of amorphous silicon films for application to silicon heterojunction solar cells. *Jpn. J. Appl. Phys.* 56: 08MB06-1–08MB06-5.

Index

Catalytic Chemical Vapor Deposition: Technology and Applications of Cat-CVD,
First Edition. Hideki Matsumura, Hironobu Umemoto, Karen K. Gleason, and Ruud E.I. Schropp.

k

l

m

q

r